# Immunology

# Immunology
## Mucosal and Body Surface Defences

Andrew E. Williams
Centre for Respiratory Research, University College London

WITH CONTRIBUTIONS FROM

Tracy Hussell
Professor of Inflammatory Diseases, Leukocyte Biology Section
National Heart & Lung Institute, Imperial College London

Clare Lloyd
Professor of Respiratory Immunology, Head of Leukocyte Biology Section
National Heart & Lung Institute, Imperial College London

WILEY-BLACKWELL
A John Wiley & Sons, Ltd., Publication

*Registered office:* John Wiley & Sons Ltd, The Atrium, Southern Gate, Chichester, West Sussex, PO19 8SQ, UK

*Editorial offices:* 9600 Garsington Road, Oxford, OX4 2DQ, UK

The Atrium, Southern Gate, Chichester, West Sussex, PO19 8SQ, UK

111 River Street, Hoboken, NJ 07030-5774, USA

For details of our global editorial offices, for customer services and for information about how to apply for permission to reuse the copyright material in this book please see our website at www.wiley.com/wiley-blackwell.

*Library of Congress Cataloging-in-Publication Data*

Williams, Andrew, E. 1975-
    Immunology : mucosal and body surface defences / Andrew E. Williams – 1st ed.
        p. cm.
    Includes index.
    ISBN 978-0-470-09003-9 (cloth) – ISBN 978-0-470-09004-6 (paper)
1. Immunology – Textbooks. I. Title.
    QR181.W67 2011
    616.07′9 – dc23

                                                                2011021727

A catalogue record for this book is available from the British Library.

This book is published in the following electronic formats: ePDF 978-0-470-09005-3; ePub 978-1-119-97988-3; Wiley Online Library 978-1-119-99864-8; Mobi 978-1-119-97989-0

Set in 9.25/11.5pt Minion by Laserwords Private Limited, Chennai, India
Printed and bound in Singapore by Markono Print Media Pte Ltd

First impression 2012

# Contents

Preface, xv

List of Standard Cells and Symbols, xvii

1    Basic Concepts in Immunology, 1
     1.1    The immune system, 1
     1.2    Tissues and cells of the immune system, 1
     1.3    Activation, regulation and functions of immune responses, 4
     1.4    Innate versus adaptive immunity, 5
     1.5    Primary and secondary immune responses, 6
     1.6    Immune cell development, 7
     1.7    Mast cells and basophils, 9
     1.8    Eosinophils, 11
     1.9    Neutrophils, 11
     1.10   Monocytes and macrophages, 11
     1.11   Dendritic cells, 12
     1.12   Natural killer cells, 12
     1.13   CD4+ T helper cells, 13
     1.14   CD8+ cytotoxic T cells, 14
     1.15   B cells, 15
     1.16   γδ T cells, 16
     1.17   Natural killer T cells, 16
     1.18   Anatomy of the immune system, 16
     1.19   Lymph nodes, 16
     1.20   Spleen, 19
     1.21   Summary, 19

2    The Innate Immune System, 20
     2.1    Introduction to the innate immune system, 20
     2.2    Innate immune receptors and cells, 20
     2.3    TLRs and pattern recognition, 22
     2.4    TLR signalling in response to LPS, 23
     2.5    Peptidoglycan and Nods, 24
     2.6    Nod-like receptors recognize PAMPs and DAMPs, 25
     2.7    Damage associated molecular patterns (DAMPs), 26
     2.8    Complement proteins perform several innate immune functions, 27
     2.9    The classical complement pathway, 28
     2.10   The lectin and alternative complement pathways, 29
     2.11   Biological properties of complement cleavage products, 29
     2.12   Opsonization by complement proteins, 30
     2.13   Phagocytosis, 31
     2.14   Fc receptors induce phagocytosis, 32
     2.15   Neutrophil function and the respiratory burst, 32

2.16 ADCC, 33
2.17 NK cells recognize missing self, 35
2.18 Activating adaptive immunity, 36
2.19 Dendritic cells link innate and adaptive immunity, 38
2.20 Summary, 40

**3 The Adaptive Immune System, 41**
3.1 Introduction to adaptive immunity, 41
3.2 T cells and B cells recognize foreign antigens, 41
3.3 Overview of antibody structure, 42
3.4 Constant region and antibody isotypes, 45
3.5 B cell receptor (BCR) diversity, 46
3.6 Genetic recombination of BCR genes, 46
3.7 Mechanism of VDJ recombination, 47
3.8 Introducing junctional diversity, 48
3.9 Somatic hypermutation and affinity maturation, 49
3.10 Immunoglobulin class switching, 50
3.11 Structure of Fc receptors, 51
3.12 Fc receptor specificity and affinity, 53
3.13 Cross-linking of antibody is necessary for Fc receptor signalling, 53
3.14 Fc receptor immune functions, 54
3.15 T cell receptor diversification, 54
3.16 T cells undergo positive and negative selection within the thymus, 55
3.17 Antigen presentation to T cells, 57
3.18 MHC class II processing pathway, 59
3.19 MHC class I processing pathway, 59
3.20 Activation requires co-stimulation, 60
3.21 Late co-stimulatory signals, 62
3.22 Activation of B cell responses, 63
3.23 CD4+ T helper cell differentiation, 63
3.24 Activation of CTLs, 65
3.25 Generation of memory T cells, 66
3.26 Summary, 67

**4 Cytokines, 68**
4.1 Introduction to cytokines, 68
4.2 Structure of cytokine families, 69
4.3 IL-1 superfamily, 71
4.4 IL-6 family, 71
4.5 IL-10 family, 72
4.6 Common $\gamma$-chain family, 73
4.7 IL-12 family, 74
4.8 Interferons, 75
4.9 TNF ligand superfamily, 75
4.10 Growth factors, 77
4.11 Functional classification Th1 versus Th2, 78
4.12 Th17, immunopathology and regulatory cytokines, 79
4.13 Cytokine receptor signalling, 79
4.14 Type I and type II cytokine receptors, 79
4.15 The JAK/STAT signalling pathway, 80
4.16 IL-2 signalling through the JAK/STAT pathway, 81

4.17   The JAK/STAT pathway is also used by IL-6, 83
4.18   Plasticity in type I cytokine signalling, 83
4.19   Suppressor of cytokine signalling (SOCS), 83
4.20   IFN-γ signalling pathway, 84
4.21   TGF-β and the SMAD signalling pathway, 85
4.22   Type III cytokine receptors and the TNF receptor family, 86
4.23   The IKK complex and the activation of NF-κB, 87
4.24   The IL-1R family of type IV cytokine receptors
       activate NF-κB, 88
4.25   Soluble cytokine receptors act as decoy receptors, 90
4.26   IL-33 and ST2 signal regulation, 91
4.27   Potential for cytokine therapy, 91
4.28   Summary, 92

5   Chemokines, 93
5.1    Introduction, 93
5.2    Structure and nomenclature of chemokines, 93
5.3    Chemokine receptors, 94
5.4    Expression of chemokines and their receptors, 97
5.5    Chemokines promote extravasation of leukocytes, 97
5.6    Chemotaxis, 99
5.7    Chemokine receptor signalling cascade, 99
5.8    Tissue specific homing, 100
5.9    Lymphocyte migration to secondary lymphoid tissues, 101
5.10   Chemokines involved in lymphoid structure formation, 102
5.11   Chemokines contribute to homeostasis, 104
5.12   Chemokine receptors on T cell subsets, 104
5.13   Redundancy in the chemokine/receptor system, 106
5.14   Chemokines in disease, 108
5.15   Chemokines as new anti-inflammatory drugs, 109
5.16   Summary, 110

6   Basic Concepts in Mucosal Immunology, 111
6.1    Introduction, 111
6.2    What is a mucosal tissue?, 112
6.3    Immune defence at mucosal tissue is multi-layered, 113
6.4    Origins of mucosal associated lymphoid tissue, 114
6.5    Concept of the common mucosal immune system, 115
6.6    How do T and B lymphocytes migrate into mucosal tissues?, 116
6.7    Special features of mucosal epithelium, 117
6.8    Toll-like receptors and NOD proteins in the mucosa, 120
6.9    Antigen sampling at mucosal surfaces, 121
6.10   Mucosal dendritic cells, 122
6.11   Secretory dimeric IgA at mucosal surfaces, 124
6.12   Regulation of J-chain and secretory component expression, 126
6.13   How does the sub-mucosa differ from the epithelium?, 126
6.14   Organized lymphoid tissue of the mucosa, 127
6.15   Cytokines in the mucosa, 128
6.16   Pathogens that enter via mucosal sites, 130
6.17   Immune diseases of mucosal tissues, 130
6.18   Summary, 132

**7   Immunology of the Gastrointestinal Tract, 133**

7.1   Structure of the gastrointestinal tract, 133

7.2   Development of the gastrointestinal tract, 133

7.3   The digestive tract as a mucosal tissue, 135

7.4   Barrier function, 136

7.5   Defensins and Trefoil factors, 138

7.6   Structure of Peyer's patches, 139

7.7   Lymphoid follicles and germinal centre formation, 140

7.8   M cells sample the intestinal lumen, 143

7.9   Dendritic cells sample the lumen contents, 143

7.10   Lymphocytes within the epithelium (IELs), 143

7.11   γδ T cells in the GALT, 146

7.12   NKT cells, 147

7.13   T cells in the lamina propria, 148

7.14   Maintenance of T cell homeostasis, 148

7.15   Sub-mucosal B cells and mucosal IgA, 149

7.16   How IgA is produced at intestinal mucosal sites, 150

7.17   Cytokines in the gut, 151

7.18   Chemokines and the homing of lymphocytes to GALT, 152

7.19   Pathogens and immune diseases, 153

7.20   Summary, 154

**8   Immunology of the Airways, 156**

8.1   The airways as a mucosal tissue, 156

8.2   Development of the respiratory tract, 156

8.3   The structure of the respiratory tract, 158

8.4   Barrier function and the mucociliary elevator, 159

8.5   Mucins and mucociliary clearance, 160

8.6   Defensins and antimicrobial peptides, 160

8.7   Structure of the tonsils and adenoids of the Waldeyer's Ring, 161

8.8   Local lymph nodes and immune generation, 163

8.9   Structure of the NALT, 165

8.10   Structure of the BALT, 165

8.11   Cells of the lower respiratory tract, 166

8.12   Surfactant proteins, 167

8.13   Immune modulation by airway epithelial cells, 167

8.14   Innate immune response, 168

8.15   Dendritic cells are located throughout the respiratory tract, 168

8.16   Alveolar macrophages maintain homeostasis, 169

8.17   NK cells in the lung, 171

8.18   T cells at effector sites in the lung, 171

8.19   Memory T cell responses within the lung, 172

8.20   Migration of circulating T cell into the lung tissue, 172

8.21   IgA production in the respiratory tract, 173

8.22   Respiratory diseases and pathogens, 174

8.23   Summary, 176

**9   Immunology of the Urogenital Tract and Conjunctiva, 177**

9.1   The urogenital tract as a MALT, 177

9.2   Epithelial barrier function, 178

9.3   Passive immunity, 181

9.4    Immunoglobulins, 181
9.5    APCs in genital tract mucosa, 182
9.6    NK cells and the semi-allogeneic foetus, 183
9.7    Pre-eclampsia is an immune-mediated disease, 184
9.8    Maintenance of foetal tolerance, 185
9.9    T cells and adaptive immunity, 186
9.10   Sexually transmitted diseases and pelvic inflammatory disease, 187
9.11   Alloimmunization and autoimmune diseases, 189
9.12   The foetal and neonatal immune system, 189
9.13   Immunity in the urinary tract, 190
9.14   Eye associated lymphoid tissue, 191
9.15   Conjunctiva associated lymphoid tissue (CALT), 192
9.16   Immune privilege of the eye, 192
9.17   Immune privilege and inflammation, 193
9.18   Conjunctivitis, 194
9.19   Summary, 195

10 Immunology of the Skin, 196
10.1   The skin as an immune tissue, 196
10.2   Barrier Immune function of the skin, 196
10.3   Cellular immune system of the skin, 198
10.4   Keratinocytes can act as immune cells, 199
10.5   Keratinocytes secrete antimicrobial peptides, 200
10.6   Langerhan's cells act as immune sentinels in skin, 202
10.7   Dermal dendritic cells and cross-presentation of antigen, 203
10.8   Mast cells and NK cells in the skin, 205
10.9   Intraepidermal lymphocytes in the skin, 206
10.10  Lymphocytes in the dermis, 206
10.11  Skin homing T cells express CLA, 206
10.12  Chemokines and migration, 207
10.13  Initiation of an immune response in the skin, 208
10.14  Cytokines, 211
10.15  Psoriasis, inflammation and autoreactive T cells, 211
10.16  Autoimmune-mediated diseases of the skin, 213
10.17  Systemic diseases that affect the skin, 214
10.18  Infectious diseases of the skin, 215
10.19  Summary, 216

11 Immunity to Viruses, 217
11.1   Introduction, 217
11.2   Structure of viruses, 217
11.3   Classification of viruses, 218
11.4   Viruses replicate within host cells, 218
11.5   Infections caused by viruses, 219
11.6   Certain viruses can infect immune cells, 220
11.7   Virus infection of epithelial cells, 221
11.8   IFN-α response, 222
11.9   NK cell response to viruses, 222
11.10  Viral evasion of NK cell responses, 223
11.11  Macrophages contribute to virus elimination, 225
11.12  TLRs and NLRs recognize virus motifs, 226

11.13  Activation of the inflammasome by viruses, 226
11.14  Dendritic cells present virus antigens to CD8+ CTLs, 227
11.15  T cell responses to viruses, 229
11.16  Evasion of CTL-mediated immunity by viruses, 229
11.17  Bystander effects of immune responses to viruses, 231
11.18  Antibody response to viruses, 232
11.19  Difference between cytopathic and non-cytopathic viruses, 233
11.20  Immune evasion by antigenic shift and drift, 235
11.21  Vaccination and therapies against viral infections, 235
11.22  Summary, 237

12  **Immunity to Bacteria, 238**
12.1  Introduction to bacterial immunity, 238
12.2  Classification of bacteria, 238
12.3  Structure of the bacterial cell, 240
12.4  Diseases caused by bacteria, 241
12.5  Mucosal barriers to bacterial infection, 241
12.6  Anti-microbial molecules, 242
12.7  Recognition of bacterial PAMPs by Toll-like receptors, 243
12.8  Complement and bacterial immunity, 244
12.9  Neutrophils are central to bacterial immune responses, 245
12.10  Some bacteria are resistant to phagosome mediated killing, 247
12.11  NK cells and ADCC, 248
12.12  The role of antibody in bacterial immunity, 249
12.13  Dendritic cells and immunity to bacteria, 250
12.14  Autophagy and intracellular bacteria, 251
12.15  T Cells contribute to protective immunity, 253
12.16  The DTH response and granuloma in TB, 253
12.17  Th17 cells in bacterial immunity, 254
12.18  Treg cells in bacterial infection, 255
12.19  Unconventional T cells, 256
12.20  Vaccination against bacterial diseases, 256
12.21  Summary, 256

13  **Immunity to Fungi, 258**
13.1  Introduction, 258
13.2  Morphology of fungi, 258
13.3  Yeasts, 260
13.4  Moulds, 260
13.5  Fungal dimorphism, 261
13.6  Diseases caused by fungi, 262
13.7  Immune response to fungi, 263
13.8  Innate immunity, 263
13.9  Mucosal barriers to fungal infection, 263
13.10  Anti-fungal molecules, 265
13.11  Recognition of fungal PAMPs by Toll-like receptors, 266
13.12  Complement and fungal immunity, 266
13.13  Dendritic cells link innate and adaptive fungal immunity, 268
13.14  DCs provide the adaptive immune response with instructive signals, 270
13.15  Macrophages are important APCs during fungal infection, 270
13.16  Neutrophils participate in the inflammatory response to fungi, 271

13.17   NK cells provide inflammatory signals to macrophages, 271
13.18   Adaptive immunity to fungi, 272
13.19   The DTH response and granuloma formation inhibit fungal dissemination, 272
13.20   The role of antibody in fungal resistance, 273
13.21   Vaccination and immunotherapies, 274
13.22   Fungal immune evasion strategies, 276
13.23   Immuno-modulatory fungal products, 276
13.24   Evasion of phagolysosomal killing, 276
13.25   Modifying the cytokine response, 277
13.26   Summary, 277

**14   Immunity to Parasites, 278**
14.1   Introduction, 278
14.2   Protozoa are diverse unicellular eukaryotes, 278
14.3   Structure of the protozoan cell, 278
14.4   Life cycle of protozoan parasites, 280
14.5   The life cycle of *Trypanosoma brucei*, 281
14.6   Life cycle of *Leishmania species*, 281
14.7   The life cycle of *Plasmodium falciparum*, 281
14.8   Helminths are multicellular, macroscopic parasites, 282
14.9   Structure of the trematode *Schistosoma mansoni*, 283
14.10   Life cycle of *Schistosoma mansoni*, 284
14.11   Structure of the nematode *Ascaris lumbricoides*, 285
14.12   The life cycle of *A. lumbricoides*, 286
14.13   Immune responses to parasites, 286
14.14   Innate immunity to trypanosomes, 287
14.15   Adaptive immunity to trypanosomes, 287
14.16   Innate immunity to plasmodium, 288
14.17   Adaptive immunity to plasmodium, 289
14.18   Immunity to *Leishmania* – Th1 versus Th2, 290
14.19   Immunity to Giardia, 291
14.20   Immunity to schistosomes, 292
14.21   Innate immunity to schistosomes, 292
14.22   Adaptive immunity to schistosomes, 293
14.23   Granuloma formation in schistosomiasis, 294
14.24   Immunity to intestinal nematode worms, 294
14.25   Innate immunity to nematode worms in the gut, 294
14.26   Adaptive immunity to nematode worms in the gut, 295
14.27   Immune evasion strategies of parasites, 296
14.28   Trypanosome variant surface glycoproteins (VSGs), 297
14.29   Plasmodium life cycle contributes to immune evasion, 298
14.30   *Leishmania* evade phagocytic killing, 298
14.31   Immune evasion strategies of helminths, 298
14.32   Summary, 300

**15   Disorders of the Immune System, 302**
15.1   Introduction to immune disorders, 302
15.2   Types of allergy, 302
15.3   Sensitization and the acute phase response, 304
15.4   Mast cell degranulation, 305

15.5 Late phase response, 306
15.6 Allergic asthma, 307
15.7 Mast cells and the early phase allergic asthma, 308
15.8 Epithelial cells can trigger allergic asthma, 308
15.9 T cells and the late phase of allergic asthma, 310
15.10 Allergic rhinitis, 310
15.11 Skin allergy and atopic dermatitis, 311
15.12 Food allergies, 311
15.13 T cell subsets in allergy, 312
15.14 Mechanisms of autoimmune disease, 313
15.15 Disregulation of tolerance and autoimmunity, 313
15.16 Inflammatory bowel disease, 316
15.17 Coeliac disease, 317
15.18 Systemic lupus erythematosus, 317
15.19 Other autoimmune diseases, 318
15.20 Immunodeficiencies, 320
15.21 Summary, 321

16 **Mucosal Tumour Immunology, 322**
16.1 Introduction, 322
16.2 Transformation into cancer cells, 322
16.3 Proto-oncogene activation, 323
16.4 Mutation in the p53 protein, 324
16.5 Mutant Ras proteins enhance proliferation, 324
16.6 Aneuploidy and colorectal cancer, 324
16.7 Tumourigenesis, 324
16.8 Angiogenesis, 326
16.9 Metastasis, 327
16.10 The immune system and cancer, 327
16.11 Immune surveillance, 328
16.12 Immunogenicity of tumour cells, 329
16.13 Recognition of transformed cells, 330
16.14 Tumour associated antigens, 331
16.15 Carcinoembryonic antigen in colorectal cancer, 331
16.16 Melanoma differentiation antigens, 332
16.17 Viral tumour associated antigens, 332
16.18 Effector molecules during tumour immune surveillance, 333
16.19 Dendritic cells modulate anti-tumour immune responses, 333
16.20 Tumour reactive T cells are activated in lymph nodes, 335
16.21 NK cell recognition – missing self, 335
16.22 NKG2D receptor on NK cells, 335
16.23 Macrophages and neutrophils phagocytose tumour cells but support tumour growth, 336
16.24 Immune cells can augment tumour growth, 337
16.25 Immune evasion strategies, 337
16.26 Darwinian selection and tumour cell escape, 338
16.27 Cytokine environment and tumour escape, 339
16.28 Tumours have disregulated MHC expression and antigen presentation, 339

16.29    Tumour escape through Fas/FasL, 340
16.30    Summary, 341

**17  Vaccination, 342**
17.1    Introduction, 342
17.2    The principles of vaccination, 342
17.3    Passive immunization, 344
17.4    Active immunization, 344
17.5    Processing of the vaccine for immune recognition, 344
17.6    Adaptive Immune response following vaccination, 347
17.7    Vaccine adjuvants, 347
17.8    Alum, 348
17.9    Freund's complete adjuvant, 348
17.10    Mucosal adjuvants and vaccine delivery, 350
17.11    Prospects in adjuvant design, 350
17.12    Th1/Th2 polarization and vaccine development, 351
17.13    Live-attenuated vaccines, 351
17.14    Inactivated vaccines, 353
17.15    Polysaccharide vaccines, 354
17.16    Peptide vaccines, 354
17.17    DNA vaccination, 355
17.18    Immuno-stimulatory complexes (ISCOMs), 355
17.19    Dendritic cell vaccines, 358
17.20    Mucosal administration of vaccines, 359
17.21    Nasally administered vaccine against genital infections, 360
17.22    New strategies for vaccine development, 360
17.23    Summary, 362

**Glossary of Terms, 363**

**Index, 374**

# Preface

For thousands of years humans have marvelled at how the body is able to protect itself from infectious pathogens. Even the ancient Chinese and Greeks acknowledged the protective effects of the immune system, noting how one is rendered resistant to catching the same disease a second time. The first empirical studies were performed by Edward Jenner, and later Louis Pasteur, who developed vaccines against smallpox and anthrax, respectively. Indeed, vaccination has become such an important aspect of human health it is sometimes easy to forget the central role the immune system plays in affording protection against so many diseases.

The vast majority of medically important pathogens infect their host across a body surface such as the skin, or across a mucosal tissue such as the respiratory tract or intestines, as these sites are the ones exposed to the external environment. Vertebrates have therefore evolved elaborate immune defence mechanisms to protect against infection across mucosal linings and body surfaces. Mucosal immune defence mechanisms are therefore integral to our survival. However, conventional immunology textbooks largely overlook this aspect of the immune system, even though it remains fundamental for the prevention of infectious disease. Many have continued to teach immunology based on knowledge of the central immune system of the blood and spleen, rather than teaching immunology from the perspective of mucosal and body surfaces. After all, these are the places where host–pathogen interactions actually take place. Therefore I have tried to redress this bias by focusing on immunity at mucosal and body surfaces. This book should therefore prove useful for science undergraduates studying immunology, medical students undertaking academic studies, postgraduate students working toward a higher degree and the broad spectrum of professional academic and clinical scientists working in the field of immunology.

Knowledge of how the immune system operates has increased extensively in the past 50 years, including our insight into mucosal immunology. The first three chapters describe the basic architecture of the immune system and the elements of innate and adaptive immunity that contribute to protective immune responses. A more focused description of the innate immune system is given in Chapter 2, including aspects of barrier, chemical and mechanical defence, components of innate immunity which are so often overlooked. A description of the effector functions of the cells of the innate immune system, such as macrophages, granulocytes and NK cells, is also given. A similar approach is used in Chapter 3 to illustrate the ways in which adaptive immune responses are orchestrated, including how B cells produce antibodies and how T cells elicit their effector functions. This includes a discussion of B cell and T cell selection and the generation of memory cells, which are key to providing long-lasting protection and is a central concept in immunology.

The next two chapters focus on two important families of signalling molecules, the cytokines and chemokines, which have fundamental roles in orchestrating the spatial and temporal mechanics of an immune response. These chapters define just how important cytokines and chemokines are to the organization of the immune system.

Chapters 6 to 10 describe the central thesis of this textbook, in that they describe the workings of the mucosal immune system. An introductory chapter outlines the central concepts of the mucosal immune system that differentiates it from the central or peripheral immune systems. The key structural and cellular components and the common themes that link mucosal tissues are explored. For example, epithelial barrier formation, aggregation of organized lymphoid tissues, the importance of secretory IgA in mucosal defence and the need to balance immunity with homeostasis, are discussed. The concept of inductive sites, where immune responses are initiated, and effector sites, where immune cell functions take place, are discussed. From there, a description of the major tissues that form mucosal associated lymphoid tissue (MALT) is described, including the gastrointestinal tract, respiratory tract, urogenital tract and the conjunctiva of the eye. In addition, the importance of the skin in body surface immunity is examined.

The next four chapters are devoted to studying immunity against the four major groups of pathogen, the viruses, bacteria, fungi and parasites, with particular emphases on those infectious microorganisms that infect mucosal or body surfaces. This discussion includes the innate and adaptive immune mechanisms that are responsible for protection and the evasion strategies that these pathogens employ in order to subvert host immune responses.

Chapter 15 focuses on immune-mediated diseases that affect mucosal and body surfaces, including hypersensitivity reactions, allergies and autoimmunity. Chapter 16 details the various aspects of mucosal tumour immunology, in particular how the tumour and the immune system are constantly competing with each other. Finally, Chapter 17 describes the process of vaccination, from the conventional strategies most commonly used today, to novel regimens that specifically target the mucosal immune system and to cutting edge technologies used in modern vaccine development.

# List of Standard Cells and Symbols

KEY –Standard cells and symbols

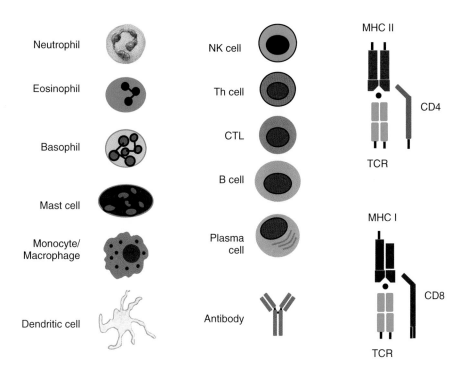

| | | | |
|---|---|---|---|
| Neutrophil | | NK cell | MHC II |
| Eosinophil | | Th cell | CD4 |
| Basophil | | CTL | TCR |
| Mast cell | | B cell | |
| Monocyte/ Macrophage | | Plasma cell | MHC I |
| Dendritic cell | | Antibody | CD8 / TCR |

# 1 Basic Concepts in Immunology

## 1.1 The immune system

The immune system evolved so as to defend our bodies against infectious microorganisms such as viruses, bacteria, fungi and parasites. Throughout history it has been observed that people who survive an infectious disease acquire protection against that disease, which is otherwise known as immunity. As far back as the fifteenth century attempts have been made to induce immunity against infectious diseases, a process referred to as vaccination. The realisation that immunity can be transferred from one person to another demonstrated that soluble factors exist in the blood and body fluids that protect against pathogens. It is now known that cellular components of the immune system are also present throughout the entire body and that these immune cells engage with any harmful substance or microorganism in order to preserve the integrity of host tissues. The defence against microorganisms is fought on many fronts and there are immune cells and innate components of the immune system within every tissue and organ. There are a multitude of cells and soluble factors that can be considered part of the immune system. For example, the barrier function of the outer layers of the skin, the mucus produced in the airways, the antibodies secreted into the gut lumen or the circulating lymphocytes that destroy virus-infected cells. The immune system comprises a number of different cell types and a multitude of secreted factors and surface bound molecules.

The immune system has a multi-layered organisation that provides immunity to infectious organisms (Figure 1.1). Each layer of the immune system can also be considered to have an increasing complexity. The first layer is provided by physical barriers such as the skin and the mucosal epithelium of the respiratory and gastrointestinal tracts. These barriers aim to prevent pathogens gaining access to underlying tissue. The next layer is the non-specific chemical barrier that consists of antimicrobial compounds and factors of the humoral immune system (soluble factors found in body fluids). Other chemical immune defence mechanisms include the acidic environment of the stomach and the proteolytic enzymes produced in the intestines. The third layer is composed of all the cells of the immune system. Therefore, if a pathogen breaches the physical barriers and chemical barriers then the immune system utilizes its immune cells.

The cellular components of the immune system can be divided into the innate immune system and the adaptive immune system. The innate immune system provides a rapid, early response and is considered to be the first line of cellular immune defence. If the innate immune response is overcome by an infectious pathogen then the adaptive immune system comes into play. Only jawed vertebrates have evolved a complex adaptive immune system, which provides highly specific immune protection against microorganisms. The immune protection afforded by the adaptive immune system is retained by the host over a prolonged period of time and is capable of generating immunological memory. It is this immunological memory that confers immunity to subsequent infections with the same pathogen.

## 1.2 Tissues and cells of the immune system

The organs and tissues of the immune system can be compartmentalized (Figure 1.2). There are certain areas that are more susceptible to infection than others and these usually correspond to areas that come into contact with the environment. Therefore the mucosal immune system has evolved over millions of years in answer to selection pressures forced upon it as a result of host-pathogen interactions. The immune system therefore comprises a series of specialized organs and tissues that function by counteracting the threat of pathogens. The areas of the body at

*Immunology: Mucosal and Body Surface Defences*, First Edition. Andrew E. Williams.
© 2012 John Wiley & Sons, Ltd. Published 2012 by John Wiley & Sons, Ltd.

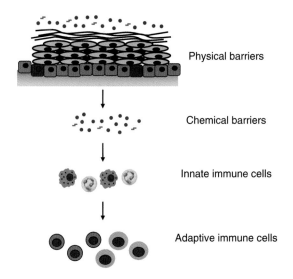

Physical barriers

Chemical barriers

Innate immune cells

Adaptive immune cells

**Figure 1.1** The multiple layers of the immune system.

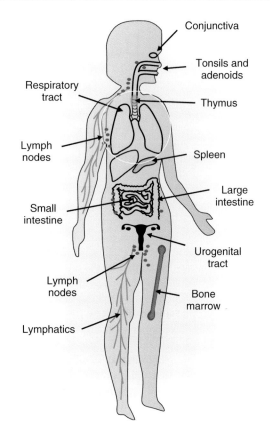

Conjunctiva

Tonsils and adenoids

Respiratory tract

Thymus

Lymph nodes

Spleen

Large intestine

Small intestine

Urogenital tract

Lymph nodes

Bone marrow

Lymphatics

**Figure 1.2** Tissues of the immune system.

most risk are the ones that are most frequently exposed to the outside, the most visually obvious tissue being the skin. Other tissues come into direct contact with the outside including the urogenital tract, gastrointestinal tract and the respiratory tract. For example, the lungs sample in the region of 11,000 litres of air every day and with each breath there is the risk of inhaling a harmful substance or potentially pathogenic microbe. Likewise the intestines are constantly exposed to material ingested through swallowing and, in addition, the gut has to cope with the billions of commensal bacteria that reside there. These tissues have therefore developed a series of immunological barriers to prevent infectious disease. The common mucosal immune system and mucosal-associated lymphoid tissue (MALT) are phrases that have been used to describe the composition of the immune system at sites that possess a mucosal lining. The respiratory, gastrointestinal and genital tracts are the major components of the MALT and function as immunological barriers at sites that are exposed to external substances.

Other tissues, which are not classified as mucosal, also contribute to the immune system. Central to haematopoiesis is the bone marrow, which is located within cavities of the long bones and is the site where all cells of the blood are derived. The bone marrow is home to multipotent stem cells that give rise to red blood cells, platelets and all the different types of white blood cell. The thymus is another important organ responsible for the differentiation and maturation of a population

of white blood cells known as T cells and is located in the chest cavity just above the heart and surrounding the trachea.

Both the bone marrow and thymus are known as primary lymphoid organs, as these are the sites of immune cell development. There are also secondary lymphoid organs such as the spleen, which is located within the upper left hand quadrant of the abdomen. The spleen is responsible for the removal of moribund red blood cells and the initiation of immune responses directed toward blood borne antigens. Other secondary lymphoid organs include the lymph nodes, which are part of the lymphatic system. Lymph nodes are critical for the proper initiation of many immune responses. They are found throughout the body and are concentrated in regions that drain large body parts such as the neck, thorax and abdomen. They provide sites for the initiation of immune responses to antigens derived from body tissues that have been filtered into lymph nodes via the extensive lymphatic system. The tonsils and adenoids are further examples of organized secondary lymphoid organs, which play an important role in the initiation of immune responses to pathogens that enter the body through the oral cavity.

The cells of the immune system are sometimes referred to as immunocytes, the most important of which are the white blood cells (otherwise known as leukocytes; from the Greek leuko (white) and cyte (cell)). All leukocytes originate within the bone marrow from precursor stem cells and can be divided into three groups, depending on their ontogeny (Figure 1.3). The first group are the granulocytes, which include neutrophils, eosinophils, basophils and mast cells. The second group are the myeloid cells that include the monocytes, macrophages and dendritic cells (DCs). The third and final group are the lymphocytes that comprise the natural killer (NK) cells, T cells and B cells. This classification is based on developmental lineage, which will be discussed further in this chapter, as a result of a process known as haematopoiesis. However, when studying the immune system it is sometimes more helpful to classify the different cell types in accordance with cell function. To this end the immune system is often divided into the innate immune system and the adaptive immune system (discussed later in the chapter). Alternatively, the immune system can be studied in terms of the type of immune response it generates and can therefore be classified as either being antibody-mediated or cell-mediated (Figure 1.4). These terminologies will be become clearer as we proceed through the subsequent chapters.

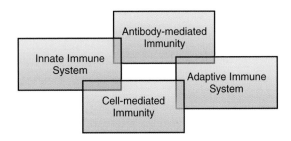

**Figure 1.4** The immune system can be compartmentalised based on function. Certain components overlap, although innate and adaptive systems are considered separate.

There are several key concepts that must be addressed when considering how the immune system works. The first is why the immune system exists at all? It can be argued that the immune system has evolved in order to defend our body against invading pathogens. It is therefore important to understand how the immune system is organized in terms of the cell types responsible for orchestrating an immune response and in terms of the tissues that provide an appropriate environment for the generation of an immune response. It is then important to understand how immune responses are initiated and, once active, how these responses are regulated. Finally, it is important to gain an understanding of how the immune system is able to provide us with immunological protection against a pathogen that we have previously encountered, a process known as immunological memory. Some

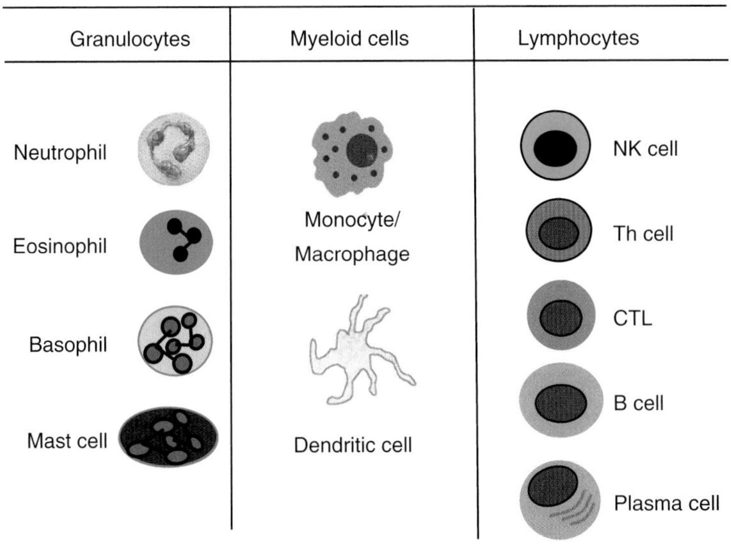

| Granulocytes | Myeloid cells | Lymphocytes |
| --- | --- | --- |
| Neutrophil | | NK cell |
| Eosinophil | Monocyte/ Macrophage | Th cell |
| Basophil | | CTL |
| Mast cell | Dendritic cell | B cell |
| | | Plasma cell |

**Figure 1.3** Cells of the immune system, which are divided into granulocytes, myeloid cells and lymphocytes.

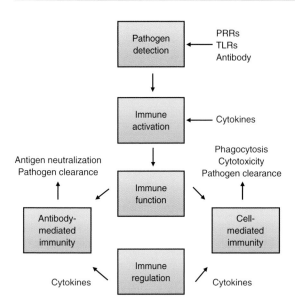

**Figure 1.5** Key concepts in the development of an immune response. First pathogens must be detected and recognised as a threat. This pathogen recognition then activates the immune system and results in immune function, which can be divided into antibody-mediated or cell-mediated immunity. The functional activity must then be regulated after the pathogen has been dealt with and is known as immune regulation.

of these key concepts and immunological terms will be introduced in this chapter (Figure 1.5), while subsequent chapters aim to provide detailed descriptions of specific immunological mechanisms.

## 1.3  Activation, regulation and functions of immune responses

One of the most important concepts in immunology is the recognition of foreign substances by the immune system. For example, microorganisms such as bacteria and viruses produce a multitude of proteins, carbohydrates and glycolipids, which can all be recognized by cellular receptors expressed on the surface of immune cells. The most widely studied family of microbial recognition receptors are known as pattern recognition receptors (PRRs), because they recognize evolutionary conserved molecular patterns produced by microorganisms. The recognition of microbial products, which are known as pathogen associated molecular patterns (PAMPs), is one of the first steps in the activation of an immune response. Numerous cells of the immune system, and also tissue

cells such as epithelial cells, express these PRRs. Probably the most ubiquitous family of PRRs are known as the toll-like receptor (TLR) family, which are responsible for the recognition of a host of PAMPs (discussed further in Chapter 2). The recognition of foreign molecules activates immune cells and results in the initiation of an immune response. For these reasons PAMPs are often referred to as danger signals. This is one way in which the immune system is able to discriminate between foreign substances (non-self) and its own molecules (self).

The interaction between immune cells, and indeed between immune cells and tissue cells, can dictate the phenotype, magnitude and duration of an immune response. It is often the case that an individual immune cell relies on signals derived from the extracellular environment in order to become activated or initiate one or more of its effector functions, for example through the recognition of danger signals. A key family of molecules, known as cytokines, play a central role in the initiation and regulation of immune responses (discussed in detail in Chapters 4 and 5). Cytokines are produced from all types of immune cells and from tissue cells such as epithelial cells of the respiratory or gastrointestinal tracts. Cytokines are normally produced and released from a cell in response to an external substance, such as an invading bacteria or virus. The released cytokine then exerts its biological effects on a target cell. For example, an epithelial cell will respond to an invading bacteria and will secrete several cytokines. These cytokines then signal to nearby immune cells and cause a functional response in those immune cells. This response may involve one or more effects including the cell activation, proliferation, migration, further cytokine secretion or initiation of effector functions. There are several families of cytokine including the interleukins, interferons, growth factors and chemokines, all of which provide a network of soluble mediators that regulate the immune response.

Specialized lymphocytes, known as B cells, can also secrete numerous proteins called antibodies, which are found in bodily fluids such as the blood serum and lymphatics. The role of antibodies within the immune system is to recognize and bind to foreign proteins derived from microorganisms. Any protein that can be bound by an antibody is known as an antigen. The interaction between an antibody and an antigen is a key principle in immunology. The binding of an antibody to an antigen has several downstream consequences, including the neutralisation and clearance of the antigen, and the activation of the effector functions of numerous immune cells. Often the word antibody is interchanged with the word immunoglobulin, as these two terms describe the very same molecule. The functional consequences of antibody and antigen interactions will be discussed further in this chapter and in subsequent chapters. Immune responses generated as a result of antibody and antigen interactions are commonly referred to as antibody-mediated immunity, which can be considered to be part of the humoral immune response. Any immune component that affords protection and is not associated with the cellular fraction of body fluids is considered humoral.

In addition to PRRs, released soluble mediators and antibody, the immune system utilizes a number of specialized cells that participate in cell-mediated immunity. The mechanisms of cell-mediated immunity are largely independent of antibody and other humoral factors (such as complement proteins). Cell-mediated immunity involves the activation of immune cells and the subsequent deployment of cellular effector functions. For example, macrophages and neutrophils participate in cell-mediated immunity by phagocytosing invading pathogens, or infected host cells, and releasing a cascade of antimicrobial products. Phagocytosis is an important process that engulfs foreign substances and microbes and clears them from the body. NK cells and T cells also participate in cell-mediated immunity by recognizing and lysing virally infected cells. This mechanism is known as cytotoxicity, which kills infected or abnormal cells. The cells that are involved in cell-mediated immunity also secrete a number of cytokines, which regulate any ongoing immune response. Historically, cell-mediated immunity has been separated from antibody-mediated immunity or humoral immunity, depending on whether immunological protection can be found in the cellular fraction or the cell free fraction of body fluids, respectively.

Therefore, the key concepts in immunology can be summarized as components that activate immune cells and initiate immune responses, soluble mediators that signal to immune cells and regulate the immune response, components of antibody-mediated immunity (and also other humoral components), and constituents of cell-mediated immunity.

## 1.4 Innate versus adaptive immunity

Components of the immune system can be conveniently grouped into either the innate immune system or the adaptive immune system (Table 1.1). The innate immune system encompasses all those aspects of non-cellular

**Table 1.1** Comparison between the innate and adaptive immune systems.

| Innate Immune System | Adaptive Immune System |
| --- | --- |
| Rapid response (hours) | Delayed response (days) |
| Non-specific response to conserved molecules | Highly specific response to antigen |
| Response fixed (not adaptive) | Response adaptive (changes over time) |
| No immunological memory | Immunological memory |
| Humoral and cell-mediated components | Humoral and cell-mediated components |
| Components found in all animals | Only found in jawed vertebrates |

immunity, including epithelial barrier defence, antimicrobial peptide secretion, chemical barriers and the complement system. The innate immune system also involves aspects of cellular immunity associated with granulocytes (neutrophils, eosinophils, basophils and mast cells), monocytes, macrophages, DCs and NK cells. Cells of the innate immune response are rapidly initiated and are considered to be the first line of cellular defence against invading microorganisms (Figure 1.6). Essentially, innate immune cells provide protective immunity against infectious microorganisms until adaptive immune responses can be initiated.

The cell receptors expressed by innate immune cells recognize evolutionary conserved molecules derived from invading pathogens or damaged host tissues, for example through the recognition of danger signals. Each of these receptors recognizes the same molecular motifs, irrespective of the cause or progression of an immune response, and is therefore considered to be part of a

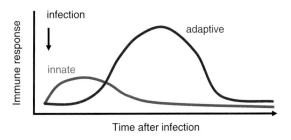

**Figure 1.6** Kinetics of a primary immune response. Innate immunity precedes the adaptive immune response.

non-specific response. For example, different bacteria will be recognized by macrophages in much the same way and will activate the same process of phagocytosis. Activation of non-specific innate immune receptors often leads to the development of an inflammatory reaction at the site of tissue damage. This also results in the release of several pro-inflammatory mediators that increase blood vessel permeability, stimulates the migration of more inflammatory cells into the area and is the first step in the activation of the adaptive immune system.

The adaptive immune system takes longer to establish itself and principally involves the T lymphocyte and B lymphocyte populations. Cells of the adaptive immune system are considered to be part of the specific immune response, due to the nature of the receptors that they express. T cell receptors (TCRs) and B cell receptors (BCRs) recognize very specific components of external substances, usually proteins, which are known as antigens. The entire T cell population is thought to consist of as many as $10^9$ different T cells, each of which displays a slightly different TCR. Similarly, the number of B cells, each capable of expressing a slightly different BCR, probably exceeds $10^9$. Therefore, the adaptive immune system has the capacity to identify a sizeable number of antigens, through the generation of antigen receptor diversity (discussed in detail in Chapter 3). T cells and B cells are therefore part of a highly evolved adaptive immune system that aims to maximize the recognition of as many different pathogens as possible. Furthermore, the mechanisms by which these receptors are produced enable the adaptive immune system to fine tune its response to antigen, so that subsequent responses are more effective. The plasticity built into the generation of antigen receptor diversity has the capacity to alter itself in response to new pathogenic challenges, hence the term adaptive immunity.

Following the stimulation of antigen specific T cells and B cells, a proportion of those cells differentiate into memory cells (Figure 1.7). These memory T cells and B cells are retained within various tissues of the immune system, until they are required at a later time point, when an individual becomes re-infected with the same pathogen. The adaptive immune system is sometime referred to as the acquired immune system, due to its ability to form populations of T cells and B cells that furnish the immune system with immunological memory.

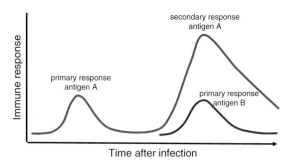

**Figure 1.7** Primary and secondary adaptive immune response. Secondary infections with the same pathogen elicits a more rapid and heightened response compared to a primary response.

## 1.5 Primary and secondary immune responses

The first time that the body encounters a new pathogen a primary immune response is triggered. It usually takes up to 7 days for a T cell response to become established, while it may take as long as 10–14 days for B cells to produce a significant amount of antibody that can be detected in the bloodstream. The reason for such a delay is that neither the B cell nor the T cell population has encountered such an antigen in the past and for this reason these cells are known as naïve B cells or naïve T cells. Therefore, these immune cells require sufficient time to recognize the antigen and start proliferating in order to produce sufficient numbers of antigen-specific clones. In the meantime, while the adaptive arm of the immune system becomes established, the innate immune system plays a vital role in controlling pathogen replication and dissemination. Once fully activated, the adaptive immune system prevents further infection and eventually eliminates the pathogen from the body. The primary immune response is then downregulated so that antigen-specific antibodies become less frequent and T cell numbers return to normal (Figure 1.7).

A secondary immune response occurs when an individual encounters the same pathogen for a second time and principally involves B and T cells of the adaptive immune system. For example, if a person has recovered from influenza infection and encounters the same strain of virus on a subsequent occasion, a secondary immune response is initiated. A secondary immune response is much quicker in establishing itself than a primary immune response is, because the antigens derived from the pathogen have been encountered before. The magnitude of a secondary immune response is also higher, meaning that more cells participate in the reaction, which results in a much more effective response (Figure 1.7). There is a rapid elevation in antibody levels within the bloodstream, which remain elevated for a longer period of time. This heightened response involves the activation of memory B cells, which can directly differentiate into antibody secreting plasma cells, without having to undergo the various stages of B cell development that a naïve B cell has to undergo. A secondary immune response is also dominated by the production of highly specific antibodies for a particular antigen. The antibodies produced during a secondary immune response are much more specific than those produced during a primary response (Figure 1.8). This is due to a process known as affinity maturation and antibody isotype switching (discussed in detail in Chapter 3), which involves a switch in IgM production to IgG production. Likewise, memory T cells are much more readily activated and they too have a heightening effector response, which is capable of responding more rapidly and with a greater magnitude than during a primary response.

The effectiveness of a secondary immune response relies on the generation of memory B cells and memory T cells (Figure 1.9), which develop following the initiation of a primary immune response. These memory cells reside

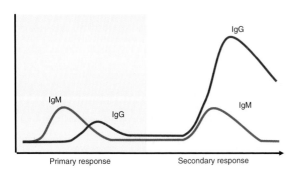

**Figure 1.8** Antibody production in primary and secondary immune responses. Primary antibody responses are initially dominated by IgM, while secondary responses are dominated by elevated IgG.

in various lymphoid tissues throughout the body and contribute to what is known as immunological memory. The term immunity was first used to describe the ability of the immune system to provide protection against infectious diseases and relies on the activation of memory B and T cells. Importantly, vaccination relies on the ability of the immune system to respond more effectively to a secondary encounter with antigen. Many infectious diseases can be prevented by vaccination (Figure 1.10), through the generation of antigen-specific memory cells that become activated in response to a challenge from the real pathogen.

## 1.6 Immune cell development

Before the array of different immune cells can exert their effector functions and prevent infection, they must first undergo a highly controlled series of developmental stages, collectively known as haematopoiesis. The bone marrow is extremely important for this process and for the continuity of the immune system. It is situated at the centre of all the long bones in the human body and consists mostly of a fatty substance surrounding a stroma of dividing stem cells. The major function of the bone marrow is to produce new lymphoid and myeloid cells, which originate from pluripotent haematopoietic stem cells. In fact, these special stem cells can give rise to any blood cell, hence the term haematopoietic, meaning blood forming. These cells then divide and develop into mature lymphocytes from the lymphoid line or monocytes, dendritic cells and granulocytes from the myeloid line (Figure 1.11). The common myeloid stem cell is also capable of differentiating into red blood cells and platelets. The differentiation of the many cells of the immune system occurs in precise developmental stages and at each of these stages a particular cell lineage is formed. This involves a complicated series of differential gene expression events that subsequently determines the commitment to a certain cell lineage. The genetic potential of a pluripotent haematopoietic stem cell is expansive and it is possible for that stem cell to differentiate into any one of a number of available cell types. As haematopoiesis proceeds, the haematopoietic stem cell becomes more and more specialized and its genetic potential becomes increasingly restricted. This eventually leads to cell fate

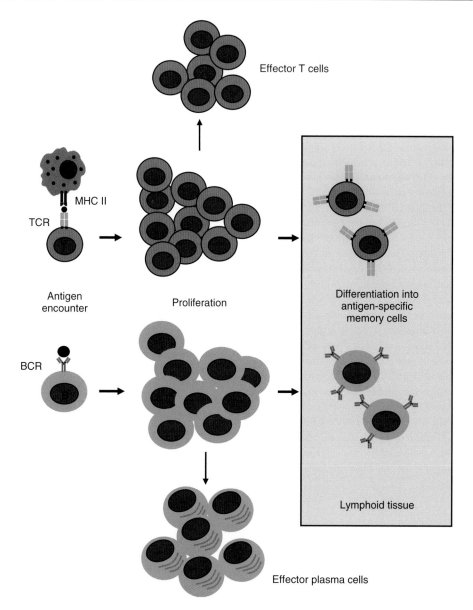

**Figure 1.9**  Generation of memory lymphocytes. Antigen stimulation leads to lymphocyte differentiation into effector cells. A subpopulation of lymphocytes differentiate into long-lived memory cells.

decisions that are irreversible and forces progenitor cells to continue down a particular cell lineage, for example the lymphocyte lineage.

The haematopoietic stem cell is driven to differentiate into a particular cell lineage based on what signals it receives from the extracellular environment within the bone marrow. One set of signals instructs the stem cell to differentiate into a common lymphoid progenitor cell and the other into a common myeloid progenitor cell. In other words, certain signals favour lymphocyte development, while other signals favour myeloid cell development. The lymphoid progenitor cell is capable of

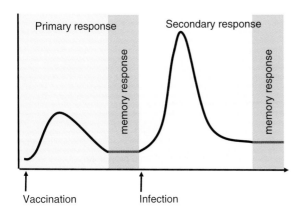

**Figure 1.10** Vaccination induces immunological memory. Pathogen-specific memory cells are generated following vaccination. Infection with the wild type pathogen reactivates these memory cells and a rapid and robust secondary immune response follows.

either differentiating into an NK cell precursor, or into a lymphocyte precursor that eventually gives rise to T cells and B cells. Several soluble factors are involved in driving this differentiation. For example, stem cell factor (SCF) is associated with both NK cell precursor and lymphocyte precursor development, while IL-7 signalling is specifically associated with T and B cell differentiation. T cell precursors leave the bone marrow and migrate to the thymus, where the final stages of T cell development take place. This results in the differentiation of immature T cells into either CD4+ T cells or CD8+ T cells. These two specialized T cell subsets will be discussed further in this chapter and extensively in Chapter 3. Following the migration of B cells out of the bone marrow, further B cell maturation takes place in other lymphoid organs such as the lymph nodes, spleen and organized lymphoid follicles associated with MALT. Within these lymphoid structures B cell maturation occurs whereby they mature into antibody secreting plasma cells.

Haematopoiesis also gives rise to all the cells of the granulocyte and myeloid lineages. For instance, the common myeloid progenitor cell can differentiate into erythrocytes, thrombocytes (platelets), granulocytes and monocytes, thereby demonstrating its pluripotent capacity within the haematopoietic system. The precursor cells of erythrocytes (red blood cells) are called reticulocytes, which leave the bone marrow and complete their maturation in the circulation. Myeloid progenitor cells also give rise to megakaryocytes that are the precursors for thrombocytes. The main precursor for all the cells of

the granulocyte lineage is the myeloblast, while the main precursor that gives rise to monocytes and macrophages is the monoblast. Mast cells, eosinophils, neutrophils and basophils diverge from the myeloblast lineage via independent precursor cells, while differentiated monocytes can subsequently mature into macrophages, once resident in tissues, or into myeloid DCs. Again, a number of growth factors and cytokines are involved in granulocyte and macrophage differentiation, including granulocyte/macrophage-colony stimulating factor (GM-CSF), G-CSF and M-CSF.

## 1.7 Mast cells and basophils

Mast cells and basophils are both granulocytes that are capable of rapidly releasing pro-inflammatory mediators into the extracellular environment, through a process known as degranulation. The release of pro-inflammatory mediators is a key process that initiates an inflammatory reaction. Examples of pro-inflammatory mediators include histamine, prostaglandins and cytokines, which will all be discussed in detail throughout the proceeding text. It was once thought that mast cells and basophils belonged to the same cell lineage; the circulating basophils giving rise to the mature, tissue residing mast cell. It is now clear that mast cells are derived from a separate precursor cell in the bone marrow, although they only fully mature once they reach their target organ. Mast cells can be detected in most tissues where they usually reside adjacent to connective tissue. They are also present in mucosal tissues such as the digestive, respiratory and urogenital tracts and the skin. Mucosal mast cells have slightly different characteristics to tissue dwelling mast cells, as they require the help of T cells to become fully activated, while tissue dwelling mast cells do not.

The primary function of mast cells is to provide an early response to the presence of microbial antigens. In order to recognize the presence of microorganisms, mast cells rely on antibodies interacting with their specific antigen. The interaction between an antibody and an antigen is then detected by the mast cells through the expression of a cell surface receptor that recognizes antibody, known as an immunoglobulin Fc receptor (FcR). When enough antigen is present, many antibody molecules cross link several adjacent FcRs. This antigen cross-linking is essential for receptor activation and in turn subsequent mast cell activation, degranulation and release of pro-inflammatory

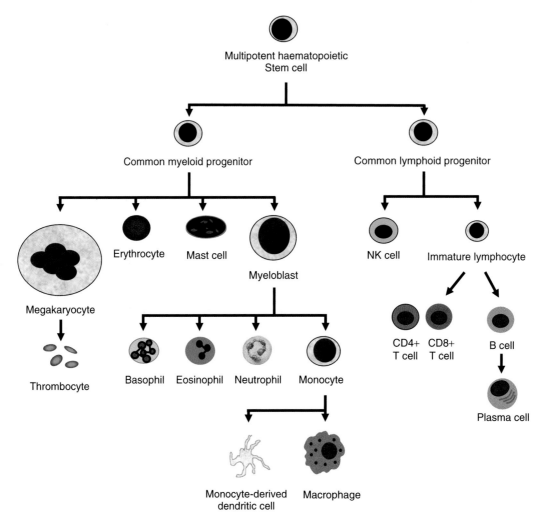

**Figure 1.11** Formation of blood cells through the process of haematopoiesis.

mediators. The biological importance of receptor cross-linking should be emphasized, as numerous receptors rely on cross-linking for proper activation. FcR cross-linking on mast cells results the release of an active mediator called histamine, which causes the dilation of blood vessels and stimulates lymphocyte migration into sites of inflammation. Mast cells and histamine have been implicated in the pathology associated with allergic asthma, whereby an allergen (an antigen involved in allergic reactions) cross links FcRs on the cell surface and causes bronchial constriction of the airways. Mast cells also release cytokines and chemokines, which attract eosinophils and other inflammatory cells.

Basophils have a similar function to mast cells and also participate in the release of histamine following FcR cross-linking. Although associated with allergic reactions, basophils and mast cells are thought to have evolved to combat parasitic infections. The degranulation of basophils results in the recruitment of other immune effector cells, following the release of histamine, leukotrines and the cytokine interleukin-4 (IL-4). The cytokine IL-4 is a critical cytokine for the development of T cell responses associated with both parasitic and allergic immune reactions. Although they are the least common of the leukocytes in the blood, basophils can constitute a significant proportion of cells migrating into sites of allergic inflammation. Therefore basophils, and mast cells, are important cells of the innate immune system capable of an immediate response to an invading pathogen, and are key initiators of subsequent immune responses.

## 1.8 Eosinophils

Eosinophils are granulocytes that reside in the circulation, MALT and lymphoid organs and are derived from precursor cells in the bone marrow. Under normal homeostatic conditions, they are not frequently observed in healthy mucosal tissues but do increase in number during diseases such as asthma, atopic dermatitis or during helminth infections. The primary function of eosinophils is the release of their granule contents (degranulation), which contain histamine, leukotrines, prostaglandin, reactive oxygen species (including eosinophil peroxidase and superoxide), growth factors and cytokines such as IL-4, IL-5 and tumour necrosis factor (TNF). Eosinophil degranulation therefore induces a local inflammatory microenvironment that attracts other immune effector cells and also results in direct damage to an invading parasite and subsequently to surrounding tissue. A number of factors result in the activation of eosinophils including the cross-linking of FcRs on the cell surface, but also in response to chemokine or cytokine receptor ligation. The chemokines RANTES (regulated upon activation, normal T-cell expressed and secreted) and eotaxin are potent inducers of eosinophil degranulation, while the cytokine IL-5 is important for maintaining eosinophils at sites of inflammation. Eosinophils are also highly phagocytic, particularly in response to antigen-antibody complexes. However, against larger parasites, such as schistosomes, eosinophils rely on the release of reactive oxygen species directly onto the surface of the parasite.

## 1.9 Neutrophils

Neutrophils are granulocytes that are predominantly found in the blood and may account for as much as 70 per cent of the total circulating leukocyte population. They respond rapidly and effectively to inflammatory signals that originate from sites of infection in tissues, particularly in response to bacterial infection and tissue damage. They are able to exit the circulation through the endothelium of blood vessels and enter sites of inflammation through a process known as chemotaxis. In particular, IL-8 and the complement factor C5a (discussed in Chapter 2) are potent chemotactic factors that recruit neutrophils from the blood and into the area of inflammation. Neutrophils adopt several effector mechanisms that aim to combat the spread of infection or kill the invading pathogen. The first of these is degranulation in response to activation signals such as IL-8 and the cytokine interferon-$\gamma$ (IFN-$\gamma$). Neutrophils release a number of antimicrobial peptides, such as defensins, cathepsins and lactoferrin, which have direct toxic effects on bacteria. In addition, neutrophils release reactive oxygen species and enzymes such as superoxide, hydrogen peroxide, nitric oxide synthase (NOS) and NADPH oxidase, a process known as the respiratory burst that is directly toxic to pathogens and surrounding cells (detailed in Chapter 2). A second effector function is the phagocytosis of bacteria by neutrophils, which places the pathogen inside a phagosome (intracellular vacuole) where more reactive oxygen species are released. Lastly, neutrophils secrete a number of pro-inflammatory cytokines, including IL-12, IFN-$\gamma$ and TNF, which are involved in the initiation of subsequent immune responses.

## 1.10 Monocytes and macrophages

Monocytes are derived from the myeloid lineage of haematopoietic stem cells in the bone marrow and constitute approximately 5 per cent of the circulating leukocytes found in the blood. In addition, almost half of all monocytes are stored in the red pulp of the spleen, where they act as a reserve of innate immune cells. In response to infection, monocytes can quickly migrate into tissues where they mature into macrophages or into monocyte-derived DCs. Within tissues, differentiated macrophages are the primary phagocytosing cell of the immune system. They are responsible for phagocytosing a multitude of foreign substances including toxins, macromolecules, cell debris, whole dead cells and microorganisms. Macrophages become active phagocytes following recognition of antibodies or complement factors that stick to the surface of pathogens or infected cells in a process known as opsonisation. Antigens present on the surface of pathogens are opsonized by antibodies, which is enhanced by complement binding to cell membranes. Macrophages are able to recognize these bound antibodies, which signal to the macrophage to induce phagocytosis.

Macrophages are also considered to be professional antigen presenting cells (APCs). The process of phagocytosis involves the digestion and processing of proteins derived from the ingested microorganisms. These proteins are presented on the surface of macrophages and are recognized by T cells. The presentation of antigen by macrophages and recognition of antigen by T cells results in the activation of T cells and the initiation of an adaptive immune response. The process of antigen processing and presentation will be discussed at length in Chapter 3. In addition, activated macrophages are important sources of the pro-inflammatory cytokines IL-1, TNF and IL-12, which are also important in generating adaptive T cell responses.

## 1.11 Dendritic cells

Dendritic cells (DCs) are derived from precursor cells in the bone marrow and are characterized by a particular morphology involving long cellular extensions known as dendrites. There are several sub-populations of DC, located within different tissues throughout the body, which can be broadly defined as either having a myeloid or lymphoid lineage. Myeloid DCs (mDCs) are thought to be derived from monocytes and can be differentiated from peripheral blood mononuclear cells (PBMCs). Lymphoid-derived DCs are known as plasmacytoid DCs, due to their resemblance to differentiated plasma cells, and are located throughout mucosal tissues. Both populations of DC are professional APCs and are potent activators of the immune system. In particular, mDCs are the most effective APC capable of activating an antigen-specific T cell response. Other DC populations include Langerhan's cells predominantly found in the epidermis of the skin and interstitial iDCs that reside in the dermis of the skin and the stromal compartments of mucosal tissues. However, the distinction between these subsets is often blurred and may reflect patterns of migration or functional specialisations within certain tissue microenvironments.

The primary function of DCs is the presentation of antigen to T cells and the initiation of an adaptive immune response. DCs are often considered to be the critical link between the innate and adaptive immune systems. In addition, they are effective at recognizing the presence of potentially harmful microorganisms through the system of cell surface molecules called pattern recognition receptors (PRRs). These PRRs, including the TLR family, are able to recognize evolutionarily conserved molecular motifs (PAMPs), on the surface of bacteria, viruses, fungi and multicellular parasites. The recognition of foreign substances is a crucial concept in the initiation of any immune response and is often referred to as a danger signal. Without such a danger signal the immune system is not activated, which is how we prevent our immune cells responding to harmless environmental substances or commensal microorganisms. Activation of danger signals through the recognition of PAMPs is one of the first events during an adaptive immune response and results in the release of IL-12 from DCs and migration into regional lymph nodes. The cytokine IL-12 primes T cells for activation, while the architecture of lymph nodes provides an ideal location for antigen presentation to T cells.

Antigen presentation is another fundamental process by which adaptive immune responses are initiated. Exogenous antigen is endocytosed by DCs and processed into small peptides. These peptides are then loaded onto a carrier protein known as major histocompatibility complex (MHC) class II. The processing and presentation of antigen in context with an MHC class II molecule is essential for the initiation of an adaptive immune response. T cells are able to recognize the peptide-MHC complex via its T cell receptor (TCR), which provides the first signal for T cell activation. Antigen presentation by DCs is a central process during the initiation of an immune response and will be discussed at length in Chapter 3.

Another subset of DCs, referred to as follicular DCs (FDCs), resides inside lymphoid follicles within lymph nodes and organized lymphoid structures associated with mucosal sites. They are dendritic in morphology and are responsible for presenting soluble antigen to B cells during B cell development and maturation. However, they should not be confused with mDCs or pDCs as they do not have their origins in the haematopoietic system but rather are derived from mesenchymal cells.

## 1.12 Natural killer cells

Natural killer or NK cells are members of the innate immune system that are derived from a common lymphoid precursor and are therefore classed as a lymphocyte. However, they do not express TCRs or immunoglobulins like T and B cells do, and therefore their cell receptors do not give them the capacity to adapt to external antigens.

They are known as NK cells due to their potent cellular cytotoxicity against tumour cells and virus infected cells. They are capable of releasing granule components onto the surface of infected cells to induce cell death. NK cell granules contain the cytotoxic proteins perforin and granzyme that induce target cells to undergo apoptosis. There are several mechanisms by which NK cells become activated, including a process known as antibody-dependent cellular cytotoxicity (ADCC). Antibodies that bind to the surface of infected cells are recognized by receptors on NK cells, specifically FcRs that recognize antibody molecules (Figure 1.12). The cross-linking of multiple FcRs activates the NK cell and induces the release of its granule contents. Perforin punches tiny holes in the surface membrane of target cells, while granzyme enters the cell and provokes it to undergo apoptosis.

An alternative means by which NK cells exert their cytotoxicity is through a mechanism known as the recognition of missing self. NK cells express surface receptors that recognize several molecules expressed by host cells. These receptors recognize normally expressed molecules on host cells, which delivers a signal to NK cells that informs them not to kill the healthy cells. This signal is known as an inhibitory signal and is recognized by particular NK cell receptors called killer cell immunoglobulin-like receptors (KIRs). However, when a cell becomes infected with a virus, for example, molecules that are normally expressed on the cell are often downregulated. In effect, this takes away the inhibitory signal from the NK cell and allows the NK cell to exert its cytotoxic functions, thereby killing the infected cell. Therefore, in normal circumstance KIRs recognize molecules on healthy cells and leave them alone. However, when these molecules are downregulated in response to infection, the lack of an inhibitory signal allows the NK cell to initiate effector functions, a term known as the recognition of missing self.

## 1.13 CD4+ T helper cells

CD4+ T helper (Th) cells are lymphocytes that originate from the bone marrow but complete their development in the thymus (the name T cell is derived from the stages of thymic development). Thymic development is a critical period for the differentiation of T cells and involves certain selection processes that only allow functional T cells to enter the periphery (positive selection). This selection process also prevents self-reactive T cells from leaving the thymus (negative selection). The details of thymic selection will be discussed in Chapter 3, although it should be noted that T cell selection is another way the immune system discriminates between self and non-self. Thymic selection is important for the prevention of self reactive T cells and the development of autoimmune reactions. The selection process in the thymus ensures that T cells are tolerant to self-antigens. T cell tolerance can be considered to be the opposite of autoimmunity.

Th cells are characterized by the expression of the CD4 molecule, which is a co-receptor for the αβTCR. CD4+ T cells are activated when they recognize a peptide antigen that is bound to the surface of an antigen presenting cell such as a DC or macrophage. In particular, CD4+ Th cells recognize peptide antigen bound to an MHC class II molecule on the APC (Figure 1.13). The function of CD4+ Th cells is to assist with the initiation and activation of other cells of the immune system. They provide the necessary signals for CD8+ T cells to exert their cytotoxic function (discussed in the next paragraph); they interact with B cells within secondary lymphoid organs and induce antibody production and

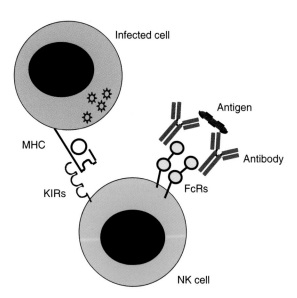

**Figure 1.12** Receptor expression on NK cells, which includes killer cell immunoglobulin-like receptors (KIRs) and FcRs. KIRs recognize abnormal expression of MHC molecules on the surface of infected cells, while FcRs bind to cross-linked antibody/antigen complexes.

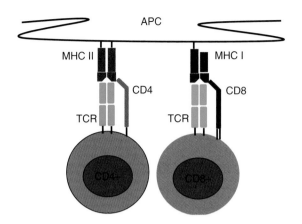

**Figure 1.13** CD4+ T helper cells express TCRs that recognize MHC class II molecules, while CD8+ cytotoxic T cells recognise MHC class I molecules. Receptor binding is stabilised by the co-receptors CD4 and CD8.

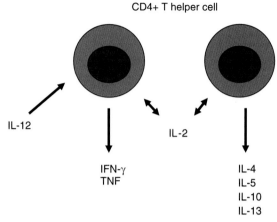

**Figure 1.14** Th1 versus Th2 differentiation and cytokine production. IL-12 induces naïve Th cells to differentiate into Th1 cells that express IFN-γ. The lack of IL-12 causes naïve T cells to differentiate into Th2 cells that express IL-4, IL-5, IL-10 and IL-13.

they stimulate macrophage phagocytosis. Therefore, they have acquired the term helper T cell (Th cell). A major part of CD4+ Th cell function is the release of stimulatory cytokines that promote proliferation, differentiation or effector functions. For example, IL-2 is produced by all CD4+ T cells and is important for the proliferation and maintenance of both CD4+ Th cell and other T cell populations.

Depending on which cytokines CD4+ Th cells produce, they can be further differentiated into Th1 and Th2 cells (Figure 1.14). This is a widely used nomenclature that helps to determine the type of immune response that an individual develops. For instance, viral infections tend to induce Th1-type immune responses, while intestinal parasites and allergic reactions are associated with Th2-type immune responses. The reason for possessing different types of immune responses is that each response is better suited to clearing a particular pathogen. Therefore, Th1-mediated immune responses are ideal at fighting intra-cellular viruses, while Th2-mediated immune responses are better at clearing extracellular parasites. The cytokines produced by Th1 and Th2 cells differ (Figure 1.14), so that Th1 cells predominantly express IFN-γ and TNF, while Th2 cells predominantly express IL-4, IL-5, IL-10 and IL-13. The differences in cytokine production by Th cells have a significant effect in the type of immune response that is initiated. Th1 and type-1 cytokines are associated with the development of cell-mediated immunity. For example, they stimulate cytotoxic functions in CD8+ T cells and NK cells and

induce macrophages to undergo phagocytosis. On the other hand, Th2 responses are associated with antibody-mediated immunity, enhanced antibody production by B cells and allergic-type reactions.

Another important CD4+ T cell population exists that are known as T regulatory cells (Tregs), which are characterized by the production of IL-10 and transforming growth factor (TGF)-β. Treg cells are responsible for the regulation of immune responses, so that any immunopathological consequences of inflammation are minimized. IL-10 and TGF-β are thought to be the principle immunoregulatory cytokines that act to dampen down excessive immune responses.

In summary, CD4+ T cells can be divided into several sub-populations whose diverse functions are mainly attributed to the cytokines that they produce. They are key mediators of immunity that instruct and orchestrate the pattern of immune cell development and differentiation, either through activation or regulation of the immune response.

## 1.14 CD8+ cytotoxic T cells

CD8+ T cells are lymphocytes that are commonly known as cytotoxic T lymphocytes (CTLs) due to their cytolytic capabilities against virally infected cells and tumour cells. Like CD4+ Th cells, CD8+ CTLs originate in the bone marrow and differentiate in the thymus. CTLs also undergo a similar selection process in the thymus, whereby only those CD8+ T cells with a functional αβTCR are allowed to survive (positive selection), while those that are reactive to self antigens undergo apoptosis (negative selection). The CD8 molecule expressed by CTLs acts as a co-receptor with the αβTCR, which specifically recognizes MHC class I molecules expressed on the surface of all somatic cells (Figure 1.13). The recognition of peptide: MHC class I complexes by TCRs on the surface of CD8+ CTLs activates their cytolytic machinery. The cytolytic factors are similar to those released by NK cells and include perforin and granulysin, which have the desired effect of killing virally infected cells through the induction of apoptosis. In addition, CTLs may be able to kill target cells through the surface expression of FasL, which binds to its receptor Fas on the target cell. The ligation of Fas initiates an apoptotic pathway within the target cell, ending in cell death. Therefore, CD8+ CTLs are important lymphocytes that kill infected or abnormal cells.

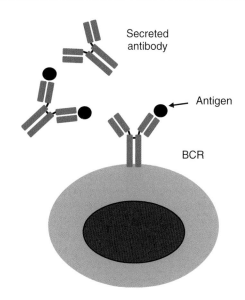

**Figure 1.15** Immunoglobulin expression by B cells can take two forms. When bound to the cell surface of B cells it forms part of the B cell receptor complex (BCR). When secreted by plasma cells it is known as antibody. Both bind to antigen.

## 1.15 B cells

The lymphocytes that are responsible for the production of antibodies are called B cells, so named as they were originally characterized in the avian lymphoid organ the bursa of Fabricius. In most mammals, including humans, B cells are derived from the bone marrow and therefore the nomenclature remains appropriate. The term antibody is also interchangeable with immunoglobulin, which when bound to the surface membrane is known as the B cell receptor (BCR). In general, an immunoglobulin that is secreted by a B cell is known as an antibody, while one bound to the surface of a B cell is known as the BCR (Figure 1.15). B cells that leave the bone marrow continue to mature in the spleen, lymph nodes and secondary lymphoid aggregates of MALT. Each B cell produces antibody that selectively binds to a particular antigen and is known as a B cell clone. Within secondary lymphoid tissues, B cell clonal selection takes place, whereby only those B cell clones that produce antibody with the highest affinity to antigen are preferentially selected. Those B cell clones with low affinity to antigen are deleted. The clonal selection theory is an important concept in immunology that allows a better understanding of how antibody responses are generated. Clonal selection ensures that those antibodies that are selected have the highest possible affinity to an antigen. This favours antigen clearance and the removal of infectious microorganisms from the body.

Mature B cells differentiate into plasma cells that produce significant amounts of secretary antibody. Successful B cell clones also divide and produce more B cells that express antibody of the same specificity, thereby effectively amplifying an antibody response. Antibody production is an important mechanism for the clearance of antigens from the body. This may include the clearance of whole pathogens following antibody binding to antigens on the surface of microorganisms. In addition, B cells can also act as professional APCs under certain circumstances, for example during interactions with CD4+ Th cells within secondary lymphoid tissues.

The majority of antigens recognized by the membrane bound BCR are not sufficient to induce B cell activation alone, but rather secondary signals are required from CD4+ Th cells. The antigens that these B cells recognize are therefore known as T-dependent antigens. On the other hand, some antigens are able to stimulate B cell activation in the absence of T cell help and are therefore known as T-independent antigens. These antigens tend to be formed of multiple repeating units that are able to cross-link BCRs and provide a sufficiently strong signal to induce activation, differentiation and subsequent antibody secretion. However, it is thought that the vast majority of antigens are T-dependent antigens.

## 1.16  γδ T cells

Conventional T cells, such as CD4+ Th cells and CD8+ CTLs, express the αβTCR that comprises one α-chain and one β-chain. However, a population of unconventional T cells express a different TCR called the γδTCR, made up of one γ-chain and one δ-chain. Instead of recognizing peptide:MHC complexes, the γδTCR is thought to bind to certain non-classical MHC molecules, which present phosphorylated or lipid antigens rather than normal peptide antigens. γδ T cells are far less common than αβ T cells and are predominantly located in mucosal tissues such as the gut, where they tend to reside in the intra-epithelial compartment. Therefore, they are often known as intra-epithelia lymphocytes (IELs). They possess some characteristics of adaptive immune cells, such as a TCR, while at the same time resemble innate immune cells. Although it is not entirely clear what their precise role is in the immune system, it is thought that they may be important during the early phase of an immune response or as regulatory sub-population of T cells.

## 1.17  Natural killer T cells

NKT cells represent a small population of T cells that share properties of both NK cells and conventional T cells. The TCRs expressed by the conventional T cell population are highly variable, so that they have the potential to recognize many different peptide antigens. However, the TCRs of NKT cells are far less variable and can only recognize a limited number of antigens, mostly glycolipids. The most characterized population of NKT cells express a restricted αβTCR repertoire that recognizes the non-classical MHC molecule CD1d, and are known as invariant NKT cells (iNKT cells). The functional significance of NKT cells is only just beginning to be appreciated. They are effective producers of IFN-γ, TNF and IL-4 and therefore are capable of contributing to pro-inflammatory signals. However, they may also play a role in regulating the immune system, through the recognition of lipid antigens derived from both pathogens and cellular sources resulting from tissue damage.

**Table 1.2**  Primary and secondary lymphoid tissues.

| Primary lymphoid tissues | Secondary lymphoid tissues |
| --- | --- |
| Bone marrow | Spleen |
| Thymus | Lymph nodes |
| Bursa of Fabricius (birds) | Peyer's patches (gut) |
| | Tonsil and adenoids (airways) |
| | Appendix |

## 1.18  Anatomy of the immune system

The physical structure of the immune system comprises many organs and tissues of the body, from the bone marrow, which is responsible for generating new cells, to the localized lymph nodes where lymphocytes are primed to mount their attacks on invading pathogens. The organs of the immune system can be divided into primary lymphoid organs and secondary lymphoid organs (Table 1.2). The bone marrow and the thymus are examples of primary lymphoid organs, where immune cells are generated and undergo differentiation. Secondary lymphoid organs include the lymph nodes, Peyer's patches and tonsils and are sites where immune responses are initiated. Secondary lymphoid organs are known as inducer sites, because they provide the correct environment for immune responses to be induced. Sites of inflammation or infection, where the actual immune response takes place and where leukocytes exert their biological functions, are known as effector sites. Examples of effector sites might include the dermis of the skin, the epithelium of the intestine or the bronchial lining of the airways. Without the presence of inducer sites, immune responses can not be effectively mounted and therefore secondary lymphoid tissues are crucial for the normal function of the immune system.

## 1.19  Lymph nodes

Lymph nodes are encapsulated, organized secondary lymphoid tissues that provide the immune system with several important functions. They are situated at various sites throughout the body, often in clusters in the abdomen, chest, neck, under the arms and in the groin area, and are integrated with the vascular lymphatic system. One function of lymph nodes is the non-specific filtration and accumulation of particulate antigens and microorganisms. Lymphatic fluid that drains from peripheral tissues enters a lymph node via the afferent lymphatic vessels. Within the lymph node any particulate antigens or microorganisms carried within the lymph are deposited, where the phagocytic activity of macrophages removes them from the system. Another important function of lymph nodes is the provision of an anatomical microenvironment where antigen is filtered and presented to lymphocytes by APCs. Lymphocytes and APCs enter lymph nodes via the bloodstream, through specialized capillaries called high endothelial venules. The structure of the lymph node is such that the likelihood of antigen encounter is maximized, thereby providing a greater opportunity for an interaction between an APC, T cell and B cell. Lymphocytes leave lymph nodes via the efferent lymphatic vessels, enter the circulating lymph and eventually rejoin the bloodstream via the large lymph vessels

of the thoracic duct or right lymphatic duct. Lymphocytes activated within a lymph node either migrate to peripheral tissues where they exert their effector functions, or they continue to circulate and eventually re-enter another lymph node.

The anatomical organisation of lymph nodes (Figure 1.16) also enables them to function as sites of B cell and T cell activation and proliferation. Most lymph nodes are bean-shaped structures only a few millimetres long but can considerably increase in size and cellular mass following infection. Lymph nodes are encapsulated by a highly collagenous membrane, which is punctuated by afferent lymphatic vessels that bring draining lymph into the lymph node. The inside of a lymph node can be divided into a highly cellular outer layer known as the cortex and a less cellular inner layer called the medulla. Afferent lymphatic vessels drain into sinuses within the cortex, which are channelled toward the medulla region where efferent lymphatic vessels converge and drain lymph away from the node. Blood vessels enter lymph nodes via the medullary cords and form extensive

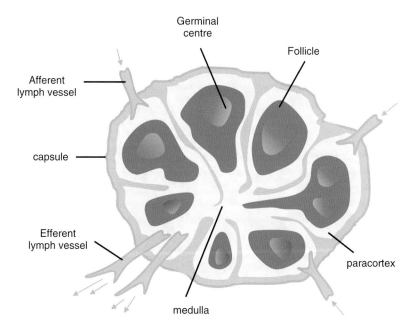

**Figure 1.16** Anatomy of a lymph node. A lymph node receives lymph from the afferent lymphatics, while lymph exits a lymph node via the efferent lymphatics. Lymph node architecture consists of T cell areas called the paracortex, which surround B cell areas known as lymphoid follicles. Mature follicles contain germinal centres where B cell differentiation takes place. A central medulla contains migrating lymphocytes.

capillary beds within the cortex. Therefore, the anatomy of a lymph node is such that lymph entering a node via the afferent lymphatics is filtered within the cortex, while lymphocytes enter the cortex in the opposite direction and migrate into areas where antigen is trapped and cells become aggregated. This greatly enhances the chances of a lymphocyte encountering antigen and initiating an immune response.

The outer regions of the cortex contain areas particularly rich in densely packed lymphocytes known as lymphoid follicles, which are in turn surrounded by a less cellular area called the paracortex. The lymphoid follicles represent regions of B cell proliferation, while the paracortex is dominated by proliferating T cells. Although the majority of T cells are located throughout the paracortex, some can also be detected within lymphoid follicles.

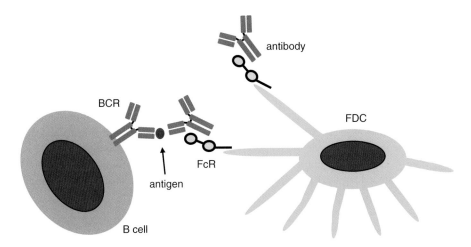

**Figure 1.17** Follicular dendritic cells (FDCs) present antigen to B cells in germinal centres. The antigen is bound by pre-existing antibody and attached to FDCs via FcRs. B cell proliferation and maturation requires the presentation of antigen in this way.

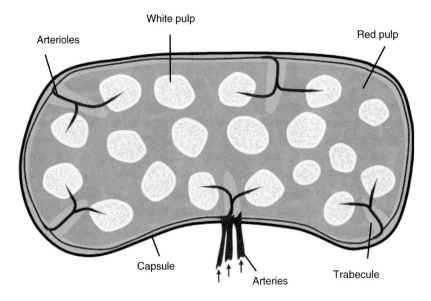

**Figure 1.18** Anatomy of the spleen. The spleen consists of a lymphocyte-rich white pulp and a red blood cell-rich red pulp. The spleen receives antigens derived from the blood.

These are thought to be Th cells involved in the activation of B lymphocytes. Within some lymphoid follicles a less dense area of cells can be observed, termed the germinal centre, which is thought to be an area where B cell and T cell interactions take place and where B cell proliferation occurs. Follicles without germinal centres are known as primary follicles, while those with a germinal centre are classified as secondary follicles. Specialized stromal cells also exist within germinal centres called follicular dendritic cells (FDCs), which are responsible for trapping and retaining antigens (Figure 1.17). Presentation of antigen by FDCs induces B cells specific for that antigen to undergo proliferation and start producing antibody. Activated B cells leave lymphoid follicles and migrate through the paracortex into the medulla, where they mature into antibody secreting plasma cells. The medulla is also the site where efferent lymphatic vessels leave the lymph node and eventually drain into the bloodstream via the thoracic duct or right lymphatic duct, thereby transporting activated lymphocytes into the circulation.

## 1.20 Spleen

The spleen is a large organ of the reticuloendothelial system that functions like a large lymph node, but rather than sampling material from the lymphatic system, the spleen filters components from the blood. It is situated in the upper left quadrant of the abdomen where it receives blood from the splenic artery. Efferent lymphatics within the spleen drain tissue fluid, although no afferent lymphatics are present. Instead blood is drained into the hepatic portal system via the splenic vein. The spleen has two main functions, the first of which is to filter particulate material and defunct red blood cells (erythrocytes) from the blood and second, to provide a site for the initiation of immune responses against antigens derived from the blood. It can be divided into the red pulp and the white pulp, which are responsible for the filtration of red blood cells and the activation of immune responses, respectively (Figure 1.18). Although the primary function of the red pulp is the removal of red blood cells, it also acts as an important reservoir for immature monocytes. The splenic artery branches into arterioles that split into capillary beds that in turn drain into the red pulp sinuses. Within the red

pulp sinuses are many macrophages that contribute to the clearance of particulate matter and dying red blood cells. Blood flows into the venous sinuses that give way to larger blood vessels which eventually enter the splenic vein.

Most of the spleen is composed of red pulp, while the white pulp can be divided into the B cell-rich lymphoid follicles and the peri-arteriolar lymphoid sheaths that act as sites for T cell activation. As the splenic artery enters the spleen it branches into many arterioles that drain into the sinuses of the red pulp. Surrounding these arterioles are the peri-arteriolar lymphoid sheaths, although in humans these sheaths are less well defined than in other mammalian species. Nevertheless, large T cell areas are located around arterioles that merge into the red pulp. The B cell-rich lymphoid follicles are situated adjacent to the peri-arteriolar lymphoid sheaths, often where an arteriole branches. Splenic lymphoid follicles also contain germinal centres, which are similar to those within lymph nodes. Certain areas adjacent to arterioles are composed mainly of plasma cells and are also reminiscent of the medullary zone found in lymph nodes. The area of red pulp that immediately surrounds lymphoid aggregations is known as the perifollicular zone. Blood flow through the spleen is much slower within the perifollicular zone, which is thought to maximize the interaction between blood-borne antigen and the lymphocytes present within the spleen.

## 1.21 Summary

**1.** The immune system evolved to defend the body against infectious diseases and can be divided into non-specific innate immunity and specific adaptive immunity.
**2.** The immune system is able to recognize self (its own molecules) from non-self (foreign molecules).
**3.** Some immune responses are characterized by antibody-mediated immunity, while others are characterized by cell-mediated immunity.
**4.** An immune response involves a number of key stages: pathogen recognition, activation and initiation, regulation and the generation of immunological memory.
**5.** All leukocytes (white blood cells) are derived from a process of haematopoiesis in the bone marrow, which gives rise to granulocytes, myeloid cells and lymphocytes.

# 2 The Innate Immune System

## 2.1 Introduction to the innate immune system

The innate immune system provides the first line of immunological defence against infection. The innate immune system is distinct form the adaptive immune system and is characterized by several key factors. Innate immunity relies on generic protection using molecules and receptors that are somatically expressed and phylogenetically conserved. In other words, the molecules and receptors of the innate immune system are considered to be non-specific, providing a broad range of protection. For this reason the innate immune system is sometimes referred to as providing natural immunity. Elements of innate immunity are found throughout the entire animal kingdom, from simple invertebrates to the more complex vertebrates. This differs from the adaptive immune system, which is only found in higher vertebrates and utilizes receptors that are highly specific for a particular antigen and allow adaptation and increased specificity to an infectious pathogen. Unlike the adaptive immune system, the innate immune system does not afford immunological memory or provide long lasting protection against infection. The principal components of the innate immune system include physical barrier defence (e.g. skin and mucosal epithelia), chemical barriers (e.g. antimicrobial peptides and reactive oxygen species), innate immune cells (e.g. granulocytes, monocytes, DCs and NK cells), components of humoral immunity (e.g. complement factors and innate antibodies) and associated cytokines. Although innate and adaptive immunity are readily separated on the basis of differing functionality, there is considerable interaction between the two systems.

The functions of the innate immune system are therefore diverse. This includes the prevention of pathogens from entering the body through the formation of physical barriers and the release of antimicrobial mediators; the prevention of the spread of infections by the activation of the complement cascade and other humoral factors; the removal of pathogens from the body through mechanisms of phagocytosis and cytotoxicity; and finally the activation of the adaptive immune system through the synthesis of cytokines and the presentation of antigens to T cells and B cells (Table 2.1). These innate immune functions can vary between different tissues. Therefore, rather than discussing individual detail within this chapter, the reader is directed to the relevant sections describing each mucosal tissue in turn. For example, the epithelial layers at mucosal surfaces are such an important aspect of innate immunity, and mucosal biology in general, that they are given detailed analysis in subsequent chapters. The remainder of this chapter will therefore focus on those elements of innate immunity that are generic and are shared by the various cells that comprise the innate immune system.

## 2.2 Innate immune receptors and cells

The reason why cells of the innate immune system are considered to provide non-specific immunity is due to the nature of their receptors. Cells of the innate immune system are able to recognize distinctive molecular structures that are synthesized by foreign pathogens. Moreover, these molecular structures are only produced by viruses, bacteria, fungi or parasites and are not found on cells derived from the host. Innate immune cells can therefore determine the difference between foreign organisms and self tissue, based on the general structure of pathogen molecules. The response of innate immune cells to foreign molecules is much the same, so that the recognition of the same molecular patterns results in the activation of similar signalling pathways and the induction of innate immune mechanisms, hence the non-specificity (Figure 2.1). Immunological specificity arises from receptors that are expressed on cells of the adaptive immune

*Immunology: Mucosal and Body Surface Defences*, First Edition. Andrew E. Williams.
© 2012 John Wiley & Sons, Ltd. Published 2012 by John Wiley & Sons, Ltd.

**Table 2.1** Innate immune mechanisms.

| Innate immune mechanism | Cell type | Function |
| --- | --- | --- |
| Barrier defence | Epithelial cells | Prevent pathogen entry |
| Humoral immunity | Soluble factors | Opsonisation |
| | Complement proteins | Cytotoxicity |
| | Acute phase proteins | Systemic effects |
| | Antibodies | |
| Phagocytosis | Macrophages | Engulfment of foreign particles and cell lysis |
| | Neutrophils | |
| Cytotoxicity | NK cells | Cell lysis |
| | Neutrophils | |
| | Eosinophils | |
| Cell signalling | All innate immune cells | Activation of innate and adaptive immune responses |
| Antigen presentation | Macrophages | Activation of T and B cells |
| | Dendritic cells | |

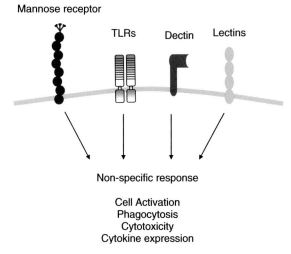

**Figure 2.1** Innate immune effector function resulting from non-specific receptor binding. Innate immune receptors recognize conserved molecular patterns derived from pathogens. Receptor binding leads to a non-specific response.

system, such as T cell receptors and B cell receptors. Importantly, there is a link between the innate and adaptive immune systems. Induction of innate immunity results in the recruitment and activation of adaptive immune cells. The immune response as a whole therefore provides a systematic means of defending against invading pathogens. Recognition of pathogens results in the activation of innate immune cells and the induction

of innate defence mechanism, while at the same time it induces the expression of mediators that activate the adaptive immune system. Innate immunity can therefore be divided into recognition mechanisms, effector mechanisms and induction mechanisms.

The cells of the innate immune system are derived from the granulocyte, myeloid and lymphoid lineages. Therefore, all members of the granulocyte family are considered to be part of the innate immune system, along with monocytes, macrophages, dendritic cells and NK cells. Together these cells supply the innate immune system with three separate mechanisms: the recognition of foreign molecules and discriminating between self and non-self; killing pathogens or infected cells and removing foreign particles through phagocytosis or cytotoxicity; and providing signals for the induction of inflammation and the recruitment of adaptive immune cells (Figure 2.2). These fundamental mechanisms of innate immune cells will be discussed in turn. In addition to receptors expressed on the surface of cells, soluble proteins contained within blood serum and tissue fluids also bind to microorganisms. Collectively, these soluble proteins are components of the humoral immune system and include complement proteins and acute phase proteins. Antibodies are also considered to be part of the humoral system, although they originate from B cells, which are part of the adaptive immune system, and will therefore be discussed in detail in Chapter 3. When these proteins bind to the surface of pathogens or infected cells they are known as opsonins, while the foreign particles or cells are said to have been opsonized. The mechanism

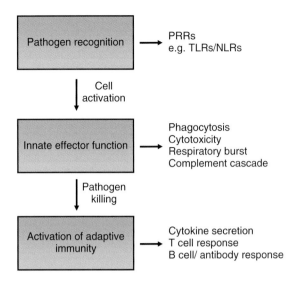

**Figure 2.2** Functions of the innate immune system. Activation of effector functions requires pathogen recognition. Innate effector functions are diverse and include cell-mediated and humoral components. Innate immune cells also activate the adaptive immune response.

of opsonization is important for the induction of effector mechanisms such as phagocytosis, cytotoxicity and release of pro-inflammatory mediators.

## 2.3 TLRs and pattern recognition

Toll-like receptors (TLRs) play a significant role in innate immune defence by recognizing evolutionarily conserved microbial structures and providing downstream signals for cell activation. They were first discovered in the fruit fly *Drosophilia melanogaster*; the original toll gene being initially implicated in development and later in immunity against certain fungi. Homologues were soon discovered in mammals, of which there are a total of 13 that have so far been identified, 10 recognized in humans and 12 in mice, of which TLR1-9 are conserved between the two species. These TLRs are all encoded by the germline and are known as pattern recognition receptors (PRRs). Each TLR specifically recognizes one type of microbial component, or pathogen associated molecular pattern (PAMP), such as lipopolysaccharide (LPS), lipopeptides, RNA or DNA. The expression of several different TLRs enables a number of different PAMPs to be recognized (Figure 2.3), allowing the detection of viruses, bacteria, fungi and protozoa. Therefore, TLRs are often expressed

by cells at the front line of immune defence, such as epithelial cells and DCs, so that a response to infection can immediately be initiated.

TLRs also represent an important means of discriminating between one's own molecules and foreign molecules. This enables a distinction between self and non-self, a concept that is often referred to as the danger signal hypothesis. Considering that TLRs only recognize components derived from microbes and not host factors, they provide a danger signal to the immune system when a pathogen enters the body. The ability of the immune system to distinguish between potentially harmful microorganisms and innocuous host material is an important concept in immunology.

TLRs are members of the immunoglobulin gene superfamily, due to the presence of extracellular immunoglobulin-like domains that function in the recognition and binding of specific microbial ligands. Most TLRs are formed of homodimers, which consist of two identical molecules, or heterodimers, which contain two different TLR members. TLRs also have a short transmembrane domain and an intracellular domain that is capable of initiating an intracellular signalling cascade. The dimerization of TLR molecules enhances the signalling capacity of these receptors by bringing all the necessary proteins into a functional complex. Signalling through TLRs normally initiates an inflammatory response and the transcription of several pro-inflammatory cytokines. TLRs are members of the TLR/IL-1 receptor (IL-1R) superfamily because they possess a Toll/IL-1R (TIR) domain, which is important for the recruitment of intracellular signalling proteins. The classical pathway induced by TLR/L-1R family members activates the transcription factor NF-κB and the transcription of cytokines such as IL-1, IL-6 IL-12 and TNF.

TLRs are expressed on numerous cell types, including macrophages, DCs, epithelial cells, T lymphocytes and B lymphocytes. Some TLR family members are expressed on the cell surface plasma membrane, while other members are expressed on intracellular membranes, within phagosomes and endosomes. The location of TLR expression reflects the type of PAMP they recognize and as a consequence the type of pathogen likely to be encountered. For example, TLRs expressed on the cell surface are likely to recognize extracellular pathogens, such as bacteria and fungi, while TLRs expressed within endosomes are likely to recognize intracellular pathogens that infect cells, such as viruses and certain bacteria. Therefore, TLRs expressed

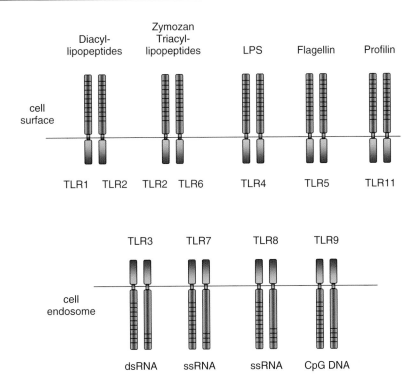

**Figure 2.3** Recognition of PAMPs by TLRs. Several TLRs are expressed on the cell surface and recognise a diverse array of PAMPs contained in the extracellular environment. Other TLRs are expressed on the membranes of endosomes and recognize PAMPs that have been endocytosed.

on the cell surface include TLR1 (that recognizes lipopeptides from bacteria), TLR2 (that recognizes lipoteichoic acid from bacteria and zymosan from fungi), TLR4 (that recognizes LPS from bacteria), TLR5 (that recognizes flagellin from bacteria) and a more recently discovered murine TLR11 (which is thought to recognize protozoa via profilin).

The TLRs expressed within the vesicular compartments of host cells include TLR3 (that recognizes dsRNA from viruses), TLR7 and TLR8 (that both recognize ssRNA derived from viruses) and TLR9 (that recognizes DNA with a high unmethylated CpG content, including DNA derived from bacteria). The PAMP-binding domain is positioned so that it can recognize molecules within the endosomal environment, rather than the cytoplasmic compartment. The recognition of CpG DNA is an important example of how the immune system can distinguish between self and non-self. Mammalian genomic DNA has a low frequency of CpG sequences, while those that are present are highly methylated. This allows for effective

discrimination between bacterial and self-DNA. The discrimination process is assisted by TLR9 being restricted to endosomal compartments, as self DNA never enters these vesicles.

## 2.4 TLR signalling in response to LPS

LPS is a highly inflammatory product of Gram-negative bacterial cell walls. LPS is a complex molecule that spans the bacterial membrane and is formed of sugars and a large extracellular domain known as lipidA. Only the lipidA is recognized by the innate immune system but is so inflammatory that excess lipidA in the bloodstream can lead to septic shock. For this reason lipidA (or LPS as a whole) is known as endotoxin. The receptor for LPS is the Toll/IL-1R family member TLR4, ligation of which ultimately leads to activation of the transcription factor NF-κB and the expression of several pro-inflammatory mediators such as TNF and IL-6. TLR4 is actually part of a larger protein complex, known as the LPS receptor, that includes the membrane bound CD14 and the extracellular

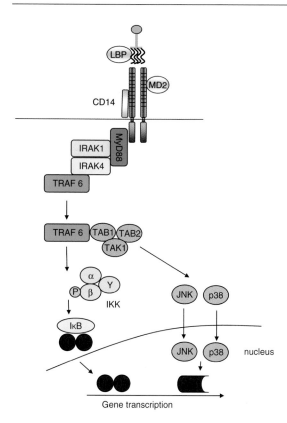

**Figure 2.4** TLR signalling pathway, exemplified by TLR4 binding to LPS. The adaptor protein MyD88 and the kinases IRAK1 and IRAK4 are recruited to the intracellular domain of TLR4 following recognition of LPS. The signalling molecule TRAF6 is phosphorylated allowing it to activate the NF-κB and the AP-1 transcription factor pathways.

MD2 glycoprotein (Figure 2.4). The role of CD14 is to bind LPS and bring CD14-LPS complexes to the LPS receptor formed of TLR4 and MD2. Uptake of LPS is further enhanced by the soluble LBP (LPS-binding protein), which is an acute phase protein found in plasma. LBP converts large LPS micelles (lipid complexes) into a monomeric form that CD14 can bind. LPS-induced signalling through TLR4 is dependent on the adapter protein MD2, which may augment the recognition of LPS. Signalling is further enhanced by the homodimerization of TLR4 molecules and the subsequent signalling cascade is mediated by the intracellular domain of TLR4.

Binding of LPS to the TLR4 complex leads to the recruitment of the adaptor protein MyD88, which contains a Toll/IL-1R (TIR) domain and a death domain. The TIR domain on MyD88 interacts with the TIR domain on TLR4 and initiates the signalling cascade. MyD88 then recruits IL-1R-associated kinase (IRAK) family members IRAK1 and IRAK4 through interaction with its death domain. Once IRAK1 and IRAK4 become phosphorylated they dissociate from the MyD88-TLR4 complex and interact with tumour necrosis factor receptor-associated factor 6 (TRAF6). TRAF6 then phosphorylates the IκB kinase complex (IKK), which is composed of IKB-α, IKB-β and IKB-γ, in association with the TAB1/TAB2/TAK2 complex, which assists in the activation of IKK. IKK functions by causing the degradation of the NF-κB inhibitor IκB that otherwise blocks nuclear transport and activation of NF-κB. Activation of IKK by TRAF6 leads to the degradation of IκB and the release of NF-κB, which enters the nucleus and initiates gene transcription. This MyD88-dependent pathway is shared by all the TLRs.

LPS can also initiate alternative MyD88-dependent pathways as a result of TLR4 activation. TRAF6 phosphorylation results in the activation of mitogen-activated protein kinases (MAPKs) including p38 MAPK, extracellular signal-regulated kinase (ERK) and c-jun $NH_2$-terminal kinase (JNK). These kinases can then activate transcription factors including AP-1 and NF-κB, leading to pro-inflammatory cytokine expression or type-I interferon (IFN-α and INF-β) release. TRAF6 can also activate PI3 kinase (PI3K), which results in the downstream activation of Akt and NF-κB. However, there is evidence of a MyD88-independent signalling pathway in response to LPS as well. This involves the activation of the transcription factor IFN-regulatory factor 3 (IRF3) through the direct phosphorylation of TRAF6, or other adaptor proteins, and leads to the expression of IFN-β. Overall, the outcome of LPS-induced signalling is the upregulation of pro-inflammatory cytokines and chemokines that initiate an inflammatory response in the presence of bacteria, firstly resulting in activation of the innate immune response and ultimately linking innate defence to adaptive immunity.

## 2.5 Peptidoglycan and Nods

Peptidoglycan is the major constituent of Gram-positive bacteria cell walls and, as LPS is unique to Gram-negative bacteria, is the most stimulatory PAMP for this group of microbes. The most widely studied receptor for peptidoglycan is TLR2, which functions as a heterodimer with TLR6 in the specific recognition of this bacterial product. TLR2 can recognize a number of other PAMPs including lipoproteins, lipoteichoic acid, lipoarabinomannan and atypical LPS, functioning either as a homodimer or a heterodimer with TLR1 or other receptor proteins. Binding of peptidoglycan by TLR2 results in a MyD88-dependent signalling cascade, NF-κB activation and inflammatory cytokine expression, in the same way as TLR4 signalling. Intracellular peptidoglycan is also recognized by nucleotide-binding oligomerization domain receptors (Nod1 and Nod2) independently of TLR signalling. Nod1 and Nod2 are members of the family of nucleotide-binding oligomerization domain–like receptors (NLRs) that recognize PAMPs within the cytoplasm, rather than in endosomal compartments. Activation of Nod1 and Nod2 results in the activation of NF-κB and MAP kinases and the expression of pro-inflammatory genes. Nod1 and Nod2 signal through the receptor-interacting protein 2 (RIP2), which is recruited via their CARD domains (caspase-recruited domain) and possibly through an interaction with the NF-κB-inducing kinase NIK (Figure 2.5). RIP2 then binds to the IKKγ subunit of the IKK complex, the inhibitor of NF-κB,

initiating activation of IKKα and IKKβ and the downstream phosphorylation of IκB. This releases NF-κB from its inhibitor so that it can enter the nucleus. However, it is still unclear whether Nod1 and Nod2 interact with peptidoglycan directly and it may be that bacterial infection is necessary for the activation of intracellular Nod1 and Nod2.

## 2.6 Nod-like receptors recognize PAMPs and DAMPs

Nucleotide-binding oligomerization domain–like receptors or Nod-like receptors (NLRs) are a family of approximately 20 intracellular proteins that recognize several PAMPs within the cytoplasm and endosomal compartment of cells. They are characterized by a C-terminal leucine-rich repeat (LRR) domain and can be categorized into four distinctive families (NLRA, NLRB, NLRC and NLRP) based on the structure of their N-terminal domain (Table 2.2). Numerous bacterial and viral-derived molecules have been shown to induce the activation of NLRs. In addition, self-derived molecules are recognized by NLRs following cellular stress or damage, usually following phagocytosis of cell debris. The self-derived molecules are often referred to as damage-associated molecular patterns (DAMPs). NLRs are also thought to be activated following disruption of the plasma membrane, thereby acting as detectors of cell distress. Several immune cells including macrophages, and non-immune cells, particularly epithelial cells in the gastrointestinal tract, express NLRs. The activation of NLRs can result in the activation of the NF-κB pathway, activation of the caspase-1 dependent inflammasome or the induction of apoptosis, depending on which NLR is activated.

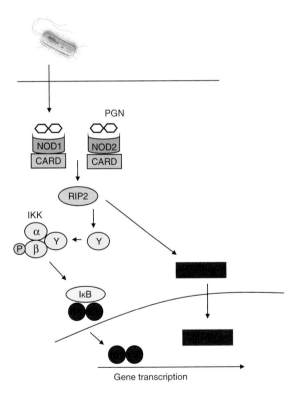

**Figure 2.5** NOD/Nod-like receptor signalling pathway is activated following the recognition of intracellular PAMPs such as peptidoglycan (PGN). The NOD pathway results in the activation of NF-κB and MAP kinases (MAPKs).

Table 2.2 Nod-like receptors and their ligands.

| Nod-like receptor | Alternative name | Ligands |
|---|---|---|
| NLRA | CIITA | IFN-γ |
| NLRB family | NAIP family | Flagellin |
| NLRC family | Nod family | Peptidoglycan |
| | | Flagellin |
| NLRP family | NALP family | Viral DNA/RNA |
| | | Bacterial toxins |
| | | Endogenous |
| | | products |

For example, activation of NLRC1 and NLRC2 (Nod1 and Nod2 in humans) results in the activation of the NK-κB pathway (Figure 2.5) and the transcription of pro-inflammatory cytokines such as TNF and pro-IL-1β, while NLRP1 and NLRP3 activation results in the activation of the inflammasome and the release of IL-1β.

The inflammasome is a term used to describe an intracellular protein complex that is involved in the processing and release of pro-inflammatory cytokines. The ligation of NLRP1 or NLRP3 with any of a number of PAMPs (such as LPS, bacteria nucleic acid or endotoxin) or DAMPs (in particular cathepsin B derived from phagolysosomes), activates NLRs and initiates the formation of inflammasomes. Activation of NLRP1 or NLRP3 causes the recruitment and processing of pro-caspase-1 into the active form of caspase-1. Rather than functioning as a pro-apoptotic enzyme like other caspase proteins, caspase-1 functions by cleaving the inactive pro-IL-1β molecule into the active, mature form of IL-1β (Figure 2.6). Indeed, another name for caspase-1 is IL-1β

converting enzyme (ICE). Similarly, activated caspase-1 cleaves the immature forms of IL-18 and IL-33, which are secreted along with IL-1β and contribute to the pro-inflammatory cytokine network. Therefore, NLRs are important receptors that are capable of detecting the presence of microorganisms or responding to cellular stress signals.

## 2.7 Damage associated molecular patterns (DAMPs)

In addition to the danger signals that are provided by microbial-derived PAMPs, host cells also produce molecules, known as DAMPs, which act as ligands for various receptors of the immune system (Figure 2.7). DAMPs are able to signal to the immune system during time of cellular stress or tissue injury in the absence of any microbial stimulus. Such danger signals derived from host cells have also been termed alarmins, to denote their endogenous origins as opposed to the exogenous origins of PAMPS. Several DAMPs have been identified that are either proteins or products of abnormal metabolism. For example, high mobility group box 1 protein (HMGB-1), heat-shock proteins (HSPs) and S100 calcium-binding proteins are examples of protein DAMPs, while uric acid, ATP and DNA are examples of metabolic or non-protein DAMPs. Although damaged cells are able to release DAMPs, macrophages, DCs, neutrophils, eosinophils and lymphocytes can all upregulate DAMPs in response to inflammatory signals. The receptors for these DAMPs, including TLRs and RAGE (receptor for advanced glycation end products), are expressed on endothelial cells and on inflammatory cells themselves.

Probably the best characterized DAMP is HMGB-1, which is normally a DNA-binding protein located within the nucleus but can also act as an inflammatory mediator. As well as activated macrophages, distressed cells that are undergoing cell death, in the absence of apoptosis, are able to release HMGB-1. During apoptotic cell death, histone proteins increase the binding of HMGB-1 to modified DNA. However, during unscheduled cell death, DNA is not modified properly and HMGB-1 is released into the extracellular environment. Release HMGB-1 can then act as a danger signal by activating endothelial cells, recruiting inflammatory cells into the site of cell damage and upregulating inflammatory cytokines such as IL-1β, TNF and IL-8. HMGB-1 is thought to signal

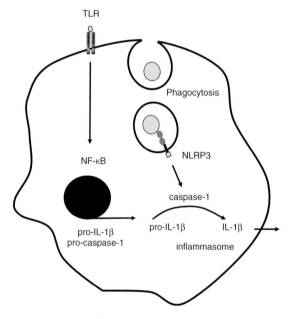

**Figure 2.6** Nod-like receptor signalling and release of IL-1β from the inflammasome. Activation of Nod-like receptor protein 3 (NLRP3) recruits and activates pro-caspase-1, which in turn cleaves pro-IL-1β and pro-IL-18. Active IL-1β and IL-18 are then released from the cell. The initial transcription of pro-IL-1β and IL-18 is induced by other PRR signalling cascades.

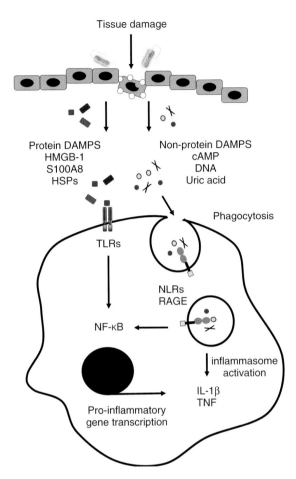

**Tissue damage**

Protein DAMPS
HMGB-1
S100A8
HSPs

Non-protein DAMPS
cAMP
DNA
Uric acid

Phagocytosis

TLRs

NLRs
RAGE

NF-κB

inflammasome
activation

IL-1β
TNF

Pro-inflammatory
gene transcription

**Figure 2.7** Damage associated molecular patterns (DAMPs) activate the NF-κB signalling pathway, upregulate pro-inflammatory cytokines and activate the inflammasome. Endogenous molecules act as DAMPs and include heat-shock proteins (HSPs), S100A8, high-mobility group protein B-1 (HMGB-1) and non-protein molecules such as DNA, uric acid and cAMP.

through TLR2, TLR4 and RAGE, thereby activating the NF-κB pathway. Similarly, S100 proteins, and in particular S100A8, S100A9 and S100A12, are released from activated epithelial cells and macrophages and upregulate pro-inflammatory cytokines through ligation with RAGE. Intracellular DAMPs, such as non-protein ATP and uric acid, are recognized by NLRs expressed within endosomes and result in the formation of inflammasomes and the release of active IL-1β. Therefore, DAMPs participate in the recognition of danger signals and the initiation of inflammatory reactions. Furthermore, DAMPs are also thought to stimulate the recruitment of cells and the upregulation of cytokines involved in the wound healing response.

## 2.8 Complement proteins perform several innate immune functions

The complement cascade is an evolutionary conserved system of secreted proteins that is central to immune defence against invading microorganisms. The function of the complement cascade is two fold: firstly it is involved in the recognition of invading microorganisms, and secondly it signals to innate immune cells and induces their effector functions (such as phagocytosis). Although the complement cascade is chiefly a part of the innate immune system, certain complement proteins regulate the function of adaptive immune cells. Most complement proteins are produced by hepatocytes in the liver, although some components are synthesized by macrophages and mucosal epithelial cells. Complement proteins are involved in several immune processes, including opsonization of invading pathogens, formation of receptor complexes, direct cell lysis and chemotaxis. Complement proteins therefore bridge pathogen recognition functions and effector functions within the innate immune system.

Opsonization is a term used to describe the way in which the binding of proteins to the surface of pathogens is able to enhance the ability of immune cells to recognize that pathogen and execute effector functions. For example, opsonization by complement proteins enhances the ability of macrophages to phagocytose bacteria. Complement proteins also assist in NK cell-mediated antibody-dependent cellular cytotoxicity (ADCC), a mechanism that benefits from the clustering of antibody and complement complexes on cell membranes. Opsonization, in the context of phagocytosis and ADCC, is discussed in a later section of this chapter. The second and equally important function of the complement cascade is the direct lysis of bacterial cells following the formation of a protein complex known as the membrane attack complex (MAC). The series of protein interactions that eventually culminate in the formation of the MAC are the result of one of three pathways: the classical complement pathway, the lectin pathway or the alternative pathway. In addition, a number of complement proteins can function as chemotactic molecules, thereby contributing to cell recruitment into sites of inflammation.

## 2.9 The classical complement pathway

The initial phase of the classical pathway relies on the binding of immunoglobulins to the surface of bacteria. Activation of the classical complement cascade involves the interaction with antibodies; in particular the antibody isotypes known as IgM and IgG (refer to Chapter 3 for structural details). Therefore, the classical complement cascade provides an important link between molecules of the adaptive and innate immune systems. Both IgM and IgG can initiate the cascade, although only one molecule of IgM is required while multiple IgG complexes are needed. This is because the complement protein C1 requires antibody/antigen complexes in order to become activated. IgM occurs as a pentamer and can therefore form antibody/antigen complex by itself, while multiple monomeric IgG molecules are required for antigen cross-linking. This is also part of the reason why the complement cascade is not activated by soluble antibodies, together with certain complement proteins requiring a cell membrane to act as a substrate for downstream enzymatic reactions.

The first complement component to bind an immunoglobulin is C1q, which forms a complex with two serine proteases C1r and C1s (Figure 2.8). The interaction between C1q and the immunoglobulin causes a conformational change in C1q that allows for two molecules of C1r and two molecules of C1s to form a complex $C1qr^2s^2$. Activation of C1r, following binding to C1q, in turn causes the activation of C1s, which is responsible for the next step in the complement cascade, the formation of the C3 convertase. The protease activity of C1s is responsible for the cleavage of C4 and C2. The splitting of C4 into C4a and C4b results in C4b binding to the C1q complex and allows the active binding component of C4b to be liberated. On the cell surface C2 binds to C4b, while the enzymatic activity of C1s cleaves C2 into C2a and C2b. The final step in the initiation phase of the classical pathway is the binding of C4b and C2a to form the C4b2a complex, otherwise known as C3 convertase. The other components C4a and C2b are released and act as potent chemotactic proteins known as anaphylotoxins.

The formation of the C3 convertase (C4b2a) is the pivotal step in the activation of the classical complement

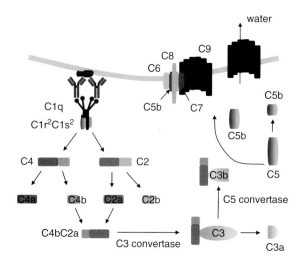

**Figure 2.8** Classical complement pathway. The binding of C1q to antibody results in the formation of the C3 convertase (C4bC2b complex). Cleavage of C3 leads to the formation of the C5 convertase. Cleavage of C5 ultimately leads to the formation of the membrane attack complex (MAC).

pathway, which allows C3 convertase to start converting numerous C3 proteins into C3a and C3b. This cleavage results in the release of C3a and the binding of C3b to C4b2a on the cell surface to form the C4b2a3b complex, otherwise known as C5 convertase. The pathway continues with the cleavage of C5 into C5a, which is released, and C5b, which binds to the cell surface and initiates the formation of the MAC. Another complement protein C6 binds to C5b, to form the C5bC6 complex, and then C7 binds to C6. The binding of C7 to the C5bC6 complex exposes a hydrophobic domain that allows C7 to insert itself into the phospholipid membrane of the bacteria. The insertion of C7 into the membrane acts an anchor for the binding of C8 and C9, which forms the MAC proper. C8 is responsible for guiding as many as 18 molecules of C9 onto the cell surface, resulting in the formation of a ring-like structure that punches a hole through the membrane. The C9 proteins possess a hydrophobic external surface that allows interaction with the phospholipid bilayer and a hydrophilic internal surface that allows the passage of water through the MAC. The process of osmosis causes water to enter the bacteria from the outside and results in cell lysis and killing of microbes. The complement cascade can be regulated at the level of the MAC by CD59 expressed on T cells. CD59 binds to C8 and C9, thereby preventing C9 interacting with the C5bC6C7C8 complex and inhibiting the formation of the MAC. This inhibition helps to prevent unnecessary lysis of host cells as a result of complement binding.

## 2.10 The lectin and alternative complement pathways

The lectin pathway is similar to the classical pathway but, instead of utilizing immunoglobulin as the opsonin, mannose-binding lectin (MBL) binds to mannose residues on the surface of bacteria (Figure 2.9). The activation of the lectin pathway is independent of C1q, which is substituted with MBL-associated serine proteases (MASP-1 and MASP-2) that function as alternatives to C1r and C1s. MASP-1 and MASP-2 are able to cleave C4 into C4a and C4b, and C2 into C2a and C2b, resulting in the formation if the C4b2a complex and the formation of the C3 convertase. In addition, MASP1 is able to cleave C3 directly. Therefore, by a slightly different route the lectin pathway activates the complement cascade through the formation of the same C3 convertase. The remainder of the complement pathway and the formation of the MAC occur as in the classical pathway.

The alternative pathway dispenses with the need for immunoglobulin or MBL opsonization and is achieved

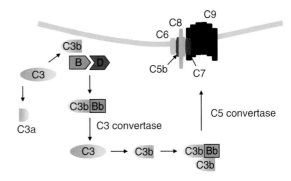

**Figure 2.10** The alternative complement pathway is activated following spontaneous cleavage of C3, which forms a C3 convertase with Factor B, which in turn cleaves more C3 molecules. C3b and Factor B form a complex that acts as the C5 convertase, ultimately leading to the formation of the membrane attack complex.

by the spontaneous binding of C3b to the surface of a pathogen (Figure 2.10). In aqueous body fluids the C3 protein is mildly unstable and dissociates into C3a and C3b. In the presence of bacteria C3b binds to the phospholipid cell membrane, otherwise C3a and C3b are reunited and the enzyme becomes deactivated. In order to generate a C5 convertase, a protein known as factor B must bind cell bound C3b to form a C3bB complex. Another protein, factor D, cleaves factor B into Ba and Bb to form C3bBb. This complex effectively hydrolyses further C3 molecules, thereby releasing more C3b, which then adds to the existing membrane-bound complexes to form C3bBbC3b, which functions as the C5 convertase for the alternative pathway. Therefore, C3bBbC3b converts C5 into C5a and C5b and the pathway concludes with the formation of the MAC by the same mechanism as the classical and lectin pathways.

## 2.11 Biological properties of complement cleavage products

In addition to the formation of the MAC, both C3b and C5b are effective opsonizing proteins (Figure 2.11). For example, C3b binds to complement receptor-1 (CR1) on macrophages, dendritic cells and B lymphocytes, which increases the phagocytic activity of these cells. The effectiveness of this opsonization is enhanced by factor B binding to C3b, which stabilizes C3b on the bacterial cell

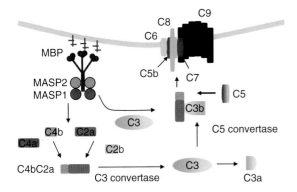

**Figure 2.9** The lectin complement pathway. Mannose binding protein recognises pathogen-derived mannose motifs. Mannose-associated serine proteases (MASP) function in a similar way as C1q resulting in formation of the C3 convertase (C4bC2a complex). Cleavage of C3 forms the C5 convertase and ultimately leads to the formation of the membrane attach complex (MAC).

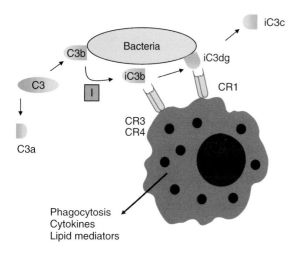

**Figure 2.11** Complement proteins opsonise pathogens and are recognized by complement receptors expressed on the surface of effector cells. Several cleavage products of C3 can act as opsonins.

surface. There is a danger, however, that C3b binding to host cell surfaces may inadvertently activate phagocytes or the downstream complement cascade. Therefore, regulatory proteins exist that either compete for C3b binding or deactivate C3b and prevent C5 convertase activity. For example, a protease known as factor I is able to bind to C3b on the cell surface and deactivate its enzymatic activity by cleaving C3b from Bb. This breaks the positive feedback loop that attracts more C3b to the complement complexes. Another protein called factor H acts as a cofactor for factor I. Factor H is only found on host cell surfaces, not on bacteria or viruses, so that host cells are protected from the alternative pathway. In a similar way, another cell surface protein known as decay accelerating factor (CD55) inhibits the alternative pathway by interfering with the assembly of the C3bBb convertase. Indeed, any protein that is able to inhibit C3bBb assembly is considered to possess decay accelerating factor properties.

Other complement proteins are released as soluble mediators, such as C5a, and act as potent chemoattractants, especially for macrophages, that attracted leukocytes into sites of inflammation. In addition, C5a acts as a pro-inflammatory mediator by binding to receptors on numerous immune cells. For example, it causes the degranulation of mast cells resulting in the release of histamine and thereby acting to promote vasodilation and endothelial permeability. It contributes

to further cell recruitment by increasing the expression of cell adhesion molecules on the surface of endothelial cells. In addition, C5a stimulates neutrophils to initiate the respiratory burst, it can act as a pyrogen, and it increases the expression of CR1 on phagocytes. Other soluble complement factors such as C4a and C3a are also pro-inflammatory mediators.

## 2.12 Opsonization by complement proteins

Macrophages are able to recognize molecular signatures on the surface of pathogens (pathogen associated molecular patterns, PAMPs) through the ligation of several pathogen recognition receptors (PRRs). Probably the most studied is the Toll-like receptor (TLR) family of PRRs, which play a central role in the internalization of pathogens, otherwise known as phagocytosis (Figure 2.12). This process is further mediated by accessory receptors that bind to soluble proteins that have attached to the surface of microorganisms. The tagging of pathogens by numerous proteins is collectively referred to as opsonization. Opsonization of pathogens effectively enhances the process of phagocytosis and is mediated by several receptors including Fc receptors (FcRs), which bind the constant regions of immunoglobulins, and complement receptors (CRs), which recognize

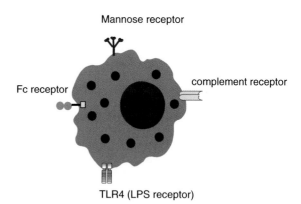

**Figure 2.12** Cell receptors on the surface of macrophages recognize several different PAMPs and opsonins such as antibody and complement factors. Receptor ligation leads to activation of the macrophages and the initiation of effector function, e.g. phagocytosis and/or cytokine release.

membrane-bound components of the complement cascade. In particular, macrophages preferentially recognize cleavage products of C3b following ligation with a CR family member, of which there are four, CR1, CR2, CR3 and CR4.

Cleavage of C3 into the small C3a subunit (also known as anaphylotoxin) and the large C3b subunit results in the exposure of a thioester bond that allows C3b to bind to the surface of microorganisms. This cleavage event also results in conformational changes in C3b, which subsequently make it possible for CR1 to recognize opsonized protein. Further proteolysis of C3b by factor I produces the degradation product iC3b. Although this cleavage event inhibits further activation of the complement pathway, iC3b is an effective opsonin and is recognized by CR3 and CR4 expressed on the surface of macrophages. Further proteolysis results in the degradation of iC3b into C3c, which is released as a soluble protein, and C3dg, which remains attached to the pathogen surface and is specifically recognized by CR2 expressed mainly on B cells. Phagocytosis is initiated by the recognition of opsonized particles, a process which is assisted by the cross-talk and synergy between different receptor families. For example, PPRs such as TLR4/CD14 specifically recognize LPS on bacteria and activate the pro-inflammatory NF-κB pathway, while FcRs (which recognize antibodies) may act in concert with CR3 by binding to different opsonins that collectively induce the physical process of phagocytosis.

## 2.13 Phagocytosis

The process of phagocytosis (from the Greek meaning *phago*, eating and *cyto*, cell) is a central mechanism for the clearance of pathogens and apoptotic cells. Phagocytosis is important in fighting infections and in maintaining tissue homeostasis. Those cells that are capable of phagocytosis are known as phagocytes and, although most cell types have a certain phagocytic capacity, those with professional phagocytic activity include neutrophils, macrophages, DCs and B cells. The process of phagocytosis is different to that of endocytosis or pinocytosis and usually involves the recognition of a ligand by one or more receptors. The receptor families responsible for initiating phagocytosis include the TLRs, which are a family of pattern recognition receptors (PRRs), scavenger receptors such as mannose receptor, which binds repeating sugar motifs on the surface of pathogens, FcRs

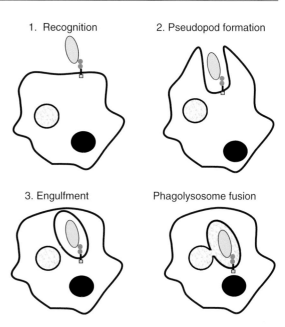

**Figure 2.13** Process of phagocytosis is a multi-step process. Receptor ligation initiates this process, leading to pseudopod formation and particle engulfment. The engulfed particle is contained within a phagosome, which fuses with a lysosome to form a phagolysosome, which contains lytic enzymes and reactive metabolites the destroy the phagocytosed particle.

responsible for the recognition of immunoglobulins, and complement receptors (CRs) that recognize components of the complement system.

Activation of phagocytosis results in the engulfment of particles through a restructuring of the plasma membrane and the creation of pseudopodia that extend around the pathogen (Figure 2.13). These pseudopodia eventually engulf the entire particle, a process that consequently leads to the formation of a large intracellular vesicle known as a phagosome. The phagosome then fuses with another vesicle called the lysosome, which contains reactive oxygen species, antimicrobial factors and lytic enzymes such as lysozyme. The fusing of a phagosome with a lysosome forms a phagolysosome. The internalized microorganism is then killed within the newly formed phagolysosome. The generation of intracellular reactive oxygen species, otherwise known as the respiratory burst, is referred to as oxygen-dependent killing, while the production of enzymes and other protein mediators is called oxygen-independent killing. The internalization of a pathogen by phagocytosis, and the release of microbicidal substances within the phagolysosome, kills the microorganism. At the same time, the retention of cytotoxic molecules within the phagolysosomal compartment prevents these harmful substances entering the cytoplasm of the phagocyte, or being released into the extracellular environment, thereby preventing any damage to host tissues. In cases of extreme inflammation, however, the contents of the lysosome may be released into the local microenvironment, which can cause tissue damage.

## 2.14  Fc receptors induce phagocytosis

Antibodies are excellent at neutralizing invading pathogenic microorganisms. They are able to bind to the surface of viruses or bacteria, block the interaction between molecules on the pathogen and cell surface and thus prevent them from entering cells. Vaccines tend to be excellent at inducing an antibody response, thereby providing protection by neutralizing the pathogen before it can cause disease. In so doing, antibodies bound to the surface of pathogens are recognized by phagocytic cells, a process known as opsonization. Again, this illustrates the close association that components of the adaptive immune system have with those of the innate immune system.

The Fc receptors (FcRs) on phagocytes, such as macrophages and neutrophils, recognize antibodies that have bound to the surface of bacteria or infected cells. The bacteria are therefore referred to as being opsonized, a process that activates a number of innate effector mechanisms. In order for FcRs to become activated, antibodies must cross-link antigens on the surface of bacteria or infected cells. This cross-linking is important for the activation of FcRs and also for the prevention of soluble antibodies unwittingly activating immune cells. The FcRs become activated through the phosphorylation of ITAMs (immunoreceptor tyrosine-based activation motifs), which initiates a downstream signalling cascade that results in the mobilization of the phagocytic machinery and the subsequent phagocytosis of the bacteria or infected cell. Antibody-mediated opsonization is enhanced by the cumulative binding of complement proteins. Internalization of the bacteria initiates the bactericidal activities of the phagocyte and results in its degradation within lysosomes. Further processing leads to presentation of bacterial antigens on the surface of the phagocyte in conjunction with MHC molecules, which is an important event for the activation of T cells and the initiation of adaptive responses. Phagocytosis

also induces the production of pro-inflammatory cytokines, such as TNF, which trigger the induction of an inflammatory response.

## 2.15  Neutrophil function and the respiratory burst

Another important family of phagocytic leukocytes are the neutrophils, which are the most abundant of the professional phagocytic cells. Following neutrophil activation and phagocytosis, external material is compartmentalized inside intracellular vacuoles called lysosomes. In addition to the cocktail of enzymes that are released into the lysosome, phagocytes synthesize superoxide anions via a process known as the respiratory burst. Neutrophils are a significant mediator of the respiratory burst, which results in the production of reactive oxygen species (ROS) that damage and kill phagocytosed microorganisms. One of the key stimulators of neutrophil phagocytosis is LPS, which initiates the assembly of enzymatic complexes that catalyse the conversion of oxygen into the various ROS. The principle ROS is superoxide ($O_2$-), which forms antimicrobial radicals when combined with other molecules. The enzyme nicotinamide adenine dinucleotide phosphate-oxidase (NADPH-oxidase) is central to the conversion of oxygen into $O_2$- (Figure 2.14). Under normal conditions that NADPH-oxidase complex remains dormant in neutrophils and only after stimulation (with LPS or complement for example) do the various subunits assemble into a catalytic complex. Another enzyme called superoxide dismutase (SOD) catalyses a dismutation reaction to produce hydrogen peroxide ($H_2O_2$), which in turn possesses antimicrobial properties. Other systems contribute to the rapid production of $O_2$-, such as the mitochondrial electron transport chain, which involves the enzyme cytochrome $c$ oxidase (COX).

Another important free radical with antimicrobial properties is nitric oxide (NO), which is synthesized by numerous cell types including neutrophils, macrophages, dendritic cells and NK cells. NO is also produced by subsidiary cells such as epithelial cells, endothelial cells and fibroblasts. The production of NO in phagocytes involves the enzyme inducible nitric oxide synthase (iNOS), which is upregulated in response to microbial stimulation. The generation of NO results in a chemical reaction with $O_2$- and the generation of peroxynitrite ($ONO_2$-) that is toxic to microbes. In addition to being antimicrobial, NO has other functions including the dilation of airways through its effects of smooth muscle cells, the vasodilation of blood

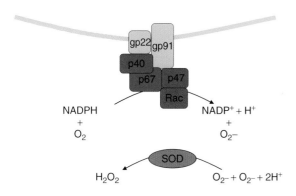

**Figure 2.14** The respiratory burst. NADPH-oxidase is formed of several subunits. Firstly a heterodimer of the glycoproteins gp91$^{phox}$ and gp22$^{phox}$ form at a cellular membrane and recruit three other proteins p40$^{phox}$, p47$^{phox}$ and p67$^{phox}$. In addition a GTPase called Rac (Rho-related C3 botulinum toxin substrate), which associates with GTP (guanosine triphosphate), also becomes recruited to the membrane complex. The NADPH-oxidase complex, once assembled, converts oxygen into superoxide (O$_{2-}$). Furthermore, the enzyme superoxide dimutase (SOD) can convert O$_{2-}$ and H$^+$ into hydrogen peroxide (H$_2$O$_2$).

vessels and contributes to the adhesion of leukocytes to endothelial cells during extravasation. It is therefore not surprising that defects in NO synthesis have been associated with several diseases such as hypertension, arthrosclerosis and cardio-vascular disease.

Activation of the respiratory burst is an important way in which neutrophils and macrophages kill microbes. However, excessive production of ROS and the release of oxidants from the cell can cause significant damage to host tissues. The damage due to ROS is often referred to as oxidative stress, an event that has been associated with many diseases including sepsis, acute lung injury, cardiac ischaemia and organ failure. This is the reason that the body has in place a number of antioxidant strategies that aim to eliminate tissue damage. For example, vitamin A, vitamin C and vitamin E are effective antioxidants that interfere with the effects of O$_{2}$-, while the enzyme SOD can also have antioxidant effects through the conversion of O$_{2}$- to the less toxic H$_2$O$_2$ (which is further broken down into H$_2$O and H$^+$).

## 2.16 ADCC

Antibody-dependent cellular cytotoxicity (ADCC) is a term used to describe the mechanism of cell lysis that occurs in response to target cells opsonized with surface bound antibodies. Leukocytes that participate in ADCC include neutrophils, eosinophils and NK cells, although they recognize antibody through different FcRs. The antibody isotypes that are particularly effective at opsonizing infected cells are IgG, IgE and IgA. These are recognized by cells expressing the appropriate Fc receptors (FcγRs, which recognize IgG, FcεRs, which recognize IgE and FcαRs, which recognize IgA). For example, NK cells express FcγRs, which bind to surface bound IgG molecules, while eosinophils express FcεRs, which specifically recognize IgE. ADCC mediated by different immune cells is also effective against different kinds of pathogens. For instance, ADCC mediated by NK cells is particularly effective at killing virally infected cells, while ADCC mediated by eosinophils is more effective against extracellular parasites such as helminth worms.

With regard to NK cell ADCC, antibody cross-linking and the formation of antibody/antigen complexes is essential for signalling via its FcR. This is to bring the essential receptor subunits into close proximity so that effective signalling events can take place (Figure 2.15). NK cells recognize opsonizing IgG via FcγRs, in particular FcγRIII (CD16). However, FcγRIII cannot signal

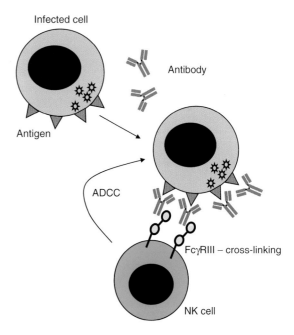

**Figure 2.15** NK cells participate in antibody-dependent cellular cytotoxicity (ADCC). Antigen bound to the surface of a target cell is recognized by FcRs expressed by NK cells. Cross-linking of these receptors leads to NK cell activation and initiation of ADCC.

through its own intracellular domain and requires the recruitment of a common γ-chain molecule that contains an ITAM. The cross-linking of FcγRIII with antibody and the formation of a receptor complex with the common γ-chain initiates an intracellular signalling cascade that leads to the release of cytotoxic granule contents. Cross-linking of these receptors by multiple antibody molecules leads to the recruitment of Src-family kinases and the phosphorylation of ITAMs (FcR activation is discussed further in Chapter 3). This in turn results in the recruitment of SH2 domain-containing Syk kinase family members or the signalling molecule Zap-70. This recruitment results in the activation of phosphatidylinositol-3 kinase (PI3K) and eventually causes the activation of the transcription factor NFAT. NK cells activated in this way produce pro-inflammatory cytokines such as TNF and IFN-γ and also initiate mechanisms of cell killing. Activation results in the orientation of intracellular granules to face toward the target cell and the formation of a synapse between the target cell and NK cell. The controlled release of granules into the synapse prevents the escape of destructive enzymes into the microenvironment.

NK cell cytotoxicity involves the formation of an immunological synapse with its target cell and the release of cytotolytic effector molecules from specialized lytic granules. The compartmentalization afforded by the synapse restricts the release of cytotoxic molecules so that neighbouring cells are unaffected. The formation of an immunological synapse requires the rapid polarization of the NK cell, which involves intracellular reorganization of molecules such as actin. The formation of a synapse allows the membranes of the NK cell and target cell to fuse, resulting in the formation of a cleft into which the cytolytic effector molecules can be released in order to kill the target cell (Figure 2.16). This process also involves the recruitment of lytic granules to the site of the immunological synapse, which involves further reorganization of the actin cytoskeleton. The two main components of lytic granules are perforin and granzymes (for example, granzyme B). Once the lytic granules fuse with the NK cell plasma membrane, the granule contents are released through exocytosis into the synaptic cleft. Perforin than embeds itself into the target cell membrane and punches holes through it so that granzymes can freely enter the cell. Granzymes are essentially serine proteases that cleave a variety of proteins that function in maintaining cell viability. For example, granzymes cleave caspases, in particular caspase-3 that functions by activating the enzyme DNase, which in turn enters the nucleus and starts to degrade target cell DNA. The degradation of DNA induces the cell to undergo apoptosis. In addition, granzyme B activate the pro-apoptotic proteins Bax and Bak, which further activate caspases and therefore accelerate the apoptotic process.

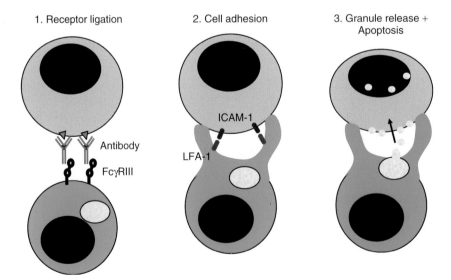

**Figure 2.16** ADCC is assisted by the formation of an immunological synapse between the plasma membranes on the NK cell and the target cell. Interaction between ICAM-1 and LFA aids cell adhesion. NK cell granule contents are then released into the synapse, which prevents cytotoxic factors causing bystander damage to neighbouring cells.

## 2.17 NK cells recognize missing self

An important mechanism by which NK cells detect infected or diseased cells is through the monitoring of surface expression of MHC class I proteins. Certain viruses and tumour cells evade detection by CTLs by downregulating MHC class I molecules (known as human leukocyte antigen, or HLA in humans). However, NK cells are capable of detecting this downregulation through a specialized set of molecules known as killer immunoglobulin-like receptors (KIRs). The detection of a lack of MHC class I molecule expression is known as the 'missing self' hypothesis (Figure 2.17). Under normal circumstances, the KIRs recognize their specific MHC class I molecules expressed on the cell surface, which sends an inhibitory signal to the NK cell in order to suppress cytotoxicity. This is achieved through the activation of the KIR receptor, which utilizes an ITIM (immunoreceptor tyrosine-based inhibitory motif) within in its cytoplasmic tail. However, in the absence of inhibitory receptor ligation (i.e. MHC class I downregulation), continued KIR ligation with MHC class I molecules is interrupted and activation of the NK cell can take place. This is equivalent to releasing the brakes on an NK cell and results in target cell lysis. NK cell lysis in response to missing self is an important mechanism for the control of intracellular infections and the removal of damaged or abnormal cells.

Not all tissue cells that lack MHC class I protein are susceptible to NK cell lysis. For example, red blood cells and nerve cells are refractory to NK cytotoxicity. This led to the discovery that NK cells require a specific activation trigger in addition to missing self. NK cells express several natural cytotoxicity receptors, such as NKp44, NKp46 and NKp30, and the C-type lectin NKG2D (Figure 2.18). These receptors are able to recognize foreign molecules

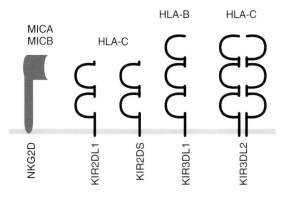

**Figure 2.18** NK cell receptor recognition of MHC class I (HLA) molecules.

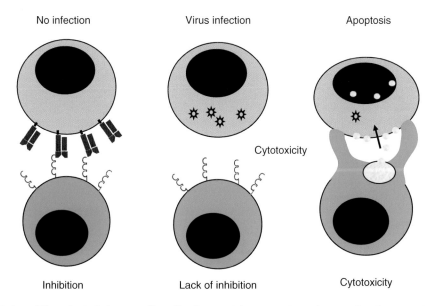

**Figure 2.17** Missing-self hypothesis during NK cell-mediated cytotoxicity. Many viruses downregulate the expression of MHC class I molecules on infected cells. NK cells identify this as being abnormal through the removal of an inhibitory signal provided by KIRs. The lack of inhibition then allows NK cells to initiate cytotoxicity.

expressed on the surface of infected cells or they can recognize self proteins that are upregulated in response to cellular stress (known as DAMPs). For example, NKG2D recognizes a family of molecules that are closely related to MHC class I molecules that are termed MICA and MICB (major histocompatibility complex (MHC) class I chain-related). MICA and MICB are upregulated in response to viral infection or on the surface of tumour cells. Therefore, NK cell cytotoxicity is regulated by a balance of activation and inhibitory signals, although the exact regulatory mechanisms responsible for this complex signalling have yet to be fully elucidated.

## 2.18 Activating adaptive immunity

Many different cell types are capable of secreting pro-inflammatory mediators that recruit and activate cells of the adaptive immune system. These include non-immune cells such as epithelial cells, endothelial cells and fibroblasts. However, innate immune cells, including neutrophils, macrophages and DCs, are the principal cell types responsible for subsequent activation. In particular, phagocytic cells are effective activators of adaptive immunity, as they come into direct contact with pathogenic stimuli. For example, TLR ligation with the relevant

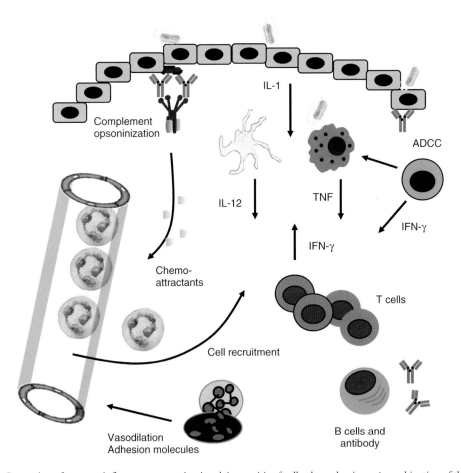

**Figure 2.19** Generation of an acute inflammatory reaction involving positive feedback mechanisms. A combination of the release of cytokines and chemokines causes the recruitment and activation of effector cells. These effector cells in turn release more pro-inflammatory mediators, thereby acting in a positive feedback loop by attracting and activating even more leukocytes.

PAMP results in an intracellular signalling cascade that ultimately leads to the expression of the pro-inflammatory cytokines IL-12 and TNF. Indeed, phagocytes are prolific producers of inflammatory cytokines, which stimulate an acute inflammatory reaction through a positive feedback mechanism (Figure 2.19). The more inflammatory cytokines that are produced the greater the number of immune cells that enter a tissue and the larger the inflammatory reaction. For example, activated DCs secrete significant quantities of IL-12, which stimulates the differentiation of T helper (Th) cells and the development of cell-mediated immunity. As well as being highly phagocytic, macrophages secrete large quantities of TNF in response to activation, which drives inflammatory reactions and reinforces the differentiation of effector cells, including cytotoxic T cells (CTLs). Specialized populations of lymphocytes, known as unconventional T cells (Box 2.1), are also able to influence and modify adaptive immune responses through the release of various cytokines. These unconventional T cells include NKT cells and γδ T cell, which possess properties associated with both innate and adaptive immune cells. However, due to their superior ability to prime the effector phase of an immune response, DCs are thought to provide the most potent activating signals during the initiation of an adaptive immune response.

---

### Box 2.1. Unconventional T cells

The lymphocyte lineage is conventionally divided into NK cells, T cells and B cells. These three subpopulations comprise the majority of lymphocytes found throughout the human body. However, it has become evident that other lymphocyte populations exist that share some of the characteristics of NK cells and some of conventional T cells. These unconventional T cells are less easily placed within the innate or adaptive immune systems. Firstly, a lymphocyte population known as natural killer T (NKT) cells express αβ T cell receptors (TCRs), which are usually associated with conventional T cells, but also express receptors that are normally only found on NK cells such as NK1.1 (CD161). The second unconventional lymphocyte population are the γδ T cells, which are most commonly located in the epithelial compartments of the gut and skin. These γδ T cells express γδ T cell receptors, instead of αβ T cell receptors. It is difficult to classify NKT cells and γδ T cells as part of either the innate or adaptive immune systems as they share properties of both. On the one hand they express TCRs that have undergone genetic recombination but they are not restricted by conventional MHC molecules. On the other hand, the TCRs that they do express are often invariant. This means that NKT cells and γδ

T cells are capable of recognizing only a limited number of antigens, thereby making them more akin to innate immune cells. It is also thought that intsead of recognizing peptide antigens presented to them by classical MHC molecules, NKT cells and γδ T cell recognize lipid or glycolipid motifs presented by non-classical MHC molecules such as CD1d.

A certain population of NKT cells are known as invariant NKT (iNKT) cells, due to the expression of an invariant αβTCR (e.g. Vα24–Jα18). This TCR is only able to recognize glycolipid antigens presented in conjunction with CD1d, which is expressed by monocytes and DCs. NKT cells are known to contribute to immune defence against infection by virus and bacteria. Indeed the invariant TCR of NKT cells has been shown to recognize certain bacterial glycolipids. NKT cells have also been implicated in several diseases, including asthma and psoriasis, and their dysregulation is associated with the development of cancer and autoimmune disease (such as type 1 diabetes, rheumatoid arthritis and multiple sclerosis). Therefore, NKT cells are associated with host defence against infection, provide a stimulus for driving excessive inflammation and can regulate autoimmune reactions and tumour development. How NKT cells perform such varied immune functions remains unclear, although it is apparent that they are capable of secreting a variety of cytokines. For example, they can express the pro-inflammatory cytokines IFN-γ, TNF and IL-17 under certain conditions but also express IL-4, IL-10 and IL-13. Indeed, NKT cells predominantly express the Th2/regulatory cytokines, which may contribute to their regulatory capacity in autoimmune disease. It is likely that signals from the microenvironment will determine whether NKT cells become inflammatory or regulatory.

The γδ T cell population has a more diverse TCR repertoire than the NKT cell population. It is thought that γδ T cells are produced during foetal life and migrate to specific mucosal tissues in order to provide first line defence for the newborn. In mice, γδ T cell subsets have been tracked to various tissues. It appears that certain tissues preferentially recruit γδ T cell that express particular γδ TCRs. For example, Vγ7γδ T cells preferentially migrate to the gut, while Vγ1γδ T cells migrate to the skin. The reasons for this preferentiality are not clear, but may relate to the order in which different γδ T cell subsets are released from the thymus during T cell development. Like NKT cells, γδ T cells recognize glycolipid antigens that are restricted by CD1c or the related MICA/MICB molecules. Therefore, although γδTCRs undergo gene rearrangement like αβ T cells and can form populations of memory γδ T cells, each receptor is only able to recognize a limited number of molecular patterns and therefore γδ T cells are often considered to be part of the innate immune system. In the gut, γδ T cells comprise a large percentage of the intra-epithelial lymphocyte (IEL) population. This makes them ideally located for the detection of cellular stress, through the recognition of microbial or host-derived glycolipids. Activation of γδ T cells results in the rapid release of the pro-inflammatory cytokines IFN-γ and IL-17 and the chemokines CCL5 (RANTES) and XCL1 (lymphotactin). This has the effect of recruiting lymphocytes into sites of infection and activating their effector functions once there. In the skin a subset of γδ T cells known as dendritic epidermal γδ T cells (DETCs) play a similar role in immunosurveillance. Skin homing γδ T cells are thought to interact closely with keratinocytes and play a major role in wound healing, in part by releasing TGF-β and KGF. Therefore, γδ T cells provide an important link between innate immune recognition and the initiation of adaptive immune responses.

The primary functions of DCs can be divided into two phases (Figure 2.20). The first phase involves immature DCs, which are actively phagocytic. It is at this phase that immature DCs receive signals from the external environment, which provide activation signals and drive DCs to undergo maturation and release pro-inflammatory cytokine such as IL-12. As a result of the phagocytosis of foreign matter, immature DCs also participate in antigen processing. The correct processing of foreign antigens is essential for the subsequent activation of T cell responses. The activation of DCs causes them to migrate out of

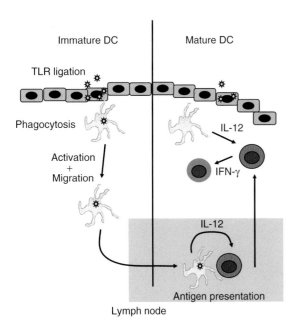

**Figure 2.20** Maturation of DCs following antigen encounter. Within primary sites of infection DCs act as antigen processing cells. Activation results in the migration of DC into regional lymph nodes, where they mature into antigen presenting cells and activate adaptive immune responses.

the site of pathogen encounter and into regional lymph nodes where they come into contact with T cells. The now mature DCs enter a second phase as they revert from being phagocytic, antigen processors to antigen presenting cells (APCs). The molecular mechanism of antigen presentation to T cells will be discussed in the next chapter. At this stage it is sufficient to appreciate that the mechanism of antigen presentation is a central event during the initiation of adaptive immunity and forms a central paradigm in immunology. The activation of T cells as a result of antigen presentation by DCs occurs in combination with other signals, known as co-stimulatory signals, and the secretion of cytokines, such as IL-12 and IL-2. The consequence of activation is the migration of T cells out of lymph nodes and into the site of inflammation, where they can exert their effector functions and prevent further infection.

## 2.19 Dendritic cells link innate and adaptive immunity

Dendritic cells (DCs) have a central role to play in the induction of adaptive immunity and as such several sub-populations of DCs have evolved in mammals (Figure 2.21). These DC populations also perform subtly different functions within the immune system, although they all share a similar morphology characterized by typical dendritic protrusions. DCs can broadly be divided into myeloid DCs (mDCs), which are also referred to as conventional DCs (cDCs), and plasmacytoid DCs (pDCs), a division based on cellular lineage. Both mDCs and pDCs are derived from haematopoietic stem cells in the bone marrow. It was traditionally thought that mDCs were derived from the myeloid precursor cells, while pDCs were derived from lymphoid precursors. This idea was further substantiated by the finding that a large population of mDCs are derived from monocytes, while pDCs lack conventional myeloid markers such as CD11c. However, it is now known that populations of both mDCs and pDCs can differentiate from both common myeloid progenitor cells and common lymphoid progenitor cells. It may be that the tissue microenvironment and the cells that they interact with influence the differentiation of DC populations.

The main subset of myeloid DCs are the monocyte-derived DCs, which are considered to be the main pro-inflammatory DC subset as they are rapidly recruited to sites of inflammation and are active stimulators of T cell immunity. Monocyte-derived DCs differentiate from

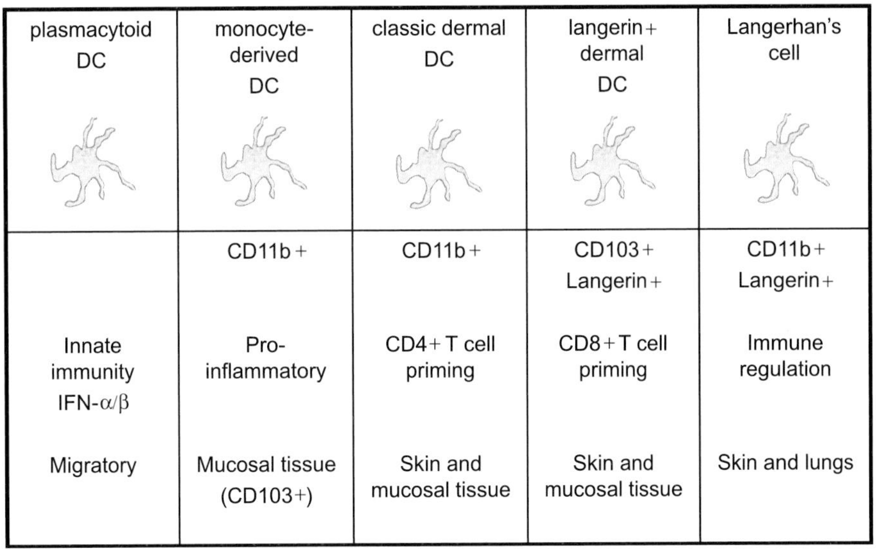

| plasmacytoid DC | monocyte-derived DC | classic dermal DC | langerin+ dermal DC | Langerhan's cell |
|---|---|---|---|---|
| | CD11b+ | CD11b+ | CD103+ Langerin+ | CD11b+ Langerin+ |
| Innate immunity IFN-α/β | Pro-inflammatory | CD4+ T cell priming | CD8+ T cell priming | Immune regulation |
| Migratory | Mucosal tissue (CD103+) | Skin and mucosal tissue | Skin and mucosal tissue | Skin and lungs |

**Figure 2.21** DC subsets found throughout mucosal tissues and the skin.

monocytes that have migrated from the circulation. They are significant producers of IL-12 and are effective at presenting antigen to CD4+Th cells. Another widely studied group of mDCs are the Langerhan's cells (LCs), which are found throughout the epidermis of the skin. LCs form an extensive immunosurveillance network in the epidermis where they can detect the presence of potentially harmful pathogens. Another mDC subset located in the skin is the dermal DC (DDC) population, which express CD11b and are thought to stimulate T cell help and initiate primary and secondary adaptive immune responses.

A third subset of mDC has been isolated from the skin, similar to DDCs but also expressing the cell surface marker CD103. Furthermore, equivalent CD103+ DC subsets can be isolated from mucosal tissues of the gut and airway. The CD103+ DC subsets may have dual roles in the immune system. On the one hand they are capable of inducing CD8+ CTL responses, while on the other hand they enhance Treg cell function. Research performed on mouse DC subsets has revealed that the CD103+ DC population can be further divided based on the expression of CD11b. Therefore, CD11b+CD103+ DCs function by stimulating CD4+ T cells (including Treg cells), while CD11b-CD103+ DCs are involved in the stimulation of cytotoxic T cells at sites of inflammation. It remains to be elucidated whether equivalent DC subsets occur in humans.

In addition, a population of mDCs, known as interstitial DCs, are found within the circulation. When activated, interstitial DCs express IL-12 and stimulate Th1-mediated immune responses. Slightly different DC subsets have been described in lymph nodes and spleen. These DCs are also derived from the myeloid lineage and resemble mucosal DCs to a certain extent. For example, CD11b+ DCs in lymph nodes and spleen express CD4, while CD11b- DCs express CD8 (there is also a CD11b+ double negative population). In a similar way to mucosal DCs, CD11b+ DCs are responsible for stimulating CD4+ T cell responses, while CD11b- DCs predominantly stimulate CD8+ T cells. Despite all the different subsets of mDC, it is clear that these specialized APCs are central for the induction of inflammatory immune responses and, depending on the DC subset, for the regulation of immune responses.

The population of pDCs are less frequent in number than the other more conventional mDC subsets and are characterized by the lack of the myeloid markers CD11b and CD11c. They are thought to be highly migratory and are found in the circulation, lymphoid tissues and sites of active inflammation. The main function of pDCs is the secretion of IFN-α and IFN-β. Therefore, pDCs are referred to as natural interferon-producing cells. They are characterized by the high expression of TLRs, in particular TLR7 and TLR9. Therefore, they are capable of responding to viral and bacterial infections. The primary function of pDCs is to activate other DC populations, in particular antigen presenting mDCs, and to stimulate T cell and B cell responses. Therefore, mDCs and pDCs play a significant role in the initiation of host immune defence mechanisms and provide an important link between the innate immune system and the adaptive immune system.

## 2.20 Summary

**1.** The innate immune system provides a first line of defence against infectious pathogens and comprises physical, chemical and cellular components.

**2.** The non-specific recognition of foreign substances and/or danger signals activates innate immune cells.

**3.** Innate immune cells are able to discriminate between self and non-self.

**4.** Effector mechanisms of innate immunity include humoral and cell-mediated components.

**5.** Innate immune cells are capable of activating cells of the adaptive immune system.

# 3 The Adaptive Immune System

## 3.1 Introduction to adaptive immunity

The adaptive immune system consists of highly specialized cells that have the capacity to adapt to the threat of new and divergent pathogenic microorganisms. Adaptive immunity is thought to have evolved in jawed vertebrates, as it is not found in evolutionarily distant species, which only possess an innate immune system. The principal cell types that constitute adaptive immunity are thymus-derived T cells and bone marrow-derived B cells. These cells are capable of recognizing a multitude of different foreign antigens in a very precise way, which has led some to refer to the adaptive immune system as the specific immune system. T cells and B cells also have the capacity to generate immunological memory, otherwise known as immunity, against a particular pathogen. This allows T cells and B cells to recognize a pathogen that they have encountered before and mount a stronger, more effective immune response against a subsequent exposure. For this reason secondary immune responses are normally much more effective at clearing an infection than primary immune responses are. Not only does adaptive immunity provide an elegant means of protecting against future infection, the principles of vaccination (synonymous with immunization) also rely on cells of the adaptive immune system to generate immune memory (primary response) and to ensure that an individual is protected from infection by the real pathogen (secondary response).

Central to the initiation and development of an adaptive immune response is the specific recognition of foreign antigens by T cells and B cells. Antigen recognition by T cells and B cells is mediated by the expression of T cell receptors (TCRs) on T cells and B cell receptors (BCRs) on B cells (Figure 3.1). Each person is able to generate an enormous number of TCRs and BCRs, each having unique binding properties to a specific antigen. It is thought that the adaptive immune system is capable of recognizing any pathogen that it could potentially encounter because such a vast number of TCRs and BCRs are generated. For example, it has been estimated that as many as $10^7$ different T cell clones exist, each capable of recognizing a slightly different antigen. However, the human genome is not large enough to encode $10^7$ separate TCR and BCR molecules. Therefore, TCRs and BCRs are synthesized by a process known as genetic recombination (discussed in detail later), which is responsible for the generation of receptor diversity. The TCRs and BCRs utilized by cells of the adaptive immune system are therefore much more sophisticated than the receptors of the innate immune system, which only recognize common molecular families. The receptor diversity of T cells and B cells also allows for the generation of immunological memory, as antigen-specific T cells and B cells are maintained following initial infection.

## 3.2 T cells and B cells recognize foreign antigens

The lymphocytes of the adaptive immune system provide both cell-mediated immunity and antibody-mediated immunity (the latter is part of the humoral immune response). T cells either differentiate into CD4+ T helper (Th) cells or CD8+ cytotoxic T cells, the latter being more commonly known as cytotoxic T lymphocytes (CTLs). CD4+ Th cells and CD8+ CTLs are both components of cell-mediated immunity. The TCRs expressed on the surface of Th cells and CTLs enable them to recognize foreign antigen that has been enzymatically processed into short peptides. However, these peptides are not able to be recognized *de novo* but must be presented to T cells bound to specialized carrier molecules known as major histocompatibility complex (MHC) molecules. The way in which antigen is processed means that CD4+ Th cells recognize exogenous antigen that is presented on the

*Immunology: Mucosal and Body Surface Defences*, First Edition. Andrew E. Williams.
© 2012 John Wiley & Sons, Ltd. Published 2012 by John Wiley & Sons, Ltd.

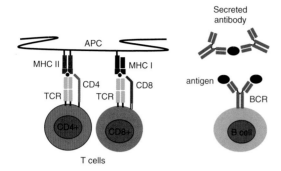

**Figure 3.1** T cells express T cell receptors (TCR), which recognize peptide antigen presented by MHC molecules expressed on APCs. B cells express B cell receptors (BCR) and secrete antibodies, which bind to antigen.

surface of professional antigen presenting cells (APCs) in the context of MHC class II molecules, while CD8+ CTLs recognize endogenous antigen presented on the surface of all nucleated cells in the context of MHC class I molecules (Figure 3.1).

Th cells, as their name implies, impart assistance for the activation of other immune cells by providing important stimulatory signals that are required for the activation of CTLs and B cells. While CTLs function by lysing and killing infected cells, activated B cells differentiate into plasma cells that are responsible for the production of antibodies. Antibodies are soluble proteins that form a major component of the humoral immune system. Essentially, an antibody molecule is a soluble form of the BCR and can be interchanged with the term immunoglobulin. Th cells therefore provide stimulatory signals that drive both cell-mediated and humoral immune responses. Antibodies are found throughout the body, including the circulating blood and lymph vessels, within tissue fluids and at mucosal surfaces. Antibodies recognize antigen in a slightly different way to TCRs. Rather than recognizing short peptides, antibodies recognize specific shapes that are part of whole proteins. As well as the neutralization

of antigens, antibodies initiate downstream effector functions. For example, antibodies are major activators of the classic complement system (discussed in Chapter 2) through the binding of the C1 complement protein. Antibodies that are bound to a specific antigen therefore direct the complement system in an antigen-dependent manner. Antibodies also direct macrophages in an antigen-specific manner and induce phagocytosis through the antibody binding to antibody receptors. Antibody cross-linking to antigen also activates mast cells and eosinophils and causes the release of pro-inflammatory and vasoactive mediators. Therefore, antibodies act as immune system adapter molecules by influencing a spectrum of immune effector mechanisms.

Thus, T cells and B cells together form a substantial adaptive immune defence network involving various effector mechanisms (Table 3.1). The following chapter will therefore describe how such immune receptor diversity is generated by B cells and T cells and how antigen is presented to T cells in the context of MHC molecules. It will include an overview of antibody and TCR structure and the generation of receptor diversity through a mechanism known as genetic recombination. The principles of B cell and T cell activation will also be discussed, as will the generation of immunological memory.

## 3.3 Overview of antibody structure

Antibody molecules are an essential part of the adaptive immune response. These immunoglobulins are synthesized by specialized B cells known as plasma cells and have two basic functions, which are directly related to their structure. The first function is binding to antigen through the recognition of molecular shapes, known as epitopes, by the antigen-binding site of the antibody. The second function is to induce an effector response through the ligation of antibody molecules with cell surface receptors, known as Fc receptors, on effector cells (FcRs were briefly introduced in the preceding chapter in

**Table 3.1** The main features of the principal cells of the adaptive immune system.

|  | CD4+ T cell | CD8+ T cell | B cell |
|---|---|---|---|
| Effector cell | T helper cell | Cytotoxic T lymphocyte | Plasma cell |
| Function | Co-stimulation | Cytotoxicity | Antibody secretion |
| Principle Receptor | T cell receptor (TCR) | T cell receptor (TCR) | B cell receptor (BCR) |
| Antigen Recognition Receptor | MHC class II | MHC class I | Antibody/BCR |

## Box 3.1. Generation of the Fc Fragment

The function of different regions of an immunoglobulin can be separated by partial cleavage of the molecule by proteases (see figure accompanying this box). The enzyme papain proteolytically cleaves the immunoglobulin within the hinge region of the molecule, thus producing three fragments. Two identical fragments are able to bind antigen and therefore they are known as fragments of antigen binding (Fab). The Fab fragments are composed of an entire light chain plus the $V_H1$ and $C_H1$ domains of the heavy chain. The remaining fragment can be easily crystallized and is therefore known as the Fc fragment. The Fc fragment is responsible for eliciting the various effector functions and is composed of bound $C_H2$, $C_H3$ and $C_H4$ domains of the two heavy chains. Pepsin is another protease that also cleaves the immunoglobulin within the constant region but below the di-sulphide bonds that link the two heavy chains together. This cleavage forms an F(ab')2 fragment as the two antigen-binding domains remain covalently bonded together. The remaining constant domains are cleaved a number of times to produce several small fragments and a larger pFc' fragment. These cleavage products enabled researchers to identify the role that each part of an immunoglobulin plays during an immune response, in particular the effector functions of the Fc domain.

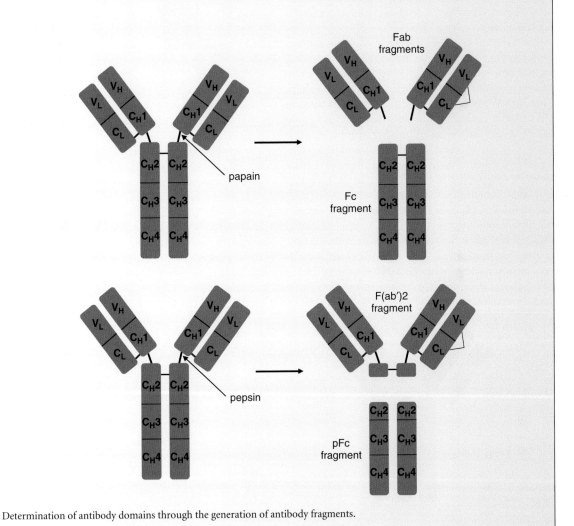

Determination of antibody domains through the generation of antibody fragments.

the context of stimulating innate effector mechanisms). The antibody-binding sites on antibodies are extremely diverse so that many antigen epitopes can be recognized by the antibody repertoire. The region of the antibody molecule responsible for antigen binding is therefore known as the variable region (V region). This region is structurally separate from the region that binds to Fc receptors (Box 3.1), which is known as the constant region (C region). The constant region has very little variation. Antibody molecules are composed of two identical heavy chains ($C_H$) and two light chains ($C_L$) that are covalently linked together by disulphide bonds (Figure 3.2). This results in each antibody molecule possessing two antigen-binding sites. Therefore individual antibody molecules are said to have a valency of two.

The V region is responsible for the antigen-binding capacity of antibodies. The actual antigen-binding site is formed of three hypervariable loops in the heavy chain and three smaller hypervariable loops in the light chain. These hypervariable regions are also referred to as complementarity determining regions (CDR1, CDR2 and CDR3). Each variable region together with the constant heavy 1 ($C_H1$) or constant light ($C_L$) region is composed of β-sheets. The hypervariable loops are formed from the peptide loops that connect the anti-parallel β-strands that form each β-sheet of the variable region. The amino acid sequence within the hypervariable loops largely determines the shape of the antigen-binding site. This will determine the tertiary structure and the shape of the immunoglobulin. This amino acid sequence is extremely variable between different immunoglobulins and therefore the antigen-binding capacity of the entire

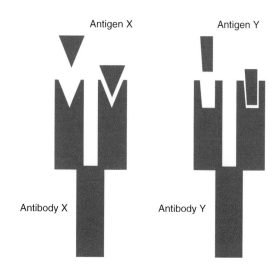

**Figure 3.3** The lock and key hypothesis. Antibody binding to antigen is largely dependent on the shape of the antigen. Different antibodies recognize different shapes, which forms the basis of antibody specificity.

immunoglobulin repertoire is diverse. This diversity is accentuated when taken into account that both the amino acid sequences of light and heavy variable regions combine to form one antigen-binding site.

A simple analogy is often used to describe the way in which antibodies recognize certain antigens, which is referred to as the lock and key hypothesis (Figure 3.3). The same concept can be applied to enzyme/protein interactions and ligand/receptor interactions that are dependent on structural identity. The lock and key hypothesis states that the three dimensional shape of the antibody-binding site will determine its antigen-binding properties. Antigen X will fit into the antigen-binding site of antibody X, while antigen Y will fit into the antigen-binding site of antibody Y, which has a different three-dimensional shape. However, other factors are also involved in antigen-binding specificity. For example, hydrogen bonds are formed between the antibody and antigen, which are stabilized by water molecules and Van der Vaal forces. Hydrophobic and electrostatic forces also contribute to the stability of the binding. Often the binding between antibody and antigen is so strong that the interaction is irreversible.

The structure of the C region also dictates the isotype of antibody and depends largely on the number of C region domains and covalent bonding between the two heavy chains (Figure 3.2). Alternative effector functions can be attributed to certain antibody isotypes due to the slight structural differences between them. These differences are

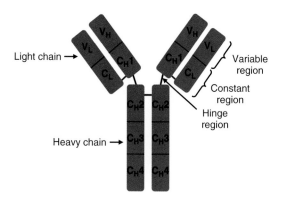

**Figure 3.2** Generalized structure of an antibody molecule. Antibodies are composed of two light ($_L$) and two heavy chains ($_H$), linked together by disulphide bonds and divided into variable (V) and constant regions (C).

**Table 3.2** Effector functions of different antibody isotypes.

| Antibody Class | $C_H$ Domains | Molecular Form | Effector mechanism |
|---|---|---|---|
| IgM | $C_H1-C_H4$ | Pentamer | Complement fixation (classical)<br>Antigen neutralization<br>Immune complex formation |
| IgA | $C_H1-C_H3$ | Dimer | Complement fixation (mannose)<br>Antigen neutralization (mucosal surfaces)<br>Secretry (binds to secretory component)<br>Induces phagocytosis and ADCC |
| IgG | $C_H1-C_H3$ | Monomer | Complement fixation (classical)<br>Antigen neutralization<br>Induces phagocytosis and ADCC |
| IgE | $C_H1-C_H4$ | Monomer | Mast cell and eosinophil degranulation<br>Anaphylaxis |
| IgD | $C_H1-C_H3$ | Monomer | Function unknown (developmental?) |

related to the type of Fc receptor (FcR) each isotype binds and in turn to the activation or inhibition of various cell types. A general outline of the various effector functions of each antibody isotype is shown in Table 3.2.

## 3.4 Constant region and antibody isotypes

The antigen-binding specificity of an antibody is dependent on the variable region but the effector function of an antibody is determined by the heavy chain constant region. Antibodies can be classified into five broad types based on the characteristics of their heavy chain constant regions. These fives classes are known as antibody isotypes and include IgM, IgD, IgG, IgA and IgE (Figure 3.4) and their corresponding constant regions are termed $C_\mu$, $C_\delta$, $C_\gamma$, $C_\alpha$ and $C_\varepsilon$. They are specifically written in that order because that is the order in which each constant region gene segment is positioned within the immunoglobulin gene locus and it reflects the temporal expression of antibody isotypes over the lifetime of a B cell clone. The number of heavy chain domains also differs depending on the antibody isotype (Table 3.2) so that IgG, IgA and IgD have three constant heavy chain domains ($C_H1-C_H3$), while IgM and IgE have four ($C_H1-C_H4$). Within the IgG and IgA isotypes further subclasses have been identified. IgG has four separate isotypes designated IgG1, IgG2, IgG3 and IgG4 in humans, while IgA has two isotypes known as IgA1 and IgA2.

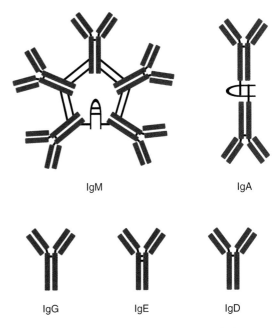

IgM          IgA

IgG          IgE          IgD

**Figure 3.4** The structural differences between antibody isotypes. Secreted IgM forms a pentamer, while IgA forms a dimer, which are linked by a structural protein known as the J-chain. IgG, IgE and IgD all form monomers.

Each isotype has a slightly different tissue distribution or subtle functional differences, which were elucidated by studying antibody fragments in solution (Box 3.1). For example, the biggest difference between IgG subclasses lies within the hinge region, which is a flexible domain between the antibody-binding domain (Fab fragment) and the constant domain (Fc fragment). The hinge region allows considerable molecular flexibility so that the two variable regions can bind to disparate antigenic epitopes. IgG1 and IgG3 both have large hinge regions that allow their variable regions to extend away from the constant regions. On the other hand, IgG2 and IgG4 have relatively small hinge regions meaning that the variable regions have far less flexibility. This may also explain why IgG1 and IgG3 are more efficient activators of complement and FcR-mediated ADCC compared to IgG2 or IgG4.

Differences in the heavy chain constant region of antibody isotypes also determine the structure that each antibody eventually forms (Figure 3.4). For instance, IgA and IgM exist as polymers, while IgG and IgE both exist as monomers. IgA is most commonly found as a dimer, with the two IgA monomers linked together by a third protein called the J-chain. The J-chain links together the constant regions of the two IgA monomers and in so doing confers a degree of flexibility to the dimer. IgM is normally found as a monomer when bound to the B cell plasma membrane but when it is secreted it forms a pentamer. The pentamer is formed by the J-chain linking two IgM monomers via $C_{H4}$ ($C_{\mu 4}$ with regard to IgM) and the linking of all five monomers by disulphide bonds formed between the $C_{\mu 3}$ domains. Due to the multiple antigen-binding sites that multimeric IgA and IgM possess, they are able to bind to several antigen epitopes at once. For this reason they are said to possess multivalency, with IgA having a valency of four and IgM a valency of ten. This enables them to cross-link antigens so that they can form antibody-antigen complexes. In other words they can agglutinate antigens, which make antigens more susceptible to phagocytosis and removal from the body. Although IgG and IgE do not form polymers, their monomeric structure means they are much more proficient at binding to Fc receptors on the surface of effector cells. Furthermore, IgM is the first antibody isotype to be produced during a primary immune response, while IgG and IgE synthesis is associated more with secondary immune responses.

## 3.5 B cell receptor (BCR) diversity

The initial stages of B cell development occur in the bone marrow (or the bursa of Fabricius in birds), which is where the term B cell is derived. Each stage of B cell development

is related to a change in the state of the genetic locus of their BCR. The major components of the BCR are formed of two heavy chains and two light chains (Figure 3.2). Actually a BCR and a secreted antibody are the same protein and are therefore derived from the expression of the same immunoglobulin gene. The antibody repertoire in any given individual is vast, so that many different kinds of antigen are able to be bound. It has been estimated that on average the antibody (or B cell) repertoire may well exceed $10^7$ different specificities. This poses a problem for the genome, as it is unable to accommodate so many individual immunoglobulin genes. Instead, BCRs and immunoglobulin genes generate immunological diversity by a mechanism known as genetic recombination. The mechanism of genetic recombination allows individual genetic components to be stitched together to form a whole gene. In this way, many different combinations of genetic components can recombine in order to generate combinatorial diversity. This has an advantage, in that B cell development can proceed according to the effectiveness of the antibody that the B cell produces. Antibody diversity is enhanced even further by another mechanism known a somatic hypermutation. This occurs when random nucleotides are added when the other genetic components are recombined. The random addition of nucleotides into V regions of antibodies means that the overall diversity of BCRs certainly exceeds $10^7$. In addition, antibodies are expressed as several different classes (e.g., IgM, IgG, IgA), resulting in functional diversity as well as diversity in antigen specificity.

## 3.6 Genetic recombination of BCR genes

Antibodies are formed of two identical heavy chains and two identical light chains, both of which contribute to antigen-binding specificity. Antibody heavy chains and light chains are transcribed from separate genes and, in addition, two forms of light chain genes known as kappa (κ) and lambda (λ) can be used. Therefore, the combination of two different immunoglobulin chains already encompasses a certain amount of combinatorial diversity. However, the largest element of combinatorial diversity arises from the random recombination of three different gene segments, which occurs in the bone marrow during early B cell development. The heavy chains of BCR (and TCR) genes are comprised of a variable region (V), diversity region (D) and joining region (J), while light chain genes are comprised of a V and J region. Therefore, three separate gene segments recombine to form heavy

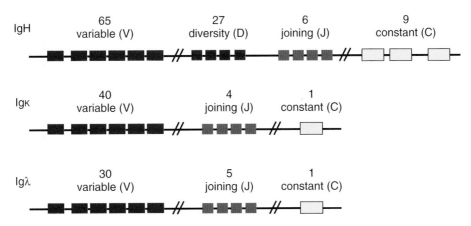

**Figure 3.5** Generalized structure of the immunoglobulin heavy chain (IgH) and immunoglobulin kappa and lambda light chain (Igκ and Igλ) gene loci. The IgH loci contains variable (V), diversity (D), joining (J) and constant (C) gene segments, while the light chain loci only contain V, J and C gene segments.

chains, while two gene segments recombine to form light chains.

Each immunoglobulin heavy and light chain locus has numerous V, (D) and J gene segments (Figure 3.5). In the human there are 65 different heavy chain V gene segments, 27 different functional D segments and 6 different heavy chain J segments. In human light chain genes there are 75 different light chain V gene segments and 9 different light chain J segments. The immunoglobulin gene segments get their names due to their positioning within the translated protein, so that V-regions are situated in the antibody-binding domain where antigen specificity is important. D-regions also contribute to the variability in antigen binding, while J-regions join the antigen-binding domain to the constant heavy domains ($C_H$) of immunoglobulins. The combination of V-(D)-J gene segments is therefore a vital process in generating immunological diversity and is referred to a V(D)J recombination. This process is also tightly controlled at the molecular level and can be used to distinguish individual stages of B cell development.

## 3.7 Mechanism of VDJ recombination

Within the heavy chain locus, located on human chromosome 14, each genetic region is in a particular order so that V gene segments precede D segments, which precede J segments (reading from 5′ to 3′; V-D-J). At the 3′ end of the variable regions are the constant region gene segments, which are also in a particular order (discussed in section 3.10). Within immature B cells the first recombination event to occur is between a D and J region within

the heavy chain locus, which is closely followed by V region recombination (Figure 3.6). This early recombination event forms a newly rearranged V-D-J gene. The remaining V-D-J gene segments located between the rearranged segments are deleted from the genome. The initial immunoglobulin gene combines with both a constant Cμ and Cδ region. Subsequent translation into protein results in the expression of a heavy chain Cμ peptide. Immunoglobulin light chains rearrange in a similar way so that the first event is V-J recombination (remember no D-region) and then the subsequent combining of the constant region. The assembly of heavy chains and light chains takes place in the cytoplasm and results in the formation of an immunoglobulin molecule.

There are various developmental checkpoints that an immature B cell undergoes in order to ensure that V(D)J recombination is proceeding correctly. The first checkpoint occurs at the pre-B cell stage, which follows VDJ recombination of the heavy chain and involves the expression of a surrogate light chain VpreB. The surrogate light chain VpreB binds to the heavy chain in the same way as a functional light chain would and ensures that the heavy chain is able to be expressed on the cell surface. If this happens correctly then the cell signals for the inhibition of further heavy chain recombination events and the initiation of light chain recombination. If heavy chain expression fails then the pre-B cell will attempt recombination at the other allele and if that also fails then the pre-B cell will die. Successful expression leads to light chain rearrangement, first by attempting to successfully recombine the kappa locus, and if that fails, then the

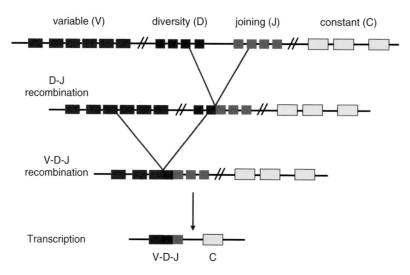

**Figure 3.6** Genetic recombination of the immunoglobulin heavy chain gene. Diversity (D) and joining (J) gene segments recombine before variable (V)-D-J gene segment recombination. Recombination of the V-D-J segment with the constant (C) region occurs last.

lambda locus. Again, if light chain rearrangement fails completely then the pre-B cell will die.

The enzymes responsible for V(D)J recombination are the recombination activation genes (RAG1 and RAG2), which utilize signal sequences at the ends of each V-D-J gene. RAG1 and RAG2 are only expressed in B cells and T cells where they are vital for gene rearrangement and the maturation of a functional adaptive immune response. Without these enzymes an individual would be severely immunosuppressed. The signal sequences that RAG1 and RAG2 recognize are known as recombination signal sequences (RSS), which act as specific cleavage sites (Figure 3.7). Therefore RAG1 and RAG2 use the RSS at the ends of V, D or J genes in order to direct cleavage of both strands at these genes. Each RSS consists of a heptamer and a nonamer with either 12 base pairs or 23 base pairs separating them, which corresponds to one (12 bp) or two (23 bp) helices of double-stranded DNA. These RSS motifs are important for genetic recombination, as only 12 bp RSS will recombine with 23 bp RSS sequences. This is known as the 12/23 RSS rule and it ensures that only the correct gene segments are recombined (Figure 3.7). For example, V region genes are flanked by one 23 bp RSS (3′), while D region genes are flanked by two 12 bp RSS (5′ and 3′). This allows VD recombination and also DJ recombination because J region genes have a 23 bp RSS (5′). Although RAG1 and RAG2 mediate this lymphocyte-specific genetic recombination, ubiquitous DNA splicing machinery is used to join the ends of V-D-J segments.

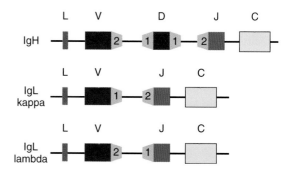

**Figure 3.7** Recombination signal sequences following the 12/23 pair rule. Only a 12 signal sequence can match with a 23 sequence (corresponding to 1 or 2 turns of a DNA helix). A leader sequence (L) is also required for recombination. This ensures that the correct V-D-J gene segments recombine with each other.

## 3.8 Introducing junctional diversity

Combinatorial diversity is increased further by the introduction of additional nucleotides, referred to a junctional diversity (Figure 3.8). The process by which RAG1 and RAG2 select V-D-J region genes leads to the formation of hairpin loops, which are subsequently spliced by DNA polymerases. This sometimes introduces nucleotide overhangs in the double-stranded DNA, which need to be filled before V-D-J segments are joined. As well as filling

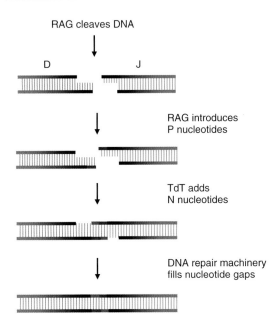

RAG cleaves DNA

D          J

RAG introduces
P nucleotides

TdT adds
N nucleotides

DNA repair machinery
fills nucleotide gaps

**Figure 3.8** RAG and TdT enzymes introduce additional nucleotides to the end of gene segments. This mechanism is responsible for junctional diversity. Exonucleases remove unpaired nucleotides, while DNA repair enzymes close the double-stranded DNA end.

in the gaps, extra nucleotides are sometimes added or exonucleases cleave the overhang so that nucleotides are actually lost. Either way, the result is the addition of extra BCR receptor diversity by the introduction of genetic alterations at the ends of V-D-J gene segments. The enzyme responsible for the addition of extra nucleotides is terminal deoxynucleotidyl transferase (TdT). The additional diversity during recombination tends to be larger in heavy chains than the light chains due to the presence of D-regions. The antibody repertoire that results from such diversity is therefore capable of potentially recognizing any antigen encountered.

## 3.9 Somatic hypermutation and affinity maturation

Somatic hypermutation refers to a mechanism by which added BCR diversity is introduced into the B cell repertoire following antigen encounter. This mechanism also enhances the binding specificity of the BCR (or antibody) through a selection process that selects those B cell clones expressing BCRs with the highest affinity to

antigen. Somatic hypermutation and selection occurs in the germinal centres of secondary lymphoid structures such as lymph nodes and Peyer's patches, a process which is commonly referred to as affinity maturation. The default pathway for B cells within germinal centres is apoptosis, unless a positive signal is received in the form of high affinity antigen presented on the surface of follicular DCs. Therefore, only those B cells expressing BCRs with high affinity for antigen are rescued from apoptosis and are able to mature into antibody secreting plasma cells or enter the memory B cell population. Affinity maturation represents a type of Darwinian selection and is commonly referred to as the clonal selection theory.

The clonal selection theory was formulated by the Nobel Prize–winning immunologist Frank Macfarlane Burnet. This theory is central to the way in which certain B cell clones are selected on the basis of antigen stimulation (Figure 3.9). Each B cell clone expresses an antibody that is specific to a particular antigen. It is this antibody/antigen interaction that provides the stimulation for proliferation and expansion of that B cell clone within the entire B cell population. Those B cell clones that have a higher affinity for antigen will undergo more proliferation compared to those B cell clones with a low affinity for antigen. Therefore, high affinity clones persist and begin to dominate the B cell repertoire, while those of low affinity get deleted from the B cell repertoire. The mechanism that introduces antibody variability, and therefore introduces higher affinity antibodies to the B cell repertoire, is the process of somatic hypermutation. B cells that have already encountered antigen undergo somatic hypermutation, which allows for the introduction of mutations within antibody variable regions and provides the driving force for the selection of B cell clones that express the high affinity antibody. This selection process occurs mainly within lymph nodes and other secondary lymphoid tissues and is commonly known as affinity maturation. Clonal selection and somatic hypermutation are also the reasons why antibody produced during a secondary immune response tends to be of a higher affinity than during a primary immune response. In addition, clonal selection is important for the generation of memory B cells, so that high affinity antibodies are secreted during a recall response. Clonal selection theory can also be applied to the expansion and proliferation of antigen-specific T cells.

The mechanism of somatic hypermutation involves a nucleotide substitution, and to a lesser extent an insertion

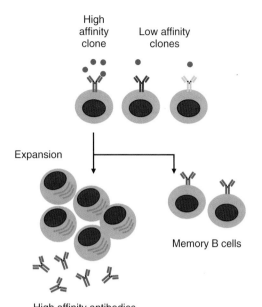

High affinity antibodies

**Figure 3.9** Clonal selection theory states that antigen drives the expansion of B cell clones that express antibody that is of the highest affinity for antigen. Clonal selection is important for both effector responses and for the generation of memory B cells.

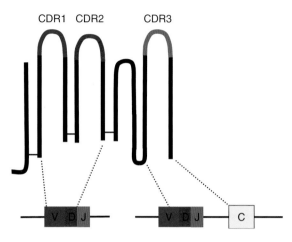

**Figure 3.10** The V-D-J gene segments form hypervariable regions, known as complementarity determining regions (CDRs), of the immunoglobulin heavy chain. CDR1 and CDR2 correspond to V and D gene segments, while CDR3 corresponds to V-D-J gene segments.

or deletion, at specific genetic sites within immunoglobulin genes known as hypervariable regions. These hypervariable regions coincide with the antigen-binding site of the antibody, otherwise known as the complementarity determining region (CDR). This hypervariable region is contained within the V-D-J gene segments and can be separated into CDR1 and CDR2, which span V-D gene segments, and CDR3, which spans the V-D-J and some of the constant region (Figure 3.10). The tertiary structure and the particular folding of immunoglobulin domains largely determine where these CDRs are located. The actual process of replacing a nucleotide within a hypervariable region is guided by an enzyme called activation-induced (cytidine) deaminase (AID). This enzyme replaces a cytosine with a uracil following a deamination reaction. Normally uracil is not present in DNA molecules and results in a uracil-guanine mismatch pair. In order to maintain the correct integrity of the DNA a DNA-mismatch repair enzyme is recruited during replication, which substitutes a thymine to replace the uracil and an adenine instead of a guanine. Therefore AID forms a thymine:adenine (T-A) base pair instead of a cytosine:guanine (C-G) base pair. Therefore, somatic

hypermutation involves a C-G to a T-A substitution known as a transition mutation. Somatic hypermutation can occur randomly across hypervariable regions and therefore introduces a significant amount of antibody diversity.

## 3.10 Immunoglobulin class switching

Immature or naïve B cells only express IgM or IgD on their cell surface. However, activation in the germinal centres of secondary lymphoid tissues can induce B cells to express a different antibody isotype. The mechanism by which B cells achieve this is called immunoglobulin class switching and occurs through a process known as class switch recombination (Figure 3.11).Class switching occurs in response to a specific set of external signals that are usually driven by the cytokine microenvironment and by the ligation of the co-stimulatory molecule CD40. For example, Th1 immune responses are associated with increased levels of IFN-γ, which can induce B cells to class witch from IgM to IgG. In mucosal tissues such as the intestines and airways the cytokine environment is dominated by TGF-β and IL-4 and therefore the predominant immunoglobulin is IgA. In another example, allergic reactions are associated with IL-4 and IL-5, which results in B cell class switching to IgE. The differences in these antibody isotypes reflect their divergent functions.

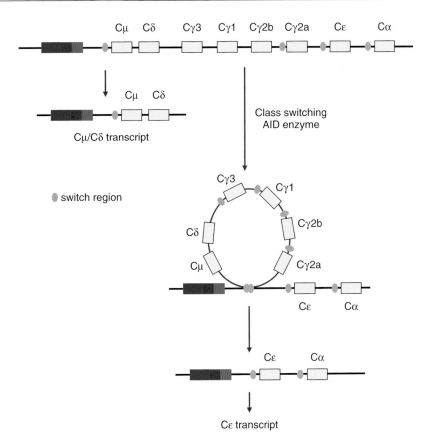

**Figure 3.11** Immunoglobulin class switching involves a process of genetic recombination. Class switching occurs in an ordered fashion corresponding to the order of constant (C) region gene segments. The enzyme activation-induced (cytidine) deaminase (AID) guides class switch recombination via specific switch regions.

The particular immunoglobulin isotype is determined by the constant region. The order in which B cells can undergo class switching is also influenced by the order in which the different constant gene regions are arranged. For example, going from 5′ to 3′ within the human immunoglobulin locus, IgM (Cμ) precedes IgG1 (Cγ1), which in turn precedes IgA (Cα) and IgE (Cε). Therefore, B cells tend to produce IgM before any other isotype, as this is the first constant region exon, then IgG before IgA and so on. During class switching any preceding constant region exons are deleted from the chromosome and replaced with a downstream constant region. For example, Cμ and Cδ exons are deleted and replaced with a Cγ, Cα or Cε exon. An individual B cell cannot then go back and produce IgM but can go on to produce IgG, IgA or IgE. Each constant region gene segment possesses a repetitive sequence at the 5′ end known as

the switch region. During class switch recombination the AID enzyme recognizes the switch region and induces the formation of a hairpin loop. The free ends of the DNA are then joined to the gene segments from the rest of the immunoglobulin by a process called non-homologous end joining (NHEJ), which involves various DNA polymerases and DNA ligase enzymes. Class switch to a particular antibody isotype often reflects the functional requirement of antibody during that particular immune response.

## 3.11 Structure of Fc receptors

Antibody molecules are an essential part of the adaptive immune response and have two basic functions, which are directly related to their structure. The first function is

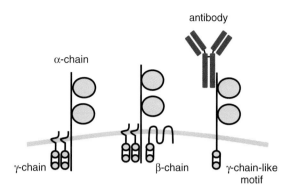

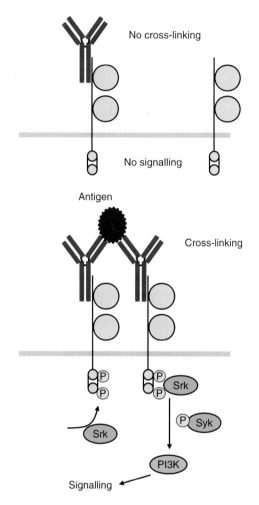

**Figure 3.12**  Fc receptors recognize antibody via the α-chain, which binds to the constant region of antibody molecules, and signal via the γ-chain, or γ-chain-like motifs.

binding to antigen through recognition of molecular epitopes by the antigen-binding site of the variable region. The second function is to induce an effector response through binding to Fc receptors on the surface of effector cells via their constant region. The majority of Fc receptors belong to the immunoglobulin gene super-family (Figure 3.12). They are transmembrane proteins with a dominant extracellular domain responsible for Fc binding and a short intracellular domain. Certain Fc receptors possess an accessory protein that mediates intracellular signal transduction. The main part of the Fc receptor, which binds antibody, is therefore known as the α-chain. The α-chain is responsible for receptor specificity. The accessory protein, which is known as the Fc receptor common γ-chain, is invariable and relays the receptor signal. Certain Fc receptors are able to dispense with the common γ-chain as they possess a γ-chain-like motif as part of the intracellular domain of the α-chain (Figure 3.12), while others also possess a β-chain that enhances receptor signalling.

Both the common γ-chain and the γ-chain-like motifs contain an activation domain, which is known as an immunoreceptor tyrosine-based activation motif (ITAM). Antibody binding to the extracellular domain of the α-chain causes activation of the ITAM within the

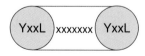

**Figure 3.13**  Immunoreceptor tyrosine-based activation motif (ITAM).

**Figure 3.14**  Cross-linking is required for Fc receptor (FcR) signalling and induces the phosphorylation of the kinases Srk, Syk and PI3 kinase.

γ-chain and results in a cascade of intracellular signalling events, gene expression and activation of the effector cell. ITAMs possess a specific amino acid sequence composed of two tyrosine-leucine repeats separated by 6 to 8 unspecified amino acids with a consensus sequence YxxL 6–8 YxxL (Figure 3.13). Receptor ligation induces phosphorylation of the ITAM, translocation of Syk family kinases to the ITAM and the initiation of signalling events leading to gene transcription (Figure 3.14). Some Fc receptors do not possess an ITAM but rather an inhibitory motif called an ITIM (immunoreceptor tyrosine-based inhibitory motif). Therefore, instead of activating the cell, they inhibit its effector functions. In other words, antibody binding to Fc receptors can either activate or inhibit a cellular response.

## 3.12 Fc receptor specificity and affinity

Fc receptors are specific for the Fc portion of antibody molecules. Each Fc receptor α-chain binds only to its particular antibody isotype so that Fc receptors specific for IgA only bind to the IgA isotype, for example. A general nomenclature is followed so that the Fc receptor that binds IgA is termed the Fcα receptor (FcαR), while FcγRs bind IgG and FcεRs bind IgE (Table 3.3). Fc receptors are also expressed on a variety of cell types. In this way effector functionality can be controlled by isotype specificity and by the pattern of receptor expression. For example, the FcαR (CD89) is specific for IgA and is predominantly expressed on macrophages. The resultant effector activity of macrophages, involving phagocytosis and killing, is therefore restricted to sites of elevated IgA production.

Not all Fc receptors have the same affinity for their particular antibody isotype. Several FcγRs exist that are either classified as high-affinity receptors or low affinity receptors. For example, FcγRI (CD64) is a high affinity receptor, particularly for IgG1 and IgG3, while FcγRIIA and FcγRIIB are low affinity receptors. As a general rule, the nomenclature dictates that all type I receptors are high affinity while all type II and III receptors are low affinity (Table 3.3). This adds another level of complexity to Fc receptor-mediated control of cellular effector function. Furthermore, different Fc receptors activate different effector mechanisms. This is partly due to the expression pattern on different cell types but also due to the antibody isotype and the specific signalling events following receptor ligation. In addition, certain Fc receptors are inhibitory and therefore result in a cessation of cellular stimulation.

## 3.13 Cross-linking of antibody is necessary for Fc receptor signalling

Free antibody is present in the serum and tissue microenvironment all the time. However, a mechanism exists that prevents the activation of effector cells as a result of free antibody binding to Fc receptors, thereby avoiding possible tissue damage. In order for antibody to successfully activate intracellular signalling cascades the Fc receptors must be crosslinked (Figure 3.14). The concept of cross-linking is a ubiquitous phenomenon regarding the initiation of receptor-mediated signalling events. Cross-linking only occurs in the presence of an antigen, as two or more antibody molecules bind the same antigen. The

**Table 3.3** Effector functions of different Fc receptors.

|  | FcαRI | FcεRI | FcγRI | FcγRIIA | FcγRIIB | FcγRIII |
|---|---|---|---|---|---|---|
| Antibody isotype | IgA | IgE | IgG1/IgG3 IgG4 IgG2 | IgG1 IgG3/IgG2 IgG4 | IgG1 IgG3/4 IgG2 | IgG1 IgG3 |
| Affinity | High affinity | High affinity | High affinity | Low affinity | Low affinity | Low affinity |
| Signalling motif | γ-chain ITAM | γ-chain ITAM β-chain | γ-chain ITAM | α-chain ITAM | α-chain ITIM | α-chain ITAM |
| Cell Expression | Macrophages Neutrophils Eosinophils | Mast cells Eosinophils Basophils | Macrophages Neutrophils Eosinophils DCs | Macrophages Neutrophils Eosinophils DCs | B cells Macrophages Neutrophils Eosinophils Mast cells | NK cells Macrophages Neutrophils Eosinophils Mast cells |
| Effector mechanism | Phagocytosis Killing | Release of granules | Phagocytosis Killing Respiratory burst | Phagocytosis Release of granules | Inhibition of phagocytosis | ADCC |

opsonized antigen effectively acts as a bridge for multiple antibodies and results in the simultaneous ligation of multiple receptors on the cell surface. Cross-linking causes the aggregation of receptors thereby enhancing the avidity of receptor binding. These interactions also cause the localization of the intracellular signalling machinery required for appropriate signal transduction.

## 3.14 Fc receptor immune functions

Antibodies play a vital role in the killing of infected cells and in the eventual clearance of pathogens from the body. A virus for example, enters a cell across the plasma membrane and in so doing leaves viral-associated proteins on the surface. In addition, replicating viruses may produce viral-associated proteins that are transported to the cell surface. The viral proteins are recognized by the Fab fragment of antibody molecules, which bind to the plasma membrane and opsonize the infected cell. The cells responsible for effecting antibody-dependent cell-mediated cytotoxicity (ADCC) are predominantly NK cells. The binding of antibody to FcRs also induces other immune effector functions such as phagocytosis by macrophages and neutrophils and the release of granule contents by mast cells, basophils and neutrophils. Although antibodies are produced by B cells of the adaptive immune system, they exert the majority of their effects through FcR binding on cells of the innate immune system. It can therefore be appreciated that the boundaries between innate and adaptive immunity are not always so rigid.

The archetypal inhibitory Fc receptor is FcγRIIB, which has significant extracellular homology to the activation Fc receptors but differs in that it possesses an immunoreceptor tyrosine-based inhibition motif (ITIM) instead of an ITAM. FcγRIIB is expressed on neutrophils, macrophages and mast cells but is noticeably absent from NK cells. Cross-linking of FcγRIIB with an activation receptor leads to the phoshorylation of the ITIM by lyn kinase. This phosphorylation event results in the recruitment of SHIP, which is able to block other kinases involved in activation and prevents calcium mobilization. The recruitment of SHIP inhibits certain effector functions that are dependent on calcium signalling such as proliferation, cytokine production and phagocytosis. In addition, the FcγRIIB has inhibitory functions that are independent of the ITIM sequence. Homogenous aggregation of FcγRIIB leads to pro-apoptotic signalling events and cell death.

## 3.15 T cell receptor diversification

The proteins that comprise the T cell receptor (TCR) are structurally similar to immunoglobulins. Each αβTCR is composed of a TCR α-chain (TCRα) and a TCR β-chain (TCRβ), while each chain can be divided into a constant region and a variable region. The α-chain and β-chain combine to form the variable domain proper, which is responsible for antigen specificity, in much the same way as the light and heavy chains of immunoglobulins form the antigen-binding site. It is no surprise therefore that the generation of TCR diversity uses the same mechanism as BCR diversity, through a process of V-(D)-J recombination. The TCRα is equivalent to the immunoglobulin light chain as the Vα domain is formed from V-J recombination, while the TCRβ is equivalent to the immunoglobulin heavy chain as it is formed from V-D-J recombination. The junctions between the V-(D)-J regions also undergo genetic alterations through the same process of junctional diversity. However, TCRs do not undergo somatic hypermutation in the same way as immunoglobulin molecules. This may have evolved so as to prevent autoimmunity by avoiding previously selected T cells from acquiring a mutation that could have self-reactive potential. The generation of γδTCRs also occurs by the same mechanisms of V-(D)-J recombination.

Considering that TCR diversity is generated by a very similar mechanism of V-(D)-J recombination to BRC diversity, and both TCRs and BCRs are members of the immunoglobulin supergene family, the actual genetic architecture of TCR gene loci are also similar to that of BCR genes (Figure 3.15). The genes that encode the TCRα and TCRγ chains are known as *TCRA* and *TCRG* and contain V and J gene segments and a constant gene segment (similar to immunoglobulin light chains). The genes *TCRB* and *TCRD* encode the TCRβ and TCRδ chains and comprise V, D and J gene segments and a constant gene segment (similar to immunoglobulin heavy chains). However, the *TCRB* gene loci is organized so that two clusters of D, J and C gene segments proceed the variable gene segments (Figure 3.15). In humans there are 42 Vα gene segments and 48 Vβ gene segments, while there are only 6 Vγ genes and 8 Vδ gene segments (the *TCRDV* genes are interspersed among the *TCRAV* segments). Therefore, it is already clear the αβTCRs have a much more diverse antigen-binding repertoire than γδTCRs. In addition, there may be as may as 60 Jα and, 2 Dβ and 13 Jβ gene segments, further increasing the

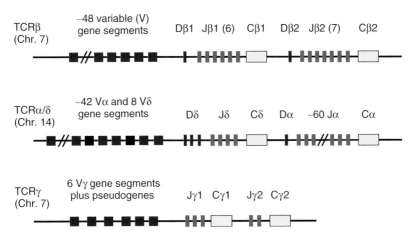

**Figure 3.15** Diagrammatic reproduction of the arrangement of V-(D)-J gene segments within the TCR gene loci.

potential αβTCR repertoire. The numbers of Vγ, Dδ and Jδ gene segments are far fewer.

The functions of TCRs differ considerably from BCRs or secreted immunoglobulins. While immunoglobulins are responsible for recognizing antigen based on the shape of the antigen, TCRs recognize peptide antigens that are bound onto MHC molecules. Despite these functional differences, the actual shape of the antigen-binding groove is remarkably similar to that of immunoglobulins. For instance, the predominant region responsible for the majority of antigen-binding variability is the CDR3 loop, which is composed of the V-(D)-J regions. The remainder of the αβTCR variable region (CDR1 and CDR2 loops) is less variable, as this is the part that makes contact with the MHC molecule. TCRβ chains have an extra area of hypervariability, known as HV4, but it is not involved in peptide binding and therefore not considered a CDR. Although TCRs exhibit less structural variability across the antigen-binding region compared to immunoglobulins, the TCR loci possess more J gene segments. This means that the diversity of TCRs may be even higher than immunoglobulins. Unlike immunoglobulins, the constant region of the TCR does not perform an effector function as it is anchored into the surface membrane of T cells. The constant domain also has a transmembrane region and a cytoplasmic tail, which aggregate with several accessory molecules to form a TCR complex. The various accessory molecules of the TCR complex are known collectively as CD3, which is responsible for generating receptor signalling following peptide:MHC binding to the TCR. The CD3 complex is composed of membrane

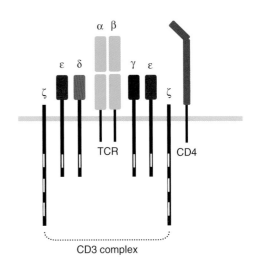

**Figure 3.16** The CD3 complex is composed of at least two ζ subunits and usually ε subunits and either a γ or a δ subunit. The TCR complex is therefore formed of a TCR, CD3 and a CD4 molecule.

bound CD3γ, CD3δ and CD3ε chains and cytoplasmic CD3ζ chains (Figure 3.16).

## 3.16 T cells undergo positive and negative selection within the thymus

The thymus is a primary lymphoid organ located in the chest, proximal to the trachea and just above the heart. The main function of the thymus is to provide a suitable

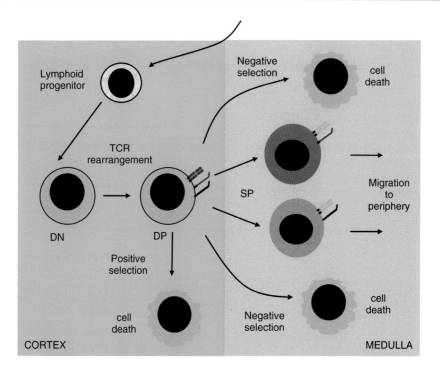

**Figure 3.17** T cell development in the thymus involves positive and negative selection of immature T cells. The stages of T cell development can be distinguished based on the expression of the surface markers CD4 and CD8. Immature T cells go through double negative (DN), double positive (DP) and single positive (SP) stages.

microenvironment for the development and differentiation of immature T cells (Figure 3.17). Anatomically it can be divided into an outer cortex and a central medulla. The early stages of T cell development occur within the cortex, which coincides with TCR gene rearrangement, positive selection and the first phase of negative selection. Within the less densely populated medulla, further rounds of negative selection take place, which results in the removal of autoreactive T cells. Fully differentiated T cells exit the thymus and migrate to secondary lymphoid structures such as the spleen, lymph nodes and Peyer's patches in the gut. Most T cell differentiation and selection occurs within the late embryo and during the neonatal period. By the time early adulthood is reached the peripheral T cell repertoire is complete and the thymus undergoes a process of involution. Although the thymus is almost undetectable in adults, it still functions as an endocrine gland that is thought to stimulate certain aspects of the immune system.

Cells derived from the haematopoietic lineage that reside in the thymus are known as thymocytes. These thymocytes, or immature T cells, undergo several phases of differentiation, which can be distinguished based on the

expression of CD4 and CD8 surface markers (Figure 3.17). Thymocytes begin life as double negative (DN) cells, as they express neither CD4 nor CD8. This is the stage that TCR gene rearrangement begins. Firstly, V-D-J rearrangement occurs at the TCRβ gene loci, which also coincides with the start of positive selection. The positive selection process ensures that a functional TCR is produced that can interact correctly with MHC molecules. The first checkpoint in positive selection therefore, is to ensure that the TCRβ gene has successfully rearranged. Surface expression of the TCRβ chain is assisted by an invariant pre-Tα chain, the dimerization of which forms the pre-TCR. If a correct TCRβ chain is expressed V-J gene rearrangement of the TCRα loci can then proceed, which coincides with the transition to the second double positive (DP) stage of thymocytes development, whereby immature T cells express both CD4 and CD8. The synthesis of both TCRα and TCRβ chains results in the expression of an αβTCR on the cell surface. A further round of positive selection then ensures that the fully formed αβTCR is functional and that a positive rearrangement has taken place. The same process of positive selection also occurs for γδTCR gene rearrangement. Due to the randomness of V-(D)-J recombination, there is always a high likelihood that an unsuccessful TCR gene rearrangement will take place. This results in a high percentage of immature thymocytes undergoing apoptosis and dying in the thymus. Furthermore, TCRs that fail to recognize cognate MHC molecules also undergo apoptosis during positive selection, which represents an important checkpoint in thymocyte development. Successfully rearranged DP thymocytes that have positively rearranged their TCR begin to undergo negative selection and migrate from the cortex into the medulla.

Migration from the cortex and into the medulla also coincides with the single positive (SP) stage of thymocyte development, as only SP CD4+ or CD8+ T cells are detected in the medulla. The thymic medulla also contains mature T cells that are either exiting the thymus or mature T cells that are migrating into the thymus form the periphery. Negative selection involves the removal of T cells that are reactive to self-antigens. DCs and macrophages that express high levels of both MHC class I and MHC class II molecules are abundant at the corticomedullary junction, which immature T cells must pass through, and throughout the medulla itself. DCs and macrophages present self-peptides to immature T cells undergoing negative selection. Epithelial cells within the medulla are also thought to present peptides to immature T cells during negative selection. Medullary epithelial cells express the transcription factor AIRE (autoimmune regulator) that drives the expression of genes that would otherwise not be expressed in the thymus. The overall effect of negative selection is the removal of self-reactive T cells from the peripheral T cell pool and to prevent any possibility of autoimmune reactions. This process is known as central tolerance and is a key concept in T cell immunology. Central tolerance is also supplemented by peripheral tolerance (discussed in Chapter 15), which is driven by the action of regulatory T cells and the induction of anergy following the lack of co-stimulatory signals.

## 3.17 Antigen presentation to T cells

The principle function of T cells is the recognition of foreign antigens through the binding of their TCR to MHC molecules loaded with peptide (Figure 3.18). The TCRs of T helper (Th) cells recognize MHC class II molecules due to the association of the co-receptor CD4, which specifically binds only MHC class II molecules. Importantly, MHC class II molecules are only expressed on professional APCs such as DCs, macrophages and to a lesser extent B cells. Antigen enters the MHC class II processing pathway only if it has been taken up from an exogenous source by an APC, through a process of endocytosis. Therefore, CD4+ Th cells recognize exogenously derived peptides that have been derived from an extracellular environment. Conversely, CTLs express the co-receptor CD8, which specifically binds to MHC class I molecules. Every somatic cell within the body is capable of expressing MHC class I molecules. Antigen that enters the MHC class I processing pathway is derived from intracellular sources, in particular the cytoplasm. Therefore, CD8+ CTLs recognize endogenously derived peptides. CTLs are a crucial component of the adaptive immune system responsible for defending against intracellular pathogens. Every somatic cell within the body is capable of expressing MHC class I molecules, which represents an evolutionary strategy intended to defend the entire body against intracellular microorganisms. To summarize, the TCRs of CD4+ Th cells recognize exogenously-derived antigen expressed in conjunction with MHC class II molecules on the surface of professional APCs, while CD8+ CTLs recognize endogenously-derived antigen expressed in conjunction with MHC class I molecules on the surface of somatic cells (see also Box 3.2).

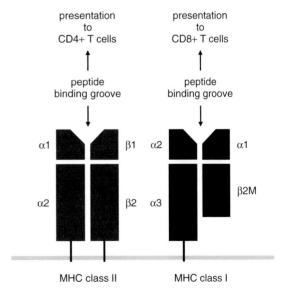

**Figure 3.18** Structure of MHC class II and MHC class I molecules ($\beta$2M, $\beta$2-microglobulin). MHC class II molecule present exogenous peptide antigen to CD4+ Th cells, while MHC class I molecules present endogenous peptide antigen to CD8+ CTLs.

## Box 3.2. MHC Gene Loci in Humans and Mice.

MHC class I molecules are formed from one α-chain and one β-chain. The α-chain is responsible for binding to antigenic peptides and is encoded from a region within the large MHC loci, which is located on chromosome 6 in humans and chromosome 17 in mice and contains more than 120 different genes. In humans MHC molecules are referred to as the HLA (human leukocyte antigen) molecules and in mice as H-2 (histocompatibility-2) antigens. The β-chain is formed from a small, globular protein called β2-microglobulin and is encoded from a different chromosome (chr. 15 in humans and chr 2 in mice). MHC class I molecules are responsible for presenting peptide to CD8+ CTLs. MHC class II molecules are also composed of an α-chain and a β-chain, which both contribute to peptide binding. They are both encoded from within the MHC gene loci and are responsible for presenting peptide to CD4+ Th cells. MHC class I and MHC class II molecules are known as classical MHC molecules because they function in presenting peptide to T cells. The MHC gene locus also encodes non-classical MHC molecules that are not involved in peptide presentation. Instead, non-classical MHC molecules perform other functions within the immune system such as peptide loading in the ER and the presentation of unconventional molecules to specialized populations of T cells (e.g. MICA presents structurally related antigens to intestinal γδ T cells). Another region within the MHC locus encodes MHC class III molecules, which are also non-classical MHC molecules, some of which have immunological functions including TNF, LTα/β and complement proteins.

In humans the MHC loci is assembled in a particular order with MHC class II genes first followed by MHC class III genes and then MHC class I genes. The gene order in mice is similar apart from some MHC class I genes are found at the beginning of the loci as well as at the end (see figure accompanying box). Three human classical MHC class II genes have been described and are known as HLA-DP, HLA-DQ and HLA-DR. Mice only have two MHC class II genes known as I-A (the counterpart of HLA-DQ) and I-E (the counterpart of HLA-DR). The genetic region encoding MHC class II genes also encodes non-classical molecules such as HLA-DM, HLA-DO, TAP and LMP proteins, which are involved in endogenous antigen processing and peptide loading onto classical MHC molecules.

The human MHC locus encodes three classical MHC class I molecules called HLA-A, HLA-B and HLA-C. The equivalent genes in the mouse are known as H-2D, H2-L and H2-K, which are separated at either end of the mouse MHC locus. The MHC class I region in humans also encodes the non-classical HLA-E, HLA-F, HLA-G, MICA and MICB, which are all thought to be capable of presenting unconventional antigens to specialized subsets of T cells.

An important property of classical MHC class I and class II molecules is that they are highly polymorphic within the human population. Firstly, each person inherits two copies of each gene, one maternal copy and one paternal copy. For example, an individual will inherit one copy of HLA-A from his mother and one copy from his father, resulting in the expression of two different HLA-A molecules on the cell surface. The same is true for the other classical MHC molecules. The polymorphism within MHC genes means that the peptide-binding capacity of different alleles is slightly different. It is therefore likely that an individual will inherit two different alleles of each MHC molecule from his parents. This is evolutionarily advantageous, as each person will inherit an MHC repertoire that is capable of presenting different peptides and therefore inheriting an immune system capable of recognizing a diverse range of pathogens. Inbred populations, which inherit identical MHC molecules, will have a smaller MHC repertoire and therefore be more susceptible to new infections.

Human MHC class I

Mouse MHC class I

Human MHC class II

Mouse MHC class II

Diagrammatical representation of the human and mouse MHC gene loci.

## 3.18 MHC class II processing pathway

MHC class II molecules are heterodimers composed of an α-chain and a β-chain. Both α-chain and β-chain are structurally similar and both participate in forming the peptide-binding groove (Figure 3.19). Although MHC class II molecules present exogenous antigen, the assembly of peptide: MHC complex occurs intracellularly. A mechanism therefore exists that prevents endogenous peptides from entering the MHC class II processing pathway. The MHC class II α-chain and β-chain are synthesized in the endoplasmic reticulum (ER) along with a smaller trimeric protein called the invariant chain (Ii). The Ii chain prevents intracellular peptides from loading onto MHC class II molecules by binding over the peptide-binding groove. The Ii chain remains bound to MHC class II until it enters an endocytic vesicle that contains the peptide products of phagocytosed exogenous antigen. Degradation of the Ii occurs until only a small fragment of it is left, called CLIP. Finally, CLIP is replaced by an antigenic peptide by a non-classical MHC class II molecule known as HLA-DM. Only a stable MHC class II: peptide complex is able to be transported to the cell surface where it can then present peptide to CD4+ Th cells.

## 3.19 MHC class I processing pathway

MHC class I molecules are heterodimers composed of an α-chain, which is responsible for peptide binding, and an invariant protein called β2-microglobulin. These two proteins are linked non-covalently (Figure 3.20) during their synthesis within the ER. Peptides that are loaded

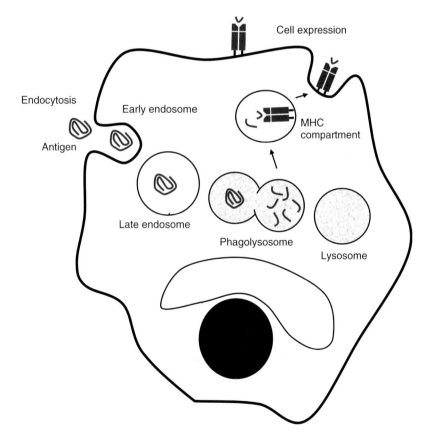

**Figure 3.19** The MHC class II processing pathway. Antigen is derived from endogenous sources through a process of endocytosis. Protein is degraded following phagolysosomal fusion and peptide fragments enter the MHC class II compartment where they are loaded onto MHC class II molecules. Exocytosis results in the presentation of peptide on the cell surface in the context of MHC class II molecules.

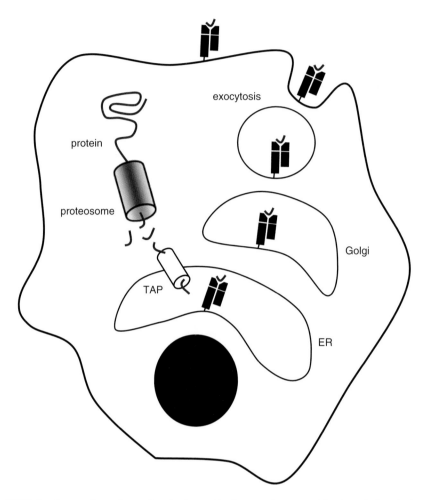

**Figure 3.20** The MHC class I processing pathway. Proteins located in the cytosol are tagged for degradation via the proteosome. Peptide fragments are transported to the endoplasmic reticulum (ER) via TAP (transporter associated with antigen processing) proteins and loaded onto MHC class I molecules. Peptide is the transported to the cell surface by exocytosis where it is presented on the cell surface in the context of MHC class I molecules,.

onto MHC class I molecules are derived from endogenous proteins that have been ubquitinated and degraded into peptides 8–10 amino acids in length. Protein degradation occurs with a structure known as the proteosome located within the cytosol. Therefore, in order for peptide to be loaded onto MHC class I they must be translocated form the cytoplasm and into the lumen of the ER. Two proteins known as TAP1 and TAP2 (transporter associated with antigen processing) are responsible for mediating this translocation. The actual loading of peptide into the binding groove of MHC class I involves TAP and several other proteins, including tapasin, calreticulin, calnexin and ERP57, that form a mutimeric complex. Once peptide

is bound onto MHC class I molecules they can enter the default secretory pathway and be expressed on the cell surface where they can present peptide antigen to CD8+ CTLs.

## 3.20 Activation requires co-stimulation

In addition to binding to peptide: MHC complexes expressed on the surface of APCs, effective TCR signalling requires the help of other stimulatory molecules. A three signal model has been proposed for T cell activation that involves firstly TCR binding to MHC molecules, secondary co-stimulatory molecule ligation and a third signal provided by IL-2. T helper cells express CD4, which specifically binds MHC class II molecules and prolongs the engagement between the TCR and peptide antigen/MHC complex. In a similar way CTLs express CD8, which specifically binds to MHC class I molecules on the surface of somatic cells. In addition, CD4 and CD8 participate in the recruitment of protein kinases to the intracellular region of the CD3 complex in order to initiate downstream signalling events (Figure 3.21). An important event is the activation and recruitment of the tyrosine kinase Lck, which phosphorylates ITAM motifs on CD3$\zeta$. These ITAM motifs recruit another important protein tyrosine kinase called ZAP-70, which

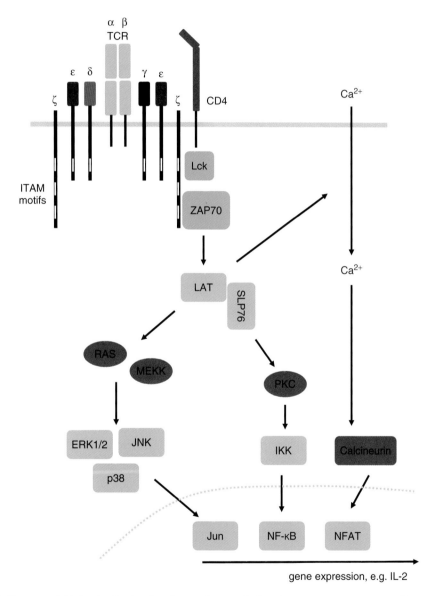

**Figure 3.21** T cell receptor (TCR) activation involves the recruitment of Lck and ZAP-70. Downstream signalling results in the phosphorylation of several kinases, NF-$\kappa$B activation and calcium (Ca$^{2+}$) mobilisation. The transcription factor Jun, NF-$\kappa$B and NFAT result in gene transcription.

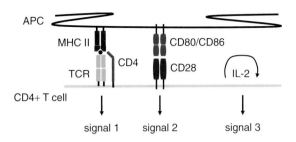

**Figure 3.22** T cell activation requires three separate signals. The first signals following TCR ligation with MHC/peptide complexes. The second signal involve activation of the co-stimulatory molecule CD28 on T cells following ligation with CD80/CD86 on APCs. The third signal follows autocrine signalling via the cytokine IL-2.

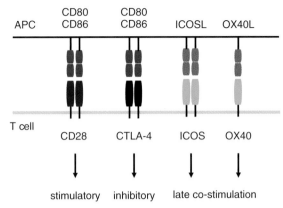

**Figure 3.23** Late co-stimulatory signals influence the progression of a T cell response. CD28 provides the initial co-stimulatory signal to T cells following binding to CD80/CD86 on APCs. CTLA-4 also binds to CD80/CD86, with higher affinity than CD28, and provides an inhibitory signal to T cells. ICOS and OX40 provide late co-stimulatory signals that promote proliferation and differentiation of T cells.

is also phosphorylated by Lck. Activated ZAP-70 in turn phosphorylates downstream signalling molecules including LAT (linker for the activation of T cells) and SLP-76 (SH2-domain-containing leukocyte protein of 76 kDa) amongst others. Ultimately, this signalling cascade results in the production of IL-2 by the T cell, which then acts in an autocrine manner and provides the T cell with a third signal for activation. IL-2 is crucial for T cell survival and proliferation. Activation of LAT also leads to the mobilization of $Ca^{2+}$ stores, which is a key event during the initiation of T cell effector functions and proliferation events.

In order for a T cell to become fully activated, secondary co-stimulatory signals are required (Figure 3.22). This secondary signal is provided by the co-stimulatory molecule CD28, which binds to CD80 or CD86 expressed on the surface of APCs. The kinases Lck and PI3K are recruited to the cytoplasmic domain of CD28, which in turn cause the phosphorylation and activation of Akt. This then acts as a pro-survival factor for T cells. Stimulation of CD28 also activates Lck so that it can be recruited to CD4 where it induces the phosphorylation of ZAP-70 associated with the CD3 complex. Furthermore, activation of Akt and other downstream kinases causes the degradation of the IκB complex and the release of the transcription factor NF-κB, which in turn initiates gene transcription events necessary for T cell activation including expression and release of IL-2 (the third signal).

## 3.21 Late co-stimulatory signals

Co-stimulatory signals are essential for the activation of T cells and for the initiation of effector functions. In the

absence of co-stimulatory signals T cells are induced to undergo anergy, whereby they remain in an inactivate state, or they undergo apoptosis. Although CD28 ligation with CD80/CD86 provides a crucial secondary signal for the early events in T cell activation, several other co-stimulatory signals take part in the late phases of activation and provide important signals for survival, proliferation and differentiation into various T cell subsets (Figure 3.23). One such receptor is ICOS (inducible T cell co-stimulator), which is a member of the CD28 family of receptors and an immunoglobulin supergene family member. ICOS binds to B7H (ICOS ligand) on APCs and is induced following T cell activation and remains expressed on the surface of memory T cells. The function of ICOS is to reinforce the express of IL-2 by T cells, which has the effect of stimulating continued survival, proliferation and effector function. ICOS is also thought to play a role in the differentiation of Th cells into Th2 cells and is associated with the upregulation of the Th2 cytokine IL-4. In addition, it has been suggested that ICOS can substitute for CD28 in activated T cells, which is supported by the shared use of a common signalling pathway.

Another CD28 family member is CTLA4 (cytotoxic T lymphocyte antigen 4), which has a negative effect on T cell activation and actually inhibits proliferation. CTLA4 is upregulated on activated T cells following TCR: MHC ligation. This inhibitory effect is firstly due to competition with CD28 for access to their common ligands CD80/CD86. CTLA4 actually has a higher affinity for CD80/CD86 compared to CD28. It therefore provides effective regulation of T cell proliferation and can inhibit certain lymphoproliferative disorders. However, it is unclear how CTLA4 initiates inhibitory signals to the T cell as it does not possess ITIMs (immunoreceptor tyrosine-based inhibitory motifs) within its intracellular domain. One possibility is that CTLA4 interferes with lipid rafts within the plasma membrane of T cells, which are important for the accumulation of receptor complexes and positive signalling events.

Another important late co-stimulatory molecule expressed by activated T cells is OX40, which is a member of the TNF receptor family. OX40 is expressed on T cells 24–48 hours after CD28-dependent activation. Its ligand OX40L is not constitutively expressed on APCs either but is upregulated following activation. Following initial activation, OX40 appears to play a significant role in maintaining T cell responses after the first 4 days and therefore contributes to generation of a memory T cell population. As well as providing signals for continued cytokine secretion, ligation of OX40 downregulates the expression of the inhibitory receptor CTLA4. OX40 signalling may also be important for the differentiation and maintenance of Treg cell populations.

## 3.22 Activation of B cell responses

The interaction between Th cells and professional APCs is crucial for the development of cell-mediated immune responses. The interaction between Th cells and B cells is also important for the effective generation of B cell responses and the production of antibodies. Like other professional APCs, B cells are capable of presenting peptide antigen in conjunction with MHC class II molecules to CD4+ Th cells. In a similar way that co-stimulatory signals are necessary for T cell activation, specific co-stimulatory signals are also required for the activation of B cells and for the effective production of antibodies. Recognition of peptide MHC class II complexes causes the upregulation of CD40L on T cells, which binds to the constitutively expressed CD40 co-stimulatory molecule on resting B cells. This receptor ligation has two principle effects. The first is that CD40L gives a signal to the Th cell and induces the expression of IL-4. Secondly, CD40 signalling results in the activation of resting B cells and acts in conjunction with IL-4 signalling to cause B cells to undergo proliferation, immunoglobulin class

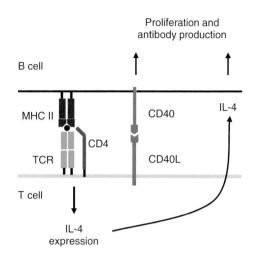

**Figure 3.24** B cell activation requires co-stimulation provided by CD4+ Th cells. In particular, binding of the co-stimulatory molecules CD40 and CD40L, together with IL-4 secretion, activates resting B cells.

switching and antibody synthesis (Figure 3.24). Without CD40:CD40L interactions B cells are unable to proliferate and cannot produce T-dependent antibodies. The interaction between CD4+ Th cells and antigen-specific B cells usually occurs in secondary lymphoid structures such as lymph nodes and specialized mucosal sites including Peyer's patches in the intestine and the tonsils.

## 3.23 CD4+ T helper cell differentiation

The differentiation of a naïve T cells into one of the several T cell subsets is largely the result of the surrounding cytokine milieu (Figure 3.25). Following activation by TCR ligation (signal 1), co-stimulation (signal 2) and proliferation induced by IL-2 (signal 3) undifferentiated T cells, known as Th0 cells, are receptive to further cytokine stimulation. The production of IL-12 by DCs, TNF by macrophages and IFN-γ by NK cells causes Th0 cells to differentiate into Th1 cells. This results in the induction of type 1 immunity and is usually the result of an infection with an intracellular pathogen. In the absence of IL-12 and in the presence of IL-4, IL-5 and IL-13 a type 2 immune response is initiated and Th0 cells differentiate into Th2 cells. This is usually the result of an infection with an extracellular pathogen, although Th2-driven immune responses are also characteristic of allergic reactions (chapter 15).

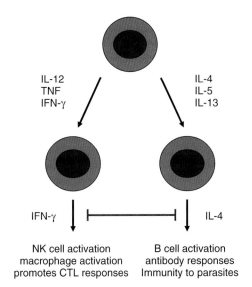

IL-12
TNF
IFN-γ

IL-4
IL-5
IL-13

IFN-γ | ⊢——————⊣ | IL-4

NK cell activation
macrophage activation
promotes CTL responses

B cell activation
antibody responses
Immunity to parasites

**Figure 3.25** Naïve CD4+ T cells (Th0) can differentiate into either Th1 cells or Th2 cells depending on the cytokine environment. Differentiated Th1 cells further secrete type-1 cytokines (IFN-γ), while Th2 cells secrete type 2 cytokines (IL-4). In addition, IFN-γ inhibits IL-4 production, while IL-4 inhibits IFN-γ production, thereby reinforcing T cell differentiation.

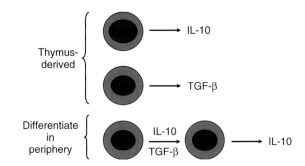

Thymus-derived

IL-10

TGF-β

Differentiate in periphery

IL-10
TGF-β

IL-10

**Figure 3.26** T cells are capable of differentiation into regulatory T cells (Tregs). Thymus-derived Tregs are known as natural nTregs, while those that differentiate in the periphery are known as induced iTregs. The regulatory cytokines IL-10 and TGF-β inhibit the production of pro-inflammatory cytokines such as IFN-γ and IL-4.

The differentiation state of Th cells is reinforced by the cytokines produced by the T cell itself. Th1 cells express IFN-γ, which activates other leukocytes involved in cell-mediated immunity and promotes effector functions. For example, IFN-γ enhances the cytolytic activity of CTLs and promotes macrophage phagocytosis and killing mechanisms. IFN-γ also reinforces Th1-driven immune responses by interfering with the expression of the type 2 cytokine IL-4. Likewise, IL-4 is able to inhibit the production of IFN-γ by T cells and IL-12 by DCs, thereby supporting and maintaining Th2-driven immune responses. IL-4 is also crucial for the activation of B cells and the initiation of antibody responses. Although the Th1 versus Th2 paradigm is an important concept regarding the differentiation of naïve Th cells, in reality there is much more plasticity in T cell development as other cell lineages have been identified in recent times.

A particular CD4+ T cell subset has been identified that plays a fundamental role in regulating an immune response and preventing unnecessary inflammation. These cells are known as T regulatory (Treg) cells, which are characterized by the expression of the IL-2 receptor α-chain (CD25) and the transcription factor FoxP3 (forkhead box P3). Like other CD4+ T cell lineages, Treg cells develop in the thymus and are known as natural Treg cells (nTregs). The main function of Treg cells is to suppress the activities of effector cells, primarily through the inhibition of cytokine production (Figure 3.26). In this way, the activity of CD4+ Th cell, CD8+ CTLs, macrophages, B cells and DCs are all suppressed by Treg cells. The mechanism by which Treg cells exert their immune regulation is through the release of the modulatory cytokines IL-10 or TGF-β. A distinction has been made between those nTreg cells that express IL-10 (Treg1 cells) and those that express TGF-β (Th3 cells). In addition to thymus-derived nTreg cells, naïve T cells in the periphery can differentiate into Treg-like cells in response to antigenic stimulation and the correct signals provided by cytokines such as TGF-β and are referred to as adaptive Treg cells or inducible Treg cells (iTregs). Importantly, the action of Tregs, through the production of IL-10 or TGF-β, suppresses inflammatory immune responses. In addition, Treg cells are thought to be pivotal in maintaining peripheral tolerance and preventing the onset of autoimmune diseases.

In contrast to Treg cells, another CD4+ T cell subset has recently been identified that is thought to be important in the defence against pathogens but also may significantly contribute to the immunopathology observed in autoimmune reactions. These CD4+ Th cells are known as Th17 cells due to their production of IL-17 and to a lesser extent IL-22. The cytokines required for the differentiation of naïve T cells into Th17 cells include TGF-β, IL-6, IL-23 and IL-1. Although TGF-β is normally involved in suppressing immune responses, the combination of the other cytokines results in Th17 cell differentiation.

This induces the expression of the transcription factor RORγt, which is sufficient to upregulate IL-17. It has been demonstrated that Th17 cells are important for host defence against extracellular pathogens. IL-17 is effective at inducing the production of antimicrobial peptides such as β-defensins that protect against invading bacteria. However, Th17 cells also participate in autoimmune diseases such as inflammatory bowel disease (IBD) and experimental autoimmune encephalitis (EAE), and contribute to the immunopathogenesis observed in immune-mediated diseases such as psoriasis and allergy.

## 3.24 Activation of CTLs

One of the principle cell types involved in cell-mediated immunity are cytotoxic T lymphocytes (CTLs), which are characterized by the surface expression of CD8 and are restricted by MHC class I antigen presentation. Naïve CTLs predominately reside in secondary lymphoid tissues such as lymph nodes or Peyer's patches. An initial priming event occurs within secondary lymphoid tissue that initiates proliferation of antigen specific CTLs and migration into the periphery (Figure 3.27). In a similar way to Th cell activation, a CTL requires a primary signal from TCR engagement with peptide:MHC class I complexes

and a secondary signal derived from the co-stimulatory molecule CD28. Professional APCs provide the necessary antigen stimulation to CTLs, either through being directly infected with a pathogen or through a process known as cross priming. Remember that peptide loaded onto MHC class I molecules should be derived from endogenous sources. However, in the absence of direct infection, DCs are capable of taking up antigen derived from infected cells and can process this antigen via the endogenous MHC class I pathway. In this way, CD8+ CTLs can be primed by migratory DCs in regional lymph nodes. In order for antigen-specific CTLs to proliferate a third signal, provided by IL-2 stimulation, is also necessary.

Once primed and stimulated naïve CTLs take on an effector T cell phenotype and migrate to the site of inflammation. The primary function of CTLs is the induction of cell lysis of infected cells. αβTCRs expressed on the surface of CTLs recognize peptide-antigen presented on the surface of infected cells, in conjunction with MHC class I molecules. Further stimulatory signals, such as IFN-γ and TNF, are provided by macrophages and CD4+ Th cells that have also migrated into the site of inflammation. These signals are enough to fully activate primed CTL effector functions. CTLs also respond to type 2 cytokines, such as IL-4 and IL-5 and can therefore be classified as being either Tc1 or Tc2 cells, depending on the surrounding cytokine microenvironment. CTLs use a

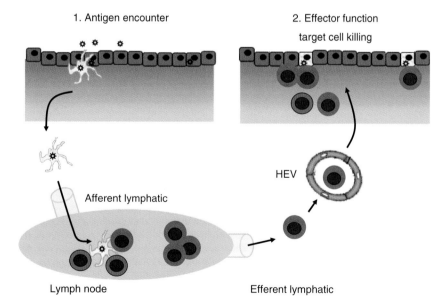

1. Antigen encounter

2. Effector function
target cell killing

HEV

Afferent lymphatic

Lymph node

Efferent lymphatic

**Figure 3.27** Activation of a CTL response against an intracellular pathogen. CTLs are primed in secondary lymphoid tissues and migrate to sites of infection where they exert their cytotoxic effector functions.

similar mechanism to NK cells during cell lysis (described in Chapter 2), which involves the formation of an immune synapse. The secretion of perforin and granzyme B by the CTL results in cell membrane permeability and the induction of cell death pathways within the infected cell, which ultimately leads to the lysis and killing of that cell.

## 3.25 Generation of memory T cells

One of the key features of the adaptive immune system is the generation of populations of memory T cells and B cells. Memory cells are essential for the induction of a secondary immune response following subsequent infection with a pathogen and provide an individual with protective immunity. With regard to CD4+ Th cells, current understanding suggests that a memory cell population emerges from the effector T cell population. Therefore, following activation and co-stimulation naïve T cells differentiate into effector cells. When the infection is cleared most of the effector cells die and those that emerge differentiate further into memory CD4+ T cells. The characteristics of these memory T cells include a heightened capacity to respond to a secondary challenge with a corresponding increase in functionality, which results in a much more vigorous immune response to subsequent pathogen infection. In addition, memory CD4+ T cells

have the ability to maintain a self renewing, long-lived population that persists throughout lymphoid and/or non-lymphoid tissues. The CD4+ memory T cell population is derived from a heterogeneous population of effector cells that require very little stimulation in order to persist, although repeated TCR signalling and IL-15 stimulation are thought to be essential for survival.

CD8+ memory T cells are also thought to be derived directly from the effector CTL cell population. It is likely that once infected cells are cleared there is a corresponding reduction of peptide:MHC class I signalling to the CD8+ T cells. This reduction in inflammatory stimuli causes the effector cells to differentiate into a long-lived memory cell. Other factors may also contribute to memory cell differentiation including a downregulation of inflammatory cytokines and growth factors. However, not all effector CTLs differentiate into memory cells, as the majority of these cells die after they have performed their effector function. Differentiation of memory CD8+ T cells appear to be more tightly controlled than CD4+ T cells, perhaps due to their cytotoxic capabilities. They are also shorter lived than memory CD4+ T cells, again reflecting the potential immunopathogenic properties of excessive CTL reactions, and may even require continued TCR engagement with peptide:MHC class I complexes in order to maintain homeostasis. Memory CD8+ CTLs also required the help of CD4+ Th cells during the

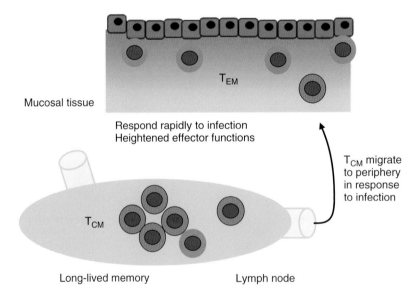

**Figure 3.28** Central memory T cells ($T_{CM}$) and effector memory T cells ($T_{EM}$). $T_{CM}$ cells are mainly CD4+ T cells that reside in secondary lymphoid organs. $T_{EM}$ cells are mainly CD8+ T cells and are a located in mucosal tissues where they respond rapidly to secondary infections.

induction of a secondary immune response, although they are activated more rapidly and possess heightened effector functions once they are activated.

Two distinct memory T cell populations have been recently described known as effector memory ($T_{EM}$) and central memory ($T_{CM}$) cells (Figure 3.28). $T_{EM}$ cells comprise mostly CD8+ CTLs and $T_{CM}$ are predominantly CD4+ Th cells, although both $T_{EM}$ and $T_{CM}$ populations contain CD4+ and CD8+ T cell populations. The distinction between these two populations lies in the responsiveness of each memory cell subtype. Whereas $T_{EM}$ cells are capable of rapidly responding to a secondary infection, $T_{CM}$ have little or nor effector function. For example, $T_{EM}$ cells express high amounts of perforin and granzyme B, while $T_{CM}$ do not. However, $T_{CM}$ cells are thought to provide a long-lived pool of memory cells that can proliferate in secondary lymphoid tissues and subsequently give rise to $T_{EM}$ cells. Therefore, $T_{CM}$ cells act almost like antigen-specific stem cells that replenish $T_{EM}$ cells that have been lost in the periphery. The distribution of these cells also varies so that $T_{EM}$ cells are located predominantly in mucosal tissues such as the lung and intestines where they can exert effector functions, while $T_{CM}$ cells are found in lymph nodes and the tonsils where they can undergo antigen-stimulated proliferation. Both $T_{EM}$ and $T_{CM}$ cells represent important memory T cell populations that are central for the dynamic function of the adaptive immune system.

## 3.26 Summary

**1.** The adaptive immune system comprises T cells and B cells and involves both cell-mediated and antibody-mediated immunity.

**2.** T cell and B cell receptors are highly specific for antigen and are capable of recognizing a diverse number of antigens.

**3.** Antibody and TCR diversity is generated by a process of genetic recombination.

**4.** Somatic hypermutation and clonal selection enhance the affinity of antibody and increase the effectiveness of secondary immune responses.

**5.** Positive and negative T cell selection in the thymus deletes self-reactive T cells but allows functional T cells to enter the periphery.

**6.** CD4+ Th cells recognize peptide antigen presented in the context of MHC class II molecules, while CD8+ CTLs recognize antigen present by MHC class I molecules.

**7.** CD4+ Th cells assist in the activation of B cells and CTLs through co-stimulation and the secretion of cytokines.

# 4 Cytokines

## 4.1 Introduction to cytokines

The immune system involves the interplay between a diverse array of cells, which need to be directed both spatially and temporally, at important checkpoints during haematopoiesis and throughout the duration of an immune response. Even within tissue microenvironments, immune cells need to be controlled at the molecular level, so that differentiation proceeds correctly and effector cell function takes place within acceptable boundaries. Therefore, a complex network of molecules exists that are responsible for the ultimate control of the immune system, known as cytokines (from the Greek meaning; *cyto*, cell and *kinin*, move). Cytokines are small soluble proteins that are secreted by various leukocytes and non-immune cells. Cytokines can alter the behaviour of the cell from which it was produced (autocrine function), of a neighbouring cell (paracrine function) or of a distant cell (endocrine function) (Figure 4.1). Together with a network of specific receptors, cytokines provide a means by which cells can communicate with each other. Once bound to their receptor, cytokines exert their effects through the initiation of an intracellular signalling cascade, which is important in controlling the development, execution and resolution of immune responses. Therefore, cytokines are a fundamental component of the immune system, without which immune cells would cease to function.

Cytokines normally operate as part of a network in order to control immune responses. In other words, a particular immune response will be coordinated by a number of different cytokines, each of which has a defined role in that particular response. Inflammation within a tissue leads to the synthesis of many cytokines, which act on neighbouring cells and elicit a complex series of cell signalling events that instruct nearby immune cells.

The induction of inflammation induces a repertoire of cytokines, the precise composition of which depends on the type of infection and the tissue in which the inflammation is taking place. For example, the cytokine repertoire initiated in response to a bacterial infection in the skin may differ considerably from the immune response initiated by a virus in the airways. The so-called cytokine storm that is elicited in response to infection will significantly influence the phenotype and magnitude of the subsequent immune response. Cytokines function by controlling the differentiation status of cells and directing immune effector functions. In addition, cytokines often act synergistically, meaning that two or more cytokines act in combination to achieve the desired effect. Furthermore, certain cytokines may affect various cells in different ways. The particular combination of cytokines, and the cell type that they are affecting, will determine the outcome of cytokine signalling. This often results in a cytokine being attributed several different functions. A multi-functional cytokine is said to exhibit pleiotropy.

Cytokines are produced primarily by the cells of the immune system, including monocytes, DCs, granulocytes and lymphocytes; although it is recognized that structural cells and stromal cells also secrete a wide range of cytokines. Cytokines that are specifically produced by lymphocytes are known as lymphokines. Actually, cytokines are capable of being synthesized by every cell of the body, given the correct stimulation. Moreover, the secretion of cytokines by structural cells, such as epithelial cells, fibroblasts and endothelial cells, means that these cells have a significant influence on the development of an immune response. For example, epithelial cells located at the surface of mucosal tissues can respond to infection by releasing several important pro-inflammatory cytokines. This early cytokine release can rapidly direct immune responses to the site of infection and protect the host from microbial attack.

*Immunology: Mucosal and Body Surface Defences*, First Edition. Andrew E. Williams.
© 2012 John Wiley & Sons, Ltd. Published 2012 by John Wiley & Sons, Ltd.

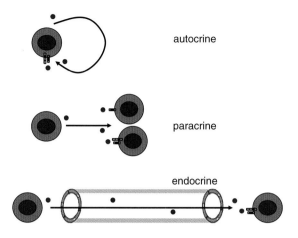

**Figure 4.1** Cytokines act in an autocrine, paracrine or endocrine manner. Autocrine signalling involves signalling to the same cell which secreted the cytokine. Paracrine signalling involves signalling to a different, nearby cell. Endocrine signalling involves signalling to a cell that is distant to the source of the cytokine.

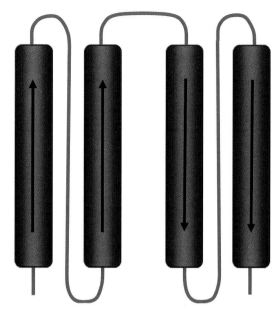

**Figure 4.2** Many cytokines, particularly type 1 cytokines, contain an α-helical bundle, which forms an up-up-down-down configuration.

## 4.2 Structure of cytokine families

The vast network of cytokines can be loosely organized into families of molecules that share similar functionality. However, there are many caveats using functional classification and many cytokines could conceivably belong to more than one of these functional groups. Therefore, it is often helpful to group cytokines according to structural similarities, as this takes into account evolutionary conservation among family members and preserves the nomenclature of cytokine homologues across species. The structure of many cytokines forms an α-helical bundle (Figure 4.2), which is constructed from four anti-parallel α-helices (arranged in an up-up-down-down configuration) and includes members of the interleukin (IL) and interferon (IFN) families. IL-1 and its related family members are composed of a β-trefoil structure, while tumour necrosis factor (TNF) family members are formed from the arrangement of β-sheets. Alternatively, in addition to structural similarities, cytokines can be grouped together based on common cytokine receptor usage, or even a combination of the two.

Cytokines broadly belong to one of five families (Table 4.1), namely the interleukins, interferons, TNF ligands, growth factors and chemokines. The interleukin family contains the largest number of cytokines that have been described to date and includes IL-1 to IL-35. Interleukins can be further classified into several

**Table 4.1** Cytokines can be classified into five broad families.

| Cytokine family | Example members |
| --- | --- |
| Interleukins | IL-1 to IL-35 |
| Interferons | IFN-α |
| | IFN-β |
| | IFN-γ |
| TNF ligands | TNF |
| | Lymphotoxin |
| | BAFF |
| Growth factors | TGF-β |
| | G-CSF |
| | GM-CSF |
| Chemokines | CCL1 |
| | CXCL1 |
| | CX$_3$CL1 |
| | XCL1 |

smaller families based on structural similarity or shared receptor binding properties. Therefore, the interleukins are divided into the IL-1 superfamily, the IL-6 family, the IL-10 family, the IL-12 family and the common γ-chain family. In a similar way, the interferon family can be further divided into type-I interferons (IFN-α and IFN-β),

**Box 4.1 Cytokine Families.**

| CYTOKINE | CELL EXPRESSION | FUNCTION |
| --- | --- | --- |
| **INTERLEUKINS** | | |
| IL-1 | Mono, MΦ, DC, NK, B | Pro-inflammatory, T cell, B cell, NK cell activation and proliferation, fever |
| IL-2 | Th | T cell proliferation |
| IL-3 | Th, NK, MC | Haematopoiesis |
| IL-4 | Th, MC, NK, NKT | B cell activation (IgE), inhibits Th1 and promotes Th2 |
| IL-5 | Th, MC | Promotes eosinophil survival, IgA production |
| IL-6 | Mono, MΦ, DC, T | Lymphocyte differentiation and proliferation |
| IL-7 | BM, thymus | Haematopoiesis |
| IL-9 | Th | MC proliferation, IgE production |
| IL-10 | Th, Treg, Tc, B, Mono, MΦ | Immune suppression, inhibits Th1 cell differentiation, inhibits antigen presentation |
| IL-11 | BM | Haematopoiesis |
| IL-12 | Mono, MΦ, DC, B | Th1 differentiation, promotes IFN-γ production |
| IL-13 | Th, MC | IgE production, B cell poliferation |
| IL-14 | Th | B cell differentiation |
| IL-15 | Th, NK, Mono, MΦ, DC, B | Promotes proliferation and cytotoxicity in NK and T cells |
| IL-16 | Th, Tc | Chemoattractant |
| IL-17 | Th | Stimulates pro-inflammatory cytokines |
| IL-18 | MΦ, DC | Promotes NK cell cytotoxicity and IFN-γ |
| IL-19 | Mono, B | Upregulate TNF, induce apoptosis |
| IL-20 | Mono | Activate keratinocytes |
| IL-21 | Th | Haematopoiesis, activate T cells and NK cells |
| IL-22 | Th | Pro-inflammatory, inhibit IL-4 |
| IL-23 | DC, Th | Pro-inflammatory, synergizes with IL-17 |
| IL-24 | Th1, MΦ, Mono | Anti-tumour activity, inhibits proliferation |
| IL-25 | Th, MΦ, MC | Promotes Th2 responses |
| IL-26 | Th, NK | Chemoattractant, pro-inflammatory |
| IL-27 | DC, Mono | Pro-inflammatory |
| IL-28 | Mono, DC | Immunoregulatory, inhibit viral replication |
| IL-29 | Mono, DC | Immunoregulatory, inhibit viral replication |
| IL-31 | Th | Allergic responses in skin |
| IL-32 | Th, NK | Pro-inflammatory, AICD |
| IL-33 | DC, MΦ | Promotes Th2 responses |
| IL-34 | Mono, MΦ | Differentiation of monocytes and macrophages |
| IL-35 | Treg | Immunoregulatory |
| **INTERFERONS** | | |
| IFN-α | ECs, most leukocytes | Inhibit viral replication |
| IFN-β | ECs, fibs | Inhibit viral replication |
| IFN-γ | Th, Tc, NK | Promotes Th1 responses, inhibits Th2 responses |
| **TMOUR NECROSIS FACTORS** | | |
| TNF | Th, Mono, MΦ, DC, NK | Pro-inflammatory, fever, anti-tumour activity, activates macrophages |
| LT | Th, Tc | Lymphoid development, inhibit viral replication, anti- tumour activity |
| **GROWTH FACTORS** | | |
| TGF-β | Treg, B, MΦ, MC | Immunoregulatory, wound healing, promotes IgA synthesis |
| G-CSF | Fibs, endo | Haematopoiesis |
| M-CSF | EC, fibs, endo | Haematopoiesis |
| GM-CSF | Th, MΦ, fibs, endo | Haematopoiesis, activates macrophages |

type-II (IFN-γ) and type-III (IL-28 and IL-29), depending on the type of receptors they signal through. The TNF ligand family includes the important cytokine TNF and other structurally related ligands such as lymphotoxin and B cell activating factor (BAFF). Growth factors are a large family of molecules that include a number of hormones involved in organogenesis but also several cytokines such as transforming growth factor (TGF)-β and granulocyte-colony stimulating factor (G-CSF). Lastly, the chemokines are a subgroup of cytokines that primarily function as chemoattractants and these will be discussed in detail in the next chapter.

## 4.3 IL-1 superfamily

The IL-1 superfamily contains the pro-inflammatory cytokines IL-1α and IL-1β, along with IL-1RA, IL-18 and IL-33, which all share structural homology. Both IL-1α and IL-1β are produced from a larger pro-IL-1α or pro-IL-1β protein through enzymatic cleavage by a specific protease called IL-1-converting enzyme (ICE/caspase-1). IL-18 and IL-33 are also cleaved from a pro-IL-18 or pro-IL-33 protein. Even though IL-1α and IL-1β share structural homology and function, they are actually encoded by separate genes. However, it is often the case that the name IL-1 is used to describe both because IL-1α and IL-1β bind to the same receptor (IL-1R). Only the cleavage products of the mature IL-1α and IL-1β are successfully transported to the cell surface and exert biological activity. The structure of IL-1 family members are formed from 12–14 anti-parallel β-strands, and in the case of IL-1α and IL-1β six β-strands fold into a β-barrel shaped configuration. It is also thought the β-barrel contains the receptor-binding domain.

The functions of IL-1α and IL-1β are considered to be pleiotropic, meaning that they have multiple effects on various cell types. The biological function of IL-1α and IL-1β are almost identical, while the principal differences between the two exist only in the cell types that produce them. IL-1 expression can be induced by a number of cytokines, including TNF, IFN-α and IFN-γ, by various PAMPs such as LPS, and even by itself. It is primarily produced by macrophages, DCs, epithelial cells and fibroblasts in response to an inflammatory insult. The primary function of IL-1 is to promote CD4+ Th cells and induce the upregulation of the cytokine IL-2, which causes T cells to proliferate (Figure 4.3). Therefore IL-1 is one of the first cytokines to be produced after pathogen recognition or tissue damage and is instrumental in establishing

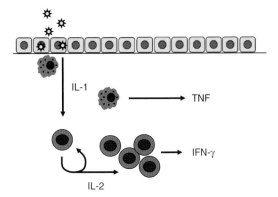

**Figure 4.3** IL-1 induction of an inflammatory immune response. IL-1 is secreted by epithelial cells in response to infection and causes the upregulation of IL-2 and the proliferation of CD4+ T cells.

an immune response. It also stimulates the production of many other cytokines including IL-2, IL-3, IL-4, IL-5, IL-6, IL-8, GM-CSF, IFN-γ and TNF from a variety of cell types. IL-1 also enhances the adhesion and extravasation of leukocytes across endothelia and into sites of inflammation through the upregulation of vascular cell adhesion molecule (VCAM)-1 and intercellular adhesion molecule (ICAM)-1. Other functions of IL-1 include enhancing B cell proliferation and antibody production and contributing to the signals necessary for DC maturation.

The IL-1 superfamily member IL-18 also possesses pleiotropic functions. It is produced mainly by monocytes, macrophages, DCs, T cells and B cells. IL-18 is a pro-inflammatory cytokine that is synthesized in response to microbial stimulation and therefore contributes to the early augmentation of an immune response. It synergizes with IL-12 to induce the expression of IFN-γ and enhance the proliferation of Th1 cells. In addition, IL-18 inhibits the production of IgE and enhances the production of IgG2a by B cells, partly through its upregulation of IFN-γ. The other IL-1 family member discussed in this section is IL-1RA, which acts as an antagonist to both IL-1α and IL-1β by competing for receptor binding. In effect IL-1RA inhibits IL-1 isoforms from activating their receptor and is therefore involved in modulating IL-1-dependent inflammatory reactions.

## 4.4 IL-6 family

Members of the IL-6 cytokine family include IL-6, IL-11, IL-27, IL-30 and IL-31, which all share structural similarity and use the same receptor subunits to instigate

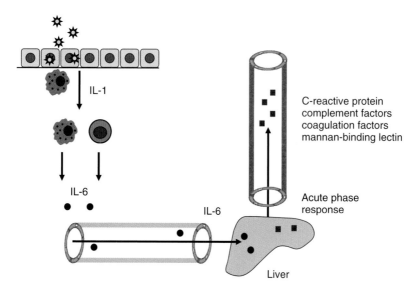

**Figure 4.4** IL-1 and IL-6 combine to induce an acute phase response, which involves the release of acute phase proteins from the liver.

cell signalling. IL-6 family members all have a helical structure that consists of four anti-parallel α-helices. There is a certain amount of functional redundancy among IL-6 family members; due mainly to the sharing of receptor subunits, in particular gp130, and therefore their biological activity is comparable. The most widely studied member of this cytokine family is IL-6 itself, which is produced by a multitude of different cell types including monocytes, macrophages, epithelial cells, smooth muscle cells, fibroblasts, T cells and B cells. IL-6 is usually synthesized in response to a pro-inflammatory signal such as IL-1, TNF or microbial products such as LPS or viral RNA. IL-6 has many functions within the immune system. It provides one of the most important signals for the induction of the acute phase response (Figure 4.4) and is instrumental in stimulating fever, as it can cross the blood–brain barrier and affect the hypothalamus directly. Indeed, IL-6 has been shown to interact with the neuroendocrine system and is able to stimulate the production of glucocorticoids. In addition, IL-6 is a potent activator of cytotoxic T cells, as it synergizes with IL-2 to promote differentiation and proliferation. It also has a similar proliferative effect on B cells.

## 4.5 IL-10 family

The IL-10 cytokine family comprises the structurally related proteins IL-10, IL-19, IL-20, IL-22, IL-24 and IL-26. Members of the IL-10 family form a V-shaped homodimer consisting of two symmetrical helical subunits, each consisting of four α-helices, which combine to form a common helical topology found in other cytokines such as IFN-γ and IL-4. Despite the similar structure, IL-10 family members have diverse biological functions. For example, IL-10 is predominantly an immunosuppressive and immunomodulatory cytokine, while IL-20 provides proliferative signals to epithelial cells and IL-22 synergizes with IL-17 during inflammatory reactions. However, the archetypal member of this cytokine family is IL-10, which functions as a suppressor of immune responses, mainly through the inhibition of several pro-inflammatory cytokines. For example, IL-10 is capable of inhibiting IFN-γ and TNF production by T cells (Figure 4.5), although this maybe the result of inhibiting IL-12 by APCs such as macrophages and DCs. IL-10 is also capable of inhibiting T cell proliferation, as well as cytokine production, and can downregulate both Th1 and Th2 immunity. Furthermore, IL-10 is able to downregulate antigen presentation by DCs through the suppression of the co-stimulatory molecules CD80 and CD86. The primary source of IL-10 is a specialized subset of T cells known as regulatory T cells (Tregs), which are characterized by the expression of CD4 and CD25 and the intracellular expression of the transcription factor FoxP3 (CD4+CD25+FoxP3+ Treg), although monocytes also contribute to IL-10 production.

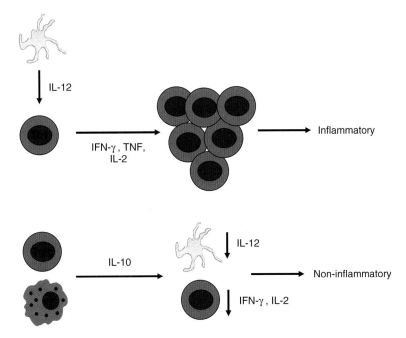

**Figure 4.5** Immune regulation by IL-10 is associated with an inhibition of IL-12, IFN-γ and IL-2 production, thereby modulating inflammation. IL-10 is predominantly expressed by T regulatory cells (Tregs) and tissue resident macrophages.

The IL-10 cytokine family member IL-20 is predominantly expressed by specialized epithelial cells in the skin called keratinocytes. Following stimulation, keratinocytes secrete IL-20, which works in an autocrine fashion to further stimulate keratinocytes and induce proliferation. IL-20 is therefore involved in the immunopathology associated with inflammatory skin diseases such as psoriasis. It also contributes to the inflammatory milieu and acts as a chemoattractant in rheumatoid arthritis. Another IL-10 family member IL-22 is also involved in the inflammatory process and is produced mainly by activated T cells. IL-22 is thought to synergize with IL-17 during inflammatory reactions and has been implicated in the pathogenesis of psoriasis, where it augments keratinocyte proliferation and activates other inflammatory cells, including CD8+ CTLs. A number of autoimmune diseases, in particular inflammatory bowel disease, are characterized by increased expression of IL-22, which is thought to synergize with other cytokines to promote immunopathology.

## 4.6 Common γ-chain family

The common γ-chain family of cytokines includes IL-2, IL-3, IL-4, IL-7, IL-9, IL-13, IL-15 and IL-21, which all signal through a shared receptor subunit known as the common γ-chain. Receptor specificity is achieved through the use of other receptor subunits specific for individual cytokines (discussed in detail later in this chapter). The common γ-chain family of cytokines do not necessarily share any structural similarities, but rather they are grouped together because of the universal use of the common γ-chain receptor subunit. Moreover, the biological function of this family of cytokine is relatively diverse. For example, IL-2 primarily promotes T cell proliferation, IL-4 and IL-13 drive Th2 cell differentiation, IL-7 is important for the development of immature T cells and B cells, IL-9 augments antibody production by plasma cells and IL-21 is a regulator of NK cell activity. This section will therefore focus on the biological function of IL-2, which is a key cytokine important for the initiation and maintenance of adaptive immune responses. The detailed function of IL-4 and IL-13 will be discussed later in the context of Th2 cell differentiation, while the other common γ-chain family members are discussed in relevant sections in other chapters.

The nucleotide sequence of IL-2 is unique among all the cytokine families, although its three-dimensional structure is similar to other α-helical cytokines (e.g. IL-4). Its tertiary structure forms an α-helical bundle composed

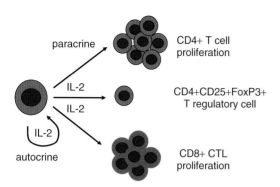

**Figure 4.6** IL-2 induces T cell proliferation in an autocrine and paracrine manner and is able to stimulate the proliferation of several T cell subsets.

of four anti-parallel α-helices, joined by another small α-helix. Activated CD4+ Th cells are the primary source of IL-2, which acts in an autocrine manner by binding the IL-2R (CD25) on the surface of the same T cell. IL-2 is a potent activator of T cell proliferation (Figure 4.6) and in its absence T cells tend to enter an inactive or anergic state. In addition, secreted IL-2 acts as a paracrine growth factor for other T cells, including CTLs and Tregs. Therefore, it is important for instigating the proliferation of T cells and in maintaining both inflammatory and regulatory T cell populations. IL-2 is also important for T cell development in the thymus and in particular for maturation of CD4+CD25+ Treg cells, although it must be remembered that CD25 is expressed by activated effector cells as well as regulatory Tregs. In addition, IL-2 can function in synergy with IFN-γ or IL-4 to promote the production of antibody by activated plasma cells.

## 4.7 IL-12 family

The IL-12 cytokine family includes IL-12, IL-23, IL-27 and IL-35. These cytokines are all heterodimeric proteins that consist of two covalently linked subunits. The heterodimer of IL-12 itself is made up of one 35 kDa protein (otherwise known as p35 or IL-12A) and one 40 kDa protein (known as p40 or IL-12B), which are transcribed from separate genes. The heterodimer forms four separate domains: the first domain is formed from the p35 subunit that contains four anti-parallel α-helices, which produce an α-bundle typical of type I cytokines, while the p40 subunit contains three separate domains formed from the folding of β-sheets and contains epitopes important

**Figure 4.7** IL-12 family members share similar structural subunits.

for receptor binding. The other family members either use an IL-12 subunit or two subunits that share similarities to the 35 kDa and 40 kDa polypeptides of IL-12. For instance, IL-23 comprises the 40 kDa subunit of IL-12 (p40) and a smaller polypeptide of 19 kDa referred to as IL-23A. IL-27 consists of one subunit that is related to p40 called EBI-3, which is otherwise known as IL-27B, and a second subunit of 28 kDa that is similar to p35 (known as IL-27p28). Lastly, IL-35 is a heterodimer formed of EBI-3 and the p35 subunit of IL-12 (Figure 4.7). The formation of a heterodimer is necessary for the biological activity of IL-12 family members.

IL-12 is an important cytokine involved in the initiation of adaptive immune responses and is secreted by DCs, macrophages and to a lesser extent B cells, following appropriate stimulation. For example, the stimulation of DCs, following microbial recognition via TLRs, will upregulate IL-12 expression and lead to its secretion. The primary function of IL-12 is to activate naïve T cells within secondary lymphoid organs, which results in their differentiation into Th1 cells. IL-12 also causes T cells and NK cells to upregulate the Th1 cytokines IFN-γ and TNF, while at the same time suppresses IL-4 expression. Within inflamed tissues, IL-12 further accentuates a Th1-type cytokine environment and contributes to enhancement of CTL and NK cell effector functions (Figure 4.8). Stimulation of an immune response in the absence of IL-12 can result in the development of a Th2-type response. IL-23 is another cytokine produced by activated DCs and, in a similar manner to IL-12, can induce the expression of IFN-γ by CD4+ Th1 cells, although IL-23 predominantly affects memory Th1 cells at sites of inflammation. IL-23 also functions in the differentiation of a subset of CD4+ T cells that express IL-17, referred to as Th17 cells. Therefore, IL-23 has been implicated as a regulator of Th17-driven immunopathology that occurs in certain immune-mediated diseases such as colitis and psoriasis.

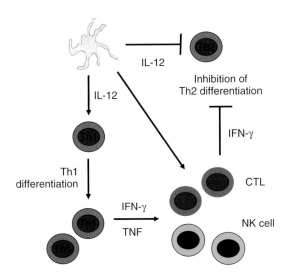

**Figure 4.8** IL-12 enhances Th1-mediated cellular immunity by inducing naïve T cells to differentiate into Th1 cells. Th1 cells secrete IFN- and TNF, which stimulate cytotoxic functions of NK cells and CTLs. The lack of IL-12 tends to result in the differentiation of Th2 cells.

## 4.8 Interferons

Interferons are so named because they were originally identified because of their ability to interfere with viral replication within host cells, although their biological functions are now known to be much broader. The interferon family is classically divided into type I interferons, which include IFN-α and IFN-β, and type II interferons, which includes IFN-γ. This classification is based on the receptor-binding capacities, as IFN-α and IFN-β both bind to the IFN-αR, while IFN-γ binds to the IFN-γ receptor. The third group, the type III interferons, is less well accepted as a standard classification and includes IL-28A, IL-28B and IL-29 (otherwise known as IFN-λ2, IFN-λ1 and IFN-λ3 respectively), which collectively bind to a receptor complex consisting of IL-10R2 and IFNλR1.

The type I interferons possess a high degree of similarity in their tertiary structure, which comprises five α-helices that form a typical α-helical bundle that is common among many cytokines. The structural similarity enables type I interferons to bind to the same receptor and therefore they also share similar biological functions. IFN-α is actually a generic name for a family of IFN-α proteins encoded by a cluster located on human chromosome 9. All the IFN-α subtypes have very similar biological functions, the only difference being in relative activity as measured by bioassays. In contrast, there are only two known human IFN-β proteins, IFN-β1 and IFN-β3, again encoded by genes located on human chromosome 9 (IFN-β2 is actually IL-6).

Type I interferons possess potent antiviral capabilities and are expressed in numerous cell types such as epithelial cells, endothelial cells, lymphocytes and macrophages. Another important source of IFN-α are plasmacytoid DCs (pDCs), which tend to reside in sub-epithelial compartments of mucosal tissues, and are sometimes referred to as natural interferon-producing cells. Binding of type I interferons to their receptor activates a number of transcription factors, such as interferon regulatory factors (IRFs) and NF-κB, leading to the expression of genes involved in immune defence. The antiviral capacity of type I interferons is numerous and includes expression of proteins that interfere with viral replication and these are discussed in more detail in Chapter 11.

The only type II interferon to be characterized in humans is IFN-γ, which is a homodimer (formed by two monomers), each containing six α-helices. Unlike the four α-helix cytokines, the IFN-γ dimer does not form an α-bundle but rather forms a globular protein with a flattened elliptical configuration. IFN-γ is primarily produced by NK cells and both CD4+ Th cell and CD8+ CTLs, following immune activation. Although IFN-γ has antiviral and antiparasitic properties, through the activation of several transcription factors and the expression of interferon-response genes, its main function is to modulate the immune microenvironment. IFN-γ is effective at promoting Th1 immunity, while at the same time inhibiting Th2 immune development (discussed further in section 4.11). This is achieved, in part, by the upregulation of the transcription factor T-bet in Th1 cells and increasing the expression of IL-2 and the IL-2R. Therefore, IFN-γ is central for immune defence against intracellular pathogens such as viruses and certain bacteria. IFN-γ also stimulates antigen presentation through the upregulation of MHC class I and co-stimulatory molecules, thereby enhancing CTL activity. Furthermore, IFN-γ enhances components of the innate immune system by augmenting NK cell effector functions and stimulating the release of reactive oxygen species by macrophages.

## 4.9 TNF ligand superfamily

Members of the TNF ligand superfamily include the cytokines TNF, lymphotoxin and B cell activating factor (BAFF or BCAF), as well as several receptor ligands including CD30L, OX40L, RANKL (receptor activator for NF-κB ligand) and FasL. Previous nomenclature recognized two TNF molecules designated TNF-α and TNF-β. It is now common usage to refer to TNF-α simply as TNF (as this was the archetypical TNF known as cachexin or cachectin) and TNF-β as lymphotoxin. Therefore, the use of TNF-α and TNF-β has become redundant, although they still exist in older texts, and are now referred to as TNF and lymphotoxin respectively. TNF and lymphotoxin are both type II transmembrane glycoproteins, which either exist as a membrane bound form or as a soluble form. The soluble form is released from the membrane by the action of a metalloprotease known as TNF-alpha converting enzyme (TACE, otherwise known as ADAM17). The soluble forms of TNF and lymphotoxin consist of a homotrimer (three identical monomers), which are made up of eight anti-parallel β-strands that form two β-sheets. The two β-sheets form what is commonly referred to a β-jellyroll. The formation of a homotrimer is necessary for the binding of TNF and lymphotoxin to their receptors (TNFR1 and TNFR2).

TNF is produced by numerous immune cells in response to microbial stimulation, for example following recognition of LPS, or following cytokine stimulation, in particular IL-1, IL-2 and interferons. The primary source of TNF is macrophages, although other cell types are significant producers, including neutrophils, mast cells, endothelial cells, fibroblasts, NK cells and CD4+ Th cells, although CD8+ CTLs are poor producers. The primary function of TNF is as an immunomodulator and an initiator of inflammation. Actually, as the name would suggest, TNF causes the necrosis of tumours through the destruction of blood vessels within the tumour mass and the activation of a number of immune mechanisms. TNF is one of the most potent pyrogenic factors induced during an inflammatory reaction. It is capable of crossing the blood-brain barrier and acting directly on the hypothalamus, causing an increase in body temperature and the symptoms that are often associated with severe infection, such as fever, sweating and cachexia (weight loss, malaise, lack of appetite). TNF also stimulates the liver to synthesize and release acute phase proteins, including C-reactive protein, which enhances the activity of complement. The activities of other immunocytes are directly influenced by TNF signalling. For example, leukocyte adhesion to endothelial cells is enhanced by TNF, which acts as an effective chemoattractant, particularly for neutrophils through the upregulation of ICAM-1 and VCAM-1. Macrophage phagocytosis, cytokine secretion and release of reactive oxygen species

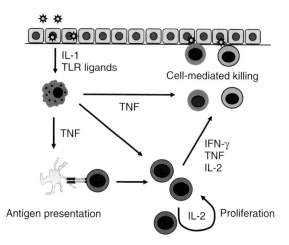

**Figure 4.9** TNF augments immune responses by acting on several immune cells during the course of an inflammatory reaction, including enhancing antigen presentation, T cell proliferation and CTL and NK cell effector functions.

are all directly augmented by TNF. Furthermore, TNF enhances cell-mediated immunity by upregulating MHC class I and II molecules, thereby promoting antigen presentation and cell-mediating killing, and increasing T cell proliferation and DC maturation (Figure 4.9). In addition, TNF acts by reinforcing and accentuating the cytokines produced during an inflammatory reaction, in particular by increasing the expression of IL-1, IL-2, IFN-γ and other pro-inflammatory mediators dependent on NF-κB activation.

Lymphotoxin (LT) is also known as TNF-β or LT-α (LTA) in order to distinguish it from LT-β, which functions to anchor LT-α to the cell membrane. The main source of LT is activated T cells, in particular CD8+ CTLs, although it is also expressed by other leukocytes, endothelial cells and epithelial cells. Therefore the pattern of expression for LT is slightly different to that of TNF. The biological function of LT during an inflammatory reaction is similar to that of TNF and contributes to the induction of apoptosis of virally infected cells. In addition, LT plays a significant role in the development of immature lymphoid organs and is produced by a specialized stem cell called a lymphoid inducer cell (LTi, discussed further in Chapter 7). In combination with LT-β, a complex (LTα1β2) is formed with the LTβR on the surface of stromal cells, which causes the release of chemoattractants that recruit other cells necessary for lymphoid organogenesis.

Another TNF ligand superfamily member BAFF is expressed by T cells, monocytes and DCs (although not by B cells themselves) and is known to be a potent growth factor and pro-survival factor for B cells, which cooperates with the co-stimulatory molecule CD40. BAFF functions by decreasing the pro-apoptotic protein BAK and increasing the anti-apoptotic molecules Bcl2 and $BCL_XL$. Other TNF family members also act as pro-survival factors for a variety of other leukocytes, examples of which include APRIL (A proliferation-inducing ligand) and RANKL, which are collectively known as TNF ligand superfamily members. There may be as many as 27 TNF ligand super-family members, although they are not bone fide cytokines as such, they do share structural homology to TNF and have significant functions within the immune system.

## 4.10 Growth factors

Growth factors represent a diverse group of molecules, usually proteins, which are capable of stimulating cellular growth, division or proliferation and are therefore classified based on function, rather than structural similarities. Although the majority of growth factors are hormones or steroids, some also fall into the cytokine family including TGF-β, granulocyte-colony stimulating factor (G-CSF) granulocyte macrophage-CSF (GM-CSF) and macrophage-CSF (M-CSF). Although other growth factors, such as the bone morphogenic proteins (BMPs), which are also members of the TGF-β sub-family, the fibroblast growth factors (FGFs) and vascular-endothelial growth factors (VEGFs), are also referred to as cytokines, they will not be discussed at depth in this section. It should be noted, however, that these growth factors significantly affect the function of the immune system by providing growth signals to structural cells of lymphoid and mucosal tissues, and function to maintain the integrity of the vasculature. Only those growth factors that directly influence the function of leukocytes will be discussed.

TGF-β is a pleiotropic cytokine that is expressed by a multitude of cell types. Its biological functions are diverse and include the active suppression of T cell responses, enhancing IgA synthesis by mucosal B cells and the induction of apoptosis pathways. Actually, TGF-β is used to describe at least three isoforms, TGF-β1-3, which share almost identical biological functions, of which the predominant form is TGF-β1. They differ only in their pattern of expression, whereby TGF-β1 is expressed by most cell types – the other isoforms have a much more restricted expression. Mature TGF-β is formed from the cleavage of a larger precursor molecule. Furthermore, TGF-β is secreted as an inactive protein, know as latent

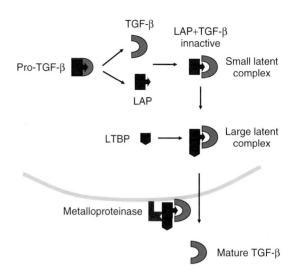

**Figure 4.10** Processing of TGF-β from its latent form. Latent TGF-β is associated with latency associated peptide (LAP) and forms an intracellular complex with latent TGF-β binding protein (LTBP). Transport of the large latent complex to the cell surface enables cleavage by metalloproteinases and release of the mature, active form of TGF-β.

TGF-β (L-TGF-β), which forms a complex with a peptide called latency associated peptide (LAP). This complex (small latent complex) is retained inside the cell until it forms a larger complex (large latent complex) with latent TGF-β binding protein (LTBP). This complex can then be secreted but remains attached to the extracellular matrix until it is processed further. The release of active TGF-β from its complex with the LTBP requires the proteolytic activity of metalloproteinases (MMP-2 for example) located throughout the extracellular matrix. The active form of TGF-β forms a homodimer, which is then free to bind to its receptor and induce a TGF-β-dependent signalling cascade (Figure 4.10). Mature TGF-β proteins contain several cysteine residues that interact to fold into a tertiary structure known as a cysteine knot, which is typical of TGF-β family members.

With regard to haematopoiesis and leukocyte function, G-CSF, GM-CSF and M-CSF represent three other important growth factors that provide important signals for the differentiation and proliferation of granulocytes and macrophages. Although they perform similar functions, they are not structurally related to each other. For example, G-CSF is more closely related to IL-6, while the structure of GM-CSF forms an α-helical bundle similar to IL-4 and other related cytokines. Bone marrow stromal cells are an important source of all three of these colony stimulating factors, which in turn provide signals to haematopoietic progenitor cells instructing them to differentiate into granulocytes or monocytes. For example, G-CSF is necessary for the differentiation of neutrophils and other granulocytes, while M-CSF is responsible for the correct differentiation of monocytes and macrophages. In addition to haematopoiesis, colony stimulation factors regulate several aspects of the innate immune response in the periphery. For example, GM-CSF is produced by tissue resident macrophages and endothelial cells in response to inflammatory stimulation. One of the functions of GM-CSF in the periphery is to recruit monocytes out of the circulation and into inflamed tissue and to contribute to the maturation of monocytes into tissue resident macrophages. In addition, GM-CSF enhances macrophage and neutrophil phagocytosis of microorganisms and induces the release of reactive oxygen species.

## 4.11 Functional classification Th1 versus Th2

One of the basic immunological paradigms of recent times has been the classification of immune responses based on the expression of predominant subsets of cytokines and the differentiation between Th1-mediated versus Th2-mediated immunity. Th1-mediated immune responses are usually initiated by intracellular pathogens such as viruses and bacteria and are characterized by the generation of cell-mediated immunity, involving macrophage and CTL activation. On the other hand, Th2-mediated immune responses are induced by extracellular parasites or in response to allergens. Th2-driven responses are characterized by the activation of humoral immunity and the production of IgE. Importantly, Th1 responses differ to Th2 responses with regard to the cytokines that are produced. This differential cytokine expression enables individual immune responses to be customized according to the type of pathogen encountered.

Th1-mediated immunity is characterized by the release of type 1 cytokines, while Th2-mediated immunity is characterized by the production of type 2 cytokines (Figure 4.11). More specifically, Th1 immunity is classically associated with the expression of IL-12 and IFN-γ, although TNF has also been related to type 1 responses. IL-12 and IFN-γ cause the differentiation of naïve T cells into Th1 cells. The predominant source of IL-12 is DCs that have been activated following microbial recognition, while the production of IFN-γ is primarily produced by

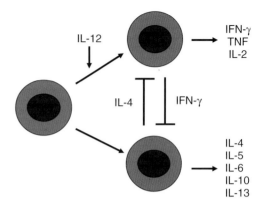

**Figure 4.11** Th1 and Th2 cytokines.

CD4+ Th1 cells. These cytokines influence the immune system in such a way as to promote various aspects of cell-mediated immunity, such as macrophage phagocytosis, NK cell activation and CD8+ CTL killing. These effector functions are amplified further by the expression of TNF, which is predominantly derived from activated macrophages and a lesser extent T cells. Therefore, in response to viral or bacterial infection, the immune response is regulated through the release of a certain subset of cytokines that direct cell-mediated immunity, which is the most appropriate means to kill intracellular pathogens. In addition, type 1 cytokines such as IL-12 and IFN-γ act by inhibiting the expression of Th2 cytokines, further enhancing Th1-mediated immunity

Th2-mediated immune responses are traditionally characterized by the expression of IL-4, IL-5 and IL-13, although the pleiotropic cytokine IL-6 and IL-10 are also considered to be a Th2 cytokine in certain situations. These cytokines are induced in the absence of IL-12 stimulation, which results in the differentiation of naïve T cells into CD4+ Th2 cells. Furthermore, Th2 cells are the main source for these type 2 cytokines. Rather than cell-mediated immune mechanisms, IL-4, IL-5 and IL-13 are more effective at inducing humoral immunity. The primary effect of inducing Th2 immunity is the stimulation of B cells to undergo class-switching and to secrete IgE antibodies, which in turn stimulates granulocytes (especially mast cells and eosinophils) and phagocytes to activate their own effector functions. Th2-mediated immune mechanisms are more effective at killing extracellular parasites. In addition, the type 2 cytokines IL-4 and IL-10 inhibit the expression of IFN-γ and other Th1-associated cytokines, thereby further supporting the differentiation of Th2 cells.

## 4.12 Th17, immunopathology and regulatory cytokines

It has long been thought that IFN-γ production and Th1-mediated immune responses are responsible for the induction of various autoimmune reactions. However, mice that are deficient in IFN-γ signalling actually develop a more severe form of autoimmune disease (e.g. rheumatoid arthritis, multiple sclerosis). It is becoming increasingly evident that IL-17, produced by Th17 and NKT cells, may be more effective at inducing excessive immune responses that lead to the immunopathology associated with autoimmune disease. These recent developments have led to the identification of a new subtype of CD4+ T cell, termed Th17 cells, which differ from the traditional Th1 and Th2 subsets. Indeed, Th17 cells have been associated with actively driving the excessive pathology observed in many immune-mediated diseases, through the secretion of IL-17. Other cytokines released by Th17 cells may also contribute to driving heightened immune responses including IL-22, which has been associated with inflammatory diseases such as psoriasis.

In contrast to the generation of excessive immunity, several cytokines actually function by downregulating inflammatory gene expression, thereby acting to modulate and regulate immune responses. For example, IL-10 functions by downregulating the expression of Th1-type cytokines IFN-γ, TNF and IL-2. Several cell types secrete IL-10 such as monocytes and macrophages, Tregs and Th2 cells. Another important regulatory cytokine is TGF-β, which is expressed by a specific subpopulation of CD4+ Tregs, known as Th3 cells. The primary regulatory function of TGF-β is to block leukocyte proliferation but it also induces apoptotic pathways in receptive cells. These regulatory cytokines play essential roles in the regulation of immune responses, on the one hand to prevent excessive inflammation and tissue damage, and on the other to coordinate the downregulation of an immune response following pathogen clearance.

## 4.13 Cytokine receptor signalling

Cytokines exert their function via interaction with specific receptors expressed on target cells. These receptors fall into distinct families of related proteins and are sometimes classified into four broad groups defined by their structural similarities. Type I and II cytokine receptors belong to the haematopoietin receptor family and are largely similar in structure. Type III cytokine receptors belong to the TNF receptor superfamily and type IV receptors include the IL-1 receptors, which are related to the Toll-like receptors (TLRs). It should be noted, however, that this classification merely forms a useful scheme for classifying different cytokine receptors based on structure. For example, Th1 cytokines bind to type I (IL-2), type II (IFN-γ), type III (TNF-α) and type IV (IL-1β) cytokine receptor family members. It should also be noted that many cytokine receptors share common components as part of multimeric receptor complexes. This therefore makes any classification of cytokine receptors more difficult and it should be remembered that such systems are often flexible.

## 4.14 Type I and type II cytokine receptors

Type I cytokine receptors all share similar structural components that allow them to bind to type I cytokines (Table 4.2). Likewise all the type I cytokines share many structural similarities such as the 'four α-helix bundle' formation, which is in an up-up-down-down configuration. It should be noted, however, that this classification is independent of the Th1 versus Th2 system, which is based on function, and the classification into cytokine families described above. These type I cytokines consist of both the short-chain and long-chain cytokines and include several interleukins, growth factors and hormones such

**Table 4.2** Type 1 cytokine receptors share common receptor subunits.

| Cytokine | α-subunit | β-subunit | γ-subunit |
|---|---|---|---|
| IL-2 | IL-2Rα (CD25) | IL-2Rβ | γC |
| IL-3 | IL-3Rα | βC | |
| IL-4 | IL-4Rα | | γC |
| IL-5 | IL-5Rα | βC | |
| IL-6 | IL-6Rα | gp130 | |
| IL-7 | IL-7Rα | | γC |
| IL-9 | IL-9Rα | | γC |
| IL-11 | IL-11Rα | gp130 | |
| IL-12 | | IL-12β1 | |
| | | IL-12β2 | |
| IL-13 | IL-13Rα | | γC |
| IL-15 | IL-15Rα | | γC |
| G-CSF | G-CSFRα | βC | |
| GM-CSF | GM-CSFRα | βC | |

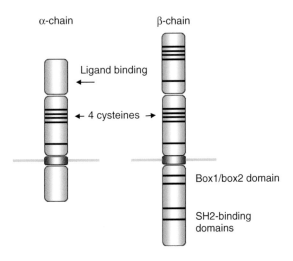

**Figure 4.12** Type 1 cytokine receptors comprise of an α-chain, which is responsible for ligand binding, and a β-chain, which contains the components necessary for signal induction.

as IL-2, IL-4, IL-5, IL-6, IL-12, IL-13, IL-15, GM-CSF, insulin and growth hormone. The receptors for type I cytokines are characterized by shared structural features and all of them are membrane spanning proteins. The type I cytokine receptors have an extracellular component, which is responsible for binding to its ligand, and an intracellular component that initiates a signalling cascade.

All type I cytokine receptors are multimeric, in that they have one α-chain and one or two secondary components, for example a β-chain (Figure 4.12). The extracellular portion of the α-chain shares two distinctive homologous domains. The first domain acts as a ligand binding region and is characterized by a barrel structure that forms a pocket. The second domain consists of four conserved cysteine residues, which maintain the structural integrity of the receptor complex. The α-chain is responsible for cytokine specificity and binds to its ligand with varying affinity that depends on the particular cytokine/receptor interaction. The α-chain is also responsible for cell specificity, as well as cytokine specificity, due to its selective expression on the surface of different cell types.

Type I cytokine receptors can be further divided into three sub-families based on the sharing of common receptor components. Type I cytokine receptors either share a common β-chain (βC), gp130 or the common γ-chaion (γC) (Figure 4.13). An example of a shared receptor component is the βC, which is shared between the IL-3R, IL-5R and GM-CSFR. Despite the number of different receptor components, all type I cytokine receptors lack intrinsic

tyrosine kinase activity. Signal transduction is therefore mediated by the recruitment of intracellular kinases. The conserved intracellular region that is important for signal transduction is referred to as the box1/box2 region. Type II cytokine receptors are distantly related to the type I cytokine receptors and share many of their structural features. Examples of type II cytokine receptors include receptors for IFN-α, IFN-β, IFN-γ and IL-10. However, type I cytokines possess a distinctive Trp-Ser-X-Trp-Ser motif (WSXWS motif) that type II cytokine receptors do not. The WSXWS motif is located within the α-chain, proximal to the transmembrane region, and is thought to be necessary for proper ligand binding and signal transduction.

## 4.15 The JAK/STAT signalling pathway

One of the key features of type I cytokine and type II cytokine receptor signalling is that signal transduction is mediated by a common intracellular pathway known as the Janus kinase/signal transducer and activator of transcription (JAK/STAT) pathway. The JAK/STAT pathway is an evolutionary conserved signal transduction pathway that directly promotes gene expression in the nucleus following cytokine binding to its receptor. Essentially JAK proteins are recruited to the activated cytokine receptor complex, where they phosphorylate tyrosine residues on the intracellular receptor domain. This results in the recruitment of STAT proteins to the receptor where they are phosphorylated and activated themselves. Dimerization of STAT proteins then allows their translocation to the nucleus where they act directly as transcription factors and induce the expression of target genes.

Type I and type II cytokine receptors do not posses their own kinase activity and therefore rely on JAKs to perform this function. There are four JAK family members present in humans, JAK1, JAK2, JAK3 and TYK2 and all are responsible for activating STATs through phosphorylation events. Cytokine binding initiates conformational changes within the intracellular domain of the receptor, which causes JAKs to bind to the box1/box2 domain of the receptor leading to activation and phosphorylation of receptor tyrosine residues. The phosphorylation of the receptor tyrosine resides opens up STAT protein binding sites. The JAKs are also responsible for phosphorylating and activating any adjacent STATs.

There are seven members of the STAT family in humans and they are characterized by several conserved protein domains. The most highly conserved STAT domain

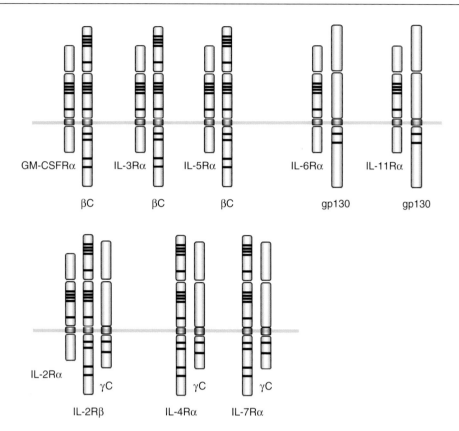

**Figure 4.13** Type 1 cytokine receptors share common receptor subunits and can be divided further into smaller families. Some type I cytokine receptors utilize the common γ-chain (γC).

is the SH2 domain, which is essential for binding to phosphorylated residues on the cytokine receptor, activation by JAKs and subsequent dimerization with other STAT proteins. The DNA-binding domain has a structure similar to the transcription factor NF-κB and therefore is responsible for the DNA binding and STAT transcription factor activity. Other domains include the linker-domain, coiled-coil domain and N-terminal domain, which have structural functions that assist in binding to other intracellular signalling molecules. Importantly, certain cytokines only activate certain STATs, thereby introducing an element of specificity to signal transduction events. This depends on receptor specificity, cell specific gene expression of a subset of STATs and the interaction between other signalling molecules that are able to modulate JAK/STAT signalling. For example, STAT1 and STAT2 are predominantly involved in IFN-α and IFN-γ signalling, while STAT4 and STAT6 are mainly associated the polarization of T cells into Th1 and Th2 cells, respectively.

Following activation STAT proteins dissociate from the cytokine receptor and form a dimer with another STAT protein (Figure 4.14). This dimerization can either be with the same STAT (homodimerization) or with another STAT family member (heterodimerization). This activation leads to the rapid translocation to the nucleus where STATs bind to corresponding DNA binding sites in order to initiate gene transcription. After transcription factor activity is complete STATs are efficiently exported out of the nucleus, a process that helps regulate STAT-induced gene expression.

### 4.16 IL-2 signalling through the JAK/STAT pathway

The IL-2 receptor is a member of a larger family of type I cytokine receptors that include the receptors for IL-4, IL-13, IL-15 and IL-21. Each of these receptors comprise a cytokine specific α-chain responsible for ligand binding

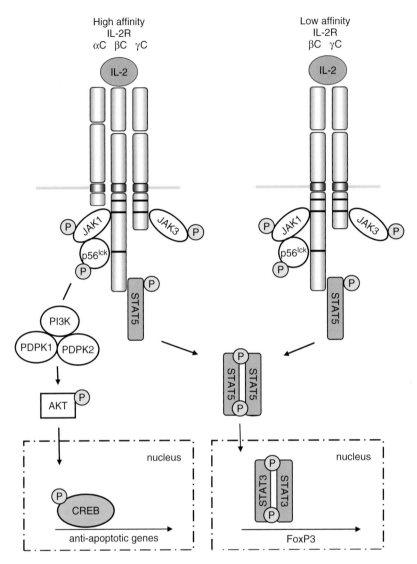

**Figure 4.14** IL-2 receptor signalling pathway involves the recruitment of Jak and STAT proteins to the intracellular domain of the receptor. A series of phophorylation events leads to the translocation of transcription factors into the nucleus and subsequent gene expression.

and a second chain that is shared by all family members known as the common gamma chain (γC). Furthermore, the IL-2 receptor has a third chain known as the β-chain (βC), which completes the high affinity receptor complex. IL-2 can also signal through a receptor with a lower affinity called the intermediate receptor, which comprises only the βC and γC. The IL-2 receptor does not possess its own kinase activity but relies on specific domains present in the βC and γC that recruit JAK proteins in order to perform signal transduction processes (Figure 4.14).

Two particular JAK proteins, JAK1 and JAK3, have been associated with IL-2 receptor signal transduction. The βC contains three important domains, the first of which is a Box1 domain that is responsible for JAK1 binding. The second is an acidic domain that associates with p56lck and the third is located at the carboxyl end, which is able to interact with STAT proteins. The γC also contains a carboxyl terminus, which is associated with JAK3 binding, and has a membrane proximal domain. In order for signal transduction to take place a heterodimerization event needs to take place by bringing the βC and γC into close proximity. JAK3 is then thought to phosphorylate JAK1 and other secondary signalling molecules such as p56lck, This phosphorylation event is essential for IL-2 receptor signalling and initiates the binding, phosphorylation and activation of STAT proteins. Secondary activation of p56lck can also lead to activation of the signalling molecule phosphoinositide 3-kinase (PI3K) and the anti-apoptotic BCL-2 protein (Figure 4.14). Phosphorylation of PI3K also causes the phosphorylation of the serine-threonine protein kinase AKT, which in turn activates the transcription factor CREB and leads to the upregulation of more anti-apoptotic genes.

The IL-2 receptor plays a vital role during the differentiation and proliferation of Treg cells. In this setting the IL-2 receptor α-chain is known as the cell surface marker CD25, the βC as CD122 and the γC as CD132. The βC is important for providing a binding site at the carboxyl terminus for the STAT5 transcription factor, which binds to the receptor via its SH2 domain. Phosphorylated STAT5 dimerizes and then translocates to the nucleus where it interacts with its DNA-binding motif within the promoter region of the FoxP3 gene. The expression of FoxP3 is essential for the maintenance of Treg cells.

## 4.17 The JAK/STAT pathway is also used by IL-6

Members of the IL-6 receptor family all share a common component, gp130, which is responsible for signal transduction. Other members of the family include receptors for IL-11, leukaemia inhibitory factor (LIF) and other receptors responsible for cell differentiation and proliferation (Figure 4.15). Like most receptors, a dimerization event is necessary for signal transduction, whereby gp130 dimerizes with a ligand specific receptor chain, in this case IL-6Rα. Although JAK3 and TYK2 are able to bind gp130, the most critical kinase for IL-6 signalling is JAK1. The membrane proximal domain of gp130 contains a box1/box2 domain, which allows for high affinity binding of JAK1. Phosphorylation and activation of JAK1 predominantly recruits and activates STAT1 and STAT3.

## 4.18 Plasticity in type I cytokine signalling

Cytokine signalling often involves more than one pathway, sometimes referred to as an alternative pathway. Therefore type I cytokine receptors such as the IL-2 and IL-6 receptors are able to initiate gene expression through the activation of signal transduction pathways other than JAK/STAT (Figure 4.16). As an example the IL-2R can also signal through PI3K/AKT/mTOR and the MAPK pathways. PI3K can be activated by the kinase activity associated with βC and γC of the IL-2 receptor. This then leads to the activation of the serine–threonine protein kinase AKT, via phosphorylation events generated by 3-phosphoinositide-dependent protein kinase 1/2 (PDPK1 and PDPK2). The activated AKT is then able to phosphorylate many other proteins, including transcription factors. However, AKT is also able to phosphorylate mTOR (mammalian target of rapamycin), which is associated with driving cell cycle progression through the activation of translation initiation factors that drive cell growth. Due to its central role in cell growth, AKT is linked with the progression of certain tumours.

Another example of cytokine signalling plasticity is the activation of the MAP kinase (MAPK) pathway following IL-6R stimulation. Adaptor proteins are recruited to the phosphorylated SH2-binding domain on gp130, which leads to the activation of RAS and initiates the MAPK signalling cascade. This includes the activation of the MAPKs JNK, ERK1/2 and p38 MAPK, which in turn activate a number of transcription factors and result in downstream gene expression (Figure 4.16). Alternative pathways involving MAPKs may also be activated following IL-2 signalling.

## 4.19 Suppressor of cytokine signalling (SOCS)

In humans a total of seven SOCS family members have been discovered, SOCS1–7. These proteins are characterized by a SH2 domain and a particular C-terminal domain known as a SOCS box. These proteins are involved in regulating cytokine expression through the inhibition of STAT-mediated signal transduction (Figure 4.17). They are also induced by the JAK/STAT pathway themselves, usually by the same cytokine signalling pathway they subsequently inhibit. Therefore, they are part of a negative feedback loop. They inhibit the JAK/STAT pathway by binding to the SH2 domain of the cytokine receptor adjacent to JAK proteins. A domain contained in the SOCS1 protein, known as the kinase inhibitory protein, binds to the enzymatic domain of the JAK and prevents kinase activity. This effectively inhibits phosphorylation events initiated by the JAKs and prevents phosphorylation of

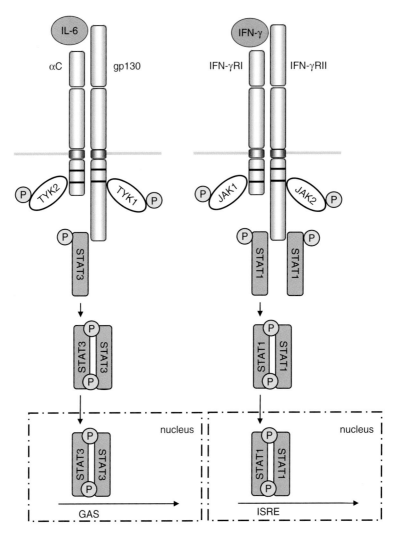

**Figure 4.15** IL-6 and IFN-γ signal through the JAK/STAT pathway, although they recruit and signal through different STAT proteins. (GAS, γ-interferon-activated site; ISRE, IFN-stimulated response elements.)

the JAKs themselves, the cytokine receptor and the STAT proteins. SOCS have been shown to inhibit different STATs, depending on the cytokine receptor combination.

## 4.20 IFN-γ signalling pathway

Both the type I IFN (e.g. IFN-α, IFN-β) and type II IFN (e.g. IFN-γ) receptors (which are members of the type II cytokine receptor family) are structurally similar to the type I cytokine receptors such as the IL-2R and IL-6R, differing only in the lack of cysteine residues in the extracellular domain. The IFN-γ receptors consist of IFN-γR1 and IFN-γR2 and predominantly utilize the JAK/STAT signal transduction pathway. IFN-γR1 and IFN-γR2 are already associated with JAK1 and JAK2 respectively, prior to ligand binding. Oligomerization of these receptor components brings JAK1 and JAK2 into close proximity with one another, so that they can phosphorylate themselves and the receptor. This opens up the SH2-binding domain on the receptor allowing STAT1 to become phosphorylated and activated. A STAT1 homodimer then translocates to the nucleus where it binds to GAP-like elements resulting in specific gene expression events (Figure 4.15).

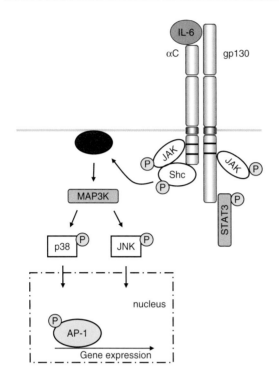

**Figure 4.16** Alternative signalling pathways induced by type 1 cytokine receptors, as exemplified by an alternative IL6R pathway. Activation of MAPK and phosphorylation of the kinases p38 and JNK leads to the activation of the transcription factor AP-1 and associated gene expression.

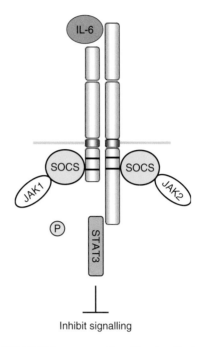

**Figure 4.17** SOCS (suppressor of cytokine signalling) proteins inhibit receptor signalling by interfering with JAK recruitment and thereby inhibiting phosphorylation of STATs.

IFN-γ signalling is known to activate many downstream transcription factors, some of which are dependent on STAT1 activation. For example, IFN-γ signalling leads to the activation of IFN-regulatory factor (IRF), which requires the cooperation of STAT1. IRF, in particular IRF1 and IRF8, bind to IFN-stimulated response elements (ISRE) in the promoter regions of interferon-stimulated genes (ISGs), which also leads to the upregulation of IFN-γ itself. In addition, it is now known that several STAT1-independent signalling cascades are associated with IFN-γ signalling, including the NF-κB pathway, MAPK pathway and the AP-1 signalling pathway. The cross-talk between different signalling pathways leads to the activation of a host of IFN-γ related genes and demonstrates the integral role that IFN-γ plays during immunity and inflammation.

## 4.21 TGF-β and the SMAD signalling pathway

TGF-β is one of the most pleiotropic cytokines and is involved in a number of important cellular and immunological processes, including cell differentiation, proliferation, angiogenesis and immune modulation. There are three isoforms of TGF-β (TGF-β1-3) having a high degree of homology and existing as homodimers but possessing overlapping biological functions. Before binding to its receptor, the latent form of TGF-β must first be activated through dissociation from its latent binding protein (LAP). The TGF-β superfamily of cytokines has more than 40 members, including various activins and bone morphogenic proteins (BMPs).

Specifically, TGF-β signals through a heterotetrameric receptor complex formed from the combination of two type I receptors (TβRI) and two type II receptors (TβRII). Both TβRI and TβRII exist as homodimers when TGF-β is not present. The first step in receptor activation is the binding of TGF-β, in the form of a homodimer, to TβRII. This results in the recruitment and phosphorylation of TβRI, forming the active receptor complex, which is then able to transduce signals to the appropriate signalling molecules. This signal transduction relies on kinase domains located within the intracellular region of the receptors. Therefore, the kinase domain on TβRII is responsible for phosphorylating TβRI, while the kinase domain on TβRI is necessary for phosphorylating

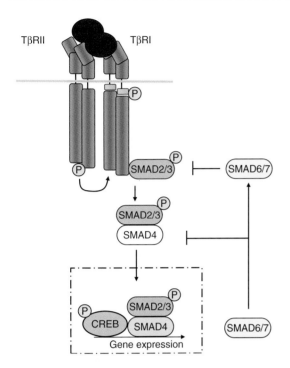

**Figure 4.18** The TGF-β signalling pathway involves the recruitment and the phosphorylation of SMAD proteins. TGF-β also upregulates the inhibitory I-SMAD6 and I-SMAD7, which inhibits receptor R-SMAD2/3 and common-partner co-SMAD4 signalling, thereby acting in a negative feedback loop.

downstream targets. The main signalling pathway activated during TGF-β signal transduction involves SMAD proteins, although SMAD-independent pathways are also activated (Figure 4.18).

In humans there are eight known SMAD proteins, SMAD1 to SMAD8. Certain SMAD proteins (SMAD1, SMAD2, SMAD3, SMAD5 and SMAD8) can directly bind to TβRI and are therefore referred to as receptor-activated or R-SMADs. Binding of TGF-β results in the recruitment of SMAD2 and SMAD3 to the TβRI, where they undergo phosphorylation at their MH1 domains. Other TGF-β family members require different SMADs for pathway activation, for example BMP ligands activate SMAD1, SMAD5 and SMAD8. Activation of the R-SMAD proteins following phosphorylation results in their release from the receptor and the formation of a heterodimeric complex with the common-partner SMAD4, otherwise known as a co-SMAD. This SMAD4/R-SMAD complex then translocates to the nucleus where it interacts with DNA-binding proteins, such as CREB-binding protein (CBP) or p300, resulting in the regulation of target gene transcription.

The structurally different SMAD6 and SMAD7 can inhibit R-SMAD signalling and are therefore known I-SMADs. They predominantly act by interacting with TβRI to prevent R-SMADs from binding to the receptor. This prevents phosphorylation and inhibits the activation of SMAD2 and SMAD3. Alternatively SMAD6 can bind to SMAD4, thereby competing with the R-SMADs and preventing nuclear translocation. Finally, SMAD6/7 can interfere with transcription factor binding; in particular SMAD6 can prevent BMPs from binding HOX transcription sites. TGF-β can also cause the upregulation of SMAD6 and SMAD7, representing an auto-inhibitory feedback mechanism for ligand induced signalling. TGF-β signal transduction is further regulated by SMAD-independent pathways, principally through the activation of MAPKs such and JNK and ERK. This can lead to an enhancement of SMAD signalling and to the activation of alternative transcription factors such as c-Fos and AP-1. Furthermore, part of the immunosuppressive properties of TGF-β signalling may be related to the inhibition of NF-κB signalling, probably through the upregulation of IκBα or through alterations in histone deacetylase proteins.

## 4.22 Type III cytokine receptors and the TNF receptor family

TNF is a key proinflammatory cytokine that plays a central role in controlling inflammation, cell proliferation and apoptosis. Following binding of TNF to either of its two receptors, TNFR1 and TNFR2, signal transduction largely results in the activation of the transcription factors AP-1 and NF-κB and the induction of proinflammatory gene expression. Through a separate signal transduction pathway, TNF can also activate caspases, which in certain cell types leads to apoptosis. However, the primary outcome of TNF signalling is the upregulation of proinflammatory genes, including auto regulation of itself through NF-κB and AP-1 binding to its promoter region.

The two TNF receptors form separate homotrimeric receptor complexes (Figure 4.19). Following TNF binding, these receptors undergo trimerization and recruit the first signalling molecules to the receptor complex, the first protein being TNFR1-associated death domain protein (TRADD). TRADD then recruits three other proteins, receptor-interacting protein 1 (RIP1), Fas-associated death domain protein (FADD) and TNF-receptor-associated factor 2 (TRAF2). Recruitment of signalling proteins to TNFR2 is not as complex, as TRAF2

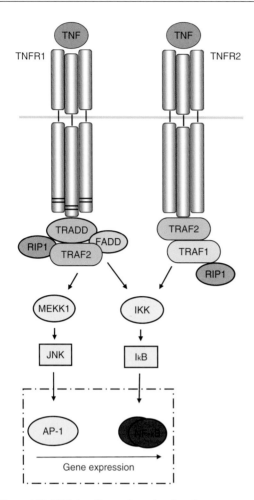

**Figure 4.19** TNF signalling pathway involves the phosphorylation and activation of the transcription factors AP-1 and NF-κB. TNFR1 signals via TRADD and TRAF2, causing the activation of MEKK1 and the phosphorylation of JNK, thereby resulting in AP-1 activation. TNFR1 and TNFR2 (which signals via TRAF2 and TRAF1) also activate IKK and lead to NF-κB translocation into the nucleus.

is directly recruited to the intracellular domain, which in turn recruits TRAF1. The recruitment of adaptor proteins to TNFR1 functions to enhance the signalling potential of the receptor, hence TNFR1 is activated more easily than TNFR2. Activation of TRAF2 then leads to the activation of the NF-κB transcription factor, which requires activation of the IKK complex in order to dissociate from the inhibitory IκB complex. Following the release of NF-κB from its inhibitor it translocates to the nucleus and regulates gene expression. NF-κB is a very important

transcription factor, involved in a host of inflammatory processes, and its activation will therefore be discussed in detail later.

The recruitment and activation of adaptor molecules to the TNFR1 complex also activates the MAPK pathway, ultimately leading to AP-1 activation and AP-1 dependent gene expression. TRAF2 is an effective activator of a number of MAPKs such as JNK and p38 MAPK. This activation is thought to be mediated by an intermediate protein called MEKK1. Cross-talk between signalling networks may also be involved in NF-κB activation. For example, MEKK1 may interact with RIP1 or TRAF2 during the activation of NF-κB, via the phosphorylation and activation of the IKK complex.

## 4.23 The IKK complex and the activation of NF-κB

The transcription factor NF-κB is actually a complex of two proteins of the Rel/NF-κB family, which are characterized by an N-terminal Rel homology domain (RHD). The RHD domain serves a multifunctional purpose as it is responsible for dimerization of the two Rel proteins, nuclear import and binding to DNA elements within the promoter regions of genes (κB binding sites). There are five members of the Rel/NF-κB family in humans: Rel A (p65), Rel B, c-Rel, p50 and p52. Of these, only Rel A and p50 are constitutively expressed, while the other family members have restricted expression patterns depending on cell type and state of differentiation. Prior to receptor activation, NF-κB is retained in the cytoplasm bound to an inhibitory protein complex known as IκB. Specific sequences (30–33 amino acids long) within IκB proteins, known as ankyrin repeats, bind to the RHD domains within Rel proteins and mask the adjacent nuclear localization signal (NLS). Phosphorylation of the IκB protein complex flags it for immediate ubiquitination and degradation within the proteosome. This degradation leads to a rapid release of NF-κB and translocation into the nucleus where it initiates gene expression. The IκB complex is composed of one or more subunits that are specific to certain Rel/NF-κB members. For example, the RelA/p50 NF-κB complex is specifically bound and inhibited by IκBα and IκBβ, the two most important proteins involved in NF-κB inhibition.

The phosphorylation events that lead to NF-κB activation require the action of an IκB kinase, referred to as the IKK complex (Figure 4.20). This complex is actually formed of three individual proteins: IKK-α and IKK-β,

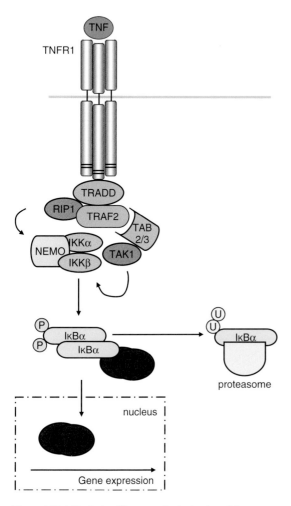

**Figure 4.20** NF-κB signalling cascade. Activation of the signalling molecules TRADD and TRAF2 cause the downstream activation of the IKK complex. This phosphorylation results in the phosphorylation and subsequent ubiquitination of IκBα and its proteasomal degradation. This releases the components RelA and p50 from the inhibitory IκBα, so that NF-κB can undergo nuclear translocation and gene expression can take place.

which possess the catalytic kinase activity, and NEMO (NF-κB essential modulator), which has an important regulatory role. It is thought that two differing pathways exist regarding the activation of NF-κB, the classical (or canonical) pathway and the alternative pathway. The classical pathway specifically involves phosphorylation of the IκB complex by IKK-β, which is able to phosphorylate IκBα, IκBβ and IκBε. The phosphorylation events in the classical pathway, mediated by IKK-β, occur very rapidly (within minutes). The alternative pathway

involves the translocation of a p52/Rel B heterodimer into the nucleus and occurs in certain developmental situations such as during the generation of secondary lymph nodes (Figure 4.21). The activation of NF-κB in this instance involves activation of IKK-α by TNF family members such as LT-α, receptor activator of NF-κB ligand (RANKL) and CD40. IKK-α then activates NF-κB by phosphorylating the p52 precursor p100, although the alternative pathway occurs much more slowly (hours rather than minutes) compared to the classical pathway.

## 4.24 The IL-1R family of type IV cytokine receptors activate NF-κB

The TLR/IL-1R family are characterized by extracellular immunoglobulin-like (Ig-like) domains and comprise two groups of structurally and functionally similar proteins, the TLR family and the IL-1R family. TLRs recognize various pathogen-associated molecular patterns (PAMPs) derived from viruses, bacteria and fungi, and function by providing early danger signals to the immune system (discussed in Chapter 2). The IL-1R family consists of receptors to IL-1α, IL-1β, IL-18 and IL-33 and share a Toll/IL-1R (TIR) domain with the TLRs. The IL-1R receptors are type I transmembrane proteins that possess three characteristic extracellular Ig-like domains that function as the ligand-specific binding site. The most widely studied IL-1R family member is the type I IL-1RI, which binds to IL-1α, IL-1β and the receptor antagonist IL-1RA. The IL-1RI requires a co-receptor for effective signal transduction called the IL-1R accessory protein, IL-1RAcP, which forms a heterodimeric complex with IL-1RI.

For signal transduction to take place both the ligand receptor IL-1RI and the IL-1RAcP need to be present, as the TIR domain on both molecules is required. Ligand binding initiates heterodimerization, bringing the two TIR domains into close proximity and forming an appropriate molecular configuration for binding to adaptor proteins. An important adaptor protein for both IL-1RI and TLR signalling is the myeloid differentiation primary-response gene 88 (MyD88), which rapidly and stably interacts with the receptor complex. MyD88 interacts with the receptor complex via is own TIR domain, although another region called the death domain is vital for signal transduction. The next step is the recruitment of IL-1R-associated kinase-1 (IRAK1) into the IL-1RI complex in association with the adaptor molecule Tollip. In quiescent cells Tollip is thought to silence IRAK1 activity. Translocation of IRAK1 and binding to the

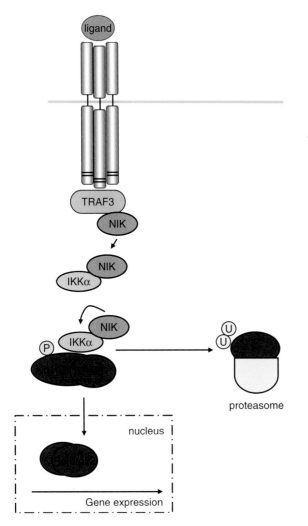

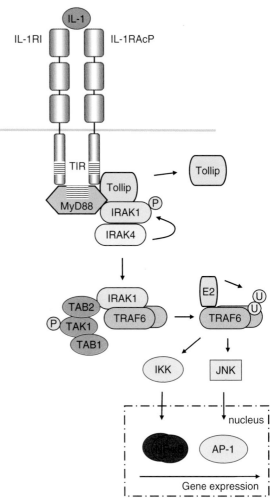

**Figure 4.21** Alternative NF-κB pathway involves the phosphorylation of p100 by IKKα, resulting in the mature p52 component. The RelB/p52 NF-κB complex then translocates to the nucleus and initiates gene transcription.

**Figure 4.22** IL-1 signalling through the TLR/IL-1R pathway. The adaptor protein MYDD88 is recruited to the TIR domains on the IL-1R complex, which initiates the recruitment and phosphorylation of IRAK 4 and IRAK1, and the release of Tollip. IRAK1 then associates with TRAF6 and the TAB/TAK complex, resulting in TRAF6 ubiquitination and the downstream activation of IKK and JNKI. This in turn activates the transcription factors NF-κB and AP-1.

IL-1RI complex is mediated via the interaction between the death domain on MyD88 and the death domain on IRAK1. IRAK4 is also recruited to the complex and undergoes autophosphorylation, which in turn phosphorylates IRAK1 causing the release of Tollip and the activation of IRAK1. The now activated IRAK1 is able to dissociate form the receptor complex allowing it to associate with TNF-receptor associated factor-6 (TRAF6). The IRAK1-TRAF6 complex then activates and forms a downstream complex with the pre-existing TGF-β-activated protein kinase (TAK1), TAK1-binding protein 2 (TAB2)

and TAB3 complex. TRAF6 is then able to undergo ubiquitination by associating with the E2 ubiquitin-conjugating enzyme complex Ubc13/Uev1A. This final step then mediates the phosphorylation and activation of IKK, which leads to NF-κB translocation to the nucleus and activation of transcription (Figure 4.22).

TRAF6 is also able to ubiquitinate the TAK1/TAB2/TAB3 complex, which in turn is able to activate alternative signalling molecules including the MAPKs JNK and p38 MAPK. This leads to the activation of the transcription factor AP-1. Further receptor cross talk may involve the association of other adaptor molecules to the IL-1RI complex, such as TRAM and TRIF. This leads to the activation of TBK1 and IKKε and the subsequent activation of the transcription factor IRF3, which initiates transcription of type I interferons.

## 4.25  Soluble cytokine receptors act as decoy receptors

Soluble cytokine receptors can either act as receptor agonists or antagonists and thereby accentuate or attenuate cytokine signalling. There are many types of soluble cytokine receptor, generated from several molecular mechanisms. They can be expressed from distinct protein coding genes, transcribed from alternatively spliced cytokine receptor transcripts or they can be derived from proteolytic cleavage of cytokine receptor ectodomains or cleaved from GPI-anchored receptors. Probably the most studied soluble cytokine receptors are those that inhibit IL-1 and IL-6 signal transduction.

The type II IL-1RII is able to form a receptor complex with IL-1RAcP on the cell surface. However, unlike IL1RI, IL1RII is not able to induce an intracellular signal because it only has a very short cytoplasmic domain without a TIR domain. IL-1RII therefore acts as a decoy receptor. Furthermore, IL-1RII can undergo proteolytic cleavage and form a soluble sIL-1RII, which is able to bind IL-1 and inhibit signalling by preventing access to the cell surface bound IL-1RI. This inhibition is enhanced by the expression of sIL-1RAcP, which is derived from an alternatively spliced version of IL-1RAcP. In this way, the pro-inflammatory properties of IL-1 can be attenuated by sIL-1RII and sIL-1RAcP (Figure 4.23).

In contrast, the soluble IL-6 receptors actually amplify IL-6 signalling, rather than acting as an antagonist. There are two forms of soluble IL-6 receptors, one derived from proteolytically cleaved IL-6Rα ectodomains and a second generated through alternative splicing to produce sIL-6Rα. Enhancement of IL-6 signalling is achieved either through sIL-6Rα binding to IL-6 and increasing its half-life, or through sIL-6Rα/IL-6 binding to the secondary gp130 receptor chain and the elicitation of an intracellular signal through a process known as trans-signalling. The

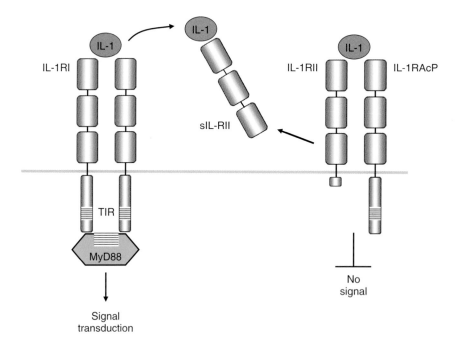

**Figure 4.23** Soluble cytokine receptors can act as decoy receptors. IL-1RII can either be membrane bound, where it forms a complex with IL-1RAcP but provides no intracellular signal, or it can be released as a soluble proteins where it binds to soluble IL-1 and prevents IL-1 binding to the IL-1R complex.

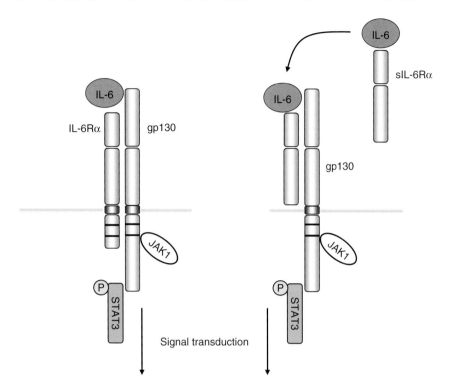

**Figure 4.24** Soluble IL-6 receptors can amplify receptor signals.

gp130 receptor is expressed in most cell types allowing IL-6Rα/IL-6 complexes to initiate signal transduction on cells that do not express membrane-bound IL-6Rα. This regulatory network is completed by a soluble form of gp130 (sgp130), which inhibits sIL-6Rα by competing for IL-6 binding (Figure 4.24).

## 4.26 IL-33 and ST2 signal regulation

IL-33 is a member of the IL-1 family of cytokines and is known to signal through the TLR/IL-1R superfamily member ST2 in association with IL-1RAcP. Moreover, there is a soluble form of ST2, which is increased in the serum of patients with inflammatory diseases such as asthma. The gene encoding ST2 contains two promoter regions that each drive the transcription of either sST2 or the membrane bound receptor ST2L. Receptor ligation and signal transduction leads to the activation of NF-κB and MAPKs and predominantly induces the Th2 cytokines IL-5 and IL-13. On the other hand, sST2 acts like a decoy receptor by sequestering IL-33 and inhibiting its binding to ST2L and preventing the expression of Th2-type inflammatory cytokines.

## 4.27 Potential for cytokine therapy

The central role that cytokines play in immune defence has led to the development of several cytokine therapies that aim to either boost the immune response (e.g. IFN-α therapy) or modulate the immune response (e.g. TGF-β therapy). Such therapeutics may be useful for the treatment of a range of different diseases including malignancies, infections and immune-mediated diseases such as autoimmunity and asthma. Despite their potential, there are a number of problematic factors that have prevented wide-scale use of cytokine therapeutics in humans. The potential for toxicity is probably the most worrying aspect. Direct toxic effects may be related to the pleiotropic nature of cytokines, so that any therapeutic will induce a number of unwanted side effects.

Toxic effects may also occur indirectly via the generation of other cytokines, or the abnormal modification of natural immune defences. One other consideration is that cytokines generally have a short half-life *in vivo*, therefore raising concern about the practicality of repeated doses, particularly with regard to treating chronic diseases. However, cytokine therapies have been successfully used in the clinic. For example, IFN-α is widely used as an antiviral agent against hepatitis C virus infection of the liver and to treat AIDS-associated Kaposi's sarcoma caused by human herpes virus-8. IFN-α has also been used as a cancer therapy, with the premise that it enhances the presentation of tumour-associated antigens, activates T and B cells and has anti-angiogenic properties. Similarly, IL-2 and IL-12 are used as cancer therapeutics due to their immunostimulatory properties. Cytokine therapy has proved to be a useful means of enhancing immunity in patients suffering from the classical immunosuppressive symptoms associated with AIDS. An example is G-CSF therapy, which is used to increase the number of circulating neutrophils, which helps to fight bacterial infections.

Cytokines also represent an attractive target molecule for the development of therapeutic drugs. A remarkable advance in the field of immunotherapy has been the development of anti-cytokine agents, which are now used in the clinic to treat a variety of different diseases. For example, the inflammatory cytokines TNF and IL-1 have both been targeted by neutralizing antibodies in order to treat the inflammation associated with rheumatoid arthritis. In the cancer setting VEGF has been target for the treatment of colorectal cancer. In a similar way, antibodies directed against cytokine receptors have also been successfully used in the clinic, particularly in the treatment of certain cancers. For example, anti-EGFR is used for the treatment of colorectal cancer, while anti-IL-2R antibodies are given to patients following organ transplant surgery in an attempt to reduce immune rejection.

## 4.28 Summary

**1.** Cytokines provide a signalling network that allows cells to communicate with each other.

**2.** Cytokines belong to one of five families: the interleukins, interferons, TNF-family, growth factors and chemokines.

**3.** Cytokines are important in regulating cell differentiation, proliferation and activation.

**4.** One important example is the control of T cell differentiation, whereby IL-12 and IFN-γ are associated with Th1 cells, while IL-4, IL-5 and IL-13 are associated with Th2 cells.

**5.** Cytokines exert their function through cytokine receptors expressed on the surface of cells.

# 5 Chemokines

## 5.1 Introduction

The immune system comprises a multitude of different effector cells located within many lymphoid and non-lymphoid tissues throughout the body. This includes the primary lymphoid organs of the bone marrow and thymus, the secondary lymphoid organs such as the spleen and lymph nodes, the mucosal lymphoid tissues of the gut, respiratory tract and urogenital tract, the extensive tissue of the skin and the circulatory system of the blood and lymphatics. Furthermore, effector cells must enter sites of infection, which could be in any non-lymphoid tissue, and exert their physiological function. With so many different immune cells and so many sites into which they can traffic, how do they all get to the right place at the right time? One important family of specialized cytokines, called chemokines, has the primary function of regulating the trafficking of leukocytes. The basic function of chemokines is to mediate the directional movement of cells, and as such they are integrally involved in orchestrating the correct migration of immune cells during health and disease. Not only are they responsible for recruiting leukocytes to sites of inflammation following infection or injury, they also maintain tissue homeostasis and direct the movement of cells during wound healing.

## 5.2 Structure and nomenclature of chemokines

The chemokine family can be considered as a group of specialized cytokines. All family members are small, being between 8 and 10 kD in size. They are distinguishable from classical chemoattractant molecules, such as complement fragment peptides C3a and C5a, and lipid molecules such as leukotriene $B_4$, due to their distinctive protein structure. The chemokine family of molecules has

rapidly expanded in recent times, due to the advances in molecular biology techniques, such as DNA sequencing, protein isolation and characterization, and the discovery of homologues in closely related mammalian species through bioinformatics methods. Indeed, these modern molecular biology techniques have made identifying new chemokine family members relatively easy. The rapid increase in chemokine family members has led to difficulties in chemokine nomenclature, due to different research groups ascribing multiple names to a single chemokine. Therefore, a universal nomenclature scheme was established, based on the configuration of cysteine residues that are vital for the correct folding of chemokine molecules (Table 5.1). The configuration of these cysteine residues largely determines the structure of the chemokine.

Chemokines have four conserved cysteine residues that form disulphide bonds, and these bonds have been shown to be critical for the tertiary structure of the protein. The chemokine family is organized into four subclasses according to the position of the first two cysteines. The two major subclasses include the CC chemokines and the CXC chemokines. All members of the CC chemokine subfamily have the first two cysteines adjacent to each other, while in all members of the CXC family, the first two cysteines are separated by one amino acid (Figure 5.1). Chemokines CCL1 to CCL28 are members of the CC family, while CXCL1 to CXCL17 constitute the CXC subfamily, where the postscript L stands for ligand. Two other subclasses have also been identified but these have far fewer members. The C class has only two cysteines instead of four and there are two members of this group, XCL1 and XCL2. Lastly, the single chemokine belonging to the $CX_3C$ subclass has three amino acids between the first two cysteines. The only member of this class, CX3CL1, is unusual in that it has a mucin stalk at the N-terminal end which is thought to enable it to stick to the cell membrane. In general, chemokines have a short

*Immunology: Mucosal and Body Surface Defences*, First Edition. Andrew E. Williams.
© 2012 John Wiley & Sons, Ltd. Published 2012 by John Wiley & Sons, Ltd.

**Table 5.1** Nomenclature of chemokines based on their structure, which is largely determined by the position of cysteine residues.

| Chemokine class | Cysteine residues | Members |
| --- | --- | --- |
| C chemokines | X-C-C Two cysteines | XCL1, XCL2 |
| CC chemokines | C-C First two cysteine residues adjacent | CCL1 to CCL28 e.g. CCL2, CCL5 |
| CXC chemokines | C-X-C First two cysteine residues separated by one amino acid | CXCL1 to CXCL17 e.g. CXCL8 (IL-8) |
| CX$_3$C chemokines | C-X-X-X-C Three amino acids between first two cysteine residues | CX$_3$CL1 |

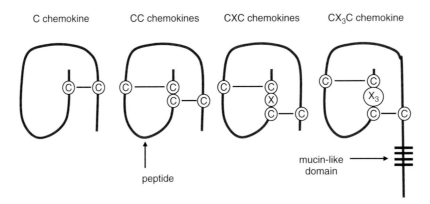

**Figure 5.1** Structure of chemokine ligands is determined by the position and number of cysteine residues. Four subclasses of chemokine have been identified.

amino-terminal domain preceding the first cysteine, a backbone made of three β-strands and the connecting loops found between the second and fourth cysteines, and a carboxyl-terminal α-helix of 20–30 amino acids.

## 5.3 Chemokine receptors

The specific biological effects of chemokines are mediated via interactions with chemokine receptors, which are G-protein coupled receptors (GPCRs) that all share a similar structure. These chemokine receptors are part of a much bigger superfamily of GPCRs that include receptors for hormones, neurotransmitters, inflammatory mediators, certain proteinases, taste and deodorant molecules and even photons and calcium ions. Chemokine receptors are approximately 350 amino acids in length and consist of a short extracellular N-terminus, an intracellular C-terminus containing serine residues and seven transmembrane domains in an α-helical configuration (Figure 5.2). The transmembrane domains are connected by 3 intracellular and 3 extracellular hydrophilic loops and a disulphide bond that links the highly conserved cysteines found in extracellular loops 1 and 2. The N terminus and the first

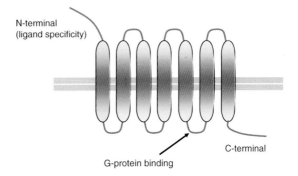

**Figure 5.2** Structure of chemokine receptors consisting of seven transmembrane domains. Chemokine receptors are G-protein coupled receptors.

extracellular loop connecting the first two transmembrane α-helices of the receptor are thought to be essential for the specific binding of chemokines. Chemokine receptors couple to heterotrimeric G-proteins through the C-terminus segment and possibly through the third intracellular loop. The intracellular C-terminus contains serine and threonine residues, which act as phosphorylation sites for receptor signalling.

Constitutively expressed chemokine receptors.

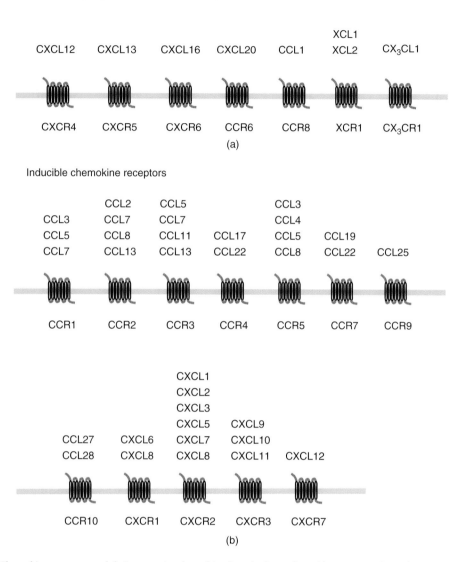

Inducible chemokine receptors

**Figure 5.3** Chemokine receptors and their respective chemokine ligands. Some chemokines are constitutively expressed (a) and tend to have homeostatic functions. Other chemokines are inducible (b) and play more of a role in inflammatory cell recruitment.

At present 19 human chemokine receptors have been identified. Chemokine receptors CXCR1 to CXCR7 bind the CXC family of chemokines, whereas the CC family consists of 10 receptors, CCR1 to CCR10, and bind members of the CC subfamily. The specific receptor for XCL1 and XCL2 is XCR1, while CX₃CL1 binds to CX₃CR1 (Figure 5.3). The chemokine receptor superfamily is unique in that there is a certain amount of promiscuity in binding to chemokine ligands. In other words, some chemokine receptors are able to bind several chemokine ligands, while certain chemokine ligands are able to bind several chemokine receptors. Receptors can be described as specific, shared or promiscuous (Box 5.1). For example, specific receptors, such as CXCR4, bind only one chemokine – in the case of CXCR4, it is CXCL12. In contrast, shared receptors bind more than one chemokine. An example of this is CCR1 which binds CCL3, CCL5, CCL7 and CCL8. Chemokine receptors are

## Box 5.1 Chemokine Receptors and their Chemokine Ligands.

| Name | Alias | Receptor |
|------|-------|----------|
| CCL1 | I-309, TCA-3 | CCR8 |
| CCL2 | MCP-1, MCAF | CCR1, CCR2, CCR3, CCR5 |
| CCL3 | MIP-1$\alpha$ | CCR1, CCR5 |
| CCL4 | MIP-1ß | CCR1, CCR5 |
| CCL5 | RANTES | CCR1, CCR3, CCR5 |
| CCL6 | C10, MRP-2 | CCR1 |
| CCL7 | MARC, MCP-3 | CCR1, CCR2, CCR3 |
| CCL8 | MCP-2 | CCR1, CCR2, CCR5 |
| CCL9/(CCL10) | MRP-2, MIP-1 | CCR1 |
| CCL11 | Eotaxin | CCR2, CCR3, CCR5 |
| CCL12 | MCP-5 | CCR2 |
| CCL13 | MCP-4, NCC-1, Ckß10 | CCR2, CCR3, CCR5 |
| CCL14 | NCC-2, HCC-1, MCIF, Ckß1 | CCR1, CCR5 |
| CCL15 | Leukotactin-1, MIP-5, HCC-2, NCC-3 | CCR1, CCR3 |
| CCL16 | NCC-4, LEC, LMC, Ckß12 | CCR1, CCR2, CCR5, CCR8 |
| CCL17 | TARC, dendrokine, ABCD-2 | CCR4 |
| CCL18 | MIP-4, PARC, Ckß7 | CCR3 |
| CCL19 | ELC, Exodus-3, Ckß11 | CCR7 |
| CCL20 | LARC, Exodus-1, Ckß4 | CCR6 |
| CCL21 | SLC, Exodus-2, Ckß9, TCA-4 | CCR7 |
| CCL22 | MDC, DC/ß-CK | CCR4 |
| CCL23 | MIP-3, MPIF-1, Ckß8 | CCR1 |
| CCL24 | Eotaxin-2, MPIF-2, Ckß6 | CCR3 |
| CCL25 | TECK, Ckß15 | CCR9 |
| CCL26 | Eotaxin-3, MIP-4a, IMAC, TSC-1 | CCR3 |
| CCL27 | CTACK, ILC, Eskine | CCR10 |
| CCL28 | MEC | CCR3, CCR10 |
| CXCL1 | Gro-$\alpha$, GRO1, NAP-3, KC | CXCR1, CXCR2 |
| CXCL2 | Gro-ß, GRO2, MIP-2$\alpha$ | CXCR2 |
| CXCL3 | GRO3, MIP-2ß | CXCR2 |
| CXCL4 | PF-4 | CXCR2, CXCR3 |
| CXCL5 | ENA-78 | CXCR2 |
| CXCL6 | GCP-2 | CXCR1, CXCR2 |
| CXCL7 | PPBP, NAP-2, RDC1 | CXCR1, CXCR2 |
| CXCL8 | IL-8, NAP-1, MDNCF, GCP-1 | CXCR1, CXCR2 |
| CXCL9 | MIG, CRG-10 | CXCR3 |
| CXCL10 | IP-10, CRG-2 | CXCR3 |
| CXCL11 | I-TAC, ß-R1, IP-9 | CXCR3, CXCR7 |
| CXCL12 | SDF-1, PBSF | CXCR4, CXCR7 |
| CXCL13 | BCA-1, BLC | CXCR5 |
| CXCL14 | BRAK, MIP-2G | unknown |
| CXCL15 | Lungkine, WECHE | CXCR2 |
| CXCL16 | SRPSOX | CXCR6 |
| CXCL17 | DMC, VCC-1 | unknown |
| XCL1 | Lymphotactin a, SCM-1a | XCR1 |
| XCL2 | Lymphotactin ß, SCM-1ß | XCR1 |
| CX3CL1 | Fractalkine, Neurotactin | CX3CR1 |

also either constitutively expressed or can be induced following an inflammatory stimulus, for example. Constitutively expressed chemokine receptors tend to be those that are involved in tissue homeostasis and basal cell trafficking. Inducible chemokine receptors tend to be more important for instructing the migration of immune cells into sites of inflammation and therefore play a vital role in coordinating host defence. Another receptor known as the Duffy antigen receptor for chemokines (DARC) expressed on erythrocytes and endothelial cells is truly promiscuous and has been shown to bind both CXC and CC type chemokines. Finally, virally encoded chemokine receptors have been described, and are thought to be a mechanism of viral evasion from the immune system. These will be discussed in more detail later.

## 5.4 Expression of chemokines and their receptors

The main function of chemokines is to promote the movement of cells, and as such chemokine receptors are predominantly expressed on the surface of leukocytes. In this way, chemokines act to attract cells that express the relevant cell surface receptors. In general, leukocytes follow chemokine gradients, from areas of low chemokine concentration to areas of high chemokine concentration (Figure 5.4). Conversely, chemokines are predominantly expressed by structural cells of tissues in response to infection or injury, such as endothelial cells of the vasculature. These structural cells are capable of secreting chemokines and retaining them on their cell surface. In this way, relevant subpopulations of leukocytes can be directed to a site of inflammation in a particular tissue. However, the pattern of chemokine and chemokine receptor expression

becomes more complex as many leukocytes themselves can secrete chemokines, while many tissue structural cells can also express chemokine receptors. The secretion of chemokines by leukocytes has an important immunological function. Leukocytes that have entered sites of inflammation produce more chemokines as a way of attracting more immune cells into sites of inflammation, thereby acting to amplify inflammatory reactions. Similarly, structural cells such as fibroblasts and epithelial cells also express chemokine receptors. Chemokine receptor expression on structural cells allows them to be recruited to sites of inflammation or tissue injury so that they can promote the repair of damaged tissues.

## 5.5 Chemokines promote extravasation of leukocytes

In general, the recruitment of leukocytes into a site of inflammation necessitates movement from the bloodstream, through the vascular endothelium, and into the tissue proper. Cells within the circulation are travelling at relatively high speed and are under large shear forces within the larger blood vessels. Therefore, the majority of cell movement out of the bloodstream occurs in specialized areas of the vasculature called high endothelial venules, where sheer forces are lower. The movement of cells out of the bloodstream and into tissues is called extravasation and involves a four-step process (Figure 5.5). Extravasation can therefore be divided into rolling adhesion, chemokine signalling (or chemoattraction), tight adhesion, and diapedesis (or transmigration). Therefore, the process of halting selective populations of leukocytes requires the carefully orchestrated expression of molecules in order to slow leukocyte movement and direct cell traffic across the endothelium and into the underlying tissue. Chemokines play a vital role in this process.

In this model of extravasation, cells travelling in the bloodstream slow down through a rolling process that involves interactions with adhesion molecules expressed on the surface of endothelial cells. At sites of inflammation, endothelial cells become activated by various inflammatory mediators, such as TNF and TLR-ligands, and upregulate the expression of P-selectins. Circulating leukocytes are able to form weak bonds with P-selectin via P-selectin glycoprotein ligand-1 (PSGL-1). The interaction between P-selectin and PSGL-1 causes leukocytes to begin rolling along the surface of endothelial cells. Shortly after P-selectin expression, another selectin known as

chemokine gradient

Direction of chemotaxis

**Figure 5.4** Chemokines stimulate chemotaxis through the provision of a chemokine gradient. Cells migrate from areas of low chemokine concentrations to areas of high chemokine concentration.

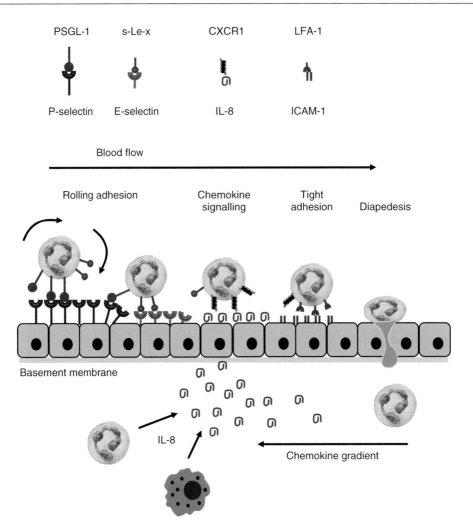

**Figure 5.5** Extravasation across endothelia involves a four-step process: 1. rolling adhesion, 2. chemokine signalling, 3. tight adhesion, and 4. diapedesis (otherwise known as transmigration). Several cell-adhesion molecules are involved in this process.

E-selectin is expressed, which interacts with PSGL-1, sialy-Lewis-x (s-Le-x) and E-selectin ligand-1 (ESL-1) on the surface of leukocytes, further slowing the leukocyte and enhancing its contact with the endothelium where they can sample the endothelial surface for activating factors.

These activating factors are the chemokines, which are secreted by endothelial cells and inflammatory cells within the tissue and presented on the endothelial cell surface. These chemokines cause the activation of leukocytes, through binding to their specific receptors, initiating the expression of high affinity integrin receptors. Integrins recognize cell adhesion molecules on endothelial cells and orchestrate the tight binding, which arrests

cell movement of leukocytes. Two important integrins expressed by leukocytes during the process of tight binding are lymphocyte function-associated antigen 1 (LFA-1) and very late antigen-1 (VLA-4), which bind to inter-cell adhesion molecule-1 (ICAM-1) and vascular cell adhesion molecule-1 (VCAM-1) on endothelial cells, respectively. The interaction between LFA-1 and VLA-4 with ICAM-1 and VCAM-1 also enhances attachment to the extracellular matrix, further securing leukocytes and allowing them to begin the process of diapedesis. During the interaction with the endothelial surface the leukocyte flattens to maximize contact with the endothelia and forms pseudopodia that penetrate through the gap between endothelial cells. The process of diapedesis involves the secretion of proteases by leukocytes and the migration of a cell across the endothelial layer and into the interstitial space. Once within the tissue, further chemokines act to recruit leukocytes to the exact site of inflammation.

## 5.6 Chemotaxis

Chemokines, unlike other molecules with chemoattractant properties, promote the movement of cells in the direction of increasing concentrations of the chemokine. In other words, they promote movement along a gradient, frequently referred to as chemotaxis. This is unique among chemoattractant molecules and is a distinctive property of the chemokine family (Figure 5.6). The interaction of chemokines with their receptors mediates a series of effects that ultimately result in the directional movement of the leukocyte. The mechanism of chemotaxis begins with a change in cell shape that occurs within seconds of contact with a chemokine. Polymerization and breakdown of actin filaments leads to the formation of lamellipodia (pseudopodia), which function as the limbs of the migrating cell. It is clear that movement of each cell necessitates a reorganization of the molecules of the cytoskeleton, so that movement takes place. This results in a polarization of the cell, where the side closest to the chemokine forms the lamellipodia (the leading edge), while the side furthest away forms a europod. The highest concentration of chemokine receptors are found at the leading edge, further reinforcing the polarized movement of the cell. Stimulation also induces the upregulation and activation of integrins, which enable the leukocyte to adhere more firmly to the vascular endothelial cell wall before migrating through to other tissues.

## 5.7 Chemokine receptor signalling cascade

Chemokine receptors initiate an intracellular signalling cascade through the recruitment of G-proteins and are therefore referred to as G-protein-coupled receptors (GPCRs). Chemokine receptor signalling may involve several signalling pathways, depending on the cell type and chemokine receptor, and includes G-protein activation of various downstream kinases and activation of the JAK-STAT pathway. It is thought that G-protein signalling is important for chemotaxis and cell motility, while the JAK-STAT pathway initiates gene transcription. Most GPCRs undergo a conformational change within their intracellular domains as a result of chemokine binding and receptor dimerization. This allows chemokine receptors to interact with G-proteins via the c-terminal and intracellular loop domains. Although there are many G-proteins capable of interacting with GPCRs, the archetypal subunit that associates with chemokine receptors is Gα. Binding of Gα to the receptor allows the recruitment of the Gβ and Gγ subunits, which form a complex at the cell membrane (Figure 5.7). Activation of the Gαβγ complex results in the dissociation of these subunits, thereby releasing Gβγ so that it can activate downstream signalling molecules. One such molecule is phospholipase C (PLC), which catalyses the formation of secondary messengers, inositol triphosphate (IP3) and diaglycerol (DAG). Firstly, the IP3 initiates the release of intracellular stores of $Ca^{2+}$, while the DAG activates protein kinase C (PKC). $Ca^{2+}$ and PKC are important for a variety of cellular functions including cell motility, polarization and adhesion. PKC is also involved in the activation of the transcription factor NF-κB. G-proteins also activate phosphatidylinositol 3-Kinase (PI3K), which possesses a diverse range of functions including the activation of several kinases such as

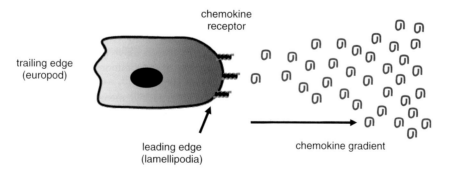

**Figure 5.6**  Cell movement up a chemokine gradient involves changes in cell shape.

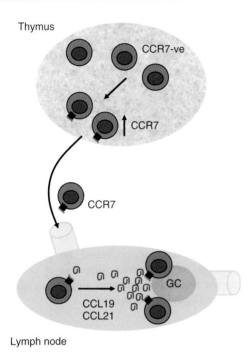

Thymus

CCR7-ve

CCR7

CCR7

CCL19
CCL21

GC

Lymph node

**Figure 5.9** Chemokines, are necessary for the migration of mature T cells from the thymus to secondary lymphoid structures. Upregulation of CCR7 on T cells enables them to migrate toward the chemokine ligands CCL19 and CCL21. CCR7 is considered to be an important chemokine receptor for the homing of leukocytes to lymph nodes.

differentiated within this microenvironment, effector or memory T cells drain into the lymph via the thoracic duct and rejoin the circulation. They can then migrate to non-lymphoid tissue or inflammatory sites in response to chemoattractants produced at those sites. A series of elegant experiments has demonstrated the importance of these chemokine/receptor interactions in the development of a fully functional immune system. For example, mice genetically deficient in CXCR5 have distorted splenic and Peyer's patch architecture and lack inguinal lymph nodes. Moreover, B cells which lacked CXCR5 could migrate to the T cell zones but not the B cell zones and thus these mice could not mount an antibody response. In a similar series of experiments, mice lacking CCR7 ligands exhibited impaired T cell homing to secondary lymphoid organs. In addition, dendritic cell migration is controlled by a similar chemokine receptor expression

pattern. Within non-lymphoid tissue, DCs express CCR5 and CCR6 (among others, depending on the tissue), both of which help to retain DCs at the site of inflammation. Following antigen stimulation and activation DCs undergo a maturation process that involves an alteration in chemokine receptor expression (Figure 5.10). During this maturation process DCs downregulate CCR5 and CCR6, while upregulating CCR7 on the cell surface. This enables them to leave sites of inflammation within peripheral tissues and migrate into the T cell zones within regional lymph nodes. In effect, DCs stop being phagocytic cells within tissues and become antigen presenting cells within lymph nodes. The expression of CCR7 by DCs, as well as T cells, is crucial for the initiation of an antigen-dependent T cell response.

## 5.10 Chemokines involved in lymphoid structure formation

The function of secondary lymphoid organs is to promote the priming and differentiation of lymphocytes in order to generate memory and effector cells. The homeostatic chemokines involved in this process include CXCL12, CXCL13, CCL19 and CCL21. The development of secondary lymphoid organs centres around a small cluster of common lymphoid progenitor cells, which are characterized by the Lin- c-kit+ IL-7Rα+ phenotype, which are called lymphoid inducer cells (see also Box 7.2, intestinal lymphoid inducer cells). During embryologic development, lymphoid tissue inducer cells express CXCR5 and CCR7, which causes them to migrate toward lymphoid stromal precursor cells, called lymph node organizers. Following stimulation with the TNF-family member lymphotoxin α1β2 and IL-7, lymph node organizer cells express the chemokines CXCL13, CCL19 and CCL21, as well as the adhesion molecules ICAM-1 and VCAM-1 (Figure 5.11). The chemokine ligands attract more lymphoid tissue inducer cells into the developing lymph node anlagen, while the adhesion molecules ICAM-1 and VCAM-1 help to retain those cells in the developing lymph node. The interaction between lymphoid tissue inducer and the lymph node stromal cells results in the signals necessary to amplify the formation of lymph node structures.

Chemokines are also important within primary lymphoid organs. For example, chemokines influence the

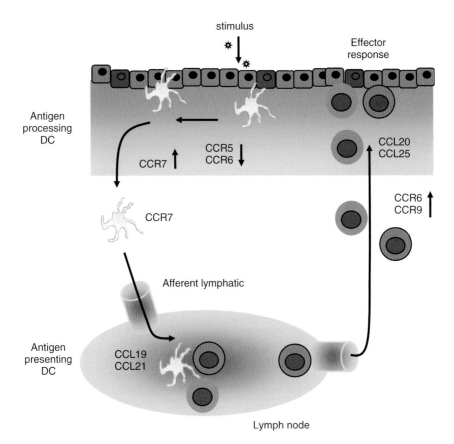

**Figure 5.10** CCR5 and CCR6 helps to retain immature DCs in tissue structures. Downregulation of CCR5 and CCR6 and the upregulation of CCR7 allows mature DCs to leave the tissue and migrate to regional lymph nodes, where they can then present antigen to T cells and initiate effector responses.

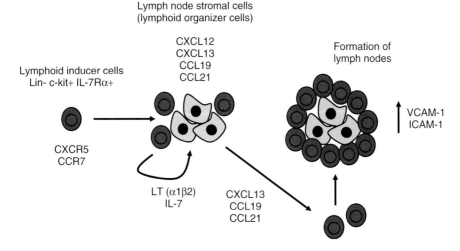

**Figure 5.11** Chemokine networks are important for the development of lymph node structures, wherein lymphoid inducer cells and lymphoid stromal (organizer) cells interact.

development of immature CD4+ and CD8+ T cells within the thymus. T cells that have undergone positive selection migrate from the cortex and into the medulla, while those cells that subsequently survive negative selection exit the thymus and enter the circulating T lymphocyte pool. Immature T cells at the CD4-CD8- (double negative) stage express CCR9 and are therefore responsive to the CCR9 ligand CCL25, the expression of which is maintained until late in the single positive stage within the medulla. Thus CCL25 functions as a retention factor, not permitting immature T cells to leave the thymus until the correct stages of differentiation and positive selection have taken place. Once this checkpoint has been cleared, immature T cells upregulate CCR7, so that positively selected T cells can migrate into the medulla. These changes in chemokine expression during thymic development are closely associated with the movement of T cells within the thymus, although their direct role in orchestrating T cell development is less certain.

Chemokines are also involved in the trafficking of haematopoietic cells in and out of the bone marrow, which is critical for normal homeostasis of lymphoid organs and for maintenance of the circulating pool of leukocytes. CXCR4 interaction with its ligand CXCL12 has been found to be essential in controlling this bone marrow traffic. The chemokine CXCL12 is expressed by bone marrow stromal cells and therefore acts as a retention factor for haematopoietic cells. Leukocytes undergoing differentiation processes express high levels of CXCR4, retaining them within the bone marrow microenvironment until maturation is complete. Emigration of cells from the bone marrow is therefore associated with a down regulation of CXCR4, enabling leukocytes to enter the circulation or complete their maturation in other lymphoid tissues.

## 5.11 Chemokines contribute to homeostasis

Although the primary function of chemokines is to promote the recruitment of leukocytes to inflammatory sites, chemokines are also important in maintaining immune homeostasis. The overriding function of the immune system is to maintain a balance of defending its host against infection, while at the same time maintaining the integrity of tissues and tolerance to self antigens. As such, cells of the immune system constantly take part in immuno-surveillance of peripheral tissue. This is of particular importance at mucosal sites

since they are the areas that are constantly exposed to environmental antigens. Therefore, the cells that survey these areas need to traffic between secondary lymphoid organs and mucosal sites, via the circulatory systems of the blood and lymphatics. Chemokine receptors play an important role in directing the migration of cells participating in immuno-surveillance. These chemokine receptors can either be expressed constitutively or their expression is inducible following appropriate stimulation. In general, those receptors expressed constitutively are developmentally regulated and are involved in the basal trafficking and homing of leukocytes, while those that are induced are involved in inflammatory reactions (Figure 5.3). However, this division should be regarded as a general guide rather than an absolute rule. One of the exemptions to the rule is CCR6 which is constitutively expressed by immature dendritic cells and T cells but is downregulated on dendritic cells upon maturation. However it is upregulated in skin diseases such as psoriasis. Furthermore, those chemokine receptors that can be regarded as constitutive tend to have monogamous pairings with their ligand (for example, CXCR5 with CXCL13), while those that are inducible generally bind multiple ligands which are themselves promiscuous (for example, CCR5 with CCL3, CCL4, CCL5 and CCL8 and CCL5 with CCR1, CCR3 and CCR5). This may be due to the need to amplify inflammatory reactions and recruit inflammatory cells to the site of infection as effectively as possible.

## 5.12 Chemokine receptors on T cell subsets

In order to promote tissue specific homing of cells to inflammatory sites, selected leukocyte subpopulations have specific homing receptor combinations that dictate the extent of their participation in an inflammatory reaction. One important example of this is the distinct chemokine receptor pattern expressed by T helper subsets. The delivery of functional subsets of T cells to particular tissues or microenvironments is a tightly controlled process involving a complex series of molecules expressed by a variety of cell types. Effector T cells can be divided into distinct subsets based upon their cytokine profiles and functional properties. Th1 cells characteristically produce IFN-$\gamma$ and contribute to host defence against viral and bacterial pathogens, whereas Th2 cells produce IL-4 and IL-5 and are associated with immunity to parasites. In addition, Th2 cells (and the cytokines they secrete) are thought to be critically

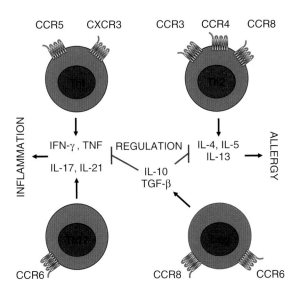

**Figure 5.12** Chemokine receptor expression differs between effector Th1, Th2, Th17 and Treg cells. The distinct pattern of expression of these receptors determines their migration into tissues and can influence whether an immune response is inflammatory (Th1/Th17) or regulatory (Th2/Treg).

important for the development of allergic reactions. The pattern of chemokine receptors seems to differ between Th1 and Th2 cells (Figure 5.12). For example, CCR5 and CXCR3 are expressed predominantly by Th1 cells, but not Th2 cells, while CCR3, CCR4 and CCR8 are expressed on Th2 cell subsets but not Th1. The expression of these different receptors may also be related to the chemokine environment in which these cells were first primed. A subsequent exposure to the same pathogen would normally trigger the release of an identical set of chemokines, thereby functioning to attract the correct subset of T cells. This is also true of T cell recruitment during allergic immune reactions, whereby repeated exposure to an allergen always recruits a Th2 cell population into the lungs. However, it is still unclear how Th2 cells specifically migrate to the lung during allergic reactions.

It has been suggested that a specific homing pathway exists for the lung, similar to that proposed for the skin or gut, whereby selected populations of T cells express a specific set of molecules to enable them to navigate principally to the airways. However, a distinct lung homing pathway involving the expression of a specific set of chemokine receptors has not yet been described, although lung T cells do express a pattern of chemokine receptors distinct from gut or skin homing T cells. Using a mouse model of allergic inflammation, the functional role of the CCR3 and CCR4 in attracting effector Th2 cells, via

CCL17 and CCL20 expression, has been investigated. These chemokine receptors/ligands were demonstrated to mediate the recruitment of antigen-specific Th2 cells to the lung, demonstrating their relevance in developing allergic responses in the airways. Indeed, in human patients a significantly greater number of CCR4+ Th2 cells were documented in airway biopsies after allergen challenge compared to pre-challenge or non-allergic patients. Moreover, CCL17 and CCL22 are upregulated in airway epithelium in atopic asthmatics, strongly suggesting that Th2 lymphocytes are recruited to the airways via CCL17 or CCL22 interacting with CCR4.

A Th1 phenotype is associated with host defence against many viral and bacterial pathogens, as well as in autoimmune diseases such as arthritis, psoriasis and multiple sclerosis. Considering that these diseases have a predominance of Th1 cells within sites of immunopathology, suggests that a specific set of trafficking molecules exists that function by recruiting Th1 cells. It has been shown that almost 90 per cent of circulating Th1 cells expresses CXCR3, which is considered to be a reliable marker for Th1 cells. However, many other chemokine receptors are thought to be associated with Th1 cells, including CCR2, CCR5, CCR6 and CXCR6, although these receptors are only expressed on Th1 cells in conjunction with CXCR3. The fact that these Th1 associated chemokine receptors show little preference for Th1 cells except when they are co-expressed with CXCR3, suggests that combinatorial expression of chemokine receptors is important for the migration of T cell subsets. Indeed, CXCR3 is an important chemokine receptor expressed on all extravasating leukocytes. The combinatorial expression of chemokine receptors allows Th1 cells to specifically migrate into inflamed tissues in response to particular inflammatory signals. For example, the skin inflammation associated with psoriasis is dominated by a Th1 phenotype and is characterized by the migration of CXCR3+CCR6+ T cells into the dermis in response to CXCL10 and CCL20 (Figure 5.13). These chemokines function in concert with the skin homing receptors CCR4 and CCR10, which are activated by CCL17 and CCL27, respectively. The chemokine system is also utilized by viruses in order to gain entry into selected tissues and to infect certain cell types. This is best exemplified by the human immunodeficiency virus (HIV), which is able to infect immune cells following binding to chemokine receptors (Box 5.2).

Treg cells also express a distinct subset of chemokine receptors, which allows them to migrate and respond to stimulation in a subtly different way to inflammatory T cells. Considering that Treg cells differentiate and

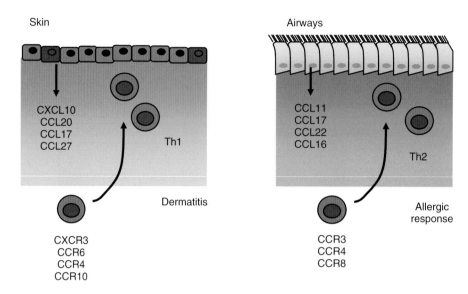

**Figure 5.13** Recruitment of inflammatory cells into different tissues utilises specific chemokine ligands and chemokine receptors. For example, Th1 mediated skin inflammation (psoriasis) is mediated by a different subset of chemokines/receptors compared to allergic responses in the airways.

proliferate in response to the same signals as Th1 cells, for example in response to IL-2, a mechanism must exist that allows these different T cell subsets to respond in separate ways. Treg cells express CCR6 and CCR8 following activation (Figure 5.12), which enables them to migrate into sites of inflammation where they can modulate the immune response through the expression of IL-10 and TGF-β. However, the immunopathogenic Th17 subset also expresses CCR6, thereby allowing the recruitment of these cells into inflamed tissues. Th17 cells contribute to the inflammatory state through the release of IL-17 and IL-21 and have been implicated in several immune-mediated diseases such as autoimmunity and allergy. Therefore, CCR6 plays an important role in balancing inflammation with immunoregulation and may be a key chemokine receptor in determining whether an immune response leads to inflammation and tissue damage or whether a state of tolerance is induced.

## 5.13 Redundancy in the chemokine/receptor system

The chemokine family performs many functions within the immune system, some of which are overlapping, while most chemokines have a promiscuous receptor usage. Analysis of animal models of inflammation, and biopsy

tissue from patients, also demonstrates that multiple chemokines are present during inflammatory reactions, many of them attracting the same populations of leukocytes. This would suggest that the chemokine system is replete with redundancy, with many chemokines performing the same tasks. However, detailed study of animal models has shown that the production of chemokines *in vivo* is organized into separate microenvironments or that chemokines with similar function are expressed at different times during an inflammatory reaction. Although the exact contribution of individual chemokines varies according to the particular model used, it is clear that chemokines function in a tightly controlled fashion, with particular chemokines operating at key stages of the response. Again it seems that the immune system has developed so that molecules with seemingly overlapping functions are secreted in an organized fashion, to ensure that leukocytes are recruited to exert the maximal functional response.

There are various aspects of chemokine receptor biology that influence the conclusions drawn from individual studies *in vitro* and *in vivo*. The reported selectivity of chemokine receptor expression is appealing in terms of anti-inflammatory therapy, since it is potentially possible to target selected leukocyte populations. However,

## Box 5.2 Viral Mimicry of Chemokines.

Viruses have evolved sophisticated mechanisms to evade detection and destruction by the immune system of their host. One of these evasion strategies adopted by large DNA viruses is to encode homologues of molecules that have a critical role in the control of the immune response. A prime example of this is the subversion of the chemokine system. Some viruses modulate the chemokine system by producing their own versions of either chemokines (vCKs) or chemokine receptors (vCKRs) or secreting chemokine-binding proteins (vCKBPs). Viral chemokine homologues function as agonists or antagonists that enable dissemination and growth of the virus. Kaposi's sarcoma virus encodes several chemokine homologues with antagonist activity against a range of CC and CXC chemokines. In contrast, the function of viral chemokine receptors is less clear. Since expression is confined to infected cells they might influence the physiological state of the cell or affect the ability of the infected cell to respond to chemotactic stimuli. An example of a chemokine receptor homologue is US28, a protein encoded by cytomegalovirus. A noteworthy feature of most viral chemokines or chemokine-binding proteins is their broad chemokine or receptor-binding capabilities, which suggests that viruses need to circumvent chemokine redundancy for effective immune subversion. The third class of viral chemokine inhibitors comprises the vCKBPs which neutralize chemokines in solution, but have no sequence similarity to chemokine receptors. These vCKBPs might interfere with binding of host chemokine to its receptor or neutralize chemokine function directly. Several of these viral chemokine inhibitors are currently being investigated for use as novel anti-inflammatory agents *in vivo*.

Some viruses use chemokine receptors to infect target cells productively, a classic example of this being the use of certain chemokine receptors by HIV. CCR5 was identified as a co-receptor, with CD4, for HIV entry into T cells. It was known that infection of T cells by HIV required interactions with two proteins on the host's cell surface. In the first interaction the viral protein gp120 undergoes a high-affinity interaction with CD4, inducing a conformational change that allows interaction with CCR5. Genetic studies have determined that around 1 per cent of the Caucasian population carry a 32-basepair deletion in the gene encoding CCR5. This results in a truncated gene that produces no functional CCR5 on the cell surface. Interestingly, this mutation was discovered by HIV researchers among a group of individuals termed 'exposed uninfected' and the $\Delta$32-CCR5 mutation seems to confer some degree of resistance to HIV infection. Since CCR5 is a cofactor for HIV entry the absence of a functional CCR5 receptor means that the interaction between the HIV protein gp120 which results in the exposure of the fusion peptide gp41 and fusion with the host cell membrane cannot occur. HIV transmission is therefore prevented in individuals homozygous for the $\Delta$32-CCR5 mutation. Heterozygous individuals belong to another clinical grouping termed the long-term non-progressors. Evidence from other studies has outlined a role for CCR5 in allograft acceptance, but so far there is no clear evidence that lack of functional CCR5 is of benefit in protecting individuals from diseases such as arthritis or MS, where CCR5 is thought to play a role.

Certain strains of HIV also use CXCR4, along with CCR5, to gain entry into T cells and macrophages. In a similar way, respiratory syncytial virus (RSV) uses CX3CR1 as a route of cellular infection. Poxviruses are also thought to require chemokine receptors, such as CCR1, CCR5 or CXCR4, for productive infection. This stimulates signal transduction pathways that the virus particle uses to gain entry into a target cell.

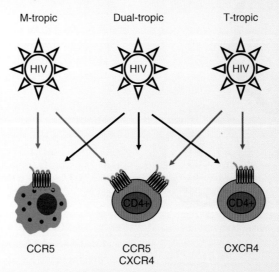

HIV utilizes chemokine receptors CCR5 and/or CXCR4 for cell entry. Different HIV strains have been identified based on their tropisms for cells expressing particular chemokine receptors. M-tropic HIV strains infect cells expressing CCR5, T-tropic strains infect CXCR4 expressing cells, while dual-tropic virus can infect both.

our current knowledge of chemokine receptor expression is based largely on studies of isolated blood leukocytes and, in the case of T cells, on artificially polarized T cell lines. There is now evidence from both *in vitro* and *in vivo* studies to show that chemokine receptor expression may be modulated temporally or by the local tissue environment. Certainly, chemokine expression on T cells can be modulated after activation, as well as during differentiation. Another example is the differential expression of chemokine receptors on airway eosinophils (from patients with eosinophilic lung disease). Airway eosinophils have lower CCR3 expression and higher CXCR4 expression compared to blood eosinophils. CCR3 is the receptor for CCL11 (otherwise known as eotaxin), which is a potent chemoattractant for eosinophils, while CXCR3 recognizes CXCL9, CXCL10 and CXCL11. It is likely that once eosinophils have been recruited from the circulation by CCL11, they upregulate CXCR3, which then acts to retain eosinophils within the airways and to promote effector functions via multiple ligand binding (Figure 5.14). However, it is likely that CCL11 also contributes to the migration of eosinophils into the airways as well as migration across the endothelia. These studies imply that the cell surface expression of chemokine receptors is a dynamic process.

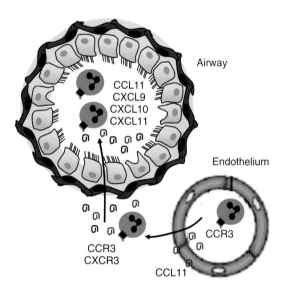

**Figure 5.14** Eosinophils are recruited from the circulation by CCL11 ligation with CCR3, which also significantly contributes to eosinophil migration into the airways. Once recruited to the airways eosinophils upregulate CXCR3 and are retained by CXCL9, CXCL10 and CXCL11.

## 5.14 Chemokines in disease

Temporal associations between chemokines and accumulation of leukocytes have been established in multiple diseases (Table 5.2). Tissue biopsies from patients have led to correlation between chemokine RNA/protein induction or upregulation and the arrival in the same tissue of particular leukocytes. This has led to the formulation of hypotheses regarding the involvement of chemokines with the recruitment of these leukocytes in the development and pathogenesis of clinical disease. For example, synovial joint fluid from arthritis patients contains the chemokines CCL2, CCL5, CCL8, CCL3 and CXCL10, which act as chemoattractants for cells associated with disease pathogenesis, such as monocytes, T cells and neutrophils. In addition, CCL2 expression at the site of atherosclerotic plaques coincides with the recruitment of monocytes. More recently, neurologists determined that chemokines are actively secreted during a number of neuropathologies, for example during multiple sclerosis (MS). CCL3 has been found in the cerebrospinal fluid of relapsing MS patients and members of the monocytes chemoattractant protein (MCP) family MCP-1 and MCP-3 (CCL2 and CCL7) are expressed in active MS lesions from autopsied brains. These associations have perhaps been easier to formulate for chemokines that bind to a single receptor that is predominantly expressed on a restricted cell population. For example, CCL11 (eotaxin) is a potent eosinophil chemoattractant that acts via CCR3. In asthma, a disease characterized by an influx of eosinophils to the airways, CCL11 expression correlates with the accumulation of CCR3-expressing eosinophils. The situation becomes more complicated with the expression of chemokines such as CCL5, which attract a number of different cell types through interactions with CCR1, CCR3 and CCR5. The complexity of chemokine receptor expression during disease states is amplified when considering that most diseases are associated with multiple leukocyte populations. Even in the case of allergic disease, which is characterized by a prominent eosinophilic infiltrate, the contribution of other cells such as Th2 cells, mast cells and basophils must be considered. Indeed, the number of chemokines expressed during allergic asthma is extensive and includes CCL2, CCL3, CCL5, CCL7, CCL8, CCL11, CCL13, CCL15, CCL17, CCL22, CCL24, CCL26 and CXCL8. However, the neutralization of a particular chemokine seems to reduce the accumulation of the target leukocyte population, with an associated abrogation of

**Table 5.2** Chemokine and chemokine receptors involved in the pathogenesis of human diseases and tumour rejection.

| Disease | Chemokine Ligands | Chemokine receptors |
| --- | --- | --- |
| Rheumatoid arthritis | CCL2, CCL5, CCL8, CCL3, CXCL10 | CC2, CCR5, CXCR2, CXCR5 |
| Asthma | CCL2, CCL3, CCL5, CCL7, CCL8, CCL11, CCL13, CCL15, CCL17, CCL22, CCL24, CCL26 and CXCL8 | CCR3, CCR4 |
| Multiple sclerosis | CCL2, CCL3, CCL7 | CCR2, CCR3, CXCR3 |
| HIV | (Viral tropism) | CCR5, CXCR4 |
| Psoriasis | CXCL8, CXCL10, CCL20 | CXCR1, CXCR2, CXCR3, CCR6 |
| Cancer (involved in tumour rejection) | CCL1, CCL2, CCL5, CCL21, CXCL10, XCL1, XCL2 | CCR2, CCR5, CCR7, CCR8, CXCR3, XCR1 |

disease and improvement in clinical symptoms. This suggests that each chemokine does indeed have a particular physiological function in complex diseases.

## 5.15 Chemokines as new anti-inflammatory drugs

Investigators are in the early stages of developing therapeutics based on chemokine research. There are several inherent difficulties in designing an effective chemokine therapeutic. These difficulties arise from the complexity of disease states and the multitude of chemokines that are involved in pathogenesis. In addition, the overlapping functions of chemokines and the redundancy contained within the system may impede the design of therapeutics directed at individual chemokines or chemokine receptors. However, it has been demonstrated that in most cases chemokine expression is differentially regulated and depends on the specific cellular source and the timing of expression during an inflammatory reaction. Understanding the temporal and spatial coordination of chemokine expression is therefore necessary for the effective design for a therapeutic and only then should it be possible to target regulatory or effector cell recruitment during disease.

The chemokine system offers multiple targets for novel therapeutic strategies, and several methods of antagonizing chemokine receptors are being investigated. These approaches include small molecule antagonists, modified chemokines, neutralizing antibodies and viral antagonists. The most direct method of interrupting chemokine function is to interfere with chemokine binding to its cognate receptor. The use of small molecule inhibitors

of chemokine receptors and modified chemokines both utilize this method. Small molecule inhibitors bind directly to the chemokine receptor and interfere with chemokine binding, thereby preventing an intracellular signalling cascade. Modified chemokines are engineered to allow them to retain binding specificity to a receptor while at the same time block intracellular signalling and therefore receptor function. An example of this strategy is a modified version of CCL5 (RANTES), whereby an additional methionine residue added to the N terminus effectively blocks CCR5 signalling. This modified chemokine has been shown to decrease airway inflammation following allergen challenge in mice.

The generation of specific monoclonal antibodies against chemokines or their receptors represents another strategy to modify chemokine function. A range of *in vivo* studies, using mouse models of allergic disease, have demonstrated the benefits of blocking chemokines. One example is the neutralization of CCL11 (eotaxin), which reduces airway inflammation by decreasing the influx of Th2 cells and eosinophils. However, the range of chemokine functions, particularly those in lymphocyte homeostasis, suggests that targeting receptors may be a more effective therapeutic prospect. For example, blocking CCR4 reduces allergic inflammation in the airway by preventing CCL17 and CCL20 recruitment of Th2 cells. Another potential source of chemokine antagonists are virally-derived chemokine homologues, such as vMIP-II, which is encoded by Kaposi's sarcoma herpes virus HHV8. This viral chemokine antagonizes many of the Th1 associated receptors such as CCR1, CCR2 and CCR5, but stimulates Th2 associated receptors such as CCR3 or CCR8. This may provide a useful strategy for downregulating Th1-induced immunopathology.

### 5.16 Summary

**1.** Chemokines play a central role in attracting cells, a process known as chemotaxis, which is vital in orchestrating the correct migration patterns of immune cells.
**2.** Cells are attracted up a chemokine gradient to the site where the chemokine is most concentrated.
**3.** The chemokine network exhibits redundancy, whereby different chemokines have similar functions.
**4.** Chemokines are considered to be pleiotropic, whereby a particular chemokine may have many functions depending on the target cell.
**5.** Many chemokine receptors exists, a number of which are able to bind several different chemokines.

## With contributions from Prof. Clare Lloyd

Head of Leukocyte Biology Section, National Heart and Lung Institute, Imperial College, London

# 6 Basic Concepts in Mucosal Immunology

## 6.1 Introduction

The practice of teaching immunology, and as a consequence being a student of immunology, has until now focused on the central immune system of the blood, spleen and lymph nodes. This is largely due to the way in which immunology is studied within the laboratory and clinic. Blood samples are much easier to acquire from human patients than internal tissues such as intestine or lung. As a result, much of the knowledge gained about how the immune system works has been based on the central immune system. The downside to this is that the majority of pathogens do not gain direct access to the central immune system; rather pathogens infect their host via mucosal sites. As a result, our mucosal immune tissues have evolved sophisticated defence mechanisms to combat such infections. An intricate network of cells and signalling molecules therefore exists throughout the digestive, respiratory and urogenital tracts, as well as other body surfaces such as the skin. A much larger immune network is found throughout the mucosal immune system than exists within the central immune system. In basic terms, the function of the mucosal immune system is to defend the body against infection at the site of pathogen entry.

The majority of cells of the immune system behave differently depending on their location within the body. The mucosal surfaces are a poignant example of how resident immunity differs from immune responses that are generated in central sites. The mucosal surfaces of the body are directly exposed to the environment and as such they encounter a vast diversity of potential antigens. The gastrointestinal tract encounters many 'foreign' antigens on a daily basis. However, to mount a vigorous immune response to food or drink items, or even to commensal bacteria, would be inappropriate. Similarly the process of respiration draws foreign antigens into the lungs, while only unfortunate individuals mount an immune response to allergens such as pollen or house dust mite faeces. The mucosal surfaces therefore face a dilemma: how to be tolerant to commensal organisms and innocuous foreign particles, whilst retaining the ability to respond to more pathogenic microorganisms. The mucosal immune system therefore has to balance its responses depending on whether the antigen is pathogenic or is non-pathogenic (Figure 6.1). A strong inflammatory immune response needs to be activated against a potentially life-threatening microbe but on the other hand inflammation needs to be tightly regulated in the absence of danger. This balance aims to maintain tissue homeostasis so that the mucosal tissue is able to perform its normal physiological functions.

The importance of the mucosal immune system has been accepted for hundreds of years. The healing and protective properties of saliva are well documented and have been used in traditional medicine for millennia. You just have to observe animals licking their wounds to realize the evolutionary significance of bodily secretions. It has long been known that feeding infants with breast milk heightens immunity to infections, through the uptake of immune components into the digestive tract. Infants acquire passive immunity from breast milk, as it contains antibodies and other antimicrobial compounds. Breastfeeding is also thought to have a beneficial impact on the intestinal microbiota, it improves immunity to intestinal infections, it is thought to reduce inflammatory diseases and it reduces the likelihood of developing allergies. There is also a link between a healthy gut

*Immunology: Mucosal and Body Surface Defences*, First Edition. Andrew E. Williams.
© 2012 John Wiley & Sons, Ltd. Published 2012 by John Wiley & Sons, Ltd.

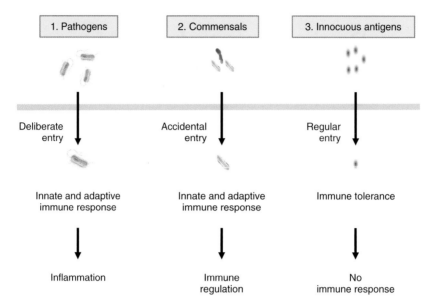

**Figure 6.1** The mucosal immune system must mount contrasting responses at the same time. How do you mount an immune response to pathogenic microorganisms that are only transiently present (1) whilst coping with accidental entry or overgrowth of non-pathogenic commensal bacteria (2) and ignoring the innocuous antigens that traverse the epithelial layer every day (3)? A balance must therefore exist that allows effective pathogen clearance but also maintains homeostasis in healthy tissue.

microflora and the development of a normal mucosal immune system. Experimental animals that lack a normal commensal microflora fail to develop organized lymphoid structures throughout the intestine and succumb to infection far more readily. The mucosal immune system of the digestive tract is also responsible for the induction and maintenance of oral tolerance to food items, while the mucosal immune system of the respiratory tract is accountable for tolerance to airborne allergens. Therefore, it can clearly be seen how important the mucosal immune system is to health and for the prevention of disease.

## 6.2 What is a mucosal tissue?

There is much debate about what constitutes a mucosal tissue. By definition, any tissue that is able to secrete mucus across an epithelial cell layer and contribute to immune defence is considered to be mucosal-associated lymphoid tissue (MALT). Generally, any area of the body exposed to the environment has to behave differently when foreign particles are present. It is useful to think of this in terms of openings to a cavity or passage in the body. The mouth and the anus are two such openings connected by a tube, the digestive tract, where the entire length is theoretically exposed to the environment. The

respiratory tract and the urogenital tract are two other examples where internal tissues are exposed to the outside (Figure 6.2). Examples of mucosal tissues are listed in Table 6.1.

The principal sites of MALT are the digestive tract, the respiratory tract and the urogenital tract (Figure 6.2) and these will therefore provide the focus of discussion in later chapters. Other sites of MALT include the conjunctiva of the eye and the exocrine glands such as the salivary and sweat glands. Organized structures such as the appendix, which is associated with the digestive tract, and the palantine tonsils, associated with the respiratory tract, may also be considered as MALT. Both obtain antigens from their respective luminal tracts, possess epithelial surfaces capable of secreting mucus and provide sites for the formation of primary and secondary immune responses. Each member of the MALT will therefore be discussed in their appropriate chapters.

By far the most controversial member of the MALT is the skin. By definition, the skin does not secrete mucus and therefore cannot be considered a *bone fide* mucosal tissue. However, the skin does provide an extremely important interface between the body and external environment. It is composed of multiple layers of differentiated epithelial cells, known as keratinocytes, and it possesses a significant immune system. Therefore, the skin performs a similar

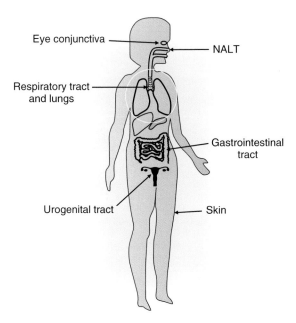

Eye conjunctiva

NALT

Respiratory tract and lungs

Gastrointestinal tract

Urogenital tract

Skin

**Figure 6.2** Mucosa associated lymphoid tissue. Mucosa associated lymphoid tissue (MALT) represents the largest surface area and contains the highest number of immune cells in the body. This includes the gastrointestinal, respiratory and urogenital tracts, the conjunctiva of the eye and the nasal associated lymphoid tissue (NALT). A MALT is defined by the secretion of mucous across an epithelial layer and the ability to contribute to immune defence. Although the skin is not a mucosal tissue, it is characterized by an epithelial layer and does provide significant immune defence at a body surface.

**Table 6.1** Mucosa associated lymphoid tissues possess innate physical, mechanical and chemical barriers.

| Site | Mechanism |
| --- | --- |
| Respiratory tract | Mucus |
| | Nasal secretions |
| | Muco-cillary escalator |
| | Sneezing |
| | Cough reflex |
| Gastrointestinal tract | Acidic pH (stomach) |
| | Peristalsis |
| | Vomiting |
| | Commensal flora |
| | Antimicrobial peptides |
| | Glycocalyx |
| | Digestive enzymes |
| | Mucous coat |
| Urogenital tract | Acidic pH |
| | Mucus secretion |
| | Antimicrobial peptides |
| Eye conjunctiva | Alkalinity of tears |
| | Eyelids and lashes |
| | Cilia |
| Skin/ducts of exocrine glands | Sweat |
| | Antimicrobial peptides |
| | RNAse enzymes |

function as MALT and is a vital component of body surface immune defence. For these reasons the skin will be discussed in depth in a later chapter.

## 6.3 Immune defence at mucosal tissue is multi-layered

Mucosal immune defence can be viewed as several layers of protection and be divided into physical or mechanical defence, chemical defence, humoral defence and cellular defence. Physical or mechanical defence mechanisms represent non-specific barrier to infection. These include the epithelial lining of mucosal tissues, the impermeable barrier of the skin, the peristaltic activity of the intestines and the mucociliary clearance mechanism within the airways. One of the features of mucosal epithelia is the presence of a layer of mucus, known as the mucous coat. This mucous layer sits on top of the epithelia on the luminal side and is composed primarily of

glycoprotein's known as mucins. These mucins interact to form a sticky mucous layer that traps antigens and prevents microbes from attaching to the epithelial cell surface. Secretions in other parts of the body also provide protection from infection, such as the saliva, tears, sweat, vaginal secretions and nasal secretions (Table 6.1). These non-specific barriers aim to exclude and expel pathogenic microorganisms from entering the body.

Other non-specific defence mechanisms include the acidic environment of the stomach and digestive tract (and to a lesser extent the urogenital tract) and the presence of large quantities of proteolytic enzymes. Collectively these defence mechanisms are known as chemical defence. These properties provide an unsuitable environment for pathogens to thrive. Other molecules, such as antimicrobial peptides, complement factors and acute phase proteins are also found in mucosal secretions and contribute significantly to host defence. Antibodies also play a significant role in host protection at epithelial surfaces. In particular, IgA is predominantly expressed by B cells and is secreted into the mucosal lumen. IgA binds to and agglutinates antigen, thereby facilitating its clearance. It is also capable of neutralizing pathogens and prevents any interaction with host cells. These soluble factors are referred to as humoral immune components.

If a pathogen is successful enough to overcome physical and chemical barriers then it is faced with a multitude of cellular defence mechanisms. Cells of the innate and adaptive immune system participate in effective mucosal immune responses. Although organized lymphoid structures are found throughout the intestines and the upper airway, most immune responses in mucosal tissues are induced in non-lymphoid sites. Firstly, innate immune cells, such as tissue resident macrophages and migratory neutrophils, respond to the initial threat of infection. Secondly, adaptive immune cells are activated by DCs in regional lymph nodes or organized mucosal lymphoid structures and are recruited to sites of inflammation. B cells respond by synthesising antibody, especially IgA, while T cells respond by secreting pro-inflammatory cytokines or instigating cytotoxic effector functions. Mucosal tissues harbour several subpopulations of T cells, including $\alpha\beta$ T cells and $\gamma\delta$ T cells. These T cells can reside in both the epithelial compartment and in sub-epithelial compartments, where they initiate and regulate effector immune responses. Therefore, mucosal immune defence has many layers, each contributing to a robust mucosal immune system.

## 6.4 Origins of mucosal associated lymphoid tissue

The immune system has been studied throughout history in a variety of animal species such as mice, rats, pigs and horses. However, it was studies performed on chickens that defined the modern immune system. For instance, the designations for T cells (thymus derived) and B cells (bursa-derived; later to refer to bone marrow-derived in mammalian systems) both originated following extensive analysis of chicken immune cells. One way in which to define what constitutes MALT, is to compare the anatomy and cellular composition to that of the bursa of Fabricius in birds (Figure 6.3). The bursa is a primary lymphoid organ associated with the terminal gut and is the site of B lymphocyte development. It is thought that the entire repertoire of chicken B cells exists at the time of hatch and therefore is independent of any contact with environmental antigens. This differs to mammalian B cell development, which seems to rely on contact with antigen. Upon hatching, however, the bursal follicles mature and the B cell repertoire can be sculpted depending on the antigenic stimulation. Chickens are

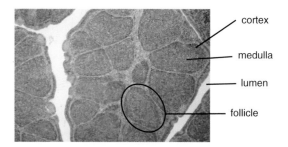

**Figure 6.3** MALT in the chicken. Avian species possess a specialized lymphoid organ known as the bursa of Fabricius. Studies in chickens revealed that the bursa is the principle site of B cell development. The bursa is the archetypal mucosal immune tissue. It is able to sample antigens directly from the bursal lumen. Each bursal follicle consists of a cortex and medulla, in a similar organization to mammalian lymph nodes, where B cell differentiation and maturation occurs.

also devoid of lymph nodes and rely entirely on the bursa to generate their antibody repertoire. Therefore the bursa has become the historical benchmark for the identification of MALT in other species.

The role of the bursa in B cell differentiation and maturation was discovered following the surgical removal of the bursa from chickens. The bursectomized birds were unable to generate immunoglobulin producing cells, demonstrating the necessity of the bursa for the generation of B cell responses. The bursa forms from an invagination of the cloaca and possesses a duct that retains contact with the luminal contents of the gut. Importantly, the bursa is able to sample antigens from the gut lumen. Like mucosal sites in other animals, an epithelial layer forms a boundary between the lumen and the inner tissues. The epithelial layer directly above lymphoid follicles is known as the follicle-associated epithelium (FAE) and at regular intervals along this cell layer there are specialized epithelial cells called microfold or M cells. The M cells are responsible for transporting antigen from the gut lumen and into lymphoid areas within the follicle, a process that is vital for the development of mature B cells. The importance of M cells for the development of gut-associated lymphoid tissue (GALT) and organized lymphoid follicles associated with other mucosal tissues will be discussed in subsequent chapters.

The bursa contains thousands of lymphoid follicles, each consisting of an outer cortex and inner medulla separated by a cortico-medullary boundary (Figure 6.3). Highly proliferative, immature B cells reside in the cortex, which is equivalent to the dark zone in mammalian lymphoid follicles. Migration into the medulla results in interactions with antigen-presenting DCs. The combination of contact with antigen and cell signalling events leads to B cell maturation, the production of high affinity immunoglobulin and the shaping of the mature B cell repertoire. The mechanism of somatic gene conversion, by which chicken antibodies gain their specificity for antigen and their enormous diversity, was also discovered in the bursa. This process differs to that in rodents and humans (detailed in Chapter 3), although the resultant generation of antibody diversity has been achieved in all vertebrate species through the co-evolution of different, but equally effective, mechanisms. Antigen primed B cells are then able to leave the bursa and enter the circulation as antibody producing plasma cells, in much the same was as plasma cells leave mammalian lymphoid follicles.

## 6.5 Concept of the common mucosal immune system

The concept of the common mucosal immune system states that immune cells that have been primed at one mucosal site are able to migrate and exert their effector functions at another mucosal site. In some way this suggests that all mucosal tissues are immunologically connected. This concept has led to the development of mucosal vaccines, whereby administration of a vaccine at one mucosal site would be able to afford protection at all the other mucosal sites. For example, a vaccine administered orally and therefore delivered to the gastrointestinal tract, in theory, should protect the recipient against a respiratory virus. A prime example of this concept is the effectiveness of the oral polio vaccine. The concept of the common mucosal immune system has also revealed the molecular mechanisms responsible for specific cellular migration patterns. In order for the

common mucosal immune system to operate as one, the mutual expression of several chemotactic and homing receptors is necessary. Therefore leukocytes express on their cell surface specific chemokine receptors and cell adhesion molecules, while specific mucosal tissues express corresponding chemokines and cell adhesion molecules. For example, endothelial cells express on their surface homing molecules that attract leukocytes out of the circulation and into mucosal tissues (discussed in Chapter 5). In effect, the expression of certain molecules on the surface of leukocytes ensures that they migrate into MALT and not other peripheral tissues or central lymphoid organs (Table 6.2). This is discussed further in this chapter with regard to B cell and T cell homing into MALT.

In reality though, different MALTs express a slightly different set of homing molecules, meaning that leukocytes primed in the gut preferentially migrate back to the gut. Although there is some cross-over in homing receptor expression, it is thought that immune cells are restricted to similar MALT and only into tissues that express the same homing molecules (Figure 6.4). This is referred to as compartmentalization of immunity and involves not just homing molecules but also distinctive cytokine, chemokine and co-stimulatory molecule expression patterns. For example, T cells primed in the intestine would normally migrate back only to the intestine but not into the lungs or urogenital tract (Figure 6.4). These patterns of cell traffic are particularly important when considering memory responses, as it is imperative for memory lymphocytes to return to the same anatomical location as the source of infection. The different homing receptors associated with each mucosal tissue is shown in Table 6.2. It is thought that leukocytes primed in one particular mucosal tissue are imprinted with a specific

**Table 6.2** The distribution of cell adhesion molecules and chemokine/chemokine receptors differs between various mucosal tissues and peripheral lymph nodes.

| Tissue type | Adhesion molecules-tissue | Adhesion molecules-leukocyte | Chemokines-tissue | Chemokine receptors-leukocyte |
|---|---|---|---|---|
| Intestines | MAdCAM-1 | $\alpha 4\beta 7$ | CCL25 | CCR9 |
| Lungs | VCAM-1 | $\alpha 4\beta 1$ | CXCL10 | CXCR3 |
| | ICAM-1 | LFA-1 | CCL5 | CCR5 |
| Urogenital tract | VCAM-1 | $\alpha 4\beta 1$ | CCL5 | CCR5 |
| | | | CCL7 | CCR2 |
| Skin | E-selectin | CLA | CCL17 | CCR4 |
| | VCAM-1 | LFA-1 | CCL27 | CCR10 |
| Lymph nodes | ICAM-1 | LFA-1 | CXCL9 | CXCR3 |
| | VCAM-1 | $\alpha 4\beta 1$ | CCL5 | CCR5 |

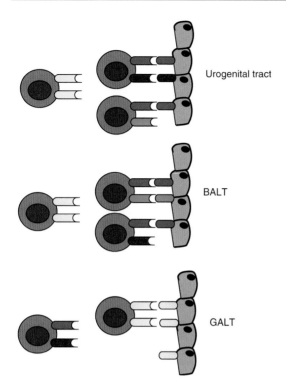

**Figure 6.4** Immunological compartmentalization in mucosal tissue. The paradigm of the common mucosal immune system suggests that all mucosal tissues are in someway linked immunologically. In reality, the migration of lymphocytes into certain MALTs is rather restricted and is defined by the specific-expression of homing receptors. For example, the urogenital and respiratory tracts express both distinct and shared homing molecules, so that lymphocytes expressing the corresponding receptors can migrate into both tissues. However, these same lymphocytes are unable to migrate into the gastrointestinal tract because this tissue expresses different homing receptors.

pattern of homing receptors, so that upon reactivation they are instructed to migrate back into the same tissue. However, cellular traffic patterns are equally important for lymphocytes involved in both antibody (e.g. B cell) and cell-mediated (e.g. T cell) immunity. The distinct requirements for leukocyte migration and the discrete compartmentalization of different mucosal tissues has even led some mucosal immunologists to abandon the concept of the common mucosal immune system.

## 6.6  How do T and B lymphocytes migrate into mucosal tissues?

Current data suggests that naïve T cells become activated in organized lymphoid tissue such as lymph nodes or the Peyer's patches (the inductive sites) and then migrate through non-lymphoid tissue where they perform their effector function (within the effector site). The specifics of inductive and effector sites are discussed later in this chapter. This non-lymphoid tissue includes the lamina propria and the intra-epithelial mucosal compartments. The requirement for cell activation by antigen within organized lymphoid tissues can be demonstrated experimentally (Figure 6.5). Transgenic mice have been bred that express a single T cell receptor on all CD4+ or CD8+ T cells. For example, the OTI transgenic mouse has CD8+ T cells that only recognize a small ovalbumin peptide. T cells that specifically express TCRs for the OTI peptide can be tracked following their transfer to another mouse. If OTI T cells from an OTI transgenic mouse are isolated and then transferred into a normal mouse we find that few of them make it into the mucosal tissues. If, on the other hand, these transgenic T cells are transferred with their specific antigen (the ovalbumin peptide) then suddenly there are large numbers of OTI T cells in mucosal tissues. This elegant experiment shows that antigen activated T cells can migrate into mucosal tissues, whereas naïve T cells tend not to.

How do immune cells know how to traffic into mucosal tissues? This is where homing receptors come into play and in particular the selective role of integrins during the process of mucosal compartmentalization. Integrins mediate binding or tethering of cells to endothelial cells present on the blood vessel wall that express the complimentary receptor. Cells that traffic into mucosal sites tend to express β7 integrins. The α4β7 integrin is common on T and B lymphocytes in the lamina propria of the gastrointestinal tract, whereas αEβ7 predominates on intraepithelial lymphocytes. The expression of integrins by antigen activated T cells may be determined by the antigen presenting cell that activates it. Dendritic cells in the gut for example have been shown to up-regulate the expression of α4β7.

T cells and B cells can therefore express different integrins. To migrate selectively into different mucosal compartments they must bind selective complimentary molecules. These complimentary molecules are called addressins. For example, α4β7 binds to mucosal addressin cell adhesion molecule 1 (MAdCAM1), which is expressed on the blood vessel walls in the gut. T and B cells expressing α4β7 will bind MAdCAM1 and migrate into the lamina propria of the gut (Figure 6.6). The importance of β7 integrins in mucosal immunity can be observed in β7 knockout mice in which lamina propria and intraepithelial lymphocytes are reduced. In addition,

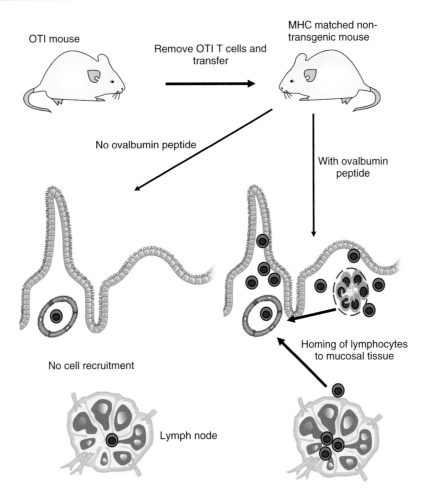

**Figure 6.5** If OTI transgenic T cells are transferred into another mouse without antigen (in this case an ovalbumin peptide) then they are unable to migrate into mucosa tissues. If these mice are given the antigen however, some T cells will be activated in the Peyer's patches or lymph nodes. These will then express the correct integrins to migrate into mucosal tissues, thus demonstrating the homing specificity of mucosal T cells.

αEβ7 is expressed on intraepithelial T cells and binds to E-cadherin that is induced on epithelial cells. E-cadherin may therefore retain αEβ7-expressing lymphocytes at this site. Skin homing T cells that express cutaneous lymphocyte antigen (CLA) are also thought to bind to E-cadherin on the surface of endothelial cells (Table 6.2).

Another example of restricted homing to mucosal sites is illustrated by plasma cells secreting IgG. Regardless of where they are activated, IgG producing B cells do not tend to populate mucosal tissues. These discerning homing properties of plasma cells are controlled by the expression of combinations of integrins and chemokine receptors, and may be the result of reducing IgG-dependent complement-mediated lysis in the mucosa

(Figure 6.7). For example, IgA secreting plasma cells express the β7 integrin α4β7 and the chemokine receptors CCR9 and CCR10, which enables them to preferentially migrate into the mucosa of the gastrointestinal tract. On the other hand, IgG secreting plasma cells express α4β1 and the chemokine receptor CXCR3, which restricts these cells to the systemic circulation.

## 6.7 Special features of mucosal epithelium

The majority of mucosal tissues are separated from the external environment by a layer of epithelial cells

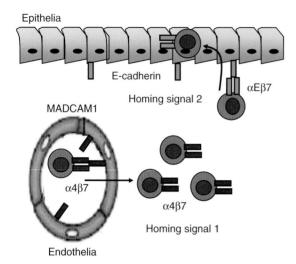

Epithelia

E-cadherin

Homing signal 2

MADCAM1

αEβ7

α4β7

α4β7

Homing signal 1

Endothelia

**Figure 6.6** Homing of lymphocytes into mucosal tissues involves the expression of specific cell adhesion molecules. Furthermore, migration across the endothelia often involves different cell adhesion molecules (1) compared to migration into the epithelium (2).

(Box 6.1). Mucosal epithelial cells form a impermeable barrier between the external environment and the tissue beneath. The epithelial cell layer prevents the unwanted movement of molecules and microorganisms across its surface. The intestine and lower airways are lined by only a single layer of epithelial cells, while the upper airways and urogenital tracts are lined with multi-layered stratified epithelia. The skin is rather different, as a thick layer of differentiated keratinocytes provides the epithelial barrier. The epithelium can be regarded as the first line of immune defence against soluble antigens since the tight junctions between epithelial cells prevents diffusion into the sub-mucosal tissues. The epithelial layer therefore functions as an effective barrier against the contents of the mucosal lumen. In addition, a mucous layer covers most mucosal surfaces (Figure 6.8), which is formed by mucins and various lipoproteins that bind the apical surface of epithelial cells.

The composition of mucin glycoproteins (of which 19 genes have been discovered in humans) differs slightly from tissue to tissue, although they perform similar functions. The production of mucins is increased following physical, chemical or microbial stress and forms a semi-permeable barrier that protects against infection and assists repair to damaged epithelium. Mucins have also been shown to bind several proteins on the surface of bacteria. In this way they trap microbes in the mucous layer,

which are then removed by the peristaltic action of the gut or by the mucociliary action of the respiratory tract. To combat these defence mechanisms several bacterial species express mucinases and colonization, adherence and invasion factors. The ability to penetrate this viscous mucous layer often distinguishes harmless commensal organisms from pathogenic varieties. The mucous layer also contains additional factors responsible for repair and digestion of antigens.

Regardless of whether microbes are harmless or harmful their density needs to be kept in check. After all, it would be difficult for oxygen to traverse from the airways into the bloodstream in the lung, and nutrients from the intestinal lumen into the circulation, even if harmless microorganisms occupied the whole surface area. However, the microbial content of mucosal tissues differs widely, with the gut containing by far the highest number of bacterial species (Table 6.3). This regulation is mediated by a variety of anti-microbial peptides that are produced by the epithelium itself. An important group of such anti-microbial peptides are the defensins. There are two main groups of defensins, α and β. In the intestine Paneth cells that are situated at the base of the intestinal villous crypts produce these peptides. In humans the two main α-defensins are human defensins 5 and 6 (HD5 and HD6), which can be found along the length of the gastrointestinal tract. Of the β-defensins BD1 is always present whereas BD2 in present in low levels but increased during times of stress or inflammation.

Epithelial cells, in addition to their barrier function, can participate in immune responses. All nucleated cells express MHC class I and can therefore present antigenic peptides to CD8+ T cells or alert natural killer cells to the presence of an intracellular pathogen. Gut epithelial cells also express the MHC class I related CD1d molecule, which is a ligand for intraepithelial lymphocytes (discussed further in Chapter 7). Interestingly when CD1d is crosslinked on epithelial cell lines they rapidly produce the immune suppressive cytokine IL-10. Therefore, CD1d expression on epithelial cells helps to maintain tissue homeostasis at mucosal sites, so that unnecessary immune reactions are prevented. However, the epithelium when under stress, such as during infection, can upregulate the expression of MHC class II. Similarly, an increase in the expression of adhesion molecules such as ICAM-1 enables them to physically interact with immune cells. They are also reported to produce a variety of cytokines and chemokines that propagate or recruit a variety of immune cells. However, expression of the T cell co-stimulatory molecules CD80 and CD86 is

## Box 6.1. Different Types of Mucosal Epithelia.

Epithelia can be categorized into two broad types depending on how many epithelial cell layers it possesses. Simple epithelia is composed of only a single cell layer, while stratified epithelia is composed of two or more layers. Common to both is the presence of a basement membrane, which is formed of connective tissue that separates the epithelial layer from the tissue parenchyma below. The mucosa of the intestinal tract and the small airways is lined with simple epithelia, while the larger airways, urogenital tract, oesophagus, anus and skin are lined with stratified epithelia. The structure of each epithelial layer reflects the physiological function of the mucosal tissue in question. For example, the gut and lower airways are only a single cell layer thick so that effective absorption of nutrients and exchange a gases can take place, while stratified epithelia tends to provide the tissue with a more rigid structure, as occurs in the oesophagus and vagina.

Simple epithelia and stratified epithelia can be further divided based on the shape of the epithelial cells, further reflecting an element of functional specialization (see table and figure accompanying this box). Simple epithelia can be formed of squamous, cuboidal or columnar epithelial cells, so named as to best describe their shape. Stratified epithelia can be divided into squamous, cuboidal or transitional and can either be keratinized (as in skin) or not. Transitional stratified epithelia can have the appearance of squamous or cuboidal epithelia depending on whether the epithelial layer is contracted or distended, respectively, as occurs in the urinary bladder. Pseudostratified columnar epithelia are also common and are classed separately for histological reasons. Even though they form a single cell layer above the basement membrane, pseudostratified columnar epithelia often resemble stratified epithelia in cross-section due to their overlapping arrangement. Certain specialized epithelial cells also exist and are distributed throughout the conventional epithelial layers. For example, goblet cells in the airways and Paneth cells in the intestine can be classified as glandular epithelial cells but are superficially columnar in structure. M cells found throughout simple epithelial layers are another example of highly specialized epithelial cells.

| Epithelial cell type | Functional example |
|---|---|
| Simple squamous epithelia | Thin, flat and tightly packed. Often known as pavement epithelia. Flattened nuclei. Suited to diffusion of soluble substances. e.g. alveolar epithelium, blood vessel endothelia. |
| Simple cuboidal epithelia | Square in cross section, cuboidal in shape. Located in secretive and absorptive tissues. e.g. exocrine glands, kidney tubules. |
| Simple columnar epithelia | Elongated and column shaped. Have many specializations. Form the majority of intestinal and airway epithelia. e.g. enterocytes, pneumocytes, mucus secreting cells (goblet cells). |
| Pseudostratified columnar epithelia | Elongated and column shaped. Overlapping appearance. Predominant in airway and nasal passages. e.g. ciliated bronchial cell. |
| Stratified squamous epithelia | Can be keratinized as in skin or non-keratinized as in the oral cavity. Often stratified epithelium is anatomically located towards the surface, while cuboidal cells are located closer to the basement membrane. e.g. oesophagus, oropharynx, vagina, cornea and skin (keratinized) |
| Stratified cuboidal epithelia | e.g. sweat glands |
| Stratified transitional epithelia | Has propensity to stretch. Appears squamous when tissue is distended or stretched and cuboidal when the tissue is contracted. e.g. urinary bladder. |

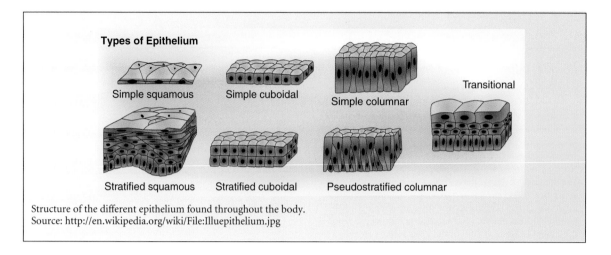

**Types of Epithelium**

Simple squamous    Simple cuboidal    Simple columnar    Transitional

Stratified squamous    Stratified cuboidal    Pseudostratified columnar

Structure of the different epithelium found throughout the body.
Source: http://en.wikipedia.org/wiki/File:Illuepithelium.jpg

uncommon. Therefore, epithelial cells maintain homeostasis in mucosal tissues but at the same time are capable of responding to infectious pathogens when required.

## 6.8 Toll-like receptors and NOD proteins in the mucosa

In Chapter 2 we introduced the Toll-like receptor (TLR) family, which are innate pattern recognition receptors (PRRs) that bind pathogen associated molecular patterns (PAMPs). Considering the microbial load in mucosal tissues it is easy to appreciate the importance of TLRs at these sites. The other family of proteins that recognize PAMPs, the nucleotide-binding oligomerization domain (NOD) proteins, are also present. Intracellular NOD1 recognizes muramyl tripeptides from Gram-negative bacteria whereas NOD2 recognizes muramyl did peptides common to all bacteria. Binding of both TLRs and NOD proteins triggers an inflammatory cascade in epithelial cells, resulting in the production of cytokines and chemokines, in particular CXCL8 (IL-8), that in turn recruit neutrophils and causes the upregulation of antimicrobial peptides. Mutations in TLRs or NOD proteins may explain certain types of intestinal diseases. For example, a loss of NOD2 function has recently been described in patients with Crohn's disease.

TLRs and NOD proteins are not uniformly expressed on all epithelial cells. In the small intestine for example, TLR4 expression is concentrated in the base of the crypts. Microorganisms are usually trapped in the mucous lining above the crypts. Only when the density of microorganisms pushes them deep into the crypts is an inflammatory

cascade initiated. Similarly, NOD2 is highly expressed in Paneth cells, which produce α-defensins. Again, this demonstrates the requirement for tolerance in mucosal sites. If TLRs and NOD proteins were expressed by all epithelial cells, we would mount an inflammatory cascade all the time. By limiting TLR and NOD expression to sites where bacteria should not venture (e.g. the base of the crypts or within intracellular compartments) ensures that infectious episodes are dealt with only when necessary and in a manner that does not require T and B cells (Figure 6.9). The concentration of TLRs is also higher on the basolateral surface of epithelial cells, rather than on the apical surface. In addition, TLRs are restricted to intracellular compartments. Therefore, inflammatory responses are only initiated if pathogenic microorganisms cross the epithelial barrier or infect cells. This expression pattern exists in order to maintain a state of tolerance to commensal bacteria in the lumen, while retaining the facility to respond to infection.

The immunoregulatory nature of mucosal tissue is further exemplified by resident macrophages. Macrophages in the lamina propria of the gastrointestinal tract are reported to lack CD14, which as you will remember from Chapter 2 is part of the receptor complex that binds lipopolysaccharide (LPS). This ensures that macrophages do not continuously respond to harmless bacteria in the gut. The immunomodulatory capacity of mucosal epithelia is also exemplified by the lung. The lung has a much lower burden of microbes than the intestine and as a consequence alveolar epithelial cells express lower levels of TLRs than their intestinal counterparts (Table 6.3). The low burden of microbes in the lung

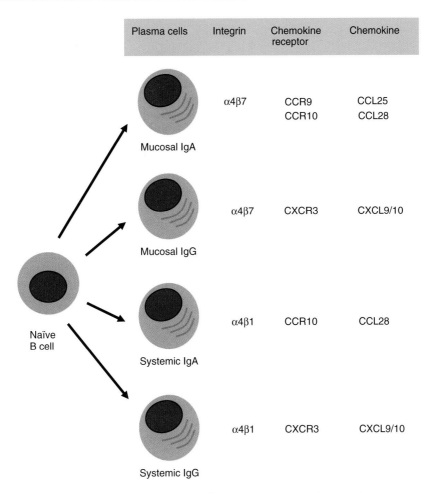

| Plasma cells | Integrin | Chemokine receptor | Chemokine |
|---|---|---|---|
| Mucosal IgA | α4β7 | CCR9 CCR10 | CCL25 CCL28 |
| Mucosal IgG | α4β7 | CXCR3 | CXCL9/10 |
| Systemic IgA | α4β1 | CCR10 | CCL28 |
| Systemic IgG | α4β1 | CXCR3 | CXCL9/10 |

**Figure 6.7** The co-ordinated expression of integrins and chemokine receptors by plasma B cells and the selective production of chemokines by cells at distinct sites orchestrates the movement of IgA secreting B cells into mucosal sites.

means that any increase in microbe numbers, as a result of bacterial infection for example, requires a lower threshold of detection than it does in the intestine. This mechanism ensures that excessive immune responses are properly controlled in order to prevent damaging inflammatory responses. Only when danger signals exceed a certain threshold are inflammatory responses initiated in response to harmful microorganisms.

## 6.9 Antigen sampling at mucosal surfaces

Due to the different organization and structure of the many mucosal epithelial barriers throughout the body, the mechanism by which antigen is sampled differs widely. Both DCs and specialized epithelial cells, called M cells, are able to sample antigen from the external environment (Box 6.2). In the intestine and lower airways, where the epithelium is only a single cell layer across, M cells perform the vast majority of antigen sampling. Tight junctions between the epithelial cells tend to prevent most cells crossing into the lumen. M cells are most frequently found in the epithelial layer above organized lymphoid tissue, such as the lymphoid follicles of Peyer's patches in the gut, or in the tonsils associated with the respiratory tract. M cells utilize a mechanism known as transepithelial transport, whereby antigen is moved from the mucosal lumen to the basal membrane. This antigen can then be further processed by resident DCs or it can be

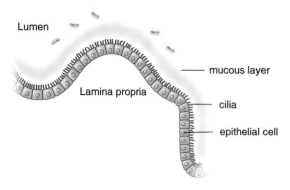

**Figure 6.8** A thick mucus layer protects the underlying epithelial surface. This mucous layers comprises mucin proteins and other lipoproteins. Its function is to trap microbes and present invasion of the epithelial layers. The mucous layers is continually being cleared from mucosal tissues.

sampled directly by lymphocytes. The basal membrane of M cells forms an invagination in which close interactions between the M cell and T cells or B cells can take place. Most lymphocytes found in association with M cells are of a memory phenotype and therefore antigen presentation by M cells may provide suitable antigen stimulation for the induction of a rapid acquired immune response. M cells seem to be most frequent in GALT and therefore it is within the digestive tract that they have mostly been studied (discussed further in Chapter 7).

The mechanism of transepithelial transport in M cells is thought to be hijacked by certain microorganisms (Table 6.4) and the digestive tract in particular is full of numerous bacterial species. Fortunately, the presence of lymphocytes, macrophages and DCs directly beneath the M cell keeps bacterial load in check. However, certain pathogenic microbes are able to exploit this antigen transport pathway and invade host tissue. Certain bacteria and viruses selectively adhere to the apical surface of M cells,

aided by the lack of a brush border, the expression of gly-colipids, which act as pathogen receptors, and the inherent phagocytic activity. The bacterial species that have been demonstrated to adhere to M cells include *Vibrio cholerae*, enteropathogenic *Escherichia coli* and *Salmonella typhi*. These bacteria tend to utilize adherence and/or invasion strategies, such as specialized pili, fimbriae or lytic enzymes, which are expressed from pathogenicity islands in the bacterial genome and act as virulence factors. Viruses are also able to gain entry to their host cells via the transepithelial transport pathway, although they are unable to manipulate M cells in the same way as bacteria. The mouse reovirus is probably the best studied, which invades M cells in the intestine and airways. It has been assumed that human viruses, such as polio virus and HIV, also exploit M cells for delivery to their target cells.

In stratified epithelium, which is many cell layers thick, DCs migrate into the epithelial layer and sample luminal contents directly, via their dendrite-like protuberances. This method of antigen sampling by DCs is thought to occur in certain areas of the urogenital tract and in the outer layers of the skin. Indeed, the skin possesses a specialized subset of DCs, known as Langerhan's cells, which form a dendritic network throughout the entire epidermis, which allows the immune system to continuous survey the skin for harmful substances. Following antigen sampling, DCs migrate into the sub-mucosal areas of MALT or to organized lymphoid structures in the mucosa or to regional lymph nodes, where they present antigen to lymphocytes.

## 6.10 Mucosal dendritic cells

Several sub-populations of DC exist within the blood, spleen, lymph nodes and throughout the various mucosal tissues. Different DC populations even exist between

**Table 6.3** Mucosal epithelia acts as a barrier to the external environment. In contrast to the blood, which is a sterile system, the airways, gut and the skin are in constant contact with microbes and numerous antigens. However, different MALTs experience differing amounts of microbial stimulation and therefore have to react in appropriate ways.

| Tissue | Microbial contact | Response |
|--------|-------------------|----------|
| Blood | None | Highly sensitive All microbial contact induces response |
| Airways/skin | Frequent | Sensitivity regulated Only high microbial load induces response |
| Gut | Permanent | Suppressed sensitivity Mostly induces tolerance Danger signals tightly controlled |

## Box 6.2. M cells and transepithelial transport of antigen.

M cells are specialized epithelial cells that occur frequently in the mucosal epithelium of the gastrointestinal and respiratory tracts. They are distributed along simple epithelium where only one cell layer forms the mucosal surface but are largely absent from stratified epithelium such as the trachea or urogenital tract. They are most abundant in the epithelium directly above organized lymphoid follicles, known as follicle associated epithelium, such as the Peyer's patches in the gut or the lymphoid follicles of the bronchioles. They are also present in the epithelium of secondary lymphoid tissues such as the tonsils and adenoids. Importantly, M cells increase in number following an inflammatory insult, for example in response to infection. Their principal function is to sample antigen from the mucosal lumen and transport it across the epithelial barrier for presentation to lymphocytes and APCs in the lamina propria. The active translocation of antigen is referred to as transepithelial transport and remains a specialized feature of epithelial M cells.

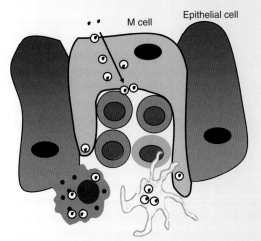

Transepithelial transport of antigen across microfold M cells. Antigen is transported from the mucosal lumen to the basolateral side of the epithelium where is can be samples by professional APCs.

M cells differ from their neighbouring enterocytes (gut) and pneumocytes (airways) in size, shape and function. Firstly they do not possess the microvilli (otherwise known as the brush border) that is so important for absorption by enterocytes and nor do they have the cilia used by airway epithelial cells to clear mucus from the lungs. Instead the apical surface of M cells is coated in glycosylated lipids and proteins that enable them to interact with substances in the lumen. The apical membrane is also adapted for phagocytosis and is characterized by a ruffled appearance under a

microscope at high magnification. M cells are capable of phagocytosing substances by clathrin-dependent endocytosis and by fluid-phase endocytosis (pinocytosis) and can even phagocytose large particles such as whole bacteria (macropinocytosis). Phagocytosis results in the uptake of material into acidic endosomes and the rapid translocation of endosomes to the baso-lateral membrane. M cells are shaped in such a way that the baso-lateral surface forms a concave pocket into which the contents of the endosomes are released through the process of exocytosis. The unique shape of M cells also means that there is only a short distance between the apical and basolateral membranes ($\sim 1$–$2\,\mu m$), which is beneficial for vesicular transport. The released antigen can then be taken up by DCs and macrophages present in the vicinity of the M cell. Although it is unclear whether M cells express MHC class II or have the potential to present antigen directly to T cells, lymphocytes are found in close association with the pocket membrane. In particular, CD4+ T cells expressing CD45RO, and therefore those with a memory phenotype, interact with the M cell pocket membrane; although a significant number of B cells have also been found to interact with the M cell pocket. This interaction may represent a mechanism for rapid T and B cell activation in response to invading microorganisms.

mucosal sites, reflecting functional diversity and tissue location. These distinct populations of cells are largely defined by the expression of cell surface molecules, which have been extensively studied in the mouse; although humans have DC subsets with similar functions and localizations. Mucosal DCs are derived from either common myeloid progenitors or common lymphoid progenitors and often do not fully differentiate until they are recruited into mucosal tissues.

It is often the case that mucosal DCs are efficient at capturing antigen but are poor activators of T cells. A process of restricted antigen presentation is extremely important for the maintenance of mucosal sites. Rather than constantly responding to the vast number of antigens present, mucosal sites have evolved mechanisms that monitor luminal contents without mounting a widespread, aggressive immune response. This highlights the balance necessary for maintaining mucosal homeostasis; potential threats need to be identified and eradicated without unnecessarily responding to inappropriate antigens. The monitoring process is so restricted and localized that the clinical symptoms typical of systemic immune responses are not apparent. Even the colon, with its vast microbial load, does not display a heavy immune cell infiltrate.

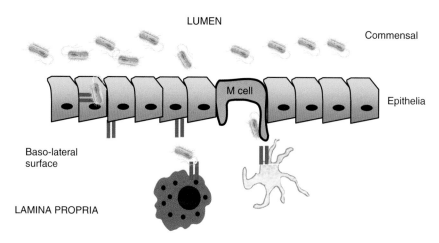

LUMEN

Commensal

M cell

Epithelia

Baso-lateral
surface

LAMINA PROPRIA

**Figure 6.9** Toll-like receptor expression in mucosal tissues is restricted to intracellular compartments and the basolateral epithelial cell surfaces. Immune cells within the lamina propria also express TLRs. The maintenance of the epithelial barrier is therefore important in order to avoid inflammation. Immune responses are only generated if this barrier is breached.

**Table 6.4** Pathogens that gain entry through intestinal M cells.

| |
| --- |
| *Vibrio cholerae* |
| *Escherichia coli* |
| *Salmonella typhi* |
| *Salmonella typhimurium* |
| *Shigella flexneri* |
| *Yersinia enterocolitica* |
| *Yersinia pseudotuberculosis* |
| *Campylobacter jejuni* |
| Roevirus |
| Polio virus |

Due to the diverse nature of the DC population and the overriding requirement for homeostasis, an element of specialization exists within the MALT. For example, some DC subsets secrete anti-inflammatory cytokines such as IL-10 and TGF-β, while other subsets secrete pro-inflammatory cytokines such as IL-12 and TNF in response to PAMP recognition. This specialization can also define the type of immune response generated, as IL-12 promotes Th1 development while a lack of IL-12 promotes Th2 development. However, mucosal DCs tend not to induce Th1 differentiation but are rather more tolerogenic, enhancing Th2, Th3 or Treg cell development. Mucosal DCs augment IgA class-switching and synthesis in B cells, again through the secretion of IL-10, TGF-β and IL-6. Mucosal DCs also function by imprinting specific homing properties on T cells and B

cells, for example gut homing lymphocytes express CCR9 and α4β7 integrin.

## 6.11 Secretory dimeric IgA at mucosal surfaces

The dominant antibody isotype secreted in mucosal tissues is IgA. This differs to systemic or peripheral antibody production, which is dominated by IgG (Figure 6.10). IgA that is secreted across the epithelium is in a dimeric form. In other words, two IgA molecules are linked together by a third molecule called the joining chain (J-chain), so that the two constant regions of each immunoglobulin are attached. The question that readily springs to mind is why IgA predominates at mucosal surfaces and why is IgA secreted in dimeric form? Dimeric IgA is secreted in vast quantities into the lumen

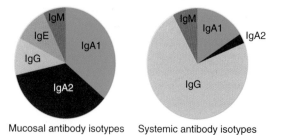

Mucosal antibody isotypes    Systemic antibody isotypes

**Figure 6.10** The dominant antibody isotypes in the mucosa are different to those in the periphery. In the mucosa IgA isotypes are prominent whereas in the periphery IgG isotypes dominate.

of most mucosal tissues. For example, approximately 3 g of dimeric IgA is transported across the gut epithelium every day. The presence of dimeric IgA is an extremely important mechanism for eliminating potential antigens in a non-inflammatory manner. In effect, dimeric IgA has four antigen binding sites (two sites on each immunoglobulin molecule) and is therefore able to cross link several antigens, resulting in antigen agglutination in the lumen. This makes the cross-linked antigen too large to traverse the epithelium and prevents access to the sub-mucosal tissues by passive diffusion. This process is termed immune exclusion and provides a first line of defence that protects mucosal epithelial surfaces.

The transport of IgA into the mucosal lumen is facilitated by the J-chain that binds the two IgA molecules together. The J-chain binds with very high affinity to a molecule known as secretory component, which is referred to as the polymeric Ig receptor (pIgR) when bound to the cell membrane on the basolateral surface of mucosal epithelial cells (i.e. the side of the epithelial cell closest to the sub-mucosal tissues or interior of the body). Once bound, the whole complex containing dimeric IgA, joining chain and secretory component is endocytosed into a vesicle and transported to the luminal membrane of the epithelial cell (Figure 6.11). Having reached the apical surface of the epithelial cell the extracellular domain of secretory component is cleaved off and secreted together with bound IgA. Secretory component is consistently trafficked in this way through the epithelial cell. As such, unloaded secretory component is also released. IgA that is transported to the mucosal lumen is often called secretory IgA (S-IgA)

Once in the lumen S-IgA is free to agglutinate antigen. Agglutinated antigen is either too big to diffuse into the body or bound antibody blocks molecules that would normally have bound to receptors on the surface of epithelial cells (pathogen binding molecules, for example). The importance of dimeric IgA can be demonstrated in patients who are deficient in this antibody isotype. Such patients suffer from repeated respiratory and digestive tract infections. This is often difficult to diagnose since IgM, though not in such abundance as IgA in mucosal tissues, can also function in a similar manner. In humans, pIgM is also transported by the same mechanism and with a similar efficiency as pIgA. However, IgA is much more effective at providing immune exclusion at mucosal surfaces and tends to have a higher affinity for antigen than IgM.

Dimeric IgA and pentameric IgM can also bind antigen that has traversed the epithelium and is present

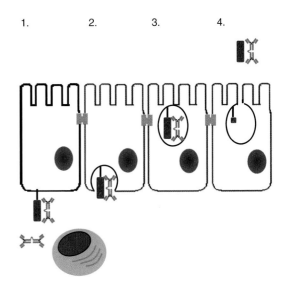

**Figure 6.11** IgA is released from plasma B cells complexed by joining (J) chain (1). IgA binds the poly Ig receptor (secretory component) on the basolateral surface of the epithelial cell and is endocytosed (2). Endocytosed IgA traverses to the luminal aspect of the epithelial cell (3) and is released with the extracellular domain of the poly Ig receptor into the lumen (4).

in sub-mucosal areas. Such antigen complexes are then transported back out into the lumen using the same mechanism of IgA transport through J-chain binding to secretory component. Dimeric IgA can even bind antigen on its journey through the epithelial cell when its endocytic vesicle merges with one present within the same cell. These three routes represent a powerful, non-inflammatory route for eliminating antigen without damaging the surrounding mucosa. pIgA and pIgM are also beneficial at mucosal surfaces because they are more stable in the harsh environment of the lumen and more resistant to protease degradation compared to IgG or IgE molecules. The presence of secretory component seems to protect secreted IgA and IgM from proteolytic degradation and prevents their clearance by forming interactions with mucus. In this way, secretory component is responsible for the localization and concentration of IgA in the mucus. Another advantage of IgA and IgM is that they do not bind complement as avidly as IgG and are therefore less inflammatory. In addition, free secretory component itself has anti-microbial properties and is able to limit infection by enteric bacteria such as *Escherichia coli*.

The migration of IgA from the basolateral to the apical surface of epithelial cells can remove pathogenic microorganisms, although it is possible that some pathogens may gain entry by binding to IgA translocated back across the epithelium. For example, reduced colonization of *Streptococcus pneumoniae* has been observed in mice lacking secretory component. However, entry of pathogens via this route is currently thought to be unlikely since apical to basolateral transfer is not efficient and IgA is cleaved from the epithelial surface. In addition, IgA lacking secretory component, or free uncomplexed secretory component itself, is likely to mop up any pathogens that bind IgA complexes. Rather, pathogenic bacteria are more likely to rely on virulence factors in order to gain entry into sub-epithelial tissue.

## 6.12 Regulation of J-chain and secretory component expression

Another question regarding the specificities of mucosal antibody production is what regulates the expression of the J-chain, which is essential for dimeric IgA and pentameric IgM formation? J-chain mRNA is expressed in early B and T cell development and switched off in T cells as they mature. J-chain transcription occurs in the presence of IL-2, which induces chromatin remodelling and allows the interaction of transcription factors with the J-chain promoter. As such, J-chain is present in most B cell lineages regardless of antibody isotype. However it is only maintained in those producing IgA and IgM. Detailed knowledge of J-chain transcriptional regulation is still missing.

The final piece in the puzzle involves the expression of the pIgR present on the basolateral surface of mucosal epithelial cells. Why is it expressed on epithelial cells in mucosal tissues? Why is there higher expression of the pIgR in the small and large intestine compared to the salivary, lacrimal and lactating mammary glands or lung? The answers to these questions are largely unknown. It may be that the higher bacterial burden in the intestine facilitates the need for a higher rate of IgA transport into the lumen of the gut compared to other mucosal sites. The requirement for immune exclusion and the control of both commensal and pathogenic microbes may be greater in the gastrointestinal tract. Constitutive expression on intestinal epithelial cells is observed 20 weeks post gestation in man but not until weaning in rodents. In man, greater then 95 per cent of IgA found in intestinal secretions is derived directly from the mucosa, whereas in rabbits and rodents it is derived from hepato-billiary transfer of circulating antibody. Important species differences therefore exist, which need to be taken into account when testing potential mucosal vaccines in animal models. Though constitutively expressed, pIgR (secretory component) can be up-regulated by a variety of stimuli including infection, cytokines such as IL-4, TNF or IFN-γ and hormones. It is therefore likely that IgA secretion into the lumen is enhanced during times of stress.

## 6.13 How does the sub-mucosa differ from the epithelium?

A vast array of leukocytes exists beneath the epithelial layer and above the muscularis mucosa in a region known as the lamina propria. In addition to smooth muscle cells, lymphatics, fibroblasts and blood vessels, the lamina propria contains T cells (predominantly CD4+ helper T cells), loosely scattered B cells, aggregates or single isolated B cell follicles and myeloid cells such as macrophages and DCs. Organized lymphoid tissue is not always present in mucosal tissues; rather it is induced when required. The stomach, for example, contains very few lymphocytes in a healthy individual whereas in those with *Helicobacter pylori*-induced gastritis B and T lymphocytes are abundant. This is also true of the healthy lung where lymphocytes are sparse but antigen presenting cells such as dendritic cells and macrophages are abundant. It is the responsibility of these antigen presenting cells and epithelial cells to recruit T and B lymphocytes should they be required, although excessive immune responses can lead to immunopathology and tissue damage (Figure 6.12). The sub-mucosa of lymphoid tissues provides an important environment in which adaptive immune responses can develop. Although, the initiation of an immune response takes place in lymph nodes or organized lymphoid structures within MALT, effector immune responses often take place within the sub-mucosa. The exception is the gastrointestinal tract, which contains organized lymphoid structures called Peyer's patches (PPs) situated along the small and large intestines, which are structurally related to lymph nodes.

Neutrophils also play a role in reducing the density of microbes at mucosal surfaces. Neutrophils are the most abundant leukocyte within circulating blood and are rapidly mobilized from the bone marrow in response to an infection. Inflammatory signals, such as IL-6 and G-CSF, contribute to the release of neutrophils from the bone marrow. Neutrophil mobilization also involves the downregulation of the bone marrow retention molecule CXCR4 (a chemokine receptor), which normally keeps neutrophils within the bone marrow in response to the

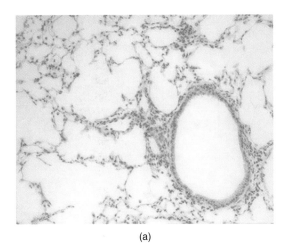

(a)

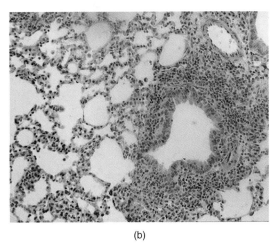

(b)

**Figure 6.12** A healthy lung contains very few lymphocytes (a). Non-immune barriers are present to prevent infection. When leukocytes are recruited to the lung the immune response is often not regulated and causes immunopathology and tissue damage (b). Sections are stained with heamatoxylin and eosin and viewed under × 200 magnification.

chemokine CXCL12 (SDF-1) produced by stromal cells. Once released from the bone marrow they are recruited to sites of inflammation or stress by the chemokine CXCL8 (IL-8) that is produced by activated epithelial cells. Neutrophils tend to locate to the basal side of epithelial cells (i.e. on the side furthest away from the lumen). Their primary function is to phagocytose and kill invading pathogens and to secrete antimicrobial peptides and inflammatory mediators. Neutrophils are also capable of migrating into the lumen in order to phagocytose and kill microorganisms, although excessive neutrophil

accumulation can result in unwanted tissue damage, for example during sepsis or acute lung injury.

## 6.14 Organized lymphoid tissue of the mucosa

In addition to the diffuse lymphocyte population that occurs throughout the epithelium and lamina propria of mucosal tissues, areas of organized lymphoid tissue are located in the digestive and respiratory tracts. These lymphoid follicles act as inductor sites that are responsible for inducing antigen-specific acquired immune responses, and include the PPs in the gut and tonsils associated with the respiratory tract. Unlike the digestive and the respiratory tracts, the urogenital tract does not possess such organized lymphoid structures (neither does the skin) but rather relies on local lymph nodes to act as inductor sites. The structure of organized lymphoid follicles throughout the MALT is similar to that described for lymph nodes and each inductor site will be discussed in subsequent tissue-specific chapters.

These organized lymphoid follicles represent sites in which T and B lymphocytes are primed by antigen presenting cells. The chronology of a mucosal immune response involves both an inductive site and an effector tissue (Figure 6.13). In general, mucosal tissue is the site where pathogen invasion takes place, for example the epithelia of the intestine where bacteria are invading or the epithelia of the lower airways where viruses are replicating. DCs recognize danger signals and process pathogen-specific antigens within mucosal tissues. These APCs then migrate to local inductive sites such as the PPs, tonsils and regional lymph nodes. Within lymphoid follicles DCs present antigen and provide co-stimulatory signals to lymphocytes. The result of B cell activation is the development of a B cell follicle, resulting in germinal centre formation, and the differentiation into antibody secreting plasma cells (Figure 6.14). Following activation, T and B cells migrate out of inductive sites into the circulation and traffic to their specific effector sites within the MALT. Within the effector tissue T cells and B cells will acquire further stimulatory signals that will allow then to perform their effector functions. In summary, inductive sites are where antigens derived from mucosal tissues stimulate and activate naïve T cells and B cells, while effector sites are where antigen-specific T and B cells perform their effector functions.

For example, CD8+ T cells will recognize antigen presented on the surface of epithelial cells in the context of

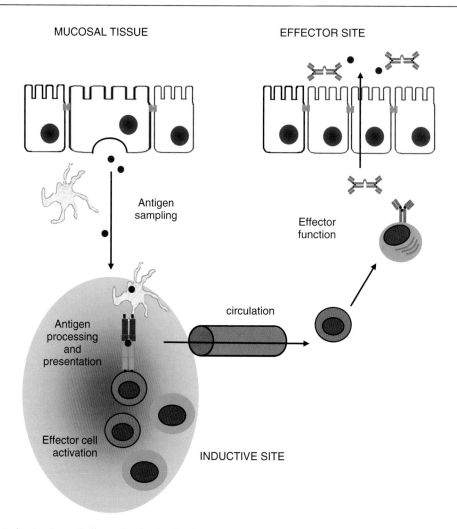

**Figure 6.13** Inductive sites and effector sites involved in the generation of a mucosal immune response. Antigen presentation and stimulation of naïve T cells nad B cells occurs within inductive sites. Antigen-specific T cells and B cell perform their biological functions within effector sites (e.g. the original mucosal tissue from where antigen was sampled).

peptide/MHC class I complexes, together with necessary co-stimulatory signals such as CD28 ligation and local IFN-γ production. These signals will fully activate the effector functions of CD8+ T cells so that they can perform their cytolytic duties. Likewise, B cells that migrate into effector tissues will receive the necessary stimulatory signals, IL-4 and TGF-β for example, to begin secreting large quantities of IgA. The anatomy of a mucosal immune response is such that lymphocyte activation and proliferation is restricted to organized lymphoid follicles, while effector functions are carried out at delicate effector sites in a controlled way.

## 6.15 Cytokines in the mucosa

The cytokine profile in mucosal tissues reflects both the requirement for vast antibody production (especially of the IgA isotype) and the requirement for tolerance to, or ignorance of, antigens by mucosal T cells. It therefore makes sense that the immunosuppressive cytokines IL-10 and TGF-β are predominantly expressed. Due to the high IgA antibody production, the B cell differentiating factors IL-4 and IL-6 are also significantly expressed, especially during episodes of inflammation (Figure 6.15). In

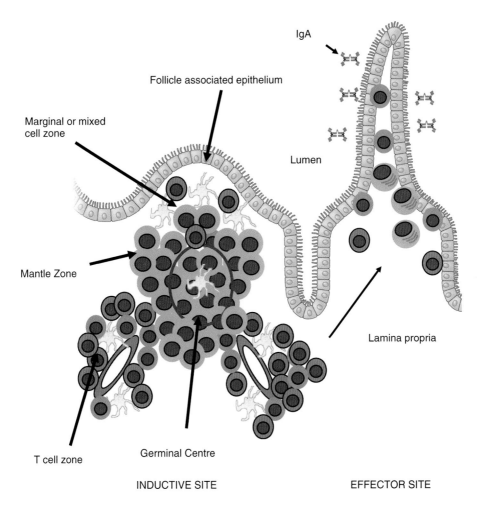

IgA

Follicle associated epithelium

Marginal or mixed
cell zone

Lumen

Mantle Zone

Lamina propria

T cell zone

Germinal Centre

INDUCTIVE SITE                    EFFECTOR SITE

**Figure 6.14** Organized lymphoid tissue of the MALT, including Peyer's patches in the gut and the tonsils in the upper respiratory tract, can act as immune inductive sites for T and B cell responses. Activated lymphocytes, including CD4+ and CD8+ T cells, as well as IgA secreting plasma cells, migrate into effector sites such as the lamina propria of the intestines.

contrast, some conventional Th1 cytokines also promote non-inflammatory immune mechanisms in the mucosa. For example, TNF is known to increase transcription of the pIgR that traffics non-inflammatory IgA through the epithelial cell and into the mucosal lumen.

IFN-$\gamma$ and TNF are not constitutively expressed in mucosal tissues, nor are they expressed in large quantities following stimulation, since these highly inflammatory cytokines are likely to induce significant amounts of immunopathology. Although IFN-$\gamma$ and TNF are central to the development of Th1-type immune responses, overexpression of these cytokines could cause that activation of cytolytic cells (macrophages, NK cells, CTLs) and

damage the delicate structures of the mucosa. Indeed, excessive cytokine production is associated with the immunopathology characteristic of serious inflammatory reactions; for example following influenza virus infection or severe acute respiratory syndrome (SARS). In addition, IFN-$\gamma$ causes antibody switching to IgG isotypes that are efficient at fixing complement, which also provides pro-inflammatory signals to leukocytes. In addition, complement has the potential of lysing cells by the formation of the membrane attack complex (Chapter 2). Therefore, the cytokine environment in mucosal tissues tends to be skewed more toward a Th2 phenotype rather than a Th1 phenotype. However, it should be noted that

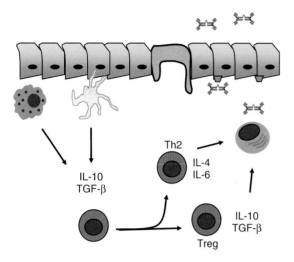

**Figure 6.15** Immunomodulatory environment in mucosal tissues is advantageous for the production of IgA and refractory to Th1-driven immunopathology. This is important for maintaining tissue homeostasis and barrier integrity.

**Table 6.5** Diseases prevalent at various mucosal sites.

| Th1 | Th2 |
| --- | --- |
| Gut | Gut |
| Inflammatory bowel disease | Food allergy |
| Crohn's disease | |
| Coeliac disease | |
| Ulcerative colitis | |
| Lung | Lung |
| COPD | Asthma |
| Bronchiolitis | Allergic rhinitis |
| Pneumonitis | |
| Skin | Skin |
| Psoriasis | Atopic dermatitis |
| | Urticaria (hives) |

even excessive Th2-mediated immune responses can lead to excessive inflammation, as is the case during allergic reactions such as during severe asthma exacerbations. Table 6.5 details common inflammatory diseases of the mucosa associated with Th1 or Th2 cytokine production. It is essential that a balance in cytokine production is achieved in the mucosa so that homeostasis and normal physiological function is maintained.

Regulatory T cells are another important cell type vital for maintaining integrity of the mucosa. They provide a mechanism for tolerance of non-pathogenic bacteria and dietary antigens present in the lumen of the gut or to potential allergens in the respiratory tract. A failure in regulatory T cells or their production of immune suppressive cytokines leads to uncontrolled inflammation. IL-10 gene depleted animals, for example, spontaneously develop colitis and a lack of tolerance may be involved in the development of inflammatory bowel disease (IBD). Several sub-populations of regulatory T cells exist (introduced in Chapter 3) and include Tr1, Th3, and CD4+CD25+ inducible iTreg cells. They are defined by their phenotype and/or their ability to produce regulatory cytokines such as IL-10 and TGF-β. Factors regulating their induction are, at present, less clear but is likely to be regulated by cytokines, dietary products (e.g. vitamin D) and the neuroendocrine system.

## 6.16 Pathogens that enter via mucosal sites

Most antigens enter the body via the mucosal surfaces. The cell type infected will depend on the pathogen and mucosal site in question. Viruses, for example, predominantly infect epithelial cells at most mucosal sites. Intracellular bacteria also initially target the epithelial layer, but may enter the sub-mucosal tissue via M cells or direct invasion across the epithelium and infect macrophages. Extracellular bacteria and fungi either replicate in the extracellular spaces close to mucosal epithelium as free organisms or specifically attach to it using receptors that recognize human cell surface molecules. The response to infection can be dramatic, such as diarrhoea, coughing or sneezing. Although these reflexes aim to clear the pathogen from its host, they often help to assist spread of the pathogen to other individuals. In addition, the mucosa may be a portal for entry but not necessarily the final target organ. For example, hepatitis C crosses mucosal surfaces but subsequently targets the liver. Examples of pathogens that infect mucosal tissues are shown in Figure 6.16 and will be discussed further in subsequent chapters.

## 6.17 Immune diseases of mucosal tissues

The importance of immune regulation at mucosal sites is no better exemplified than when tissue homeostasis breaks down, giving rise to immunopathology and disease. Both Th1 and Th2-mediated immune responses have been associated with diseases of mucosal tissues (Table 6.5). For example, inflammatory bowel disease and

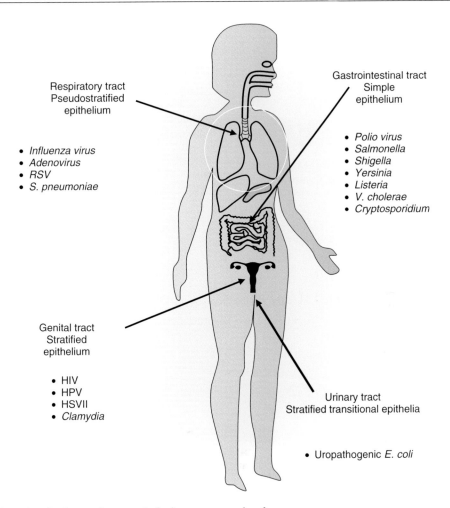

**Figure 6.16** Examples of pathogens that enter the body across mucosal surfaces.

Crohn's disease are both associated with a Th1-mediated immune response, which results in immunopathology and damage to the intestinal tract. In addition, these diseases may be the result of a lack of immunoregulation that allows excessive immune reactions to develop. A disregulation in the immunoregulatory cytokine IL-10 has been linked with disease progression. Other mucosal tissues can also be affected by Th1-dependent immune responses. For example, chronic obstructive pulmonary disease, bronchiolitis and pneumonia can all affect the respiratory tract. These diseases are associated with elevated levels of the pro-inflammatory cytokines IFN-γ and TNF and can result in cell-mediated tissue damage. Sometimes the excessive cytokines produced during an inflammatory reaction are known as a cytokine storm, as is thought

to occur during acute viral infections with influenza or severe acute respiratory syndrome (SARS) virus.

In addition to Th1-mediated immune responses, mucosal immunopathology can be associated with Th2 responses. One of the most studied examples of a Th2-mediated disease is allergic asthma, which affects the airways of susceptible individuals. It seems apparent that the incidence of asthma in Western populations has increased over the past 50 years and it has been suggested that this is due to a lack of immunoregulation (see Chapter 15, Box 15.1: Hygiene hypothesis). Allergic asthma is associated with the Th2 cytokines IL-4, IL-5 and IL-13, which in collaboration induce the production of IgE antibodies and recruit large numbers of eosinophils to the respiratory tract. This immune induction results in the release of histamine and other inflammatory mediators and causes the airway constriction that is so typical of an asthma attack. Other Th2-mediated immune reactions associated with epithelial surfaces include food allergy in the gut and atopic dermatitis of the skin.

## 6.18 Summary

**1.** Mucosal associated lymphoid tissue (MALT) provides host defence against pathogens that invade across mucosal surfaces.

**2.** Mucosal tissues are located at regions of the body that are exposed to the outside, and are characterized by an epithelial layer and the production of mucus.

**3.** Immune defence at mucosal tissues is multi-layered, consisting of physical, chemical, humoral and cellular defence mechanisms.

**4.** The secretion of IgA across epithelial layers is an important component of host defence at mucosal sites.

**5.** Specialized subsets of T cells exist within mucosal compartments, which differentiate MALT from the systemic immune system.

## With contributions from Prof. Tracy Hussell

Leukocyte Biology Section, National Heart and Lung Institute, Imperial College, London

# 7 Immunology of the Gastrointestinal Tract

## 7.1 Structure of the gastrointestinal tract

In a healthy adult human the gastrointestinal tract is approximately eight metres (28 feet) in length and stretches from one opening, the mouth, to the distal opening, the anus and can be divided into the upper and lower gastrointestinal tracts. The upper tract begins at the mouth and includes the entire length of the oesophagus, which is a muscular tube that uses the action of peristalsis to propel food down into the stomach. The oesophagus exits the chest through the diaphragm and connects to the stomach at the cardiac sphincter. The stomach is primarily responsible for the digestion and churning of food items which, through the action of muscles in the stomach wall, are pushed through the pyloric sphincter and into the lower gastrointestinal tract. The lower tract can be further divided into the small and large intestines, both located within the abdominal cavity, which eventually terminate at the anus.

The small intestine comprises the duodenum, jejunum and ileum. The duodenum is the shortest section and participates in further digestion of food. It is where digestive enzymes produced in the pancreas and digestive juices from the liver and gallbladder enter the small intestine. The jejunum links the duodenum to the ileum and is the first section to possess microvilli along its mucosal surface (Figure 7.1). The presence of villi on the lumenal surface of both the jejunum and the ileum greatly increases the surface area available for the absorption of nutrients. The ileum is the final section of the small intestine and can be as much as four metres in length. It is here that the majority of nutrients are absorbed, facilitated by the numerous villi located across its entire mucous surface. The surface area of each villus is further increased by the presence of microvilli, formed from the multiple folding of the apical membrane on intestinal epithelial cells (Figure 7.2).

Although the entire length of the small intestine contains dispersed lymphoid tissue, only the ileum possesses organized lymphoid follicles known as Peyer's patches (PPs). This is the main distinction between the ileum and jejunum. The small intestine ends and the large intestine starts at the ileocecal valve, and is where the cecum is located. Many large mammals, such as the ruminants, have extensive cecal pouches that contain large quantities of bacteria, which aid the digestion of fibrous food. In humans, the cecum is much smaller and is largely replaced by the appendix. The cecum also connects the ileum with the colon, where most of the water, vitamins and vital salts contained in the stool are absorbed. The colon can de divided into the ascending colon, transverse colon, descending colon and sigmoid colon. Lymphoid aggregates can be found throughout the cecum, appendix and colon, which are thought to be inducible rather than organized, as PPs are largely absent from the large intestine. Their presence may reflect the high density of bacteria in these sites. The sigmoid, or s-shaped, colon is very muscular as it needs to push stool into the rectum. The rectum basically provides a storage site for faeces.

## 7.2 Development of the gastrointestinal tract

The gastrointestinal tract starts to form in the early embryo at approximately 2–3 weeks of age from endodermal tissue located on the ventral surface. The gut develops from the invagination of the two distal ends of the embryo, the head (anterior) and tail (posterior). Concave folding of the ventral surface and the fusion of the anterior and posterior invaginations form a tubular structure. The

*Immunology: Mucosal and Body Surface Defences*, First Edition. Andrew E. Williams.
© 2012 John Wiley & Sons, Ltd. Published 2012 by John Wiley & Sons, Ltd.

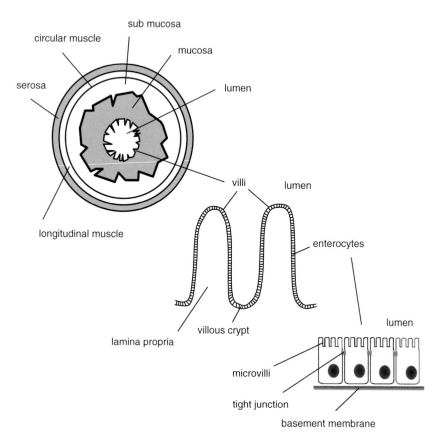

**Figure 7.1** Structure of the intestinal tract. The epithelium of the gastrointestinal tract comprises a single layer of epithelial cells and forms the immediate barrier between the lumen and the lamina propria. The gut epithelium is folded into many villi, while each epithelial cell possess microvilli on its lumenal surface. These villi and microvilli greatly increase the surface area of the gut, which aids absorption of nutrients.

endoderm of the primitive gut will eventually give rise to the epithelial layer of the intestines and airways, while the adjacent mesoderm of the embryo will give rise to the muscle layers of the digestive and respiratory tracts. From the largely uniform tube structure of the primitive gut, the more defined structures of the foregut, midgut and hindgut can be discerned by 4–5 weeks of age. The foregut gives rise to elements of the upper gastrointestinal and respiratory tracts, including the oropharynx, trachea, lungs, oesophagus, stomach and upper duodenum (up to the bile duct). The foregut also gives rise to the adjacent tissues of the liver, gallbladder and pancreas. The midgut gives rise to the lower duodenum, ileum, cecum, appendix and two-thirds of the traverse colon. An important feature of midgut development is the precise looping pattern that the small and large intestine eventually employs, which is the result of rapid elongation during embryogenesis.

Lastly, the hindgut forms the last third of the colon, the rectum and the anal canal.

In latter embryogenesis the endoderm begins to differentiate into the epithelial cells that cover the lumenal surface of the gastrointestinal tract. The specific differentiation patterns result from signalling events derived from the mesoderm that surrounds the endodermal layers of the gut. The pattern of development occurs in four planes or axes; the anterior-posterior, dorsal-ventral, left-right and rotational axis, which gives rise to the different morphological sections of the gut and their functional specialization. For example, differentiation of epithelium in the ileum gives rise to tall and thin villi, while the villi in the colon are flat and wide. Epithelial differentiation in the gut results in the formation of downward pointing crypts, which are separated from the villus by a proliferating inter-villus region. The upward pointing

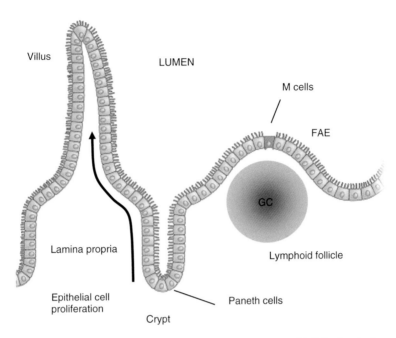

**Figure 7.2**  The villus-crypt unit forms the characteristic surface of the intestines. New epithelial cells arise from precursor cells in the crypt and eventually get sloughed into the lumen at the top of the villous. Located deep within intestinal crypts are Paneth cells, which function by secreting antimicrobial peptides into the gut lumen. Areas of organized lymphoid tissue are located beneath the follicle associated epithelium (FAE), an area rich in M cells.

villus, the inter-villus region and the downward pointing crypt are collectively known as the villus-crypt unit (Figure 7.2). In humans the villus-crypt unit of the small intestine is retained for life but is lost in the colon. This may reflect the nutrient absorptive function of the small intestine, which causes the cells of the villus-crypt unit to remain metabolically active. For these reasons epithelial cells are continuously replaced in the gut and new cells are derived from stem cell populations located in the crypts. Undifferentiated epithelial cells are found at the bottom of crypts, while fully differentiated enterocytes comprise the villus. Migration of cells out of the crypts results in morphological changes and differentiation into enterocytes, goblet cells or Paneth cells, of which the latter actually remain in the crypt. Dead enterocytes are sloughed off from the top of the villus and are continuously being replaced by new cells from the crypts, in a manner similar to a conveyor belt. Enterocytes eventually differentiate into columnar epithelial cells that possess an extensive brush border made out of microvilli, which greatly enhances their absorptive capabilities.

## 7.3 The digestive tract as a mucosal tissue

Foreign antigens are prevented from entering the body from the lumen of mucosal tissues by a variety of mechanisms. Some of these are site specific. For example, peristalsis in the gastrointestinal tract, the mucociliary escalator in the respiratory tract and the acidic environment of the stomach all prevent antigen deposition. Other features, however, are common to all mucosal sites, such as a mucosal epithelium. The immune system of the gastrointestinal tract has been the most widely studied to date, probably because tissue samples are often removed during a number of clinical procedures and the gut is an accessible organ. It is thought that there is more lymphoid tissue along the length of the gut than anywhere else in the body. This is likely to be the consequence of harbouring such a large amount of commensal microorganisms, the vast majority of which are composed of hundreds of different species of bacteria.

There is a close and very important relationship between the commensal microflora and the development of MALT along the length of the gastrointestinal tract. The commensal microflora is thought to be pivotal for proper immune development, as well as being beneficial for the physiological functioning of the gut. The immune response to these symbiotic microorganisms is characterized by tolerance, whereby innate and adaptive responses are tightly regulated so as not to cause immunopathology. The immune system therefore functions by keeping the commensal microflora in check, while at the same time preventing potentially damaging immune reactions to these friendly bacteria. However, a dichotomy exists whereby the gut immune system also has to provide protection from unwanted pathogenic microorganisms. The immune system is therefore continuously discriminating between commensal and pathogenic bacteria. The mechanisms that the immune system employs, in order to maintain homeostasis in the gastrointestinal tract, will be discussed throughout this chapter (also see Box 7.1).

Development of gastrointestinal associated lymphoid tissue (GALT) occurs primarily after birth, which coincides with microbial colonization, although some lymphoid aggregates are observed in the foetus. Within six days of birth organized follicles of B cells can be seen in the intestine, lying between the intestinal villi and beneath the dome epithelium (Figure 7.2). Such follicles are arranged at regular intervals along the duodenum, jejunum and ileum and vary in number between different species; the human and horse containing the most. In humans there are on average 239 follicles at puberty. These B cell follicles are normally isolated from each other in most species, although in others such as dogs and pigs, organized lymphoid tissue forms one long continuous array that may reach up to 2 metres in length in the ileum. The most studied mucosal lymphoid tissue is the Peyer's patch (PP), which in mammals consists of aggregates of several B cell follicles. The area between the B cell follicles (the interfollicular region), known as the lamina propria, is rich in T cells and B cells (and to a lesser extent dendritic cells and macrophages) and specialized blood vessels called high endothelial venules (HEVs). Small numbers of T cells are also present within the B cell follicle, where they provide help during B cell maturation. T and B cells within mucosal tissues have distinct phenotypes depending on where they are located. For example, lymphocytes present throughout the intestinal lamina propria are often differentiated effector cells, such as antibody-producing plasma cells or cytokine secreting Th cells.

## 7.4 Barrier function

A single layer of highly organized epithelial cells provides effective barrier defence against numerous noxious agents that enter the intestinal lumen. The contents of the gut lumen contain a vast array of soluble peptide antigens, macromolecules, electrolytes and microorganisms. Therefore the epithelial monolayer has to provide a very stable barrier in order to prevent any of the luminal contents from entering the body. Intestinal epithelial cells are polarized so that the membrane surface contacting the lumen, known as the apical surface, is different morphologically and functionally to the basolateral membrane that is adjacent to the lamina propria. For example, the apical surface possesses microvilli, otherwise known as the brush border, which significantly increases the surface area of each enterocyte and increases the efficiency of absorption. In contrast, the basolateral membrane is specialized in the transport of certain molecules, such as nutrients, for uptake into the extensive venous system found throughout the gut.

Importantly, neighbouring epithelial cells are linked together by tight junctions that provide the entire mucosal surface with a virtually impermeable barrier. These tight junctions are formed by transmembrane proteins such as occludins and claudins that are embedded in the plasma membrane of both adjoining cells. Tight junctions tend to encircle the entire epithelial cell, forming a string of beads-like structure (Figure 7.3). They function by bringing the membranes of adjacent cells into close proximity. Furthermore, the intracellular domains of tight junction proteins attach to components of the cytoskeleton and in effect link the cytoskeleton of neighbouring cells and thus prevent the passage of molecules, ions and microorganisms across the epithelium. Tight junctions are very effective at preventing the migration of microorganism and even host cells, so that the only way cells can translocate across the epithelium is through an actual epithelial cell. Tight junctions also help to maintain the polarity of cells, by preventing the diffusion of membrane proteins between apical and basolateral surfaces.

The barrier function of the single epithelial cell layer is assisted somewhat by the deposition of a glycocalyx on the apical surface of the epithelium. Epithelial cells are able to secrete a number of glycoproteins and polysaccharides that coat the surface of each epithelial cell, including the surface of the microvilli. The presence of the glycocalyx and the glycosylation of certain secreted proteins help to prevent bacteria gaining access to the

## Box 7.1. Recognition of Commensal Versus Pathogenic Microorganisms.

Commensal microorganisms are ubiquitously present at all epithelial surfaces throughout the human body. These microbes have co-evolved with their hosts, ensuring a symbiotic relationship that is often beneficial to both the commensal and the host species. The vast majority of commensal microorganisms are gram-negative bacteria found in the distal parts of the intestines. Although under normal circumstances commensal bacteria are not pathogenic, they still possess similar PAMPs as pathogenic bacteria. Considering the PAMPs recognized by TLRs and NOD proteins are shared between both pathogenic and commensal microorganisms, how does our immune system distinguish between the two?

The answer to this is not fully understood, although several hypotheses have been proposed. There is some evidence to suggest that certain commensal bacteria possess altered PAMPs that prevent or antagonize TLR recognition, thereby preventing the activation of inflammatory pathways. However, this is not the case for all commensals, while certain pathogenic species employ the same strategy to avoid immune recognition. Evidence also exists demonstrating that

commensal species actively downregulate inflammatory pathways or promote immunoregulatory responses. For example, *Lactobacillus* sp. have been shown to decrease the activity of the pro-inflammatory transcription factor NF-κB, while commensal *Helicobacter* sp. enhance Foxp3+CD4+ Treg cells and IL-10 expression. It seems likely that the commensal microflora help to shape the immunoregulatory environment of the gut.

However, consider what it takes to become a pathogenic rather than a commensal microorganism. Pathogenic microbes must traverse the mucous layer, bind to epithelial cells and invade into the sub-mucosal tissue in order to establish infection. The process of bacterial invasion will result in cellular stress and damage, causing epithelial cells to express and release several factors known as danger signals or danger associated molecular patterns (DAMPs). These danger signals act as secondary signals to those received from PRRs. The danger signal hypothesis therefore discriminates between commensal bacteria that cause no tissue damage, and pathogenic bacteria that do cause tissue damage. It is often the case that pathogenic bacteria express degrading enzymes, adhesion factors, invasins and secretion systems in order to

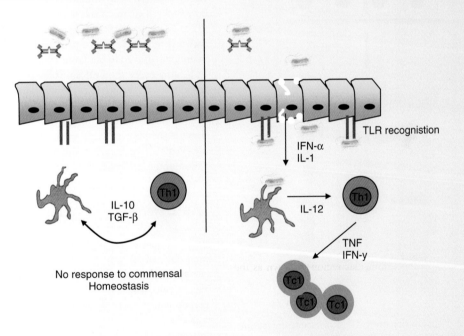

Response to commensal versus pathogenic bacteria. Compartmentalisation of commensal bacteria in the lumen and an immunoregulatory phenotype maintain a non-inflammatory state and tolerance to commensals. Pathogenic bacteria tend to disrupt the epithelial barrier leading to the recognition of DAMPs and PAMPs and the activation of an inflammatory response.

attach to and invade epithelial cells. Commensals often lack these strategies, so that the lack of such pathogenicity results in the compartmentalization of commensal bacteria in the gut lumen, rendering them invisible to immune cells. It is also possible that this sensing machinery is not fully intact at certain epithelial locations or the PRRs are expressed only on the basolateral side of the epithelial cell. In addition, certain PRRs are only expressed intracellularly, such as TLR7 and TLR9, so that only truly pathogenic microbes induce an inflammatory immune response. The absence of co-stimulatory receptors, such as CD80/CD86, on epithelial cells also contributes to the tolerogenic environment of the intestines (see figure).

The maintenance of intestinal homeostasis, in the presence of a commensal microflora, is assisted by other immunological factors. For example, the secretion of antimicrobial peptides,

the formation of a mucous layer and the extensive secretion of IgA across the epithelium and into the intestinal lumen are all important mechanisms that prevent commensal bacteria interacting with epithelial cells. Only those bacteria that break through these defences initiate inflammatory reactions. The general tolerogenic microenvironment of the intestinal mucosa is complimented by the lack of IL-12 and TNF-mediated Th1 immune responses to commensal bacteria and the production of immunoregulatory IL-10, TGF-β and IL-6. Commensal bacteria tend not to interact with host cells and are kept in check by innate defences. On the other hand, pathogenic bacteria breach host defences and have extensive interactions with host cells. This interaction is sufficient to overcome immune tolerance and initiate innate and adaptive immune responses.

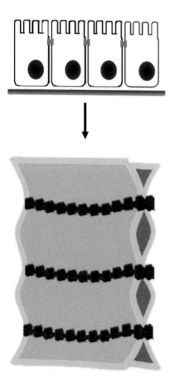

**Figure 7.3** Tight junction prevent molecules, microorganisms and cells diffusing across the epithelial layer.

the crypt epithelium. The mucous layer is principally composed of water ($\sim$ 95%), although several glycoproteins called mucins provide mucus with its viscous and elastic properties, essentially making the mucous layer very sticky. The mucous layer along the length of the gut varies from 50 to 800 μm and comprises a bottom layer, which is tethered to the apical surface of epithelial cells by mucins, and an upper layer, which is loosely adherent. In healthy tissue the bottom of the mucous layer is free from bacteria. Goblet cells can respond very quickly to a microbial insult and release mucins from pre-stored vesicles into the lumen in a matter of milliseconds. Harmful bacteria then become trapped in the viscous mucus and are eventually propelled out of the body by the peristaltic action of the gut wall.

## 7.5 Defensins and Trefoil factors

Barrier function is further augmented by the secretion of antimicrobial peptides. These are largely secreted into the gut lumen by specialized epithelial cells called the Paneth cells, which are located in the base of the intestinal crypts. Paneth cells secrete α-defensins and β-defensins into the gut lumen where they are thought to have direct antimicrobial actions that limit bacterial load and prevent the invasion of pathogenic bacteria. Another antimicrobial peptide, the cathelicidin LL37, is upregulated and secreted by Paneth cells in response to microbial infection, particularly in the colon.

In addition to the production of mucins, goblet cells express and secrete trefoil peptides, which are small, highly stable proteins characterized by the presence of a

apical membrane of epithelial cells. The barrier function of the gut mucosa is further strengthened by a layer of mucus (Figure 7.4). The cells responsible for the secretion of mucus onto the apical surface of epithelia are the goblet cells, which are preferentially located throughout

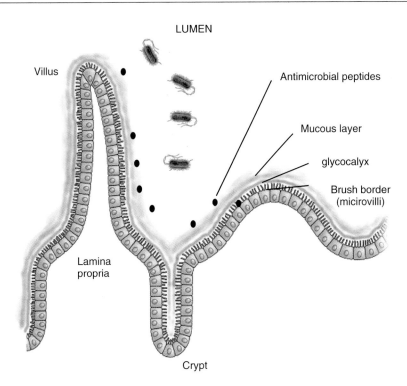

**Figure 7.4** Barrier function of the intestinal mucosa includes a brush border, glycocalyx and mucous layer. Antimicrobial peptides contribute to host defence.

trefoil motif. Through the bridging of three disulphide bonds these trefoil peptides form a clover-leaf secondary structure (the trefoil). Three trefoil peptides are known to be expressed in humans, TTF1, TTF2 and TTF3, of which the latter is also known as intestinal trefoil factor (ITF). ITF is thought to assist repair of the intestinal wall following stress, wounding or infection and is often secreted in association with mucins. Expression of ITF is upregulated in areas of tissue damage where it promotes the migration of epithelial cells in order to preserve the integrity of the epithelium. Although the mechanisms by which trefoil factors promote wound healing are not clearly understood, it is likely they are mediated through interactions with epithelial growth factor (EGF) or TGF-β pathways.

## 7.6 Structure of Peyer's patches

The organized lymphoid aggregates, known as Peyer's patches (PPs), found along the intestinal wall are named after the Swiss anatomist Joseph Hans Conrad Peyer (1653–1712). These secondary lymphoid organs represent specialized areas of lymphocyte division and differentiation throughout the intestines and share similar features to the lymphoid structures of the appendix, tonsils and lymph nodes. Smaller, isolated lymphoid follicles are also found along the small and large intestines, particularly in the colon where PPs are absent, although they are considered as inducible lymphoid tissues (tertiary lymphoid tissues). Indeed, PPs represent important sites of immune cell activation, where T and B cell expansion takes place, and in particular for the production of high affinity IgA antibody by plasma cells and the induction of memory T cell responses. The frequency of PPs differs along the length of the gut, being most abundant in the ileum, less so in the jejunum and being absent entirely from the colon. There is considerable species to species variation in the distribution of PPs along the intestine. For example, in humans, ruminants, dogs and horses, PPs develop during the foetal period and often form a continuous aggregation of lymphoid tissue along the ileum. In humans, the number of PPs increases directly after birth and are most numerous throughout the teenage years, slowly diminishing in adulthood. In other species such as

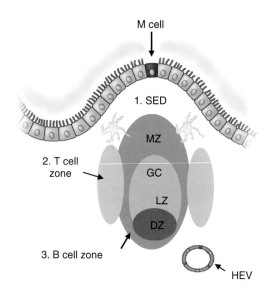

**Figure 7.5** Microanatomy of a Peyer's patch. A. The anatomy of Peyer's patches is very similar to that of other secondary lymphoid tissues such as lymph nodes. The structure can be divided into three broad areas. 1. An antigen capture area rich in APCs and often referred to as the subepithelial dome (SED). 2. A T cell zone and 3. a B cell zone where dendritic/B/T cell interactions take place in germinal centres (GC). Each germinal centre has a dark zone (DZ), light zone (LZ) and a mantle zone (MZ).

rodents and chickens, PPs develop during the postnatal period and are located at regular intervals along the entire length of the ileum and jejunum as discrete structures. The development of PPs is likely to be dependent on a population of progenitor cells known as lymphoid tissue inducer (LTi) cells (see Box 7.2). These cells arise within the developing buds of both lymph nodes and foetal PPs and drive the development of organized lymphoid structures through the production of lymphotoxin (LTα1β2).

Intestinal PPs develop directly beneath the epithelial cell layer, known as follicular-associated epithelia (FAE), while each PP is characterized by the presence of several lymphoid follicles (Figure 7.5). The FAE itself contains numerous lymphocytes, known as intra-epithelial lymphocytes (IELs), primarily CD8+ T cells. The area between the lymphoid follicle and the FAE is called the subepithelial dome and contains T cells, B cells (mostly plasma cells), DCs and macrophages and is similar to the lamina propria of other organized lymphoid tissues. Specialized epithelial cells called microfold or M cells (discussed later in this chapter) are able to sample the

contents of the intestinal lumen and transport antigen across the epithelium for subsequent DC uptake and activation. DCs within the subepithelial dome can also form long dendrites into the FAE, where they sample antigen directly. It is thought that DCs play a central role in the induction of immune responses within PPs and in the formation of lymphoid follicles. Following activation, DCs migrate from the FAE into the inter-follicular region and present antigen to Th cells. Alternatively they migrate to germinal centres within follicular regions of PPs and present antigen to B cells.

## 7.7 Lymphoid follicles and germinal centre formation

The lymphoid follicles within PPs are surrounded by connective tissue with only a small interfollicular space between adjacent follicles. Lymphocytes are also found within the inter-follicular spaces between each follicle, as is an intricate network of blood vessels and lymphatic vessels (Figure 7.5). Lymphocytes returning to PPs from the circulation enter inter-follicular areas via specialized blood vessels known as high endothelial venules (HEVs). It is thought that cells that enter the lamina propria or inter-follicular areas migrate into subepithelial areas to form immature follicles. The lymphoid follicles comprise a central zone known as the germinal centre, which contains actively differentiating B cells, DCs and reticular cells. Each germinal centre can also be separated into a dark zone, where rapidly dividing B cells (centroblasts) are located, and a light zone consisting mostly of resting B cells (centrocytes). Surrounding the germinal centre, and extending toward to the FAE is the mantle zone and the subepithelial dome that largely contains scattered T cells, DCs, macrophages and IgA-producing plasma cells. Lymphoid follicles can be further divided into a central medulla (equivalent to the germinal centre), surrounded by a cortex and an outer corona (equivalent to the mantle zone) that extends into the subepithelial dome. However, little is known about the migration of cells between the different areas of each follicle.

It is likely that antigen stimulation is required for follicle development within PPs. Once activated by antigen, lymphocytes located within the subepithelial dome area and marginal zone migrate to follicular areas and proliferate. A large number of IgA-producing B cells are contained within the germinal centres of these lymphoid follicles. Follicular DCs form an extensive network of dendrites throughout the germinal centre, enabling them

## Box 7.2. Cryptopatches and Inducer Cells.

It has long been debated whether the mucosal tissue of the intestine harbours an immature population of lymphocytes and DCs that are capable of populating the gut in response to infection. The conventional view has always been that T cells that populate intestinal lymphoid tissues are derived from the thymus. However, more recently, small clusters of lymphoid aggregates have been identified throughout the intestines that seem to contain common lymphoid progenitor cells, which are characterized by the Lin⁻ c-kit⁺ IL-7Rα⁺ phenotype. These aggregates are located just below villous crypts and have therefore become known as cryptopatches. This discovery has led to the notion that cryptopatches are the site of extra-thymic generation of intra-epithelial lymphocytes (IELs) and in particular the origin of the CD8αα sub-population of IELs, although this still remains controversial. It does seem clear though, that cryptopatches are the sites for the generation of isolated lymphoid follicles found along the length of the small intestine. In this case cryptopatches respond to a pathogenic insult and provide the necessary components required for the development of an induced lymphoid follicle.

The expression profile of common lymphocyte precursor cells is Lin⁻ c-kit⁺ IL-7Rα⁺, which is similar to lymphoid tissue inducer (LTi) cells found in foetal tissues. LTi cells are essential for the proper development of lymph nodes and Peyer's patches (PPs). They are located in the anlagen (immature bud) of foetal lymph nodes and immature PPs and have been shown to express lymphotoxin (LTα/LTβ), which is necessary for the development of these lymphoid structures. LTi cells are also characterized by the expression of the transcription factor RORγt. Mice that are deficient in RORγt (RORγt⁻/⁻ mice) fail to develop lymph nodes or PPs, thereby demonstrating the essential role that this transcription factor and LTi cells play in the development of these tissues. A similar population of cells are located in the centre of cryptopatches

that seem vital for the correct maturation of isolated lymphoid follicles in the intestine (see figure). Surrounding the central population of Lin⁻ progenitor cells are CD11c+ DCs, which may have a central function in antigen presentation and the development of lymphoid follicles. Cryptopatches remain devoid of any T cells or B cells and only when they mature into isolated lymphoid follicles are T cells and B cells recruited to these structures. It is important to note that cryptopatches have only been studied in mice and these cellular aggregates have remained elusive in humans to date, although it is known that isolated lymphoid follicles are present along the length of the small intestine and colon.

One feature of cryptopatch LTi cells is that they secrete large amounts IL-22. This cytokine has been strongly linked to the maintenance of epithelial surface integrity in the gut. However, in response to microbial stimulation epithelial cells secrete IL-7, which induces LTi cells to produce LTα/β that in turn activates neighbouring stromal cells through ligation with the LTβR. Within the cryptopatch, stromal cells upregulate chemotactic factors such as CCL19, CCL21 and CXCL13. This has the effect of recruiting T cells and particularly B cells, resulting in the formation of an immature isolated lymphoid follicle. Upregulation of LTα/β in B cells themselves augments lymphoid follicle development. Indeed, lymphotoxin-deficient mice are unable to form B cell follicles, demonstrating the importance of LTα/β signalling in lymphoid follicle development. It is likely that CD40:CD40L interactions between T cells and B cells contribute to the formation of mature lymphoid follicles and to the production of class-switched, high affinity antibodies. Presentation of antigen derived from the gut lumen appears to drive the expansion of B cells and the formation of germinal centres, resulting in the production of high affinity IgA antibodies and differentiation into memory B cell and T cell populations.

Cryptopatch    Immature lymphoid follicle    Mature lymphoid follicle

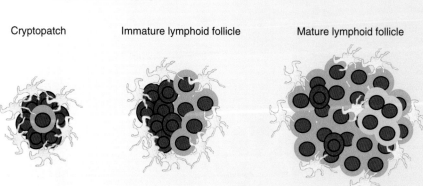

Development of lymphoid follicles from gut cryptopatches.

to capture migrating B cells. Antigen presented to B cells is retained for lengthy periods of time on the surface of dendrites. Antigen-stimulated B cells also interact with CD4+ Th cells located throughout the follicle. This interaction provides further stimulatory signals that are necessary for B cell survival and proliferation and is primarily dependent on CD40-CD40L signalling. Fully selected B cells then migrate out of the germinal centre and become IgA-producing plasma cells within the subepithelial dome region of PPs or within the adjacent lamina propria. Alternatively, plasma cells can exit the PP via the lymphatics, gain access to the blood and enter systemic sites or other mucosal tissues. In particular, lymphocytes from PPs traffic to the mesenteric lymph nodes via the thoracic duct and gain re-entry into GALT after they migrate back into the bloodstream. PPs and isolated lymphoid follicles are therefore referred to as inductive sites, as this is where lymphocytes are initially activated (Figure 7.6). Activated lymphocytes then migrate to their effector sites, such as the intestinal lamina propria, where they perform specific effector functions (e.g. S-IgA production).

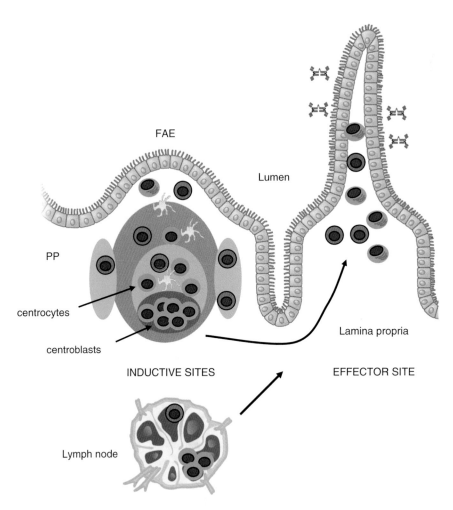

**Figure 7.6** Peyer's patches (PP) and regional lymph nodes act as inductive sites, where immune responses are initiated, while the lamina propria acts as an effector site, where lymphocytes exert their effector functions. Within germinal centres of Peyer's patches (or lymph nodes), interactions between Th cells, follicular dendritic cells and immature B cells (centrocytes) occurs. Antigen-specific T cells and plasma cells exit the germinal centre and enter effector sites, such as the lamina propria.

## 7.8 M cells sample the intestinal lumen

At intervals along the intestine the mucous and glycocalyx protective layers are absent, as is the expression of the pIgR. At these points specialized epithelial cells known as microfold or M cells are located, a cell type specifically found in mucosal tissues. M cells are distributed at intervals along the epithelial surface, most frequently above areas rich in B cell follicles and PPs. Indeed B cells may also play a role in the development of M cells. For example, B-cell-deficient mice or RAG1-deficient mice (lacking T and B cells) have very few M cells. In addition, differentiated epithelial cells can transform into M cells when cultured with B cells. The number of M cells increases rapidly after microbial challenge. The luminal surface of M cells contains short, flat and irregular microvilli and can additionally be identified by the presence of multiple endocytic vesicles. They also possess fine processes that interact with adjacent epithelial cells and lymphocytes. M cells are a portal for antigen entry as they sample the contents of the lumen and pass it via clatherin-coated vesicles to underlying dendritic cells for processing and presentation to T and B cells. They therefore play a key role in the initiation of immune responses throughout the gastrointestinal tract. This active sampling process monitors the luminal contents for pathogenic microorganisms. Unfortunately, several species of bacteria and viruses utilize this process to gain entry into the body, although the precise molecular mechanisms by which they do so are not fully understood.

## 7.9 Dendritic cells sample the lumen contents

The contents of the gastrointestinal tract lumen are also sampled by dendritic cells, which extend their dendrites through the epithelial layer and capture antigens. Alternatively, translocated antigen can be transferred to DCs that are present at the basolateral surface of M cells. These DCs are often referred to as lamina propria DCs (LPDCs), which reside directly beneath the FAE layer. Once microbes, or their microbial products, have been captured, the transepithelial dendrites contract and draw the cargo into the lamina propria. LPDCs are responsible for inducing both tolerance and effective immune responses to invading pathogens. Therefore a mechanism must exist so that LPDCs only activate an immune response in the presence of a microbial threat.

In order for LPDCs to become activated and migrate to regional lymph nodes, they must first recognize a danger signal derived from a pathogen associated molecular pattern (PAMP). Pattern recognition receptors (PRRs), such as the TLRs, recognize evolutionary conserved microbial products that are capable of activating DCs. In this way LPDCs are able to induce tolerance to commensal bacteria and the vast quantity of ingested food material, while at the same time respond appropriately to harmful microorganisms (Box 7.1). Once activated, LPDCs migrate to organized lymphoid tissues such as lymph nodes, PPs or isolated B cell follicles to initiate an immune response (Figure 7.7). During this migration LPDCs undergo a maturation process whereby they upregulate MHC class II, co-stimulatory molecules and the chemokine receptor CCR7. In contrast, other chemokine receptors such as CCR1 and CCR6, which act to retain DCs near the FAE, are downregulated, thereby allowing migration out of the FAE and into organized lymphoid structures.

The requirement for maintaining mucosal tolerance is particularly relevant in the colon, where commensal bacteria are most numerous. Here CD11b+CD103+ DCs are thought to be highly immunoregulatory and express the anti-inflammatory cytokine IL-10. These regulatory DCs help to maintain a state of tolerance to the heavy burden of commensal bacteria in the colon. Disregulation of this DC subset may contribute to the development of inflammatory bowel disease and autoimmunity, whereby excessive immune responses are directed toward self tissue following a lack of proper immune regulation.

## 7.10 Lymphocytes within the epithelium (IELs)

The lymphocytes that are distributed throughout the epithelial cell layer, at all mucosal sites, are known as intraepithelial lymphocytes (IELs), the vast majority of which are T cells (Table 7.1). The precise composition of these IELs varies between species but there are, on average, 20 IELs for every 100 epithelial cells (Figure 7.8). Considering the area of all mucosal surfaces ($300\,M^2$), this represents a vast number of lymphocytes. IELs therefore act as sentinels of the mucosal immune system and are likely to be the first lymphocyte population to encounter antigen. The majority of IELs also have a memory phenotype (CD45RO+), suggesting that they have been retained within the epithelial compartment as a result of previous antigen encounter although they exist normally in a resting state.

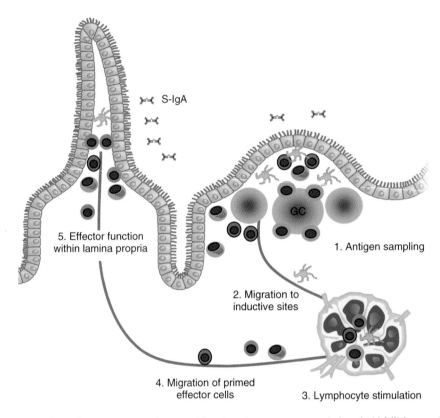

**S-IgA**

**5. Effector function within lamina propria**

GC

**1. Antigen sampling**

**2. Migration to inductive sites**

**4. Migration of primed effector cells**

**3. Lymphocyte stimulation**

**Figure 7.7** Migration of DCs from antigen sampling sites (1) activate immune responses in lymphoid follicles or migrate to regional lymph nodes (2) where they stimulate naïve T cells and B cells (3). Antigen-specific T cells and B cells then migrate (4) into effector sites where they perform their biological functions (5).

The most comprehensively studied population of IELs is located throughout the epithelia of the gastrointestinal tract. The phenotype of IELs is rather heterogeneous and consists of both conventional αβ T cells and more unconventional αβ T cells, in addition to γδ T cells. The many IEL subsets are thought to have evolved in order to cope with the dilemma of maintaining tolerance to commensal microorganisms while at the same time protecting the host against infections. In humans, the majority of IELs are αβ T cells with a small proportion being γδ T cells (γδ T cells account for approximately 10 per cent of IELs in man, although in mice the proportion may exceed >50%). Of the αβ T cell population, some IELs are of the conventional CD4+ or CD8αβ+ αβ T cell subset, which have been educated in the thymus and migrated to the epithelial compartment of the gut. The other IEL population comprises the more unconventional αβ T cell and γδ T cell subsets, which are thought to develop independently

of the thymus. It has been suggested that unconventional αβ T cells and γδ T cells develop within the intestinal tissue where they comprise a self-renewing population of T cells. These unconventional IEL populations express the more unusual CD8αα homodimer, rather than the more conventional CD8αβ heterodimer. The majority of intra-epithelial γδ T cells, and many αβ T cells, express the CD8αα homodimer. The majority of IELs are therefore CD8+ T cells. The expression of the CD8αα homodimer, rather than the conventional CD8αβ heterodimer, is often used to distinguish intestinal IEL T cells from peripheral T cells. Furthermore, small populations of αβ T cell and γδ T cell IELs are double positive, in other words express both CD4 and CD8 (CD4+CD8+), or are double negative and express neither co-receptor (CD4-CD8-). For simplicity, IELs can be divided into those that are restricted by classical MHC molecules (conventional CD4+ αβ T cells and CD8αβ+ αβ T cells) and those that are restricted by

IEL compartment

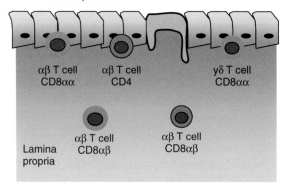

**Figure 7.8** T cells are present within the epithelial compartment of the gastrointestinal tract are heterogeneous and have multiple functions.

**Table 7.1** Phenotype of IELs. This table lists the various subsets of IELs, which are phenotyped based on the expression of the αβ or γδ T cell receptor and the expression of CD4 or CD8α/β. IELs expressing both CD4 and CD8 are known as double positive (DP) while those not expressing either CD4 or CD8 as double negative (DN).

| αβ T cells | γδ T cells |
| --- | --- |
| αβTCR+ CD8αβ+ (or CD4+) | γδTCR+ CD8αβ+ |
| αβTCR+ CD8αα+ | γδTCR+ CD8αα+ |
| αβTCR+ CD4+CD8+ (DP) | γδTCR+ CD4-CD8- (DN) |

non-classical MHC molecules (CD8αα+ αβ T cells, all γδ T cell populations and double positive/double negative populations). The various phenotypes of IELs are listed in Table 7.1.

These IELs are extremely interesting, not only for their unusual phenotype, but also because some populations can still be found in animals lacking a thymus (for example nu/nu mice or those where the thymus has been surgically removed). It is thought that some IELs do not undergo thymic education and therefore they are not subjected to the usual positive and negative selection processes that delete autoreactive T cells. Therefore autoreactive T cell

receptors are often observed on IELs, in particular the unconventional αβ T cells and γδ T cells that express the CD8αα homodimer. The expression of autoreactive TCRs may contribute to the maintenance of tolerance to self antigens in the intestine, as the T cells concerned have been shown to possess an immuno-regulatory capacity. They are also present in mice lacking the recombinase activating genes (RAG), which you will remember from Chapter 3 are essential for the rearrangement of T cell and B cell receptor gene segments. Conventional T cells possess a vast TCR diversity and are referred to as being polyclonal (allowing recognition of myriad antigens). However, this is not the case for IELs, which have a restricted TCR diversity and are therefore termed oligo-clonal. Furthermore, this oligoclonal TCR expression is stable over prolonged periods of time and is thought to account for some of the differing functional properties that exist between IELs and conventional T cells.

At present it is thought that IELs develop in, or at least migrate through, structures known as cryptopatches, which are located just beneath the crypt epithelium (see Box 7.2). Described for the first time in 1996, cryptopatches contain approximately 1000 cells of a phenotype suggesting a lymphopoietic origin (lin−1−, c-kit+, IL-7R+) and may very well be a common lymphocyte progenitor cell similar to the LTi cells found in foetal lymphoid tissues. These cryptopatches are found in germ free, athymic, SCID and TCR knockout mice, as are IELs, suggesting that their origin and development of IELs is not dependent on the thymus. However, they are absent in IL-7 knockout mice, which also have a sub-stantially under-developed thymus, suggesting that this cytokine is essential for cryptopatch formation. It should be noted, however, that a human equivalent of cryp-topatches has not yet been identified. Further doubt as to the extrathymic origin of intestinal IELs stems from studies using athymic mice, which have much reduced CD8αα+ αβ T cell and γδ T cell populations. Debate remains as to whether intestinal T cells require thymic education or whether they develop independently of the thymus.

The primary function of IELs throughout the gas-trointestinal tract is to provide early immune defence to infection. The proportion of γδ T cells within the IEL compartment is higher during the neonatal period than during adulthood. Conversely, the proportion of αβ T cells increases significantly with age. It is there-fore thought that γδ T cells provide a more primitive form of protection for the new born, while at the same time tolerating the colonization of the gut by commensal bacteria. IELs are normally in a resting state, so that the tolerance is maintained in the presence of an antigen-rich intestinal environment. However, they possess a memory phenotype, which is probably due to previous antigen encounter, meaning that they are ready to response to a previous pathogenic challenge. Resting but constitutively activated CD8+ αβ T cells and CD8+ γδ T cells are capable of initiating an immediate cytotoxic effector response to an infection. In this respect, unconventional T cells in the intestine differ to conventional T cells. While conventional T cells require a round of activation before they initiate their effector functions, unconventional intestinal T cells are already activated and immediately perform effector functions. Both γδ T cells and unconventional αβ T cells in the intestinal epithelia recognize non-classical MHC molecules such as CD1 (γδ T cells are discussed in the next section), which enables them to recognize stress response factors released from injured epithelial cells (Figure 7.9). This means that activation of γδ T cells is not restricted by specific peptide antigens, but rather they are capable of responding to host-derived danger signals.

### 7.11  γδ T cells in the GALT

In humans 10 to 15 per cent of IELs are γδ T cells, although their numbers increase in the colon (∼40%) compared to the small intestine. γδ T cells possess a restricted TCR repertoire. Within the gut the most frequently expressed

δ-chain is the Vδ1 chain (Vδ3 and other δ-chains are also expressed but are infrequent) together with a variety of Vγ-chains. This pattern of δ-chain expression is in contrast to that found in blood, which is dominated by Vγ9Vδ2 TCR expression. This specific γδTCR expression pattern is thought to be due to the selection of a population of γδ T cells within the intestine during development, which is then capable of circulating back into the intestine in adulthood or retained in the intestine as a self-renewing population. Despite the very restricted Vδ1 (Vδ3) expression, intestinal γδ T cells express a variety of Vγ-chains, the exact combinations of which are unique to each individual. Despite restricted Vδ-chain usage, γδ T cells are actually oligoclonal, meaning that this population has the potential to recognize numerous antigens, although their clonality is far less diverse compared to conventional αβ T cells.

Although it is far from clear, γδTCRs are thought to recognize antigen in the context of non-classical MHC molecules. For example, Vδ1 expressing γδ T cells recognize the stress-inducible MHC class I chain-related gene A and B (MICA and MICB), which are expressed on intestinal epithelial cells (Figure 7.9). MICA and MICB are upregulated in response to cell stress, trauma or infection. γδ T cells recognize MICA and MICB as damage associated molecular patterns (DAMPs), which allows them to respond quickly to abnormal or infected cells. Alternatively Vδ1 γδ T cells are restricted by CD1c expressed on epithelial cells, and upregulated on activated DCs and

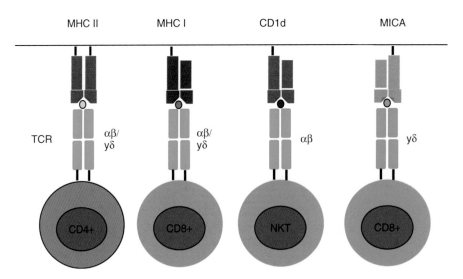

**Figure 7.9** T cell subset recognition of classical and non-classical MHC molecules. These molecular interactions are important for host defence against infection and in maintaining homeostasis and tolerance to the commensal microbiota.

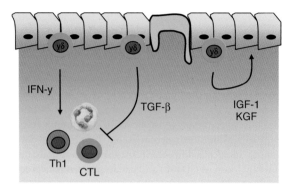

1. Inflammatory    2. Immunoregulatory    3. Homeostasis

**Figure 7.10** γδ T cells have multiple functions in the gastrointestinal tract and function in both host defence and epithelial barrier homeostasis.

macrophages. CD1c is structurally similar to MHC class I molecules, but rather than presenting peptide antigens it presents glycolipids derived from host cells or microorganisms. Therefore, intestinal γδ T cells are capable of recognizing numerous self antigens presented on the surface of abnormal cells in response to stress and can also respond to less conventional pathogen-derived antigens such as glycolipid antigens.

The function of γδ T cells may therefore involve an early response to tissue damage or cell transformation. They are capable of secreting IFN-γ, which is associated with driving Th1 immunity and enhances cytotoxic responses against virally infected cells and promotes neutrophilic responses against bacteria. It is likely that γδ T cells contribute to intestinal homeostasis by recognizing the abnormal expression of self antigens presented by non-classical MHC molecules. Although they are associated with numerous inflammatory diseases, γδ T cells are also thought to play an important role in protecting mucosal surfaces form the damaging effects of immune responses. In this respect they act as immunoregulatory T cells through the upregulation of TGF-β and the downregulation of IFN-γ and TNF. This has the effect of dampening T cell responses, particularly Th1-mediated responses, and limiting the influx of inflammatory leukocytes. In addition, γδ T cells within the IEL compartment are capable of releasing insulin-like growth factor-1 (IGF-1) and keratinocyte growth factor (KGF), both of which stimulate epithelial repair mechanisms, thereby regulating epithelial homeostasis. Therefore, within the mucosa of the gastrointestinal tract γδ T cells are multifunctional (Figure 7.10), providing protection against infectious diseases, particularly during the neonatal period, regulating mucosal immune responses, and contributing to the repair of damaged tissue.

## 7.12 NKT cells

A subset of unconventional T cells, known as NKT cells, exists in the lamina propria of the intestines. As their name suggests, they share some properties and express shared surface molecules normally associated with T cells and NK cells. For instance, they are characterized by the expression of the αβTCR found on conventional T cells, by the expression of CD4 and also by the NK cell marker NK1.1 (mice) or CD56/CD161 (human). αβTCR+CD4+ NKT cells express a limited number of αβTCRs on their surface, as they have a restricted α-chain repertoire, most notably a Vα14 chain in mice and a Vα24 chain in humans, and are therefore referred to as invariant NKT cells (iNKT cells). Vα24 iNKT cells are also restricted by the MHC class I-like molecule CD1d, which is thought to present glycolipid antigens rather than peptide antigens (Figure 7.9). NKT cells function in the early phases of an immune response and act as an interface between cells of the innate and adaptive immune systems. Following TCR recognition of glycolipid antigen presented on CD1d molecules by APCs, NKT cells can rapidly express IFN-γ, IL-17 or IL-22. It is clear that NKT cells promote Th1 immunity in certain conditions, and have even been associated with the induction of autoimmune reactions, such as inflammatory bowel disease and psoriasis. However, at other times NKT cells may play a more immunoregulatory role. For example, NKT cells express the regulatory cytokines TGF-β and IL-10, particularly in the context of controlling chronic inflammation, or can promote Th2 immunity through the production of IL-4 and IL-13 (Figure 7.11).

More recently another population of αβTCR+ invariant NKT cells have been found in GALT and are known as mucosal associated invariant T (MAIT) cells. They differ from other NKT cells in that they are restricted by the expression of the major histocompatibility molecule related 1 (MR1) molecule on B cells and are characterized as being double-negative for the CD4 and CD8 co-receptors. MAIT cells are thought to play a role in maintaining tolerance to the bacterial microflora in the gut. Furthermore, a small number of NKT cells are located in the intra-epithelial compartment of the gut and express the γδTCR receptor, rather than the αβTCR (although in mice this population may represent as many as 50 per

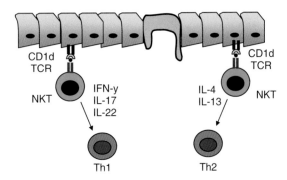

**Figure 7.11** NKT cells are early cytokine producers, they can participate in either Th1 or Th2 mediated immune responses.

cent of IEL NKT cells). Most γδ NKT cells also express the CD8αα homodimer, although the ligands recognized by γδTCR+ NKT cells have not been discovered yet.

## 7.13 T cells in the lamina propria

We described earlier the unique properties of IELs that exist within the epithelium. T cells in the sub-mucosal tissues or lamina propria are different and more analogous to those found in the periphery. However, unlike peripheral T cells, most lamina propria T cells have a memory phenotype, characterized by the expression of CD45RO, and are therefore antigen-experienced. Naïve T cells, on the other hand, which have not encountered antigen, tend to locate within lymph nodes. Most T cells within the gastrointestinal lamina propria express the αβTCR, with either the CD4 or CD8 co-receptor, although CD4+ Th cells dominate. Sequence analysis of their T cell receptors shows that they are oligoclonal (i.e. recognize a relatively small diversity of antigens). This together with their antigen-experienced phenotype suggests that constant re-encounter with mucosal antigens leads to the expansion of selected clones. Unlike the intra-epithelial compartment, the lamina propria contains few γδ T cells. A population of CD8+ αβ T cells are also found in the lamina propria but these cells display surface molecules more associated with IELs (such as CXCR4), suggesting they are merely in transit to the epithelial layer.

Interestingly, sub-mucosal T cells, especially those in the intestinal lamina propria, have a surface phenotype indicative of recent activation. For example, lamina propria T cells express CD25, which is the receptor for IL-2 (IL-2R), and CD45RO, which is a marker for memory T cells. However, activated T cells in the lamina propria

differ to memory T cells found in the periphery, in that they do not express CD29 (integrin β1), suggesting that there are genuine differences in T cell populations found in mucosal tissues. Furthermore, if these cells are cultured *in vitro* they undergo cell death or apoptosis. This activated, prone-to-die phenotype, may be of an advantage in delicate mucosal structures, so that excessive immune responses are tightly controlled and immunopathology kept to a minimum. Such activated T cells are likely to express molecules important for assisting antibody production (such as CD40L) and yet do not produce potent levels of inflammatory cytokines that could cause damage. A comparison of peripheral blood, intraepithelial and lamina propria T cells is shown in Table 7.2.

## 7.14 Maintenance of T cell homeostasis

Although conventional T cells are originally selected in the thymus, thereby generating a polyclonal repertoire of cells capable of recognizing a diverse range of antigens, the T cell pool is extensively shaped by the intestinal microflora. Commensal bacteria in the gut comprise by far the largest group of microbes that are recognized by T cells. Antigen derived from the gut microflora significantly affects the T cell repertoire located in GALT. Despite the large amount of antigenic stimulation, T cells found throughout the GALT are under strict regulation so as to maintain tissue homeostasis and prevent excessive inflammatory reactions to commensal bacteria. There are several mechanisms by which intestinal T cell homeostasis is maintained. The first relies on the intestinal barrier preventing microbial access to the lamina propria or lymphoid follicles. The second set of mechanisms involves the generation of CD4+ regulatory T cells, which produce

**Table 7.2** Percentages of T cell populations found within the intra-epithelial lymphocyte (IEL), lamina propria lymphocyte (LPL) and peripheral blood lymphocyte (PBL) compartments.

| Cell surface marker | IEL (%) | LPL (%) | PBL (%) |
|---|---|---|---|
| CD4+ | 20 | 65 | 65 |
| CD8+ | 80 | 35 | 35 |
| αβ TCR | 50 | 95 | 95 |
| γδ TCR | 50 | 5 | 3 |
| Activated | 85 | 65–95 | 30–50 |
| Naive | 15 | 10–22 | 60–90 |

anti-inflammatory cytokines such as IL-10 and TGF-β, instead of immunopathogenic effector T cells. This has a number of effects such as inducing T cell tolerance to the commensal microflora, rather than inducing effector T cell responses, and effectively inhibiting the activity of effector T cells within intestinal sites (Figure 7.12). γδ T cells and αβ T cells located in the intra-epithelial compartment contribute to the production of IL-10 and TGF-β in response to commensal bacteria. Only when tissue damage occurs do intestinal T cells become activated. The consequence of a lack of such regulation is exemplified in inflammatory bowel disease (IBD or Crohn's disease), which is characterized by excessive inflammatory reactions in the gut as a consequence of a general lack of regulatory cytokine production.

## 7.15 Sub-mucosal B cells and mucosal IgA

B cells located throughout the gastrointestinal tract are present either as diffuse cells within the lamina propria (within effector sites), or organized into B cell follicles (Figure 7.13), such as those found in PPs (inductive sites). The centre of the B cell follicle contains centrocytes and centroblasts and a network of follicular dendritic cells (FDCs). This follicle centre is surrounded by a mantle

of small lymphocytes, predominantly B cells but a few T cells are also present (Figure 7.13). In some areas of GALT the mantle zone merges into the subepithelial dome containing a mixed cell population that includes plasma cells, dendritic cells, macrophages and small B cells that extend up to the overlying epithelium. B cell follicles are the primary source of mucosal antibody production, which is also highly specialized in that the predominant immunoglobulin that is produced by plasma cells is IgA.

A large proportion of diffuse, scattered B cells are present as terminally differentiated plasma cells, found throughout the lamina propria, that produce large quantities of antibody. It has been estimated that there may be as many as $10^{10}$ plasma cells per metre of human intestine, a huge number of IgA secreting cells that could increase in frequency even more during an inflammatory insult. These intestinal B cells are present between B cell follicles, although most surround the crypts of the villi. These sites are known as B cell effector sites, where immunoglobulin secretion has a direct functional role in controlling the intestinal microflora. Unlike peripheral and circulating B cells, mucosal B cells predominantly produce IgA. Between 30 and 40 per cent of cells in the human intestinal lamina propria and 80 per cent of jejunum mononuclear cells secrete IgA. In the jejunum only 18 per cent and 3 per cent of B cells produce IgM and IgG, respectively. This is in marked contrast to peripheral

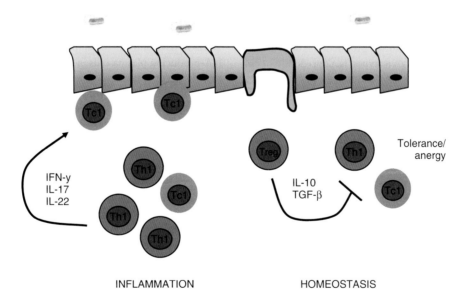

**Figure 7.12** Inflammation versus homeostasis in the gut. A balance exists whereby the gut must defend itself from pathogenic microorganisms, while at the same time maintain the integrity of the intestinal barrier. Inflammation and homeostasis are characterised by the production of different cytokines.

and stromal cells located throughout the lamina propria and the myriad leukocytes that migrate into or are resident within the GALT. As in any other tissue, the balance of cytokines is essential for the control of subsequent immune reactions and the GALT is no exception in that it can mount effective Th1 (IL-12, IFN-$\gamma$, TNF) and Th2 (IL-4, IL-5, IL-13) responses, whichever is most appropriate for the initial pathogenic insult. The type of pathogen is often a key factor in driving a particular type of immune response. For example, intracellular viruses activate Th1-driven cell-mediated immunity, while extracellular bacteria drive Th2-mediated antibody responses.

## 7.18 Chemokines and the homing of lymphocytes to GALT

Chemokines (described in Chapter 5) also play an important role in directing the trafficking of cells around the body. Epithelial cells of the small intestine produce CCL25 (TECK). CCL25 binds to CCR9, which in humans is expressed by nearly all lymphocytes in the small intestine. Interestingly, responsiveness to CCL25 is induced in T cells that also upregulate the integrin $\alpha 4\beta 7$, which in turn binds to the mucosal addressin molecule MAdCAM1. Selective lymphocyte movement into GALT is therefore controlled by the expression of selected chemokines/chemokine receptors and cell adhesion molecules. CCL25 can also bind to CCR11, which is also present on most cells found in the intestine. IELs also express restricted chemokine receptor patterns that enable them to migrate specifically to the epithelial compartment. For example, IELs express CCR9 and CXCR3, which causes them to preferentially migrate along a chemokine gradient provided by the receptor ligands CCL25 and CXCL9/CXCL10, respectively. IELs express these chemokine ligands in order to recruit cells to the epithelia. This network of chemokines and adhesion molecules also retains IELs in the epithelial compartment.

With respect to lymphocyte homing, there is a requirement for IgA secreting plasma cells to preferentially migrate to mucosal tissues. Generally B cells activated in PPs or mesenteric lymph nodes mainly populate the lamina propria of the small intestine. MAdCAM-1 is expressed on the vascular endothelium of the lamina propria and PPs, which acts as an important homing molecule for B cells expressing the integrin $\alpha 4\beta 7$. B cells activated

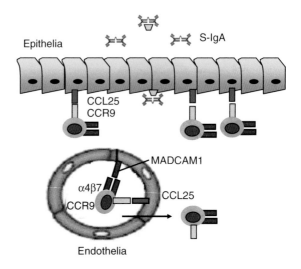

**Figure 7.15** Homing of B cells to the intestines involves the specific expression of cell adhesion molecules, such as $\alpha 4\beta 7$, and chemokines such as CCR9.

in organized lymphoid tissues within the mucosa express $\alpha 4\beta 7$, whereas those activated in the periphery express $\alpha 4\beta 1$. Therefore there exists an element of anatomical specialization. Both $\alpha 4\beta 1$ and $\alpha 4\beta 7$ integrins bind to vascular cell adhesion molecule-1 (VCAM-1), which is expressed on the endothelium of blood vessels. The expression of either of these two integrins, however, still does not allow passage of only IgA secreting B cells into mucosal tissues. This further selectivity is mediated by the expression of different chemokine receptors (Figure 7.15) and cell adhesion molecules such as MaDCAM-1 expressed preferentially by mucosal endothelia. IgA plasma B cells when induced in mucosal tissues upregulate the expression of CCR9 and CCR10, whereas mucosally induced IgG secreting plasma B cells express CXCR3. The ligand for CCR9 (CCL25) is predominantly produced by small intestinal epithelium and endothelium. Therefore for an IgA plasma B cell to enter the intestine it must express $\alpha 4\beta 7$ and CCR9. The $\alpha 4\beta 7$ binds to MAdCAM1 and the CCR9 binds the CCL25 produced by epithelial cells. Furthermore, IgG-producing plasma B cells migrate less into mucosal tissues because the ligand for CXCR3 (which is CXCL9/10) is not expressed on mucosal endothelial cells, only epithelial cells, therefore initial entry into mucosal tissues is unfavourable.

IgA-producing plasma cells are produced in non-mucosal tissues and may upregulate overlapping chemokine receptors but remember, non-mucosal induction induces the α4β1 integrin and not α4β7. The orchestrated movement of B cells therefore requires coordinated expression of both integrins and chemokine receptors. Obviously we have simplified this a little, as other chemokines and adhesion molecules are required for effective cell migration and translocation across the endothelial cell wall (discussed in Chapter 5). For example, all plasma cells express the CXCR4 chemokine receptor, which enables them to enter the bone marrow.

The migration of conventional CD4+ and CD8+ T cells into GALT is often a transient process that helps to maintain tissue homeostasis, while at the same time provide effector T cells for host defence. Most T cells located in the GALT have a memory phenotype, although a relatively small population of naïve T cells do exist. Distinctions can be made between the type of memory T cell depending on its location and function. Central memory T cells ($T_{CM}$) are primarily located in the regional lymph nodes, whilst effector memory T cells ($T_{EM}$) are located in peripheral tissues such of the intestine; although both cell types express the memory marker CD45RO (Figure 7.16). Both $T_{CM}$ and $T_{EM}$ cells can also be found in the blood and spleen, which reflects their propensity to migrate between tissues. $T_{CM}$ cells are distinguished from $T_{EM}$ cells by the expression of the homing receptors CD62L and CCR7, which allows entry into lymph nodes.

However, the distinction between $T_{CM}$ cells and $T_{EM}$ cells may be better defined based on the functional capabilities of each cell type. For instance, CD4+ $T_{EM}$ cells express a greater amount of effector cytokines than CD4+ $T_{CM}$ cells (particularly IFN-γ), while CD8+ $T_{EM}$ cells have a much higher cytotoxic capacity than CD8+ $T_{CM}$ cells. With this in mind, $T_{EM}$ cells are capable of immediate effector functions in response to antigenic stimulation within inflamed sites of the gut. On the other hand, $T_{CM}$ cells reside in lymph nodes and other lymphoid tissues where they provide an antigen-specific population of dividing memory cells. Although $T_{CM}$ cells do not have the capacity for rapid effector functions, they do upregulate CD40L, migrate to peripheral tissue in response to antigen and are capable of providing help to B cells.

## 7.19 Pathogens and immune diseases

The gastrointestinal tract is frequently subjected to pathogenic insults, whether in the form of viruses, bacteria or parasites. The mucosal immune response to each of these types of pathogen, as well as fungi, will be discussed in subsequent chapters. It is important to consider that the immune response to a certain microorganism must be tailored so as to induce the most

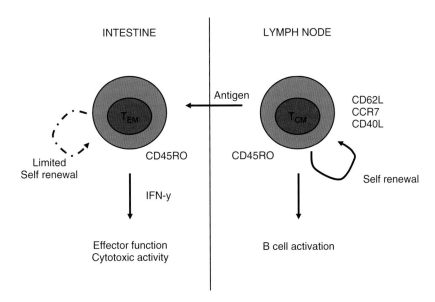

**Figure 7.16** Effector memory T cells ($T_{EM}$) and central memory T cells ($T_{CM}$) have different characteristic and effector functions within GALT. $T_{EM}$ cells are present throughout the GALT, can rapidly respond to infection and a highly cytotoxic. $T_{CM}$ cells reside within lymph nodes where they provide during B cell differentiation and can themselves differentiate into $T_{EM}$ cells following migration into effector tissues.

**Table 7.3** Immune-mediated diseases of the gastrointestinal tract.

| Disease | Immune | Phenotype |
| --- | --- | --- |
| Colitis | Inflammation of colon. Can be autoimmune (IBD), infectious, ischaemic, or idiopathic | Diarrhoea, dehydration, fever, distension, weight loss, abdominal pain. |
| Appendicitis | Inflammation of appendix | Nausea, vomiting, fever, abdominal pain, anorexia |
| Inflammatory bowel disease | Inflammation of small intestine and/or colon (autoimmune) Includes:- Crohn's disease Ulcerative colitis | Diarrhoea, vomiting, fever, abdominal pain Mainly ileum and colon Mainly colon |
| Diverticulitis | Formation of diverticula on colon and inflammation | Abdominal pain, weight loss, nausea, vomiting |
| Gastroenteritis | Inflammation of stomach or intestines caused by Infection. Examples:- Norovirus Rotovirus Enterovirus *Salmonella sp.* *Escherichia coli* *Campylobacter sp.* *Helicobacter pylori* | Diarrhoea, dysentery, dehydration, fever, stomach or abdominal pain. |

appropriate form of defence depending on the type of pathogen. For instance, viruses and intracellular bacteria induce Th1-driven cell-mediated responses, while intestinal parasites elicit Th2-driven antibody-mediated immune responses. These differences are largely a result of the method of pathogen entry. Viruses have evolved to infect and replicate inside cells and therefore the immune response to those viruses relies on cellular cytotoxicity in order to prevent further infection. Large extracellular parasites, such as nematodes, do not reside inside cells and therefore induce Th2-driven humoral immunity involving antibody opsonization, phagocytosis and granulocyte activation. The varying mechanisms of immune defence will be discussed in depth in later chapters.

Although the activation of immune cells is essential at preventing infection and eliminating intestinal pathogens, this inflammation is normally under strict regulation so as not to generate excessive immunopathology. However, a dysregulated immune response often results in unwanted tissue damage. For example, IBD is a broad term for a spectrum of immunologically-driven diseases (Table 7.3),

and is characterized by excessive immune reactions in the intestines. This may be the result of an autoimmune reaction, a lack of immunoregulatory cells or cytokines, or the result of tissue damage and microbial invasion of host tissue. As with other mucosal sites, it is crucial that the integrity of the delicate tissues is under tight immunological control, in order to preserve homeostasis and physiological function.

## 7.20 Summary

**1.** The immune system of the gastrointestinal tract (GALT) has to constantly distinguish between harmless commensal bacteria and pathogenic bacteria.
**2.** There exist mechanisms of tolerance to food items and the commensal microflora that live in the intestines.
**3.** The epithelial layer of the gastrointestinal tract plays an important part in maintaining homeostasis and preventing infectious microorganisms invading the subepithelial tissue.

**4.** Specialized epithelial cells, known as M cells, transport antigen from the gut lumen to dendritic cells for processing and presentation to lymphocytes.

**5.** Organized lymphoid tissues, known as Peyer's patches, are found along the length of the small intestine and act as inductive sites for the initiation of adaptive immune responses.

**6.** B cell differentiation occurs within germinal centres, located within Peyer's patches, resulting in the synthesis of protective IgA.

**7.** Specialized subsets of T cells exist throughout GALT, including a population of intra-epithelial lymphocytes (IELs) that function in both homeostasis and host defence.

# 8 Immunology of the Airways

## 8.1 The airways as a mucosal tissue

The airways represent an important portal of entry for pathogens, allergens and particles from the external environment. The airways, otherwise known as the respiratory tract, run from the nasal openings (nares) down to the alveoli in the lower lungs (Figure 8.1) and are primarily responsible for the transfer of gases between the bloodstream and inhaled air. It has been estimated that the average person breathes in and out 20,000 times each day and in the process samples in the region of 10,000 litres of air. The respiratory tract therefore comes into contact with a large quantity of air that has the potential of containing harmful substances such as infectious pathogens, allergens and pollutants. It is therefore necessary for the airways to have appropriate defence mechanisms in order to maintain normal structure and physiological function, while at the same time being able to defend the body from harmful agents. The importance of maintaining this delicate balance is exemplified by lung diseases such as acute lung injury (ALI) or chronic obstructive pulmonary disease (COPD), where immune regulation becomes disregulated and immunopathology ensues (Box 8.1).

The immune system within the airways is tightly regulated so as to minimize any collateral damage caused by excessive immune reactions, without allowing pathogens or other harmful material to overwhelm its host. Several tissues located within the airways can be regarded as belonging to the common mucosal immune system. These include the nasal associated lymphoid tissue (NALT) and the bronchus associated lymphoid tissue (BALT) (Figure 8.1). Specialized mucosal tissues, which contain highly organized lymphoid structures, are also associated with the respiratory tract such as the palatine, tubal and lingual tonsils and the adenoids that collectively form the circular lymphoid architecture known as the Waldeyer's ring. In addition, during inflammatory insults typically non-lymphoid areas, such as the bronchial-alveolar parenchyma, can accommodate immune reactions and are therefore host to different types of immune cells.

## 8.2 Development of the respiratory tract

The lung is a highly specialized organ that forms a diffusible interface between oxygen-rich air and the circulatory system. The developmental processes that contribute to this highly differentiated respiratory system include branching morphogenesis, alveolization, vascularization and angiogenesis. Moreover, a complex series of biochemical and genetic pathways integrate during lung morphogenesis, involving many transcription factors, cytokines, chemokines, growth factors, integrins and phylogenetically conserved signalling molecules.

The airways begin to form in the embryo, around 4–6 weeks into gestation, as a small budding from the primitive foregut that elongates to form a diverticulum located on the distal section of the trachea. The primitive airway begins to branch at points near the end of the tracheal buds, a process known as branching morphogenesis. Each branching event results in the development of a tubule with an ever-decreasing diameter. Therefore, the diameter of the mature trachea is larger than that of the bronchus, which is larger than the bronchioles and so on. The second stage of development involves further branching morphogenesis and terminal bud formation, which results in the formation of alveolar sacs within the immature lung. The developmental stage of alveolar formation continues throughout the postnatal period until 4–5 years of age. Branching morphogenesis, alveolization and the

*Immunology: Mucosal and Body Surface Defences*, First Edition. Andrew E. Williams.
© 2012 John Wiley & Sons, Ltd. Published 2012 by John Wiley & Sons, Ltd.

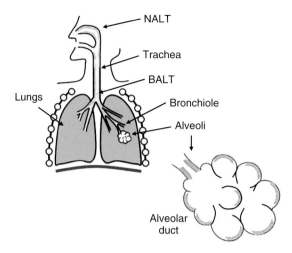

**Figure 8.1** Structure of the respiratory tract. The upper respiratory tract includes the nasal-associated lymphoid tissue (NALT), bronchus-associated lymphoid tissue (BALT), and the larger conducting airways. The lower respiratory tract consists of the small airways and the alveolar ducts.

formation of several separate lung lobes results in a very large surface area, over which gaseous exchange can take place. The molecules and signalling pathways employed during lung morphogenesis are often similar or associated with those utilized by the immune system. This had led to the notion of evolutionary-developmental (evo-devo) functional conservation and is a useful concept that can be extended to the function of the mucosal immune system.

---

**Box 8.1. Lung Diseases Associated with the Immune System.**

Acute lung injury (ALI) and acute respiratory distress syndrome (ARDS) are life-threatening lung diseases characterized by severe inflammation associated with excessive neutrophil accumulation within the alveolar spaces. Likewise, chronic obstructive pulmonary disease (COPD), which involves the destruction of the lung parenchyma and alveolar walls, is also associated with neutrophil recruitment and activation. The most common cause of ALI/ARDS is pulmonary sepsis resulting from a severe bacterial infection, although trauma and inhalation injury, as well as non-pulmonary sepsis can also trigger ALI/ARDS. The main trigger for COPD is cigarette smoke, which is thought to attract and activate neutrophils in the lung.

The early phase of ALI involves an acute inflammatory response that is followed by a fibroproliferative phase and then recovery, although protracted ALI increases susceptibility to secondary infections. The early phase of

ALI/ARDS involves destruction of the alveolar epithelium and activation of resident alveolar macrophages. Damaged alveolar epithelial cells are sloughed away from the basement membrane, causing bleeding into the air space, activation of the coagulation cascade and the deposition of a fibrotic hyaline membrane. At the same time, macrophages and epithelial cells release several pro-inflammatory mediators including TNF and IL-1 and chemoattractants such as IL-8 (CXCL8), which recruit neutrophils into the inflamed tissue. The release of these mediators is often triggered by the recognition of PAMPs, including the TLR ligands LPS and peptidoglycan. This acute inflammatory reaction also affects endothelial cells, resulting in vasculature leakage and the accumulation of fluid and protein in the alveolar space. Neutrophils accentuate the inflammation by releasing more cytokines and chemokines and initiating a respiratory burst, which involves the release of tissue-damaging reactive oxygen species. The overall consequences of ALI/ARDS are alveolar collapse, severe disruption of gaseous exchange and hypoxemia. There is a direct association between the extent of neutrophil accumulation and the severity of disease. Neutrophil recruitment and their continued activation are considered to be central to the progression of disease. Therefore, neutrophil clearance appears to be a very important process for the resolution of ALI/ARDS.

In COPD, cigarette smoke can cause the destruction of epithelial cells, which in turn release DAMPs and other stimulatory signals such as IL-8. Activated macrophages contribute to the release of TNF, IL-1 and IL-8 and accentuate local inflammation. Neutrophils are then recruited to inflamed areas of the epithelium where they release reactive oxygen species and several enzymes, such as serine proteases, elastases and matrix metalloproteinases, which degrade matrix proteins, stimulate the release of mucus and cause the destruction of alveolar walls (emphysema). This in turn attracts more inflammatory leukocytes into the lung, thereby perpetuating COPD.

The chronic inflammation associated with COPD involves the migration and activation of CD4+ and CD8+ T cells into the lung. COPD is generally considered to be a Th1-mediated disease, due to the continued release of type-1 cytokines such as IL-12, IFN-γ and TNF. The continued release of these cytokines encourages the infiltration of more neutrophils and monocytes and stimulates CD8+ T cell-mediated perforin release, which contributes to alveolar epithelial cell destruction. COPD patients are also susceptible to acute exacerbations as a result of viral or bacterial infections, which heighten inflammatory reactions and worsen symptoms. Ongoing inflammation and the release of tissue growth factors, including FGF and TGF-β by epithelial cells, enhance mucus secretion into the airways and stimulate fibroblast proliferation. This often leads to the fibrosis of the airways, further restricting respiration. COPD is a progressive disease, mediated by an uncontrolled immune response, which can lead to severe difficulties in breathing, hypoxia, tissue necrosis and death.

## 8.3 The structure of the respiratory tract

The respiratory tract can be divided into the upper and lower airways. The upper airways comprise the oral cavity, the nasal passages, the trachea and the large bronchi and bronchioles. The upper respiratory tract is known as the conducting part of the airways, where air is actively moved to the lungs from the outside. The lower respiratory tract comprises the small bronchioles and the alveoli (along with their supporting parenchymal structures) and is known as the respiratory part of the airways where gaseous exchange takes place. The upper and lower airways therefore differ in structure in order to accommodate their different physiological functions. Importantly, the upper airways are mucus secreting and participate in mucociliary clearance, while the lower airways tend not to secrete large quantities of mucus so that gaseous exchange in unhindered. Differences therefore exist in the composition of the epithelial lining between the upper and lower respiratory tracts (Figure 8.2).

The respiratory epithelium consists of four major cell types; epithelial cells, secretory cells, basal cells and neuroendocrine cells. Epithelial cells comprise the most frequent of these cell populations, although the type of epithelial cell varies according to the level of the respiratory tract. The epithelium of the upper airways, including the trachea and the large bronchi, is mostly psuedostratified and formed from ciliated columnar epithelial cells. Interspersed along the length of the large airways are secretory epithelial cells known as goblet cells, which are responsible for the synthesis and secretion of mucus. Undifferentiated columnar cells and basal cells also exist along the large airways and are thought to represent a pluripotent cell population capable of replenishing and repairing the epithelial layer. It has been estimated that the upper respiratory epithelium is replenished every 30 to 50 days.

The epithelium of the small airways is also composed of ciliated epithelial cells interspersed with undifferentiated columnar epithelial and basal cells. However, rather than containing goblet cells, the small airways contain secretory

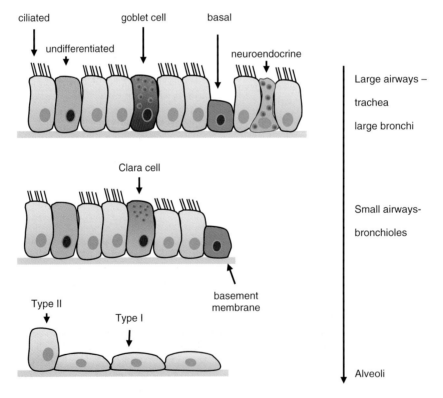

**Figure 8.2** Differences exist in the composition of epithelial cells, depending on the level of the respiratory tract. The airways consist of several key epithelial cells in varying state of differentiation. The alveoli only consist of type I and type II epithelial cells.

Clara cells, which are specialized epithelial cells unique to the bronchioles. Clara cells secrete surfactant, antimicrobial peptides, IgA and enzymes that are responsible for the breakdown of mucus produced in the upper airways. In addition, Clara cells predominantly reside at bronchiole bifurcation points and at bronchiole-alveolar ducts. The alveolar epithelium differs again and is composed of type I and type II epithelial cells. It is thought that type II epithelial cells give rise to type I epithelial cells, which constitute the alveolar duct surface responsible for gaseous exchange and are far more frequent than type II cells. Alveolar type I cells are also much flatter in shape, which aids the transfer of gases between the alveolar space and the capillary system beneath. Although rare, neuroendocrine cells are located throughout the entire length of the respiratory tract where they detect alterations in oxygen levels and respond by releasing hormones such as serotonin. Neuroendocrine cells also respond to lung injury and can stimulate tissue repair. Although their role in lung disease is only beginning to be understood, they have been linked to the development of certain lung cancers.

Leukocyte trafficking into the respiratory tract is influenced by the circulatory system. The human respiratory tract is supplied by two separate arterial systems: the pulmonary circulation and the bronchial circulation (mice only have the pulmonary system, even though they are extensively used for the study of respiratory tract immunology). The pulmonary system supplies arterial blood directly from the heart to the lung parenchyma and is therefore deoxygenated. The pulmonary system provides an extensive capillary network throughout the parenchyma in close association with alveolar ducts, where it receives oxygen from the alveolar space.

Oxygenated blood leaves the pulmonary vein and enters the heart, where it is delivered back into the arterial system via the aorta. The arterial system then delivers blood to the rest of the body, including the bronchial circulation. The pulmonary vasculature therefore provides a unique environment for the migration and extravasation of circulating leukocytes into lung tissue (discussed in section 8.20). The capillary network in the lungs is such that leukocytes do not need to undergo the classical rolling, adhesion, transmigration events as occurs within high endothelial venules (HEVs). The close proximity of pulmonary endothelial cells to alveolar epithelial cells also provides a unique environment for leukocyte transmigration.

## 8.4 Barrier function and the mucociliary elevator

The upper respiratory tract consists of the naso-pharynx, the trachea (and associated structures such as the larynx), the bronchi and bronchioles. The upper respiratory tract is considered to be the conducting portion of the airways, responsible for moving air between the lungs and the outside environment. The most numerous cell type found throughout the upper airways is the ciliated epithelial cell, which forms a physical barrier between inhaled air and internal body structures. In addition, ciliated epithelial cells act in conjunction with mucus secreting goblet cells and sub-mucosal glands to generate the mucociliary elevator. Less frequent are basal cells, which are thought to be local stem cells that are capable of replenishing cells in the airways. Goblet cells secrete mucus, which traps particles and pathogens, preventing harmful reagents from entering the lower airways (Figure 8.3). This mucous layer

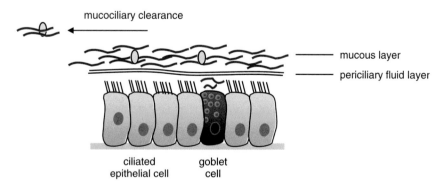

**Figure 8.3** Goblet cells, mucus secretion and mucociliary clearance. Goblet cells are responsible for the synthesis and secretion of mucus onto the lumenal surface of the airways. The action of ciliated epithelial cells results in the movement of mucus and any trapped matter out of the airways. Movement of mucus is assisted by a dynamic periciliary fluid layer.

is transported upwards and outwards by the action of ciliated epithelial cells and is eventually expelled through swallowing, coughing or sneezing. The mucociliary elevator represents an important mechanism that prevents antigens from entering the lungs and clears debris and dead cells from the airways.

## 8.5 Mucins and mucociliary clearance

Mucins are a family of glycoproteins secreted either by goblet cells found throughout the conducting airways (Figure 8.4) or by sub-mucosal glands found in the upper airways, nose and trachea. The primary function of mucins is to form a protective layer, known as mucus, which overlies the epithelial layer beneath. The mucous layer helps protect epithelial surfaces from desiccation, prevents inhaled particles from accumulating in the lungs and inhibits microbial attachment and invasion. Particles that enter the airways during the process of respiration become trapped in the mucous layer. With the assistance of ciliated epithelial cells, the mucus is continually being transported toward the mouth where it is either expelled through coughing and sneezing or is swallowed, a mechanism known as mucociliary clearance. A liquid layer, known as the periciliary fluid layer, exists between the ciliated epithelial cells and the mucous layer and is necessary for efficient mucociliary clearance. In effect, islands of mucus float on top of the periciliary layer, which assists effective transport of mucus across the epithelial sheet of the upper airways. As well as acting as mechanical barriers, both the periciliary and mucous layer contain antimicrobial peptides, such as defensins, and immunoglobulins in

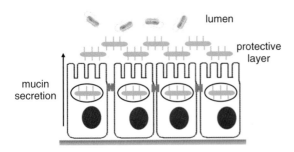

**Figure 8.4** Mucin secretion across the epithelia. Mucus is stored with goblet cell granules, which fuse with the apical membrane and release mucus into the airway lumen. Airway inflammation can trigger goblet cell hyperplasia and/or mucus hyper-secretion, which can be so severe as to obstruct the airways.

particular S-IgA. In the lower airways, a layer of surfactant exists at the epithelial cell surface, rather than a layer of mucus, which helps to maintain surface tension at the air–liquid interface. Together, these layers are known as airway surface liquid (ASL).

In addition to mucins, the aqueous mucous layer consists of lipids, proteins and other glycoproteins, although the mucins provide mucus with its elastic properties that allow ciliated transport and clearance from the airways. During episodes of infection, or inflammation of the airways, goblet cells can undergo hyperplasia resulting in mucus hypersecretion. This is beneficial for microbial clearance but can also be detrimental if associated with diseases such as asthma or COPD. The production of large amounts of mucus often relies on coughing to expel it from the airways and is symptomatic of chronic lung disease. This expectorate is often referred to as sputum.

Goblet cells are activated in response to several stimuli including oxidative stress, pro-inflammatory cytokines, microbial products and various irritants. Currently there are 16 known human mucin genes, although the most abundant mucins found in the lungs are the products of the MUC5A and MUC5B genes. These are high molecular weight, heavily glycosylated, water absorbing molecules that form large oligomeric structures. They are known as gel-forming mucins, the physical properties of which principally give surface mucus its visco-elastic properties. Other mucins, such as MUC1 and MUC4, are associated with the apical membrane of epithelial cells where they assist in mucociliary clearance. Mucins are normally contained within intracellular secretary granules, which translocate to the apical cell membrane upon activation of protein kinases (Figure 8.4). When contained within intracellular secretary granules mucins are in a highly condensed state. Exocytosis directly delivers mucins onto the epithelial cell surface where they rapidly expand, confer elasticity to mucus and enable interactions with ciliated epithelial cells.

## 8.6 Defensins and antimicrobial peptides

Several cell types along the respiratory tract are capable of producing and secreting antimicrobial peptides, including ciliated epithelial cells, alveolar epithelial cells, sub-mucosal gland cells and immune cells located in subepithelial sites. Numerous anti-microbial peptides are produced from a diverse set of cell types and include defensins, cathelicidins, collectins (such as the surfactant proteins) and protease inhibitors (Table 8.1). As

**Table 8.1** Types of antimicrobial peptides in the airway.

| Antimicrobial | Cell source |
| --- | --- |
| Lysozyme | Epithelia and neutrophils |
| Lactoferrin | Epithelia and neutrophils |
| Phospholipase A2 | Epithelia and neutrophils |
| Peroxidases | Epithelia and neutrophils |
| α/β-Defensins | Epithelia and neutrophils |
| LL37 | Epithelia and neutrophils |
| BPI | Neutrophils |
| SLPI | Epithelia and macrophages |
| SP-A and SP-D | Epithelia |

well as having direct antimicrobial properties, many antimicrobial peptides can act as inflammatory mediators or growth factors and have therefore been described as alarmins.

In humans, two families of defensins are known, α-defensins and β-defensins, of which there are six members in mice and four members in humans. Neutrophils that have been recruited to the respiratory tract are the predominant cell type that expresses α-defensins, which are stored in intracellular granules. The primary function of α-defensins is to kill phagocytosed microbes within the neutrophil phagolysosome, although these antimicrobial peptides are also secreted into the extracellular environment where they act as inflammatory mediators. The main sources of β-defenins in the airways are epithelial cells. It is likely that the detection of pathogens at the epithelial surface, particularly through the recognition of TLR ligands, triggers the production and secretion of β-defensins into the periciliary fluid. They possess broad spectrum antimicrobial properties and exert their activity by binding to the polyanionic surface membrane of bacteria, via LPS or techoic acid, for example. The hydrophobic amino acid residues on the β-defensin peptide then allow access into the bacterial membrane, which causes cellular distortion and disruption. In addition to their microbicidal activity, β-defensins can act as chemokines for DCs and T cells, thereby recruiting cells of the innate and adaptive immune systems to inflammatory sites along the respiratory tract. Cathelicidins are another family of broad-spectrum antimicrobial peptides and in humans LL37 is the only one described. LL37 is processed from its precursor protein hCAP18 (human cationic antimicrobial protein 18) into its biologically active from. Like defensins, LL37 has both microbicidal and chemotactic

properties and enhances bacterial defence in several tissues including the lung and skin (Chapter 10).

The most abundant antimicrobial proteins found in ASL are lysozyme and lactoferrin. Both of these molecules are produced by epithelial cells and neutrophils, while lysozyme is also produced by monocytes and macrophages. Lysozyme is thought to kill gram-positive bacteria through its enzymatic activity, resulting in peptidoglycan hydrolysis and cell lysis. In addition, lysozyme is able to kill gram-negative bacteria, which have a protective outer membrane preventing access to peptidoglycan, through a non-enzymatic mechanism. Lactoferrin is an iron-binding protein and has potent antimicrobial activity against iron-dependent bacteria. Furthermore, it is capable of acting as an immune modulator, polarizing responses toward a Th1 phenotype, through its recognition by receptors on DCs and macrophages.

## 8.7 Structure of the tonsils and adenoids of the Waldeyer's Ring

The location of the tonsils and adenoids makes them ideally positioned for capturing antigens that enter the upper airways. The lymphoid tissues of the Waldeyer's ring form a network that circles the naso-pharynx (Figure 8.5) and in humans represent the immune structures of the NALT. Unlike regional lymph nodes, which rely on the delivery of antigens via the lymphatic system, antigen is delivered directly to the tonsils and adenoids via their epithelial surfaces that are exposed to the outside.

The palatine tonsils are paired secondary lymphoid organs located at the entrance to the oropharynx, thereby positioning themselves at the head of the common entry site for both the digestive and respiratory tracts. Tonsils are characterized by several tubular crypts, increasing the tonsillar surface and antigen capture capabilities (Figure 8.6). Each crypt possesses an outer epithelial layer throughout which are distributed cells resembling Peyer's patch M cells. These M cells are responsible for the endocytic uptake and transport of antigen across the epithelium for delivery to subepithelial APCs such as DCs, macrophages and B cells. Beneath the epithelial layer are one or more secondary lymphoid follicles per crypt, depending on the extent of pathogen burden or antigenic stimulation. T cells and B cells can be found throughout the epithelium and, to a lesser extent, so can macrophages and DCs. B cells account for the majority of IELs in the tonsils, functioning mainly as antibody -producing cells.

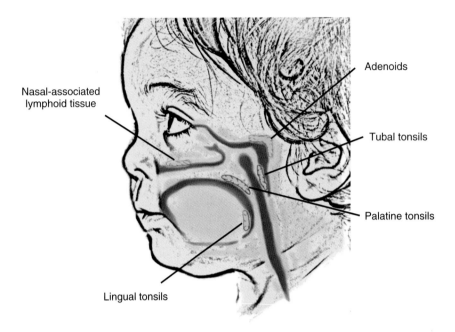

**Figure 8.5** The organized lymphoid structures of the Waldeyer's ring include the tonsils and the adenoids, located at the opening to the pharynx and oesophagus. These are important structures that sample antigens at the oropharyngeal opening and are sites of prolific IgA production.

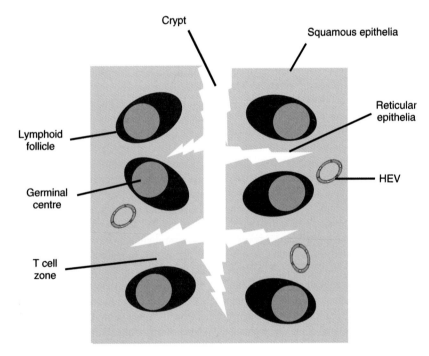

**Figure 8.6** Morphology and anatomy of the human palatine tonsil, which is characterised by deep crypts that function by trapping harmful substances. Numerous lymphoid follicles are associated with tonsillar crypts, in which B cell differentiation, germinal centre formation and antibody production take place.

Lymphoid follicles exist just below the epithelial surface and are major sites of lymphocyte activation, maturation and proliferation. These follicles are similar to those found in lymph nodes, whereby B cell rich germinal centres are surrounded by T cell rich parafollicular regions. B cells that have been activated by Th cells in organized lymphoid tissues, such as the tonsils, form primary follicles. The chemokines MIP-1α and MIP-1β are produced by B cells following BCR-antigen activation and form a chemoattractant gradient that assists in T cell migration into the B cell-rich follicles. The interaction between the antigen-specific B cell and the primed CD4+ T cell also involves ligation of the co-stimulatory molecule CD40 on B cells and CD40L on T cells. This induces the B cell to undergo proliferation and to form the secondary lymphoid follicles.

Continued proliferation of activated B cells forms each germinal centre. The rapidly dividing B cells are known as centroblasts and comprise the dark zone, while centroblasts that give rise to centrocytes translocate to the light zone, an area that contains fewer proliferating cells. In the light zone centrocytes interact with follicular DCs that retain antigen on the surface of their dendrites. This allows B cells to continue to receive positive signals through ligation and cross-linking of their BCRs. The competition for access to antigen on the surface of FDCs favours those B cells bearing receptors with the highest affinity to that antigen and forms the basis of affinity maturation and clonal selection. Antigen-stimulated B cells also interact with a specialized subset of CD4+ T cells, known as follicular helper T cells (Tfh cells), located throughout the light zone, an interaction thought to aid positive B cell selection, affinity maturation, antibody isotype class-switching and somatic hypermutation (Box 8.2). This interaction is again dependent on CD40-CD40L signalling and the production of IL-4 and IL-10 derived from T cells. Fully selected centrocytes can then migrate out of the germinal centre and become antibody producing plasma cells at secretory sites within tonsils/adenoids of the respiratory tract or become a sub-population of memory cells that are either retained in the lymphoid tissue or migrate to lymph nodes. Antibody producing plasma cells mainly secrete the IgA class of immunoglobulin. B cells that are not stimulated by antigen migrate to the mantle zone of the germinal centre.

Surrounding the germinal centres are the interfollicular areas, where T cells undergo activation through interactions with migratory DCs. The inter-follicular spaces contain numerous HEVs that act as sites of lymphocyte entry. In order for lymphocytes to cross the endothelium several cell-adhesion molecules are upregulated. In particular, lymphocyte function-associated antigen-1 (LFA-1) is upregulated by migratory cells, which recognizes inter-cellular adhesion molecules-1 (ICAM-1) on the endothelial cells. Surrounding the HEVs are T cells (mostly CD4+), B cells, macrophages and DCs. Migratory T cells that enter the inter-follicular areas scan the resident APCs for specific antigen. Ligation of the T cell receptor, plus appropriate secondary co-stimulatory signals, will activate the T cell within the lymphoid organ and initiate a primary T cell response. A subset of activated T cells exits within the tonsils via the draining efferent lymphatics and forms either an effector or memory T cell population. The effector T cells tend to migrate to sites of infection while the memory T cells migrate to regional lymph nodes.

## 8.8 Local lymph nodes and immune generation

The lung and lower airways differ from the upper airways in that they do not possess organized lymphoid structures. Therefore, in order for an effective adaptive immune response to be initiated, antigen presenting cells must migrate to sites of T cell priming within organized secondary lymphoid organs. These include the mediastinal lymph nodes, which drain the lower respiratory tract and the cervical lymph nodes, which drain the upper respiratory track (Figure 8.7). The mediastinal lymph nodes are located along the trachea, while the cervical lymph nodes are located along the muscles that form in the upper neck. Both sets of lymph node become significantly enlarged during a respiratory infection or indeed as a result of head, neck and lung cancers. Afferent lymphatic vessels drain lymph from the upper and lower respiratory tract, transporting APCs together with antigen to the regional lymph nodes (a more detailed exploration of lymphatic tissue and its role in immune cell priming is detailed in Chapter 1). Other sites of organized lymphoid tissue can also act as sites of T cell priming such as the BALT and structures of the Waldeyer's ring such as the tonsils.

Lymph nodes and the tonsils can be spatially compartmentalized into the medulla and the cortex. The medulla contains mostly proliferating B cells and is where germinal centres are formed. B cells that are activated by antigen, on the surface of follicular DCs, enter the medullary cords where they mature into antibody producing plasma cells. B cell proliferation and survival is also

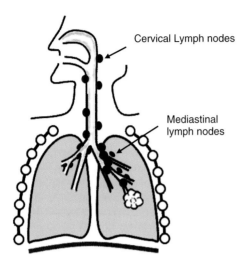

Cervical Lymph nodes

Mediastinal
lymph nodes

**Figure 8.7** Location of the regional cervical and mediastinal lymph nodes.

dependent on co-stimulation provided by a specialized T cell subset known as follicular helper T cells (Tfh cells), which are particularly abundant in the germinal centres of the tonsils (Box 8.2). Tfh cells express CD40L, which is an important co-stimulatory receptor that interacts with CD40 on B cells. Tfh cells also express the cytokines IL-4 and IL-10, which promote antibody class-switching to IgA.

---

## Box 8.2. Follicular Helper T Cells and Antibody Responses.

The activation and proliferation of B cells is a prerequisite for the synthesis of antigen-specific B cells. Although some B cell responses are independent of T cell help, requiring stimulation only with bacterial polysaccharides or TLR ligands, the majority of B cell responses are T cell dependent. The germinal centres of secondary lymphoid tissues, including the tonsils, provide an ideal environment for T cells and B cells to interact with each other. A specialized subset of Th cells, known as follicular helper T cells (Tfh cells), reside in germinal centres where they provide important stimulatory signals that support the survival and differentiation of memory B cells and antibody secreting plasma cells. Although antigen-priming of naïve B cells takes place in the periphery, Tfh cells provide the necessary signals for germinal centre formation, somatic hypermutation, class-switching and the production of high-affinity antibody.

---

Germinal centres form in the tonsils when antigen-primed B cells seed lymphoid follicles. Germinal centres can be divided into a dark zone, light zone and a mantle zone. The mantle zone contains predominately T cells and is where T cell priming is thought to take place. The dark zone comprises densely packed and rapidly dividing B cells known as centroblasts that are undergoing somatic hypermutation and antibody class-switching. Centroblasts that have successfully rearranged their immunoglobulin genes migrate to the light zone where they become non-dividing B cells known as centrocytes. It is within the light zone that interactions between centrocytes and Tfh cells takes place. In addition, follicular DCs (FDCs) reside within the light zone where they present immobilized antigen to B cells. This three-way interaction between FDCs, Tfh cells and centrocytes is an important checkpoint for the survival of antigen-specific B cells. FDCs present antigen to centrocytes, which is recognized by the newly synthesized, high affinity immunoglobulin (BCR). At the same time centrocytes process and present this antigen to Tfh cells, which recognize the antigen in the context of MHC class II molecules and induce the Tfh cells to provide B cell help. This results in the survival of centrocytes and differentiation into memory B cells or long-lived antibody-producing plasma cells.

Tfh cells express the chemokine receptor CXCR5, which is necessary for the migration into follicular germinal centres in response to CXCL13. They also express the memory marker CD45RO and the co-stimulatory receptors CD28, CD40L, ICOS and OX40. The interaction between CD40L and CD40 is crucial for the survival of germinal centre B cells, while OX40 is thought to promote the differentiation into memory B cells. Signalling via these co-stimulatory molecules also upregulates the enzyme AID, which is essential for somatic hypermutation and immunoglobulin class-switching. The co-stimulatory receptors CD28 and ICOS, expressed by Tfh cells, and their corresponding accessory molecules CD80/86 and ICOSL expressed by B cells, are also necessary for the formation of germinal centres and induce the expression of IL-10 and IL-21. These cytokines then signal to B cells, which provide further survival signals and promote antibody class-switching. Terminally differentiated plasma cells then leave the germinal centre and migrate to subepithelial areas within the tonsils. The resultant synthesis and secretion of IgA provides important immune protection in the upper airways.

The cortex of lymph nodes and tonsils consist of T cell areas where T cell priming takes places and is packed with reticular cells that trap DCs, macrophages and lymphocytes that enter the lymph node through the afferent lymphatics. The reticular network acts to maximize any interaction between APCs and T cells, effectively optimizing the chances of antigen stimulation and T cell priming. Activated T cells are then able to migrate out of the lymph nodes via the efferent lymphatics, re-enter the circulation and migrate to their effector sites in the lungs or upper respiratory tract. The homing of T cells to the appropriate sites in the respiratory tract is governed by a set of tissue-specific receptors.

## 8.9 Structure of the NALT

NALT is probably the first lymphoid site that contacts antigens from the environment and has been found to exist in several mammalian species including rats and mice. However, NALT has been less well described in humans and it may be that other lymphoid tissues such as the palatine tonsils and adenoids of the Waldeyer's ring play a more prominent role in immune defence. In animals the NALT is located at the entrance to the pharyngeal duct where it forms specific lymphoid aggregates (Figure 8.8). It has been suggested that the NALT is analogous to both the GALT and the BALT as it is structurally very similar. A layer of epithelial cells, together with M cells, goblet cells and IELs overlies the organized secondary lymphoid follicles beneath. The layer of epithelial cells directly overlying each follicle is often

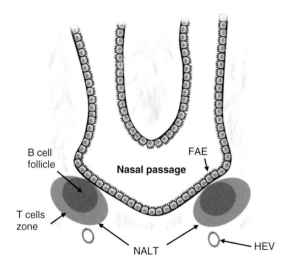

**Figure 8.8** Nasal associated lymphoid tissue (NALT) consists of organised lymphoid follicles beneath follicle associated epithelium (FAE). HEV, high endothelial venule.

referred to as the follicle-associated epithelium (FAE). As well as the presence of high endothelial venules (HEVs), the follicles are characterized by central B cell follicles surrounded by a parafollicular T cell area. As in other MALT, mature B cells (plasma cells) predominantly express IgA. The CD4+ T cells that are present are mostly in a naïve state (Th0 cells) reflecting their ability to differentiate into either Th1 or Th2 lineages. The CD8+ T cell compartments consists of conventional CD8+ $\alpha\beta$ T cells, as well as unconventional CD8$\alpha\alpha$+ $\alpha\beta$ T cells and CD8$\alpha\alpha$+ $\gamma\delta$ T cells that reside within the intra-epithelial compartment. In addition, macrophages and DCs reside throughout the lymphoid aggregates.

## 8.10 Structure of the BALT

Lymphoid tissue can be found along the entire length of the airways, referred to as BALT. The upper airways contain highly organized lymphoid structures, while the lower airways of the lung contain a diffuse immune network more characteristic of the systemic immune system. A mixture of the two types is found in the smaller airways, usually in the form of diffuse B cell aggregates without any associated T cell areas. The organized lymphoid structures of the BALT usually form at the confluence of bronchiole branches (Figure 8.9). The structure of the BALT consists of an overlying layer of epithelial cells containing M cells and IELs but lacks any goblet cells. In a normal respiratory tract very few M cells can be found, suggesting that epithelial cell differentiation into M cells occurs following antigen stimulation.

Similar to other lymphoid sites of the MALT, the follicular areas of the BALT consist of a B cell follicle and a parafollicular T cell area interspersed with macrophages and DCs (Figure 8.9). This lymphoid aggregate is similar to germinal centres found in lymph nodes and other secondary lymphoid tissues. Surrounding the follicle are interspersed T and B lymphocytes, macrophages and DCs. Follicular B cells are mostly proliferating memory B cells, while those located in the inter-follicular regions tend to be antibody secreting plasma cells. Most B cells express IgM or IgA, while the T cell population is a mixture of CD4+ $\alpha\beta$ T cells, CD8+ $\alpha\beta$ and CD8+ $\gamma\delta$ T cells. The only way that lymphocytes can migrate into the BALT is via blood thorough HEVs. Although there are draining efferent lymphatics surrounding the follicles there are no afferent lymphatics into the BALT, or indeed the NALT.

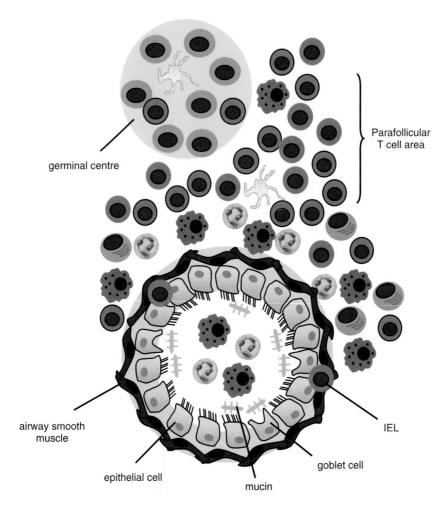

germinal centre

Parafollicular
T cell area

airway smooth
muscle

IEL

goblet cell

epithelial cell

mucin

**Figure 8.9** Development of germinal centres and lymphoid aggregates in BALT. Germinal centres form at the confluence of bronchiole branches. The germinal centre acts as B cell inductive site, wile the parenchyma surrounding the airway acts as an effector site. IEL, intra-epithelial lymphocyte.

Following an inflammatory insult or antigenic stimulus the frequency of BALT aggregates can increase along the airways, while each lymphoid focus can expand. In the absence of inflammatory stimuli, BALT is often very difficult to detect and if often found only in children and young adults, leading some to suggest that BALT is an inducible lymphoid aggregation rather than a constitutive organized lymphoid structure.

## 8.11 Cells of the lower respiratory tract

The lower airways comprise the terminal bronchi and alveolar ducts and are thought of as the respiratory portion of the respiratory tract responsible for gaseous exchange between inhaled air and the bloodstream. The lower airways are also highly vascularized in order to maximize the transfer of gases between the blood and alveolar spaces. Therefore the endothelia of capillaries are located in very close proximity to the single epithelial cell layer of the alveoli (Figure 8.10). The thin epithelial layer of the alveolar ducts consists of squamous type I epithelial cells, which comprise 95 per cent of the cells in the lower airways. Type II epithelial cells are located at the junctions between alveoli, are more cuboidal in shape and are capable of secreting pulmonary surfactant. It is thought that type II epithelial cells give rise to the more numerous type I epithelial cells. Other types

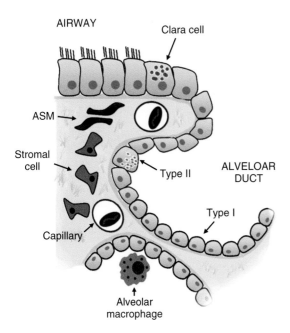

AIRWAY

Clara cell

ASM

Stromal cell

Type II

ALVELOAR DUCT

Type I

Capillary

Alveolar macrophage

**Figure 8.10** Structure of alveolar duct of the lower airways. The alveolar wall comprises mostly type I epithelial cells, which are the progeny of type II epithelial cells. Stromal cells, smooth muscle cells and an extensive capillary bed are located in the areas between alveoli (known as the alveolar septum or interstitium).

of epithelial cells, called Clara cells, are also capable of secreting surfactant and are located predominantly along the small airways and in the terminal bronchioles. Surfactant functions by reducing surface tension within the tiny alveoli and small bronchioles and consists of several phospholipids and small proteins.

## 8.12 Surfactant proteins

Multiple functions have been attributed to lung surfactant proteins including antimicrobial, immunomodulatory and in reducing lung surface tension at the air–liquid interface. Surfactant is an aqueous mixture of phospholipids and proteins, of which the surfactant proteins SP-A, SP-B, SP-C and SP-D constitute approximately 10 per cent. Surfactant proteins are produced and secreted by alveolar type II epithelial cells. In particular, SP-B and SP-C are hydrophobic polypeptides that act to lower the surface tension within the alveolus by forming complexes with phospholipids such as dipalmitoylphosphatidylcholine (DPPC), resulting in a more stable lung

microenvironment for the efficient transfer of gases. On the other hand, SP-A and SP-D have more immunomodulatory properties.

The most abundant surfactant protein in the lung is SP-A, which is a small glycoprotein ($\sim$ 30 kDa) that has both antimicrobial and immunomodulatory properties. It can directly kill microbes through the formation of reactive oxygen species (ROS) and can inhibit bacterial cell division by permeabilizing the cell membrane. It is capable of activating macrophages and enhancing the opsonization and phagocytosis of pathogens. SP-A and SP-D are both C-type lectins, also known as a collectins, which bind to various receptors on the surface of immune cells and hence act as immunomodulators. SP-D is able to bind to various molecules on the surface of viruses, bacteria, fungi and parasites, thereby enhancing phagocytosis and cell-mediated killing of pathogens in a similar manner to mannose-binding lectin (discussed in Chapters 12 and 13). Although associated with the innate immune response and pathogen-induced inflammation, SP-D also possesses anti-inflammatory properties. It can reduce the production of pro-inflammatory cytokines such as IL-1 and TNF and can reduce the number of T cells proliferating in the lung through the inhibition of IL-2.

## 8.13 Immune modulation by airway epithelial cells

Airway epithelial cells form a tight barrier at the interface between the mucosa and the environment and are in continuous contact with potentially pathogenic microorganisms but at the same time tolerate the presence of harmless microbes. Epithelial cells therefore possess sophisticated immune recognition mechanisms and have evolved strategies that recognize microbial danger signals but at the same time preserve local tissue homeostasis. They are able to secrete antimicrobial peptides, release inflammatory cytokines and chemokines and act as immunomodulatory cells that help to maintain homeostasis in the lung. In order to rapidly respond to pathogen encounters, airway epithelial cells express a number of pattern recognition receptors (PRRs) on their cell surface, in particular members of the TLR family. Although responding to the threat of pathogen invasion is crucial to the health of an individual, protecting the lungs against excessive or inappropriate inflammation is also desirable.

The expression of TLRs often varies between cell types and different mucosal tissues. The respiratory tract differs from the gut in that it does not possess such a large microbial population. Airway epithelial cells therefore do not express TLRs to the same degree as gut epithelial cells, thereby making them far less responsive to microbial stimulation. This lack of TLR expression helps airway epithelial cells to maintain the fragile microenvironment of the lung. However, TLR2 and TLR4, for example, are upregulated in response to TNF or IFN-γ, making airway epithelial cells more responsive to microbial stimulation during chronic infection or in times of heavy pathogen load. In response they produce a number of pro-inflammatory cytokines such as IL-1 and IL-6, which are mostly dependent on activation of the transcription factor NF-κB, and various chemokines such as RANTES (CCL5) and MIP−3α (CCL20). This has the combined effect of activating local DCs and recruiting neutrophils and T cells to the area. A close association between epithelial and both the innate and adaptive immune systems therefore exists in the airways.

## 8.14 Innate immune response

When an invading pathogen overcomes initial lines of barrier defence and humoral immunity, its presence must be rapidly identified in order to mount a coordinated immune response against it. Cells of the innate immune system are responsible for pathogen recognition, induction of an appropriate immune response and also the restoration of homeostasis and the prevention of tissue damage. Although the detection of exogenously derived microbial products (PAMPs) represents a crucial means of establishing the presence of a pathogen, endogenously derived molecules released as a result of tissue damage (DAMPs) are also recognized by innate immune cells. Together, these danger signals reveal to the innate immune system that the epithelial cell layer has been penetrated.

Epithelial cells, DCs and macrophages are all known to express a variety of TLRs, as well as intracellular NOD-like receptors and macrophage-borne scavenger receptors. Receptor binding to its ligand results in the release of pro-inflammatory cytokines from these cells, most notably IL−1β and IL-8. This cytokine release has the effect of activating macrophages and DCs and recruiting neutrophils to the site of inflammation. PRR ligation also activates epithelial cells to produce more antimicrobial peptides and surfactant proteins. The release

of chemokines and other pro-inflammatory mediators such as TNF and IL-12 further enhances inflammatory cell migration into the area. The release of IL-8 from macrophages is particularly adept at recruiting neutrophils form the vasculature and into the alveolar space. Although crucial for innate defence against invading pathogens, in particular bacteria, excessive neutrophil recruitment and macrophage activation can lead to unwanted inflammation and tissue damage, as exemplified during ALI and COPD (Box 8.1). Both alveolar macrophages and neutrophils phagocytose microbes, although neither is very good at presenting antigen. Rather macrophages transport microbial products to regional lymph nodes where they are taken up by DCs, which in turn present antigen to T cells and induce an adaptive immune response.

## 8.15 Dendritic cells are located throughout the respiratory tract

Although DCs are frequent throughout the BALT, NALT and the tonsils, a large number of DCs are also located throughout the respiratory epithelium and the lung parenchyma where they continuously sample antigens present within inhaled air. They are responsible for distinguishing between harmful pathogens and harmless substances. In so doing, airway DCs are central for maintaining homeostasis, either by inducing tolerance to innocuous antigens or by initiating an immune response to a pathogenic insult. Airway mucosal DCs (AMDCs) form a network along the conducting airways and are mostly mDCs, although a scattering of pDCs are present. In addition, a langerin+ population of DCs are located mostly in the mucosa of the conducting airways and are thought to play a similar role in immune surveillance as LCs in the skin epidermis. Within the lower airways and lung tissue, lung parenchymal DCs (LPDCs, or interstitial DCs) are located in the alveolar epithelium, alveolar space and the interstitial tissue between the epithelium and vasculature, and are considered similar to intestinal LPDCs in that they comprise mostly CD11b+ mDCs. Respiratory tract DCs tend to be excellent at antigen capture and processing but are weak activators of T cells, an important characteristic for preventing pathogenic immune reactions and protecting the fragile structure of the lung.

Respiratory tract DCs are able to efficiently recognize and respond to microbial PAMPs, principally through TLR ligation, although not as robustly as gut-derived DCs. It has been suggested that airway DCs are predisposed to driving Th2 responses and can effectively induce B cells to class switch to IgA through the production of TGF-β. Moreover, airway DCs tend to be more immunomodulatory, in that they can induce Th cells to express IL-10, and therefore differentiate into Treg cells, thus contributing to mechanisms of immune regulation in the airways. DCs can also migrate to draining lymph nodes and upregulate MHC class II molecules, co-stimulatory molecules (CD80/CD86) and certain chemokine receptors (such as CCR7) following microbial stimulation (Figure 8.11). In effect, DCs that are activated by danger signals undergo maturation from antigen processing cells to antigen presenting cells. Within the regional mediastinal and bronchial lymph nodes DCs are then able to transfer antigen captured in the respiratory tract to lymph nodes, where they present antigen to T cells. Primed T cells then migrate to the lungs where the original danger signal was recognized and participate in an immune response to the invading pathogen.

## 8.16 Alveolar macrophages maintain homeostasis

In healthy lungs alveolar macrophages remain quiescent, preventing unwanted inflammation and damage

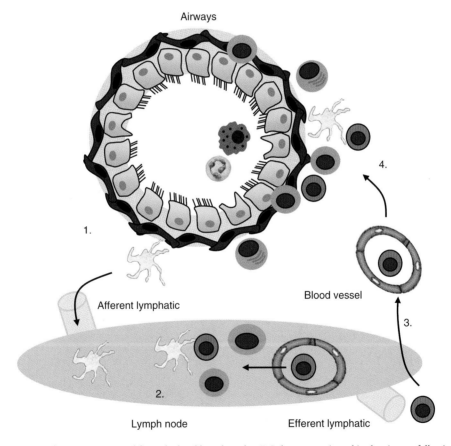

**Figure 8.11** Migration of airway DCs to and from the local lymph nodes. DCs become activated in the airways following infection and migrate to the cervical or mediastinal lymph nodes, where they present antigen to lymphocytes and activate adaptive immune responses. Primed lymphocytes then migrate out of the lymph node and are recruited to sites of inflammation within the airways, where they exert their effector functions.

to the fragile structure of the lung. They are the most abundant immunocompetent cell type found within the alveoli, alveolar-interstitial areas and the conducting airways. They are therefore one of the first cell types to encounter antigen in the lungs and hold an important balancing act between immunosuppression and immune activation. The immunosuppressive qualities of alveolar macrophages mainly stem from their ability to secrete anti-inflammatory cytokines such as IL-10 and TGF-β. The other vital duty that alveolar macrophages perform in the lung is to phagocytose invading microbes. They must respond quickly to pathogenic insults and are therefore capable of producing the pro-inflammatory cytokines IL-12, IFN-γ and TNF, which orchestrate a Th1 polarized immune response (Figure 8.12).

In addition to microorganisms, alveolar macrophages are able to phagocytose apoptotic cells in the airways, including neutrophils that have migrated into the respiratory tract in response to inflammation. Neutrophils tend to have a short lifespan and undergo apoptosis once they have phagocytosed bacteria and released reactive oxygen species into their phagolysosome. An important interaction exists between macrophages and neutrophils, whereby macrophages clear apoptotic neutrophils from the lung. Disregulation of this important process often leads to tissue damage, resulting from the release of neutrophil granule contents into the extracellular environment.

Following an effective immune response, invading bacteria are eventually cleared from the lungs. This causes a reduction in pro-inflammatory cytokine expression and the removal of pro-survival factors necessary for macrophage function. Activated macrophages therefore undergo apoptosis and are cleared from sites of

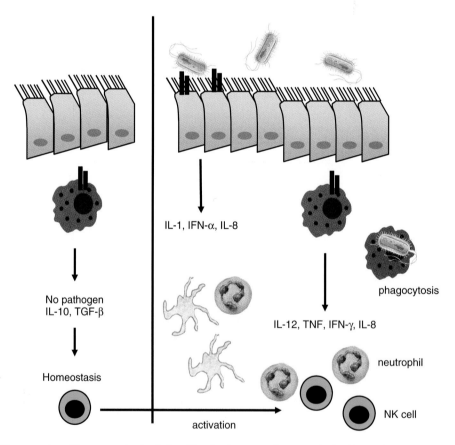

**Figure 8.12** Homeostasis and immune activation induced by alveolar macrophages. Under healthy condition alveolar macrophages have an immunoregulatory phenotype and express the cytokines TGF-β and IL-10. However, an infectious insult to the lungs can induce alveolar macrophages to be inflammatory and secrete IL-1, TNF and IL-8. These inflammatory mediators act by recruiting leukocytes into the site of infection and stimulate neutrophil, macrophage and NK cell effector activity.

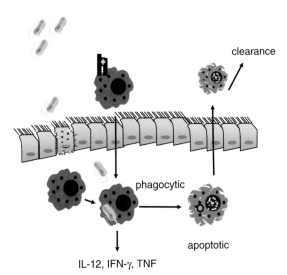

clearance

phagocytic

apoptotic

IL-12, IFN-γ, TNF

**Figure 8.13** Recognition of danger signals by alveolar macrophages causes the release of pro-inflammatory mediators and activates their phagocytic activity. Phagocytosis and killing of bacteria, for example, causes macrophages to undergo apoptosis. Clearance of apoptotic macrophages, together with the killed bacteria, contributes to pathogen clearance and downregulation of the inflammatory response.

inflammation (Figure 8.13). The induction of apoptosis in macrophages, and other inflammatory cells, is an important mechanism that limits tissue damage and restores homeostasis. Alveolar macrophages can also detect the presence of other cells undergoing apoptosis, leading to the inhibition of pro-inflammatory mediators and production of anti-inflammatory cytokines.

## 8.17 NK cells in the lung

In normal, healthy airways NK cells are maintained in a resting state, mainly through the production of IL-10 and TGF-β from alveolar macrophages. However, they are quickly activated in response to pulmonary infections and are stimulated by IFN-α and IL-2. The key ability of NK cells is to form direct synapses with infected cells and release their cytotoxic granules onto the cell surface. This is an important mechanism for the control of viral infections (discussed in depth in Chapter 11), whereby they recognize various cell surface receptors on infected cells or kill opsonized cells by means of antibody-dependent cellular cytotoxicity (ADCC). Activated NK cells are also adept at producing large quantities of IFN-γ

following exposure to IL-12, which promotes DC survival and drives the proliferation and migration of Th1 cells to the site of infection. Alternatively, in the absence of IL-12 and the presence of IL-4 NK cells can produce Th2 promoting cytokines such as IL-5 and IL-13.

## 8.18 T cells at effector sites in the lung

Considering the lower airways do not possess organized lymphoid structures, T cells are usually primed for antigen encounter within the mediastinal or cervical lymph nodes and are equipped for effector function once they migrate and enter the non-lymphoid environment of the lung. Within the lung, the precise effector site for T cells is governed by the structure of the tissue. Therefore, effector sites within the lung are usually within the lung parenchyma or the airway lumen (Figure 8.14). The phenotype of the effector response in the lung is governed by several factors. Firstly, the interaction with tissue-specific

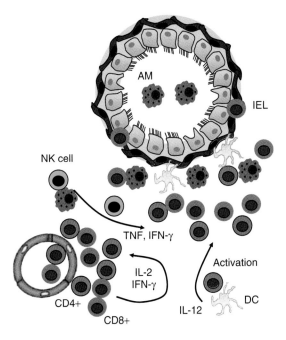

AM

IEL

NK cell

TNF, IFN-γ

Activation

IL-2
IFN-γ

CD4+

CD8+

IL-12

DC

**Figure 8.14** Induction of cell-mediated immune responses in the lower airways. Immune responses to invading pathogens can take place in the lung parenchyma and the interstitial tissue surrounding the smaller airways. Loose lymphoid aggregates can form in these areas, where primed lymphocytes exert their effector functions. However, over-exuberant cell-mediated immune responses in the airways can lead to bystander tissue damage and immunopathology. Alveolar macrophage, AM; intra-epithelial lymphocyte (IEL).

trafficking molecules and engagement with cell adhesion receptors on lung endothelium; secondly, the ligation of T cell receptors (TCRs) and co-stimulatory molecules with APCs within the lung tissue; and thirdly, the pattern of cytokines and chemokines that are expressed by various cells participating in the immune response. These interactions can largely dictate whether effector T cells that enter the lung undergo Th1, as is the case following a viral infection, or Th2 differentiation, as is the case during allergic asthma.

## 8.19 Memory T cell responses within the lung

Under normal homeostatic circumstances there is a significant population of memory CD4+ and CD8+ T cells that reside in the lung parenchyma and airways. These cells are thought to be responsible for maintaining tolerance in healthy lungs. However, under inflammatory conditions, these resident memory T cells are able to respond rapidly to a secondary respiratory infection. They mediate immunity to infection via cytokine production and direct cytotoxic activity against infected cells. They require a lower threshold for TCR engagement with MHC molecules on the surface of APCs and need less co-stimulatory signals to become activated compared to naïve T cells. Memory T cells that reside within lung tissue have low expression of the homing receptors CCR7 and CD62L and have the characteristics of effector memory T cells, including the expression of CD45RO. In addition, many T cells migrate into the lung from secondary lymphoid tissues. The phenotype of these migratory T cells is characteristic of effector memory T cells ($T_{EM}$), although central memory T cells ($T_{CM}$) can also be detected to a lesser extent. $T_{CM}$ cells express CCR7 and CD62L, which helps their retention in lymph nodes and secondary lymphoid tissue. However, upon activation $T_{CM}$ cells can differentiate into $T_{EM}$ cells and downregulate CCR7 and CD62L, which allows them to migrate into effector tissues. Such CD4+ memory T cells are able to secrete a larger amount of the cytokines IL-2 and IFN-γ, which enhance their own proliferative capacity and that of other lymphocytes. Memory CD8+ T cells have a much higher lytic capacity, allowing them to more effectively kill virally infected cells. Another characteristic of effector memory cells is the imprinted expression of lung-homing trafficking molecules. This allows for rapid migration of circulating memory cells back into the site of original antigen encounter. This includes the expression of the

chemokine receptors CCR4 and CCR5, which respond to the chemokines CCL17 and CCL5 (RANTES) respectively, both of which are upregulated in the respiratory tract during an inflammatory response.

The BALT and tonsils can also mount a secondary immune response, primarily involving the activation of memory B cells and the production of high-affinity IgA antibody. BALT and NALT may not be sites of primary B cell priming, as this is more likely to occur in the mediastinal and/or cervical lymph nodes; rather, BALT and NALT act as sites for B cell proliferation and expansion of memory B cells. The rapid expansion of plasma cells involves an interaction with T cells via CD40 on the memory B cell and CD40L on the T cell. Memory B cells produce much more antibody than naïve B cells and undergo terminal differentiation into plasma cells. This has the double effect of mounting a substantial antibody response to secondary infection and at the same time controlling the size of the memory B cell pool. During secondary immune responses, B cells can also act as effective APCs and active memory T cells. Both memory B cells and memory T cells are located in the subepithelial areas of the tonsillar crypts, directly beneath M cells. Antigen recognition by memory B cells, direct interaction with T cells and the close proximity to sites of antigen uptake all assist in the subsequent activation of memory T cells by antigen presenting B cells.

## 8.20 Migration of circulating T cell into the lung tissue

Lymphocytes that wish to enter lung tissue or the airways must express a defined set of trafficking molecules. Integrins play an important role in tissue-specific homing, in particular the expression α4 integrins (α4β1 and α4β7) on lymphocytes are thought to be necessary for migration into the lung. The ligand for α4β7 is E-cadherin and is expressed in several mucosal tissues such as the gut and lung. Another integrin (α1B1) is highly expressed on CD8+ T cells that enter the lung in response to viral infection. The relative importance of integrin expression of lymphocytes is not as clearly defined for the lung as it is for the gut. This may be due to the unique properties of the pulmonary vasculature within the lung. The small size of the capillary microvessels, and the close proximity of the endothelia and epithelia, means that lymphocyte transmigration into the lung may not follow the same rules as in other mucosal tissues. Therefore, the lung possesses a unique set of cell adhesion molecules and chemokines.

Other molecules are involved in lymphocyte trafficking into the lung. For example, leukotriene and prostaglandin receptors are involved in CD4+ T cell migration. Lung epithelial cells express the prostaglandin DP1 receptor, which interacts with the prostaglandin PGD2 expressed by T cells. Several chemokine receptors are also involved in lymphocyte homing to the lung, including CCR3, CCR4 and CCR8, which are all known to be expressed on CD4 T cells. The chemokine CCL5 (also known as RANTES) is constitutively expressed in lung tissue and is upregulated during episodes of inflammation. CCL5 is a potent lymphocyte chemoattractant as its receptor CCR5 is highly expressed on effector T cells, directly influencing the migration from the pulmonary vasculature into the lung parenchyma (Figure 8.15). However, a combination of integrin and chemokine receptor expression has not yet been identified that is specific to lung homing lymphocytes. It may also be the case that slightly different trafficking molecules and chemokines are required for homing to the lung parenchyma compared to the lung

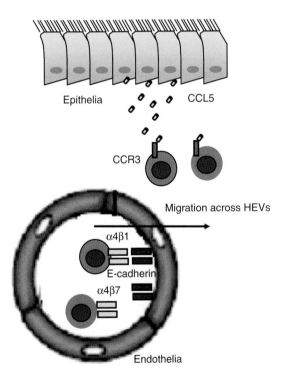

**Figure 8.15** Migration of lymphocytes into the airways from HEVs involves the expression of a specific set of cell adhesion molecules and chemokine receptors.

airways. For example, T cells located in the airways have a lower expression of LFA-1 compared to T cell within the parenchyma. It may be that as T cells migrate across the epithelial layer and into the airways they downregulate the expression of LFA-1.

Lymphocytes re-circulate from peripheral lung tissue back into secondary lymphoid structures. Migration out of the tissue and into the draining afferent lymphatics, which feed regional lymph nodes, is associated with the upregulation CCR7 and CD62L expression. The expression of CCR7 and CD62L is also associated with a $T_{CM}$ cell phenotype. Migration of lymphocytes out of lymph nodes and into the efferent lymphatics is associated with a loss of CCR7 and CD62L expression. T cells that wish to enter peripheral lung tissue express very low levels of both CCR7 and CD62L, the expression of which is associated with an effector T cell phenotype.

## 8.21 IgA production in the respiratory tract

The primary function of plasma B cells at mucosal sites is the production of secretory IgA (Figure 8.16). As in the gut and other mucosal tissues, most IgA produced in the lung is polymeric pIgA, usually in a dimeric form linked together with a joining chain (J-chain) tail-to-tail. Dimeric pIgA is a secretory immunoglobulin produced by plasma cells in sub-mucosal sites, which is translocated across the epithelial layer through an association with secretory component expressed by epithelial cells. Secretory component (otherwise known as the pIg receptor, pIgR,when bound to the basolateral surface of epithelial cells) is expressed at a much lower level in the respiratory tract compared to GALT and may reflect differences in bacterial load or differences in tissue function. In contrast to the gut, the IgA1 sub-class of antibody is more highly expressed in the lung compared to IgA2. IgM can also associate with the epithelial secretory component (pIgR,), although pIgA remains the most abundant secretory immunoglobulin in the airways. The pIgR is expressed on mucous-producing and ciliated epithelial cells throughout the airways. Other immunoglobulin isotypes can also be found in the airways, although they are not actively transported across epithelial cells. Rather, monomeric forms of IgG isotypes and IgE reach airway secretions mainly through a process of passive diffusion through epithelial cell tight junctions.

Isotype class-switching occurs in activated B cells in the germinal centres of lymphoid follicles. As well as

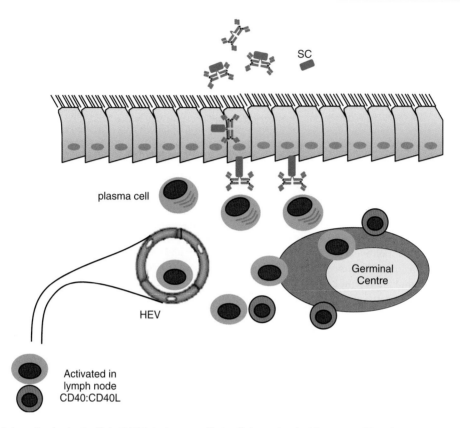

**Figure 8.16** IgA production by B cells in BALT. Antigen-specific B cells lare primed within regional lymph nodes. Migration into BALT is associated with immunoglobulin class switching and IgA secretion. IgA is transported across the respiratory tract epithelia in association with secretory component (SC).

close interactions with primed T cells, exposure to certain cytokines induces B cells to undergo class-switching. Probably the most important cytokine for class-switching to the IgA isotype is TGF-β, although IL-2, IL-4 and IL-10 may also be important, if only in providing proliferative signals for the differentiating B cells.

B cells are found throughout the lung, BALT and structures of the NALT where antigen stimulation causes them to preferentially accumulate in the germinal centres of secondary lymphoid follicles. The tonsils and adenoids contain numerous germinal centres containing IgA-producing plasma cells (Figure 8.6). Following somatic hypermutation and affinity maturation, mature plasma cells migrate out of the germinal cells and into the extrafollicular areas, particularly in areas directly beneath the tonsillar crypt epithelium. Alternatively, plasma cells will migrate to mucosal sites of the upper respiratory and the nasal mucosa. Transport of pIgA across mucosal epithelia of the upper respiratory tract forms an important

mechanism by which pathogens are prevented from invading these sites and protecting the delicate structures of the lower airways and alveolar spaces. Alternatively, plasma cells can migrate to local exocrine glands including the salivary glands of the mouth and the lachrymal glands of the eye, where they provide an important source of sIgA.

## 8.22 Respiratory diseases and pathogens

Any breakdown in the normal homeostatic mechanisms in the airways is likely to lead to a respiratory disease, which is often driven by a central immune-mediated process. Non-infectious lung diseases can be broadly categorised into obstructive lung disease and restrictive lung diseases (Table 8.2). Obstructive lung diseases are characterised by a narrowing of the airways that causes difficulties in breathing and can sometime be fatal. For

**Table 8.2** Respiratory diseases.

| Disease family | Example | Phenotype |
|---|---|---|
| Obstructive lung diseases | Asthma | Smooth muscle hyperplasia |
| | | Airway narrowing |
| | | Th2-mediated inflammation |
| | Chronic obstructive pulmonary disease (COPD) | Airway narrowing |
| | | Includes emphysema and chronic bronchitis |
| | | Th1-mediated inflammation |
| | Bronchitis | Mucus secretion, cough |
| | | Epithelial cell hyperplasia |
| | Emphysema | Alveolar and airway wall destruction |
| Restrictive lung diseases | Pulmonary fibrosis | Fibrosis of lung parenchyma |
| | | Oxygen deficiency |
| | | Includes asbestosis and idiopathic pulmonary fibrosis |
| | Acute respiratory distress syndrome (ARDS) | Inflammation of parenchyma |
| | | Alveolar damage |
| | Infant respiratory distress syndrome (IRDS) | Insufficient surfactant protein |
| | | Collapsed alveoli |
| | Sarcoidosis | Granuloma formation |
| | | Th1-polarized immune response |
| | | Autoimmune contribution |

example, COPD and severe asthma are both associated with inflammation of the respiratory tract. However, differences do exist between the two diseases in that asthma is a Th2-driven disease characterised by a reversible narrowing of the airways (discussed in detail in Chapter 15), while COPD is a Th1-driven disease characterized by a progressive, non-reversible narrowing of the airways (Box 8.1). Although the clinical outcome may be similar, the immune mechanisms that mediate these two diseases are distinct, as the combination of cytokines, inflammatory mediators and the cell types recruited during the inflammatory reaction are different.

The airways are exposed to a variety of infectious pathogens, as anyone who has recovered from the flu or a 'chest' infection will confirm (Table 8.3). The severity of these infections can be vastly different, with upper respiratory tract infections being far less deadly than lower respiratory tract infections, a consequence that maybe related to the vital gaseous exchange function of the lower airways. The most common upper respiratory tract infections are the result of viruses (discussed in detail in Chapter 11) that cause the common cold, such as

rhinoviruses, parainfluenza viruses and respiratory syncytial virus (RSV). However, as is the case with RSV, they can sometimes infect the lower airways, particularly in young infants or the elderly. Influenza virus is another potentially deadly virus that normally infects the upper respiratory tract but can cause life-threatening pneumonia in severe cases. Lower respiratory tract infections tend to result in pneumonia that is characterized by an excessive inflammation that blocks the airways, destroys the alveoli and creates breathing difficulties that prevent gaseous exchange. Bacteria such as *Streptococcus pneumoniae* and *Haemophilus influenzae* tend to be the leading causes of such life-threatening infections (discussed in detail in Chapter 12). Another bacterium that has been infecting humans for tens of thousands of years is *Mycobacterium tuberculosis*, which causes the granulomatus and potentially fatal disease tuberculosis. Although less frequent, fungal species can also infect the airways but usually only in individuals who are immunosuppressed or undergoing chemotherapy (discussed in detail in Chapter 13). Luckily, despite the constant barrage of pathogens, the vast majority of respiratory infections are dealt with appropriately and effectively by our immune system.

**Table 8.3** Infectious respiratory diseases.

| Disease family | Example | Phenotype |
| --- | --- | --- |
| Upper respiratory tract infections | Rhinovirus | Causes common cold |
| | | Sore throat, runny nose, mucus secretion, fever |
| | Respiratory syncytial virus (RSV) | Similar to common cold in adults |
| | | Major cause of bronchiolitis in infants |
| | | Can cause lower tract infections and pneimonia |
| | Parainfluenza virus | Common cold symptoms |
| | | Can cause pneumonia in infants |
| | Influenza virus | Causes the 'flu' |
| | | Fever, chills, headache, weakness, fatigue, sore throat |
| | | Can cause fatal pneumonia |
| | *Streptococcus pyogenes* | Pharyngitis or 'strep throat' |
| | | Sore throat |
| | | Toxins cause scarlet fever |
| Lower respiratory tract infections | *Streptococcus pneumoniae* | Pneumonia |
| | | Inflammation of parenchyma |
| | | Fluid filled alveoli |
| | *Mycobacterium tuberculosis* | Tuberculosis |
| | | Th1-mediated granulomas |
| | *Haemophilus influenzae* | Pneumonia |
| | | Meningitis |
| | Severe acute respiratory syndrome (SARS) | Caused by a coronovirus |
| | | Pneumonia |

## 8.23 Summary

1. The respiratory tract can be divided into the upper respiratory tract (conducting airways) and the lower respiratory tract (respiratory airways).
2. The epithelium differs depending on the level of the respiratory tract and provides barrier defence and actively participates in immune responses.
3. Mucus secretion and the mucociliary elevator are important innate defence mechanisms in the upper airways.
4. NALT, BALT and the structures of the Waldeyer's ring (such as the tonsils) possess organized lymphoid tissue and act as inductive sites for immune responses in the airways.
5. Alveolar macrophages play an important role in maintaining homeostasis within the respiratory tract.
6. Leukocyte migration into the lungs is influenced by the unique architecture of the pulmonary vasculature.
7. The lower airways can host inducible immune responses, including the recruitment of protective memory T cells.

## With contributions from Prof. Tracy Hussell

Leukocyte Biology Section, National Heart and Lung Institute, Imperial College, London

# 9 Immunology of the Urogenital Tract and Conjunctiva

## 9.1 The urogenital tract as a MALT

The excretory system of the urinary tract and the reproductive system of the genital tract are related morphologically and share a common opening to the outside, even though their respective functions are different. The development of the urogenital tract is largely completed by the end of foetal life and due to the developmental complexities, further detail is outside the scope of this textbook. The urinary system includes the kidneys, bladder and urethra, while the reproductive system includes the gonads and prostate gland in males (Figure 9.1) or the ovaries, uterus, cervix and vagina in females (Figure 9.2). In human males, the excretory system and the reproductive system utilize the same duct through which both urine and sperm pass, while the female urinary tract is separate from the genital tract. Although immune defence is of upmost importance along the male urinary tract, more attention will be given to female urogenital tract, due to the greater amount of mucosal tissue present and its vital role in protecting both the mother and developing foetus during pregnancy.

Due to the anatomical location and physiological functions, the urogenital tract (otherwise known as the genitourinary system) is often exposed to the environment and is therefore a potential entry site for pathogens. The female genital tract, in particular the vaginal canal, is especially vulnerable to infection and has therefore evolved specific mechanisms to counteract this threat. Sexually transmitted diseases, such as chlamydia, herpes and HIV (Box 9.1) are a major health issue worldwide and contribute to a significant amount of morbidity and death of both sexes. As well as having to protect the host from sexually transmitted diseases, the MALT of the female reproductive tract must handle various physiological events such

as fertilization (associated with immunogenic spermatozoa), extensive hormonal alterations and pregnancy. In doing so, the separate anatomical locations of the ovaries, cervix, uterus and vagina have evolved distinct adaptations in order to protect against pathogens and to maintain immunological homeostasis during pregnancy. Indeed, the uterus and placenta possess unique features that enable the mother and foetus to tolerate each other. It has been suggested that the various immunological mechanisms that exist within the different anatomical locations throughout the reproductive tract are designed to protect and support the survival of the mother and the foetus.

Compared to the gastrointestinal and respiratory tracts, the urogenital tract has proved to be much more difficult to study. This is due to the lack of opportunity to study specimens from the delicate tissues involved and also due to the reproductive tract being at the interface between the immune system and the endocrine system. The effect of myriad hormones on various aspects of urogenital immunity makes studying the immunology of the reproductive tract a more complex prospect compared to other MALT. This is particularly true for the female reproductive tract, which constantly responds to fluctuations in hormone levels. These endocrine changes have a significant effect on epithelial cell immune function and on cells of both the innate and adaptive immune system and therefore on subsequent responses to invading pathogens. Another very important consideration is that the MALT of the female reproductive tract differs considerably to that found in the intestinal or respiratory tracts. Unlike the latter, the female reproductive tract does not possess organized lymphoid structures equivalent to Peyer's patches or other lymphoid aggregates. Rather, loose associations of B and T cells can be found within the lamina propria of the uterus. This may be one

*Immunology: Mucosal and Body Surface Defences*, First Edition. Andrew E. Williams.
© 2012 John Wiley & Sons, Ltd. Published 2012 by John Wiley & Sons, Ltd.

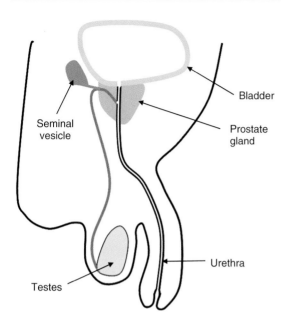

**Figure 9.1** The male urogenital tract.

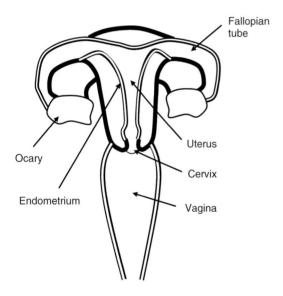

**Figure 9.2** The female reproductive tract.

of the reasons why it has proven so difficult to induce local immune responses in the urogenital tract through immunization. However, the lack of organized lymphoid aggregates may be an evolutionary consequence of providing an immunologically receptive environment for the developing foetus. The exception to this may be a region of the cervix known as the transformation zone, where the squamous epithelium of the ectocervix gives way to columnar epithelium of the endocervix. Organized lymphocyte tissue that resembles germinal centres can be induced following infection, with human papilloma virus for example, and is rich in T cells and APCs.

## 9.2 Epithelial barrier function

As a member of MALT, the urogenital tract is characterized by a continuous layer of epithelial cells that forms a protective physical barrier against pathogens. The unique anatomical function of the female reproductive tract means that there is a high degree of epithelial specialization along its length. The vagina and outside of the cervix (the ectocervix) are lined with stratified squamous epithelia, similar to the mucosa of the conjunctiva (discussed later in this chapter). The inside of the cervix (the endocervix), the uterus (endometrium) and the fallopian tubes are lined with columnar epithelial cells (Figure 9.3). The mucosa lining the uterus is particularly affected by hormonal changes during the fertility cycle, as the epithelia is sloughed off during menstruation. Epithelial shedding occurs in response to a sudden decrease in progesterone and oestrogen, following unsuccessful fertilization. The endometrium, including the columnar epithelial lining, must be rebuilt following menstruation, a process involving epithelial, immune and stromal cell interactions.

The epithelial monolayer not only provides a physical barrier between the lumen of the female reproductive tract and the cells beneath but also actively participates in adaptive immune responses. Unlike the epithelium of the intestinal or respiratory tracts, the female reproductive tract does not appear to possess the equivalent of M cells and therefore is not capable of developing lymphoid aggregates. However, epithelial cells of the reproductive tract are capable of producing cytokines and chemokines that attract or activate cells of the innate and adaptive immune systems. For example, TLRs expressed on epithelial cells recognize pathogen derived components and stimulate the production of pro-inflammatory cytokines. It is thought that uterine epithelial cells are capable of expressing most TLRs, including TLR4 (which recognizes LPS), TLR5 (which recognizes bacterial components such as flagellin) and TLR3 (which recognizes double-stranded RNA from viruses), making them important responders during first line defence. For example, stimulation of

## Box 9.1. Infection of the Reproductive Tract with HIV.

The human reproductive tract is markedly different to other mucosal tissues; in that systemic and local IgG often predominates over IgA and that it lacks organized lymphoid aggregates capable of acting as initiation sites. These unique features are thought to be the reason why it has been so difficult to design and implement a successful vaccine to pathogenic diseases that affect the reproductive tract. The development of a successful vaccine against HIV infection has also succumbed to these difficulties, especially considering that the reproductive tract is the major portal of entry for HIV, and such a vaccine would provide substantial health benefits worldwide. Despite the numerous defence mechanisms that exist in the reproductive tract, including antimicrobial peptides, mucus secretion, antibody production and cellular immunity, HIV is still able to infect its host. Several mechanisms have been proposed by which HIV invades cells and eventually infects and destroys CD4+ T lymphocytes.

HIV must first transverse the epithelial lining of the reproductive tract and in the female the simple columnar epithelia of the endocervix seems more permissive than the squamous epithelia of the vagina and ectocervix. Once entry to the lamina propria has been made HIV is capable of infecting macrophages, which then migrate into the circulation and regional lymph nodes where they come into contact with CCR5 expressing CD4+ T cells. The chemokine receptor CCR5 is also expressed on lamina propria DCs and intraepithelial Langerhan's cells, which are likely to participate in viral uptake and transmission to susceptible CD4+ T cells (see first figure accompanying this box). Interestingly, the CCR5+ DCs also reside in the epithelia of the intestine and are thought to express higher levels of CCR5 than those in the lamina propria of the vagina. Therefore, differences in DC populations, and in the structure of the epithelium, may contribute to the reasons why viral transmission is higher in the rectum compared to the vagina. HIV tropism, with regard to different chemokine receptors, was discussed in Chapter 5.

The aim of a successful vaccine against HIV is to induce both humoral and cellular immune responses against the virus. Of particular importance may be the production of virus-neutralizing antibodies, as demonstrated by the protective nature of anti-HIV IgG in the macaque model of SIV infection. In fact, generating IgG responses against HIV seems a useful approach considering that IgG is often the predominant isotype in the genital tract and that infected

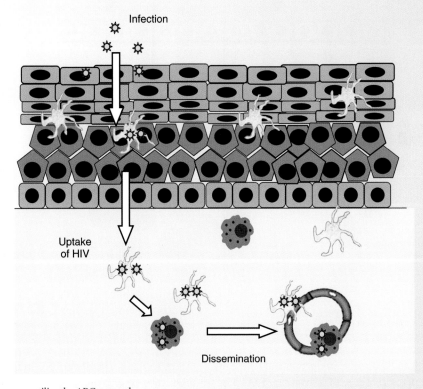

Infection

Uptake of HIV

Dissemination

HIV transmission may utilise the APC network.

individuals have HIV-specific IgG in their serum. In contrast, measurement of antibody responses in those individuals who have been exposed to the virus but who remain uninfected has revealed no detectable HIV-specific IgG in the serum. Rather, neutralizing IgA present in mucosal secretions is associated with protection, a phenotype that may rely on repeated exposure to the virus (see second figure accompanying this box). It has been proposed that in order to be protected an individual must possess neutralizing IgA antibody at the mucosal surface, which prevents cellular infection prior to dissemination to CD4+ T cells. Conversely, even though infected people have virus-specific IgG, the dissemination of HIV to T cells has already taken place and therefore they remain infected.

A protective role for cell-mediated immunity against HIV has been more difficult to prove. It is clear that HIV-specific CTLs are present in infected individuals and although CTLs help control viraemia they are unable to prevent further infection of CD4+ T cells or prevent subsequent immunosuppression and the development of full blown AIDS. However, they can be detected in the blood and genital mucosa of some exposed but uninfected people. Furthermore, HIV-specific CTL numbers decrease when these individuals are removed from continued viral exposure, a situation that has been associated with sero-conversion and consequent infection. This therefore suggests that cell-mediated immunity, driven by CTLs, does indeed contribute to protection against HIV infection, although the precise role that CTLs play remains to be elucidated.

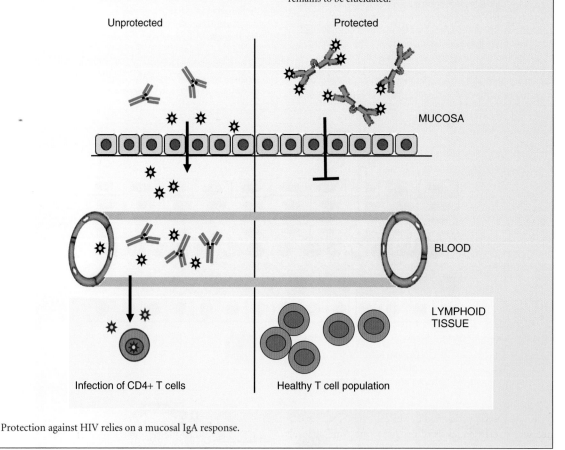

Protection against HIV relies on a mucosal IgA response.

TLR3 with its ligand causes the production and release of the type-I interferon IFN-β, while activation of TLR4 or TLR5 results in TNF and IL-6 release, via activation of the NF-κB pathway, and the secretion of the chemokines IL-8 (CXCL8) and monocyte chemoattractant protein-1 (MCP-1/CCL2). The effect of pro-inflammatory cytokine production by epithelial cells leads to the recruitment of immune cells such as neutrophils, monocytes, NK cells, T cells and B cells and causes the subsequent initiation of effector cell functions.

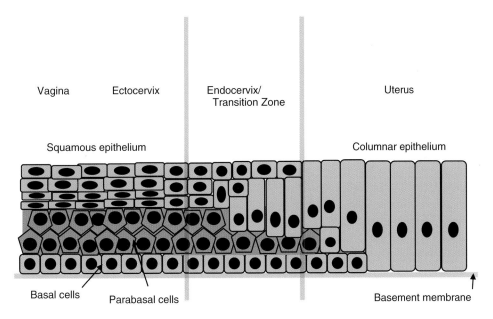

**Figure 9.3**  The epithelium of the female reproductive tract differs along it length. The epithelium of the vagina is multi-layered comprising mostly squamous epithelial cells. A transition zone exists at the endocervix, which gives way to a single layer of columnar epithelial cells that line the uterus (endometrium).

## 9.3 Passive immunity

Despite the regular shedding of the uterine epithelial layer, the uterus is capable of maintaining a sterile environment. This is largely due to the production of significant quantities of antimicrobial peptides, the secretion of which is heavily influenced by the stage of the menstrual cycle and the production of hormones (Figure 9.4). Epithelial cells produce several antimicrobial peptides including human β-defensins (HBD-1, HBD-2 and HBD-3), secretory leukocyte peptidase inhibitor (SLPI, otherwise referred to antileukoproteinase), lysozyme and lactoferrin. Indeed, the production of some of these antimicrobials, such as SLPI, has been shown to increase in the period before menstruation, thereby increasing passive immunity during the most vulnerable times of the menstrual cycle. Increased production of SLPI has also been associated with increased levels of progesterone. Similar levels of antimicrobial peptides have been detected in the mucous lining of the vagina and cervix, as well as the uterus. Increased production of antimicrobial peptides also occurs during labour, the effect of which benefits both the pregnant mother and the new-born baby during and after birth, when both are particularly susceptible to infection.

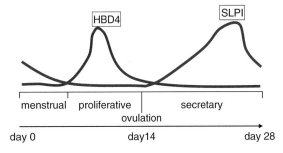

**Figure 9.4**  Antimicrobial peptide secretion is influenced by hormones during the menstrual cycle. Human beta-defensin 4, HBD4; secretary leukocyte peptidase inhibitor, SLPI.

## 9.4 Immunoglobulins

Epithelial cells of the reproductive tract express the pIgR on their basolateral surface, enabling them to effectively transport IgA to the lumenal surface. IgA plays an important role in protecting the reproductive tract from pathogen colonization and invasion. Plasma B cells reside just below the basolateral side of the epithelial layer and actively secrete IgA in response to the appropriate stimulation, thereby providing an important source of protective IgA. In the female reproductive tract an equal amount of IgA1 and IgA2 subclasses are secreted. This differs from the intestinal and respiratory tracts, which are dominated by IgA1. The transport of polymeric IgA across the epithelial layer is affected by oestrogen (in particular oestradiol) and progesterone. For example, oestrogen increases the expression of the pIgR on the basolateral surface of uterine and cervical epithelial cells and as a consequence results in enhanced transport of IgA into the mucosal lumen (Figure 9.5).

Elevated oestrogen levels are associated with two phases of the menstrual cycle, one just before ovulation and the second before menstruation. Its effect of increasing IgA transcytosis at these times therefore coincides with potential periods of heightened vulnerability to infection. In addition, significant amounts of IgG are detected in vaginal fluids and cervical secretions. Indeed, IgG often predominates over IgA, demonstrating a clear difference between the reproductive tract and other mucosal sites. It is thought that the source of this IgG is the circulation (blood plasma), although tissue residing IgG+ plasma cells may also contribute. The endocervix in particular has a high level of immunoglobulin production, as it is located at an anatomical site where it is important to prevent pathogens entering the upper reproductive tract. The secretions of the male reproductive tract (e.g. seminal fluid) are also dominated by high levels of IgG, derived from the circulation, and locally produced IgA.

## 9.5 APCs in genital tract mucosa

Antigen presenting cells (APCs) represent an important cell type for the initiation of adaptive immune responses and for the maintentence of homeostasis at the mucosal surfaces of the reproductive tract. Macrophages are found throughout the urogenital tract, where they participate in the phagocytosis of cell debris and externally derived particles. Phagocytosis of dead cells is particularly important during menstruation, as macrophage numbers increase in the pre-menstrual endometrium. As in other mucosal tissues, macrophages play an important role in tissue remodelling and in regulating the immune response through the release of modulatory cytokines such as IL-10. However, they can also be potent activators of immune responses through the release of TNF, IFN-$\gamma$ and IL-1. Activation of tissue macrophages is often the result of pathogen recognition through the ligation of TLRs or scavenger receptors.

DCs represent another important professional APC present throughout the urogenital tract and are the most effective activators of T cells. The main function of DCs is the uptake and processing of antigen for the subsequent presentation of peptide to T cells. A specialized subset of intra-epithelial, langerin+ DCs are also present within the reproductive tract mucosa, particularly in the intra-epithelial compartment of the vagina. These so-called Langerhan's cells (LCs) are thought to provide a network of DCs that survey the vaginal epithelia for the presence of pathogenic microorganisms (Figure 9.6).

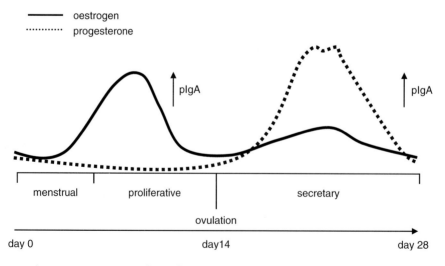

**Figure 9.5** Oestrogen increases IgA transcytosis during the menstrual cycle.

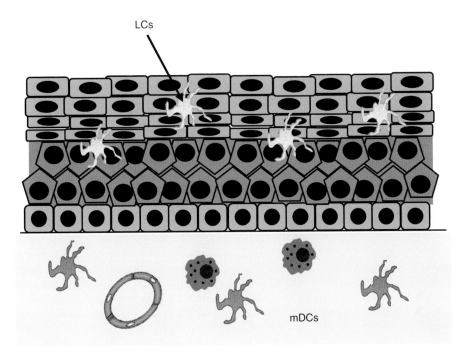

**Figure 9.6** Langerhan's cells (LCs) and DCs in the vaginal mucosa. LCs are found in the epithelial compartment of the vagina, while more conventional DC populations, for example mDCs, are located in the lamina propria along with tissue resident macrophages.

Another population of DCs are present within the lamina propria of reproductive mucosal tissue that also becomes activated following recognition of microbial components. This sub-mucosal population of DCs are thought to be similar to the dermal DC population found in skin. Microbial stimulation, in concordance with antigen uptake, causes DCs to undergo maturation and migration to regional lymph nodes (for example, the inguinal lymph nodes located in the groin). Within lymph nodes DCs upregulate MHC class II and co-stimulatory molecules (CD40 and CD80/CD86) on their cell surface, allowing effective presentation of peptide to CD4+ Th cells. Antigen-specific T cells then become activated themselves and migrate out of the lymph node and into the mucosal tissue of the urogenital tract where they can perform their effector functions.

## 9.6 NK cells and the semi-allogeneic foetus

NK cells are a relatively frequent lymphocyte population found throughout the female reproductive tract, in particular the lamina propria of the endometrium, where they function as early responders of the innate immune response. For example, it has been estimated that up to 70 per cent of all lymphocytes within the endometrium are NK cells. As well providing important immune defence functions, such as antibody-dependent cellular cytotoxicity (ADCC) and IFN-γ production, NK cells have also been associated with various pathologies including pregnancy loss and pre-eclampsia. This illustrates the central role that NK cells play in instigating innate immune responses within the reproductive tract. NK cells located in the uterus also have a unique phenotype, characterized by the expression CD90 and CD69 on their cell surface in addition to the NK cell marker CD56. These cell surface markers make them distinct from peripheral blood NK cells, which lack CD90 and CD69 expression. Hormones are also thought to influence the number of NK cells, as their frequency increases in the uterus during the menstrual cycle.

In addition to the control of infectious pathogens, NK cells are also frequent in the early stages of pregnancy, where they surround the trophoblast. These NK cells are thought to facilitate decidualization, protect foetal tissues from maternal immune responses, and protect the foetus from infectious disease. It is thought that NK cells are tightly regulated during pregnancy so that no immunopathological insults are directed toward the foetus. This regulation may be aided by the expression of the MHC class I molecule HLA-G, which is specifically expressed by trophoblasts (and the thymic medullar). Unlike the more conventional HLA-A and HLA-B MHC class I molecules, HLA-G inhibits NK cell cytotoxicity, thereby protecting the trophoblast against NK cell-mediated cytotoxicity. However, HLA-G ligation with NK cell receptors still allows the production of IFN-γ, which influences placental development and vascularization. Trophoblasts express relatively high levels of HLA-G, which is recognized by KIR2DL4 expressed on all NK cells. KIR2DL4 acts as an inhibitory receptor as it contains an ITIM (immunoreceptor tyrosine-based inhibitory motif) in its cytoplasmic tail and can therefore inhibit NK cell cytotoxicity upon ligation but still allow cytokine production. Although under normal circumstances NK cells protect the foetus from immunological damage, a mismatch between foetal KIRs and maternal HLA molecules can result in pre-eclampsia and recurrent miscarriage (Figure 9.7).

## 9.7 Pre-eclampsia is an immune-mediated disease

One of the most common diseases of the female reproductive tract is pre-eclampsia, which affects approximately 5–10 per cent of pregnancies. It is thought that pre-eclampsia is the result, at least in part, of a dysregulated maternal immune response toward the paternal alloantigens of the foetus. The characteristic signs of pre-eclampsia are hypertension, high blood pressure and protein in the urine and are associated with damage to the endothelium of the placenta. Although the precise immune mechanisms involved in the pathogenesis of pre-eclampsia are not fully understood, it is thought that the condition is the result of a lack of immune regulation. Abnormally low expression of both the regulatory cytokines TGF-β and IL-10 have been implicated in the development of an inflammatory reaction toward the placenta. This is consistent with increases in the pro-inflammatory cytokines TNF, IFN-γ and IL-12, and granulysin, which is a marker for NK cell and CTL activity.

Uterine NK cells may actually be an important cell type responsible for driving the pathology and the mechanism seems to be related to the expression of paternally-derived MHC class I molecules. As well as presenting antigen to TCRs on the surface of T cells, MHC class I molecules can also activate KIR receptors on NK cells. MHC molecules are also highly polymorphic in the human population that will affect the interaction with KIRs expressed on NK cells, depending on which alleles are inherited. Therefore, depending on which specific allele is inherited by the foetus, the interaction between MHC class I molecules and KIRs may influence the extent of NK cell reactivity against

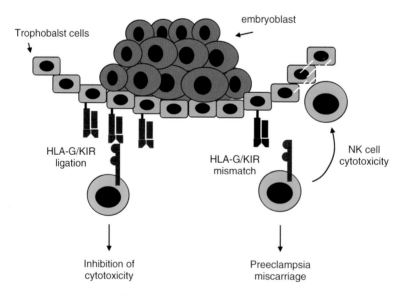

**Figure 9.7** Inhibition of NK cell cytotoxicity following KIR ligation with HLA-G. A mismatch in HLA-G signalling prevents inhibitory signals, thereby allowing NK cell cytotoxicity.

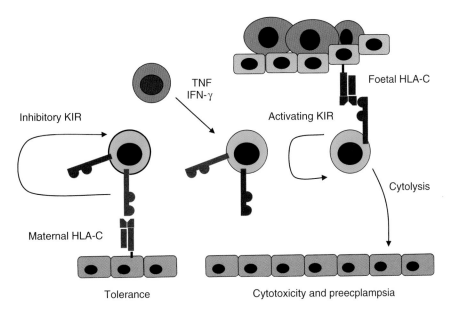

Inhibitory KIR

Activating KIR

TNF
IFN-γ

Foetal HLA-C

Cytolysis

Maternal HLA-C

Tolerance

Cytotoxicity and preecplampsia

**Figure 9.8** Breakdown of tolerance can lead to NK cell-mediated immunopathology. Depending on which HLA-C allele is inherited by the foetus, maternal NK cells can either engage inhibitory KIRs or activating KIRs.

the placenta. This appears to be particularly relevant for HLA-C alleles, which are thought to strongly interact with uterine NK cells. One set of HLA-C alleles may therefore signal through inhibitory KIR receptors, while another set of HLA-C alleles signal through activating KIR receptors that lead to an immunopathological outcome (Figure 9.8). In addition, the abnormal NK cell activity is thought to be supported by Th1-cytokine-producing T cells and possibly through the recognition of alloantigens by CD8+ T cells.

## 9.8 Maintenance of foetal tolerance

The unique biological function of the female reproductive tract means that immune responsiveness must be tightly regulated. On the one hand, an immune reaction must be allowed to develop in response to an invading pathogen but at the same time tolerance must be induced to spermatozoa and the developing foetus. Particular mechanisms therefore exist that enable the discrimination between self and nonself. This immune recognition is further complicated by the fact that spermatozoa are allogeneic (genetically different) and that the foetus is semi-allogeneic. Like other mucosal sites the reproductive tract is inherently tolerogenic, rather than pro-inflammatory. This is a crucial phenotype to have in order for the semi-allogeneic

foetus to be tolerated throughout pregnancy. Any breakdown in the immunoregulatory mechanisms that drive foetal tolerance could lead to infertility (immune reaction against spermatozoa), inhibition of implantation or even abortion.

The foetus is considered semi-allogeneic because it bears antigens from both its paternal and maternal parents. Although the maternal antigens are shared between mother and foetus, the paternal antigens are different. Lymphocytes generated by the mother are able to recognize the foetus as foreign (Figure 9.9) but instead of reacting in an inflammatory manner, the lymphocytes become tolerant. During the pre-implantation phases of pregnancy, the developing blastocyst and trophoblast are protected from the maternal immune system by the surrounding zona pellucida. Not until implantation does tissue from the embryo and mother converge. The mechanism of tolerance involves signals supplied by both the foetus and maternal tissue, primarily the placenta (Figure 9.10). For example, the endometrium receives various signals that prime it for implantation and at the same time elicits a tolerogenic microenvironment. Early pregnancy factor (EPF) is a potent immunoregulatory molecule expressed by the mother that suppresses T cell effector functions and drives tolerance to foetal antigens. Embryonic trophoblasts are highly specialized

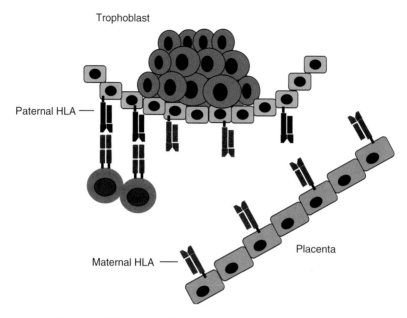

**Figure 9.9** The foetus is semi-allogeneic and is recognized by maternal CD8+ T cells through the recognition of paternal HLA molecules. The maternal CTLs essentially consider the foetal HLA molecules as foreign.

cells that are capable of inducing maternal immune tolerance through the release of several immunosuppressive factors including indoleamine 2,3-dioxygenase (IDO), which inhibits lymphocyte proliferation. Trophoblast cells also lack expression of conventional MHC class I and II molecules. Therefore, rather than expressing HLA-A and HLA-B, which are known to be important in tissue allograft rejection, they express HLA-C and HLA-G instead. Interaction between HLA-C and inhibitory KIR receptor on the surface of NK cells is thought to lead to the inhibition of cytolytic activity, while HLA-G interacts with both NK cells and CD8+ T cells and may lead to the suppression of cytotoxic function (Figure 9.10) and the promotion of T cell apoptosis. Other immunosuppressive molecules are expressed on the surface of trophoblasts including CD200, which suppresses DC activity, and FasL, which induces apoptosis of NK cell and T cells.

Due to a multitude of signals derived from maternal tissue and the embryo, many leukocytes that reside in the placenta and endometrium are immunoregulatory. This includes DCs and macrophages, the latter of which constitute the largest immune cell population within the placenta. Rather than releasing pro-inflammatory cytokines such as TNF and IFN-γ, placental macrophages produce the immunomodulatory cytokines IL-6 and IL-10. This immunomodulatory microenvironment is

further accentuated by the presence of CD4+CD25+ Treg cells, which produce TGF-β or IL-10. The important immunoregulatory cytokines accumulate in the placenta during pregnancy. The immune response is also skewed more to Th2 type response, rather than an immunopathogenic Th1 response. Th2-mediated adaptive immune responses are associated with the release of IL-4 from macrophages and Th2 cells. Furthermore, hormones significantly influence the phenotype of the immune response during pregnancy. For example, progesterone favours the development of a Th2 response by enhancing the production of IL-4 and IL-10 and reducing IL-12 and TNF activity. Therefore, the immune system is tightly controlled by several key mechanisms during pregnancy so that the developing foetus is properly tolerated.

## 9.9 T cells and adaptive immunity

Both CD4+ Th cells and CD8+ CTLs are found throughout the female reproductive tract and play an important role in protecting against pathogenic insults. The vaginal area of the lower genital tract is particularly vulnerable to pathogen invasion, due to its proximity to the outside and its susceptibility to sexually transmitted diseases.

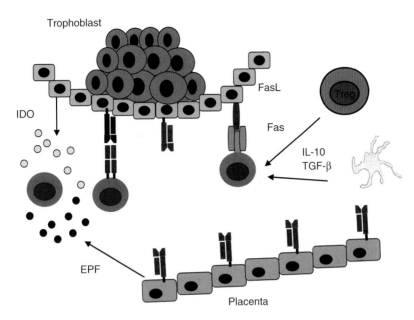

**Figure 9.10** The foetus and placenta produce tolerogenic molecules, including IDO and EPF and the immunoregulatory cytokines IL-10 and TGF-β. This induces a tolerogenic state within the placenta. Immune responses directed against the trophoblast are further inhibited by the expression of FasL.

Therefore cell-mediated immunity is a critical defence mechanism against infection. Due to the lack of well organized lymphoid structures, T cells are mainly dispersed throughout the lamina propria of the reproductive tract. In the absence of inductive sites within the mucosa, it is likely that T cell responses are initiated in local lymph nodes. The same can be assumed for local B cell responses. The frequency and recruitment of lymphocytes also differs between the mostly sterile upper reproductive tract (endometrium and fallopian tubes) and the non-sterile lower reproductive tract (vagina and ectocervix). Most reproductive tract T cells are found within the vaginal mucosa and are conventional αβTCR+ T cells, with CD8αβ+ CTLs predominating over CD4+ Th cells. Intra-epithelial T cells (IELs) are also present in the vaginal mucosa, with unconventional αβ or γδ T cells predominating (Figure 9.11), although the number of IELs is far less than is found in the intestinal mucosa. Experiments performed in mice have also revealed a CD3+αβTCR+CD4-CD8- population (double negative T cells), which are either a regulatory subset of T cells or one that may respond to non-peptide antigens in much the same way as CD1d-restricted iNKTs do.

The trafficking of lymphocytes, including T cells and B cells, into the mucosal tissue of the reproductive tract remains an area of much debate. Interestingly, intranasal immunization often results in the recruitment of T cells into the vaginal mucosa as well as the respiratory tract, suggesting that these two mucosal tissues share some common homing receptors. It is known that the gut-homing receptor MAdCAM-1 is not expressed on endothelial cells located within the lower genital tract, conferring at least one level of tissue specificity. Other homing receptors are expressed in the genital tract including VCAM-1 and ICAM-1, which are recognized by the integrins α4β1 and αLβ2, expressed on T cells, respectively (Figure 9.12). These receptors are also expressed in nasal tissue, although it is uncertain whether these receptors alone account for the homing properties of lymphocytes into the genital mucosa.

## 9.10 Sexually transmitted diseases and pelvic inflammatory disease

Pelvic inflammatory disease (PID) is caused by an infection of one or more of the organs within the female pelvis, including the uterus, cervix, fallopian tubes and ovaries. PID can result in the blockage of fallopian tubes, infertility and ectopic pregnancies and is associated with an increased susceptibility to HIV due to the loss of epithelial

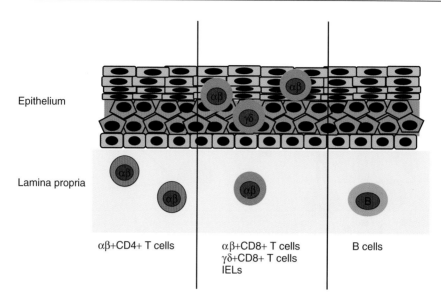

Epithelium

Lamina propria

αβ+CD4+ T cells        αβ+CD8+ T cells        B cells
                       γδ+CD8+ T cells
                       IELs

**Figure 9.11** Distribution of T cells and B cells throughout the vaginal mucosa. There is a heterogeneous population of αβ T cells and γλ T cells.

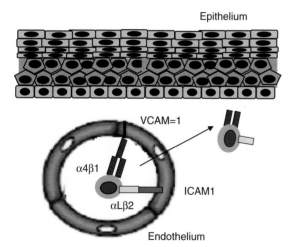

**Figure 9.12** Migration of T cell into the reproductive tract mucosa requires the expression of specific cell adhesion molecules.

integrity. The most common aetiological agents are the sexually transmitted diseases gonorrhoea and chlamydia, caused by the bacteria *Neisseria gonorrhoeae* and *Chlamydia trachomatis*, respectively. These bacteria gain access to the otherwise sterile upper reproductive tract by ascending the mucosal layer of the vagina and entering via the endocervix. Vaginal colonization is largely asymptomatic, as the thick mucous layer prevents bacteria invading and

infecting the epithelial cells. Only when the bacteria ascend into the cervix is an inflammatory reaction initiated following infection of epithelial cells that form the epithelial mono-layer. Bacterial PAMPs such as peptidoglycan and lipid A are recognized by PRRs on host cells, which initiate the inflammatory response.

The cervical area is rich in IgG and complement proteins, although *N. gonorrhoeae* and *C. trachomatis* are adept at avoiding these immune components, allowing them to gain entry into the upperreproductive tract and instigate infection and inflammation (cervicitis). *N. gonorrhoeae* is thought to avoid the complement system by inactivating C3b, following binding to bacterial lipooligosaccharide (LOS). This interaction is also thought to mediate the activation of bacterial invasion complexes, resulting in epithelial cell infection. Invasion is augmented by the binding of gonnococcal pilus to the CR3 complement receptor expressed by epithelia cells of the ectocervix and endocervix. Ascent into the upper respiratory tract is aided by the motility of the bacteria. Furthermore, the release of pro-inflammatory cytokines such as TNF and the accumulation of inflammatory cells accentuate the destruction of epithelial cells in the fallopian tubes and uterus, which contributes to PID.

Infection of epithelial cells by *C. trachomatis* also leads to the release of pro-inflammatory cytokines and chemokines, which act by recruiting inflammatory cells into the infected tissue and contributing to the pathology associated with PID. For example, macrophages release IL-12, IL-1 and TNF in response to bacterial infection, which initiates a Th1-mediated immune response. In addition macrophages and activated endothelial cells secrete IL-8, which promotes the migration of neutrophils. Although neutrophils are important for the clearance of bacteria, they also release elastases and matrix metalloproteinases that contribute to tissue damage. Activated NK cells release IFN-γ, which supports the proliferation and activation of CD4+ and CD8+ T cells. Furthermore, B cells are recruited to sites of inflammation within the genital tract. Although tissue damage is often a consequence of such a vigorous immune response, it is also necessary for bacterial clearance. For instance, IFN-γ helps to prevent bacterial replication within epithelial cells, while IgG and IgA neutralize extracellular bacteria. Once infection has been resolved the inflammatory response subsides, although tissue remodelling often persists in the form of scarring, exemplified by the fibrosis associated with chlamydia-induced oviduct disease.

## 9.11 Alloimmunization and autoimmune diseases

Considering that the foetus possesses paternal antigens that are not shared by the mother, there remains the possibility that a woman develops an immune response to those antigens. One such disease that affects the neonate is haemolytic disease of the newborn (HDN), which is most commonly the result of the mother producing alloantibodies against the rhesus-D antigen expressed on the surface of foetal red blood cells (erythrocytes). This only occurs in rhesus-D negative women, when foetal erythrocytes cross the placenta and pass into the mother's circulation. The mother becomes sensitized to these erythrocytes and produces antibodies that cross react with the rhesus-D antigen. These IgG antibodies cross the placenta themselves and enter the foetal circulation where they opsonize and cause the lysis of erythrocytes. Neonates with HDN present with severe hyperbilirubinaemia (the cause of jaundice) and anaemia and are often very sick at birth. HDN used to be associated with high rates of morbidity and mortality; that is until the 1970s when an effective treatment was developed. Rhesus-D negative women are routinely administered with anti-D antibodies during the antenatal period, which prevent sensitization to the antigen and act as an effective prophylactic for the prevention of HDN. A less severe form of HDN occurs when the mother produces IgG antibodies against ABO

blood group antigens, although this only occurs in O positive women who are sensitized to A or B antigens.

Another common antibody-mediated disease of new born babies is neonatal alloimmune thrombocytopenia (NAIT), which is the result of the mother becoming sensitized to human platelet antigens, in particular HPA-1a. Neonates with this disease have an abnormally low platelet count and a corresponding deficiency in blood clotting, which often requires the transfusion of compatible platelets. Less frequently, NAIT can affect the foetus during gestation and cause inter-cranial haemorrhaging. A higher incidence of NAIT has been associated with the inheritance of a particular HLA-DRB3 allele, which makes carriers more likely to produce anti-HPA-1a alloantibodies. Another disease that affects platelet counts is idiopathic thrombocytopenia purpura (ITP), which results from the mother having autoantibodies to her own platelets. The autoantibodies produced in ITP are of the IgG isotype and are often directed against platelet glycoproteins, which results in the opsonization and phagocytosis of platelets or megakaryocytes (platelet progenitor cells). ITP is considered to be an autoimmune disease, although it is a less severe disease compared to NAIT as the low platelet count is often asymptomatic. The most common symptom is purpura (bruising), although extremely low platelet counts can cause mucosal bleeding and even intracerebral haemorrhaging.

## 9.12 The foetal and neonatal immune system

The foetus is particularly susceptible to viral and bacterial infection within the uterus, due in part to its immature immune system. This is exemplified by a heightened susceptibility to pathogens such as HIV and *Listeria monocytogenes*. Indeed, pregnant mothers are advised not to consume unpasteurized dairy products because of the risk of contamination with the Lysteria bacterium. Although the immune system starts to develop early, approximately seven weeks into gestation, T cell and B cells remain in an immature state, although their numbers are relatively high at birth. The immature state includes poor T cell help to B cells, so that antibody production and isotype class switching is reduced. Innate immune cells, such as NK cells, macrophages and neutrophils, can also be detected in the developing foetus but they too remain in an immature state at birth compared to adults. For example, although NK cell numbers are similar in the neonate and adult, they have a reduced cytotoxic capacity. This immaturity is further reflected in the cytokine response of the foetus and neonate, in that production of the pro-inflammatory cytokines TNF and IFN-γ are low. The neonatal immune system is therefore rather immature at birth, which may account for the severity of bacterial and viral infections in this age group.

Antibody production in the foetus and neonate is characterized by a predominance of circulating IgM, as most B cells will not have undergone T cell-mediated class switching. Therefore, the amount of IgG and IgA is very low. This is partly compensated for by the transfer of maternal IgG across the placenta, so that at the time of birth the level of IgG is similar to that of the mother. The trophoblast expresses FcγRs, in particular FcγRIII, which are responsible for the transfer of maternal IgG across the placenta and into the foetal circulation. The majority of IgG is transferred after 32 weeks of gestation, which means that babies born prematurely are often more susceptible to infections. Although IgG is important for immune protection of the new born, it only has a half-life of around three weeks. Moreover, neonates do not start to actively synthesize their own IgG until approximately two months of age, resulting in a period of IgG deficiency. Infant vaccination and neonatal infections increase IgG production and stimulate B cell somatic hypermutation and isotype class switching. Passive protection afforded by maternal immune components is also provided in breast milk (discussed in Chapter 10), which contains IgA, complement components, antimicrobial peptides and leukocytes.

## 9.13 Immunity in the urinary tract

Even though the upper urinary tract is normally considered to be a sterile environment, it does have open access to the environment and therefore both the urethra and the bladder are susceptible to infection. Moreover, very serious urinary tract infections can also involve the kidneys and cause nephritis. Therefore the urinary tract has evolved several immune mechanisms that contribute to the defence against invading pathogens. In much the same way as in other mucosal tissues, the epithelium of the urinary tract functions as a passive barrier against the invasion and colonization of pathogens. Epithelial cells also secrete significant amounts of IL-6 and IL-8 in response to bacterial infection, as well as antimicrobial peptides. Cytokine release from urinary tract epithelial cells is augmented by the detection of PAMPs by various

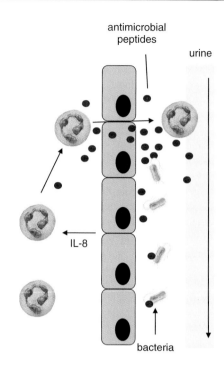

**Figure 9.13** IL-8 and antimicrobial peptides recruit neutrophils to the site of infection in the urinary tract. Activated neutrophils and killed bacteria are removed by the flow of urine.

TLRs. In particular, the detection of LPS by TLR4 and CD14 has been associated with enhanced clearance of urinary tract infections. IL-8 is particularly effective at recruiting neutrophils (Figure 9.13), and to lesser extent macrophages, from the circulation to the site of infection. Indeed, neutrophils are thought to elicit important innate immune effector mechanisms against pathogenic bacteria. These polymorphonuclear neutrophils (PMNs) are effective at phagocytosing bacteria that have breached the epithelial monolayer of the urinary tract and, in conjunction with macrophages, participate in the clearance of both bacteria and apoptotic cells.

The humoral immune response is equally important in protecting the urinary tract from infection. Antimicrobial defensins, lactoferrin, lipocalin and Tamm-horsfall protein (THP) are produced by epithelial cells and released into the urine. In fact, THP is selectively produced by the urinary tract, which can release as much as 50 mg each day. THP binds to the fimbriae of pathogenic bacteria and as a result prevents bacterial attachment to epithelial cells and enhances bacterial washout with the passage of urine. As well as having antimicrobial properties, THP also accentuates TLR4-mediated activation of epithelial cells and DCs. Another valuable component of the humoral arm of the immune system is antibody secretion by plasma cells. It has been demonstrated that S-IgA is produced by plasma cells that reside in the lamina propria of the urinary tract mucosa. IgA is easily detected in the urine of patients with a urinary tract infection and in patients with more severe nephritis. The importance of IgA in host immune defence is illustrated in children with severe infections, whereby IgA levels are significantly elevated. Furthermore, bacteria grow less well in high IgA concentrations and they seem less capable of attaching to the epithelial monolayer, thereby preventing bacterial invasion and colonization.

Although less well defined, particularly in humans, cell-mediated immunity is also thought to contribute to immune defence in the urinary tract. During infection CD4+ Th cells and CD8+ CTLs both increase in number. Following experimental bacterial infection of mice with pathogenic *Escherichia coli*, γδ T cells predominate over αβ T cells and are likely to provide substantial support to the local immune response. Whether this situation is true in humans remains to be determined, especially considering that the human γδ T cell population is not as abundant as it is in rodents.

## 9.14 Eye associated lymphoid tissue

Although not considered to be contiguous with the skin, the conjunctiva and the surface of the eye are components of the body surface and are considered part of the mucosal immune system (Figure 9.14). Like the skin, the ocular surface is exposed to the external environment and is composed of a layer of epithelial cells. However, the structure of the conjunctiva is more comparable to the female urogenital tract, as its mucosal surface comprises squamous epithelia. The conjunctiva primarily covers the inner surfaces of the eyelids, contacts the surface of the eye ball and extends over the surface of the white part of the eye ball (the sclera). It therefore seems appropriate to discuss these tissues within the context of this chapter. However, the eye is a highly specialized organ responsible for visual function and needs to remain moist, free from damaging inflammatory reactions but at the same time be protected from harmful microorganisms. The cornea of the eye, and its internal structures, remain devoid of organized lymphoid tissues. This prevents excessive inflammation that may result in immunopathology and

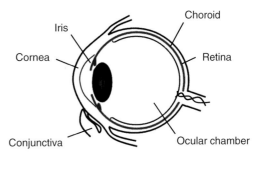

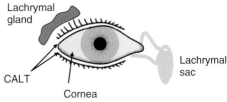

**Figure 9.14** Supporting tissues associated with the eye.

has led to the notion that the eye is an immune privileged site. However, several evolutionary adaptations do exist in order to protect the eye from infection. These include the conjunctiva associated lymphoid tissue (CALT) and the lachrymal gland and lachrymal drainage system associated lymphoid tissue (LDALT, commonly associated with tear production). The combination of these two supporting tissues is sometimes referred to as eye associated lymphoid tissue (EALT). The cornea of the eye and the CALT possess moist mucosal surfaces that are more similar to other MALT, rather than to skin.

The principal component of ocular immune defence is S-IgA, secreted by both the lachrymal gland and the CALT. The fluid of the tear film produced by the lachrymal gland contains a significant amount of IgA, as well as antimicrobial peptides such as lactoferrin and lysozyme. Plasma cells can be detected in the lachrymal gland, along with the pIgR (secretory component), and to a lesser extend in the conjunctiva. Plasma cell numbers are highest surrounding the secretory acini of the lachrymal gland but are also present in the lachrymal drainage ducts and in the conjunctiva. The position of the lachrymal gland in relation to the lachrymal ducts and sac means that tear fluid containing S-IgA flows from the naso-distal to the naso-proximal surface of the eye. This is particularly relevant when the eye lids are closed and fluid movement is assisted by the action of blinking.

## 9.15 Conjunctiva associated lymphoid tissue (CALT)

The lymphoid tissue found throughout the conjunctiva is somewhat similar to that found in other MALT, although it may be absent in some healthy individuals and only apparent during inflammatory episodes. It is more evident in infants and young children and declines with age. Diffuse lymphocytes are scattered throughout the lamina propria and the epithelium, and like the gut CD8+ T cells predominate in the intra-epithelial compartment, while CD4+ T cells a more frequent in the lamina propria. B cells (mostly IgA+ plasma cells), macrophages and dendritic cells also occur throughout the lamina propria, strongly suggesting that the conjunctiva can act as an immune effector site. The conjunctiva also contains high endothelial venules (HEVs), which are necessary for the migration and extravasation of lymphocytes into immune effector sites, and lymphatics that take cells away from the conjunctiva and into regional lymph nodes. In addition, discrete lymphoid follicles exist in the conjunctiva (Figure 9.15) and superficially resemble the PPs found along the intestine. The main population of cells within these lymphoid follicles are B cells expressing either IgM or IgA, although T cells are located in the interfollicular zones. However, there is much debate as to whether the FAE of conjunctival lymphoid follicles possess M cells and the presence of the pIgR has been difficult to determine. Recent findings do suggest that both M cells and pIgR are present in conjunctiva, making CALT a genuine component of the common mucosal immune system. It also seems likely that IgA is secreted by specialized lachrymal gland cells, positioned at regular intervals along the epithelia. It is thought that resident DCs, located within or positioned directly beneath the squamous epithelia, contribute to the important function of antigen capture, rather than relying on M cells to perform this function. Slight differences in antigen sampling are also related to the stratified epithelia found in the conjunctiva (and also virginal mucosa), as opposed to the simple epithelia found in the gut and airways.

## 9.16 Immune privilege of the eye

The delicate nature of the eye and its physiological role in providing sight make it particularly vulnerable to immune-mediated damage. Several evolutionary adaptations therefore exist that render the eye especially resistant to inflammatory reactions, a phenomenon known as immune privilege. Tissues or organs designated as immune privileged sites are defined as sites where tissue grafts have extended (if not indefinite) survival times without them being rejected, compared with conventional sites. Indeed, corneal allografts have a very high rate of acceptance in humans. Immune privileged sites are often tissues that perform specialized physiological functions and also lack the ability to regenerate following immune damage. Examples of immune privileged sites include the eye, brain and testicles. The mechanisms responsible for

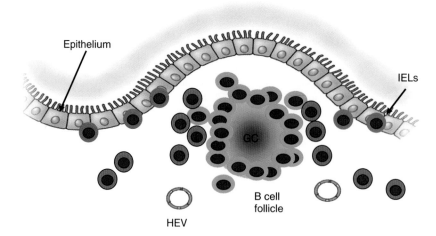

**Figure 9.15** Organized lymphoid follicles are located in the conjunctiva, including B cell follicles surrounded by a para-follicular T cell zone. The conjunctiva is also high vascularized allowing migration of leukocyte across HEVs.

maintaining immune privilege include both anatomical and immunological features.

The first mechanism relies on the blood/ocular barrier, which effectively prevents immune cells and insoluble substances from crossing into the chamber of the eye. The blockade of molecules entering the eye chamber include antibodies, complement factors, cytokines and chemokines. The anterior chamber of the eye is also devoid of draining lymphatics, meaning that no lymphocytes are activated in response to ocular-derived antigen, as this antigen is simply not delivered to lymph nodes. Instead, fluid is reabsorbed through the venous drainage system so that antigens enter the spleen rather than lymph nodes. The endothelial cells lining the blood vessels of the iris are also specialized in that they prevent extravasation of lymphocytes, while blood vessels are completely absent form the cornea. The fluid within the eye ball, the aqueous humour, also plays a role in maintaining immune privilege as it contains immunomodulatory factors such as TGF-β, while the epithelial cells of the iris and corneal endothelium are refractory to complement activation and immune stimulation. For example, the surface expression of FasL on epithelial and endothelial cells throughout the eye drives Fas-expressing cells to undergo apoptosis (Figure 9.16). Recently activated T cells upregulate Fas molecules on their cell surface, so that encounter with Fas-ligand causes apoptosis in those T cells that would otherwise be immunopathogenic. The lack of response within the chamber of the eye is known as anterior chamber associated immune deviation (ACAID).

A principal component of ACAID is the induction of systemic tolerance to antigens derived from the anterior chamber of the eye. This tolerance is thought to be driven by TGF-β-stimulated APCs that adopt an immunomodulatory phenotype. The distribution of DCs throughout the eye is rather unusual as well, in that Langerhan's cells are absent form the corneum. Although they express normal levels of MHC and co-stimulatory molecules they do not secrete IL-12. Curiously, the effect of ACAID is dependent on the spleen and it is therefore likely that immunomodulatory APCs migrate to T and B cell areas within the spleen and induce tolerance to ocular-derived antigens. Within the splenic microenvironment, these tolerogenic APCs have been shown to express low levels of MHC class II and CD40. They also express IL-10 and TGF-β themselves, further potentiating their immunoregulatory capacity. This has the effect of causing antigen-specific CD4+ T cells to become anergic (unresponsive to antigen stimulation) or to cause them to differentiate into Treg cells that express more IL-10 or TGF-β (Figure 9.16). Interestingly the CD8+ CTL response appears to remain intact, although the immunoregulatory environment prevents immunopathology. The effect of ACAID on B cells is to eliminate B cells that produce complement-fixing antibodies, such as IgG, while at the same time to enhance B cells that produce antibodies that fix complement less efficiently, such as IgA. The overall immunological outcome is the preservation of immune privilege and the continued protection of the structure of the eye, through the induction of tolerance and anti-inflammatory mediators.

## 9.17 Immune privilege and inflammation

Sensitive tissues employ various mechanisms in order to maintain immune privilege and prevent damage to delicate structures. However, what happens when pathogens invade tissues such as the eye or when an inflammatory reaction is induced by external factors? Immune privileged sites still need to defend themselves against microbes, a process that may involve the induction of an inflammatory reaction. It now appears that immune privilege in the eye is temporarily lost during episodes of inflammation. However, the principal components of ACAID remain intact following resolution of inflammation, so that immune privilege is restored. For example, APCs can be induced to drive a Th1 immune response in the eye, which involves the production of IL-12 and the recruitment of effector T cells. Once this immune response has been resolved, ocular APCs continue to produce the immunomodulatory cytokine TGF-β, thereby restoring immune privilege status.

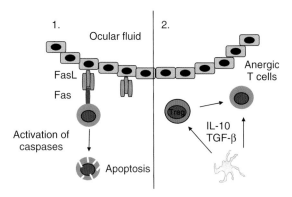

**Figure 9.16** Maintenance of immune privilege in the eye involves a number of inhibitory factors. These include expression of FasL and the secretion of immunoregulatory cytokines such as IL-10 and TGF-β.

## 9.18 Conjunctivitis

The conjunctiva of the eye is an important component of the MALT system that contains organized lymphoid tissue and protects the surface of the eye from infection. However, excessive inflammation of the conjunctiva often occurs in response to pathogenic insults, allergens or chemical irritation, and is known as conjunctivitis (Table 9.1). The most common causes of conjunctival inflammation are usually the result of a viral infection (viral conjunctivitis) or a reaction to an allergen (allergic conjunctivitis), although bacteria are also a familiar cause (bacterial conjunctivitis). Viral and bacterial conjunctivitis are collectively known as infectious conjunctivitis and are associated with redness of the eye, itching, irritation, tear formation and often discharge. Although bacterial infections can usually be treated with antibiotics, there is no treatment for viral conjunctivitis. These infections can be highly contagious and easily pass from person to person. Therefore, good hygiene is often the best practice

in order to avoid such infections. Bacterial conjunctivitis tends to be more problematic than viral conjunctivitis and is often the result of *Haemophilus influenzae, Staphylococcus aureus* or *Chlamydia trachomatis* infection (the clinical appearance is commonly known as red eye). The importance of the immune system in controlling conjunctival and other eye infections is demonstrated by the higher rate of bacterial conjunctivitis in immunodeficient patients. The production of antimicrobial factors such as lactoferrin and lysozyme, along with IgA and components of the complement system, are integral in preventing bacterial colonization and infection.

Allergic conjunctivitis is induced in response to the presence of an allergen and is normally far less severe than infectious conjunctivitis. The term allergic conjunctivitis actually refers to a collection of eye disorders that includes seasonal allergic conjunctivitis (otherwise known as rhinoconjunctivitis) and keratoconjunctivitis (Table 9.1).Allergic conjunctivitis is a transient inflammation that is largely restricted to the eye lids and

**Table 9.1** Types of conjunctivitis.

| | |
|---|---|
| Seasonal allergic conjunctivitis (SAC) | Intermittent<br>IgE-dependent<br>Mast cells, eosinophils (Th2)<br>IL-4, IL-8, IL-13, histamine |
| Perennial allergic conjunctivitis (PAC) | Persistent (chronic)<br>IgE-mediated<br>Mast cells, eosinophils (Th2)<br>IL-4, IL-8, IL-13, histamine |
| Vernal keratoconjunctivitis (VKC) | Persistent (chronic)<br>IgE-mediated<br>Mast cells, eosinophils, neutrophils, T cells (Th2/Th1)<br>IL-4, IL-8, IL-13, histamine, IFN-γ |
| Atopic keratoconjunctivitis (AKC) | Chronic<br>IgE-mediated<br>Mast cells, eosinophils, neutrophils, T cells (Th2/Th1)<br>IL-4, IL-8, IL-13, histamine, IFN-γ |
| Giant papillary conjunctivitis (GPC) | Persistent (chronic)<br>Independent of IgE<br>T cells, macrophages (Th1/Th2)<br>IL-1, TNF, IFN-γ |
| Contact dermatoconjunctivitis (CDC) | Chronic<br>Independent of IgE<br>T cells, macrophages (Th1)<br>IL-1, TNF, IFN-γ |

conjunctiva and does not normally affect the cornea. On the other hand, keratoconjucntivitis is often complicated by inflammation of the cornea. Allergic conjunctivitis is often associated with hay fever and rhinitis and primarily involves the activation of mast cells. The number of mast cells increases throughout the conjunctiva during an allergic response, as does the density of FcεRα expressing cells. Allergic conjunctivitis is associated with the development of a Th2-type immune response and the production of IgE. The specificity of IgE to environmental allergens results in cross-linking of FcεRs on the surface of mast cells, which causes the subsequent release of pro-inflammatory mediators such as histamine, chemokines and cytokines (including IL-4, IL-13 and IL-8). This results in the recruitment of eosinophils and changes the conjunctival vasculature, both of which are characteristic of the acute phase of an allergic response. Infiltrating T cells then contribute to a more chronic state of allergic conjunctivitis associated with persistent inflammation. Keratoconjunctivitis is a more severe form of the disease and involves damage to the keratin layer of the cornea. Like allergic conjunctivitis, keratoconjunctivitis is driven by Th2 cytokines such as IL-4 and IL-13 but also involves the production of the Th1 cytokine IFN-γ, which, together with a significant T cell infiltration, may contribute to the enhanced immunopathogenesis of this disease.

## 9.19 Summary

1. The anatomy of the urogenital tract differs between males and females, although both are susceptible to infection with sexually transmitted diseases.

2. The epithelium of the female genital tract differs along its length, with stratified squamous epithelium of the vagina giving way to a monolayer of columnar epithelium at the cervix.

3. Unlike other mucosal sites, IgG often predominates over IgA, although secretory IgA is still important for host defence.

4. The immune system of the reproductive tract is highly specialized, as it needs to maintain tolerance to the foetus during pregnancy.

5. The eye can be considered an immune privileged site, through a mechanism known as anterior chamber associated immune deviation (ACAID).

# 10 Immunology of the Skin

## 10.1 The skin as an immune tissue

Although not classified as a mucosal tissue *per se*, the skin possesses a number of immunological features that are similar to MALT; appreciably the shared need to maintain tissue homeostasis, while at the same time promoting effective immunity. In contrast, the skin has unique qualities, which are borne out of its distinct biological function and anatomical location. Importantly, the skin is considered the largest organ of the body and provides a robust barrier to the external environment, effectively preventing pathogen entry through a combination of structural and immunological features. In addition, the skin is able to respond to injury (following cuts and abrasions) in both a wound-healing and an immunological capacity. Wounded skin is an easy conduit for pathogens to enter the body and therefore effective immune defences are in place to combat such adverse events. The constant exposure to environmental toxins, solar radiation, a range of pathogens and the daily instances of wounding, make the skin a unique organ that has evolved its own identifiable immune system.

Structurally, the skin is composed of two main layers, the outermost epidermis and the inner dermis, separated by a basement membrane (Figure 10.1). The epidermis is much thinner than the dermis and largely comprises specialized squamous epithelial cells called keratinocytes that form the durable outer layers of the skin. Keratinocytes produce large quantities of the protein keratin, which is the main constituent of nails and hair and also forms the tough, watertight surface of the skin, the stratum corneum. As well as keratinocytes, the epidermis contains specialized DCs called Langerhan's cells (LCs), which act as sentinels in the skin by detecting the presence of potentially harmful foreign substances (Figure 10.2), and intra-epidermal lymphocytes (equivalent to intra-epithelial lymphocytes in MALT). The dermis is much thicker than the epidermis and contains more complex structures such as sweat glands, sebaceous glands, hair follicles, blood vessels and nerve endings. The dermis mainly consists of dermal fibroblasts and contains many more immune cells than the epidermis, such as T cells, dermal DCs, NK cells, mast cells and macrophages. Throughout the dermis post-capillary venules (equivalent to high endothelial venules in other tissues) act as the main sites of leukocyte entry into the dermis and can influence the extent of inflammatory cell infiltration through the expression of cell-adhesion molecules and chemokines. Below the dermis is a thick subcutaneous layer called the hypodermis, which is connected to the dermis by collagen and elastin fibres. The hypodermis consists mostly of fat cells known as adipocytes, which collect fat and effectively act as energy reserves.

## 10.2 Barrier Immune function of the skin

The principal function of the skin is to act as an effective structural barrier to chemical, physical and microbial challenges. The outer most epidermal layer of the skin, the stratum corneum, is primarily responsible for providing this passive barrier. The epidermis can be divided into several distinct layers that are largely characterized by the migration and differentiation state of keratinocytes (Figure 10.3). Actively dividing keratinocytes are located in the inner-most stratum basale from where they are able to regenerate the upper layers of the epidermis. Migration upwards through the stratum spinosum and stratum granulosum is accompanied by morphological changes and the upregulation and secretion of keratins. The outermost layer of the stratum corneum comprises dead keratinocytes that have undergone apoptosis and lost their nucleus. These dead keratinocytes, referred to as corneocytes, are arranged in tightly packed layers that form the organized layers of the stratified epithelium of the skin. These corneocytes are linked together by

*Immunology: Mucosal and Body Surface Defences*, First Edition. Andrew E. Williams.
© 2012 John Wiley & Sons, Ltd. Published 2012 by John Wiley & Sons, Ltd.

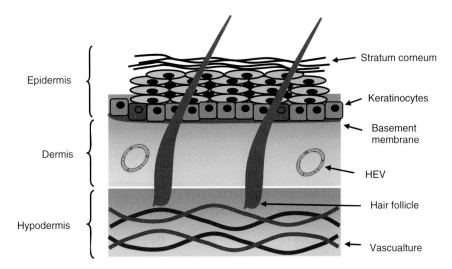

**Figure 10.1** Structure of the skin, which comprises an outer layer known as the epidermis and an underlying area called the dermis. A highly vascular area lies beneath the dermis, called the hypodermis.

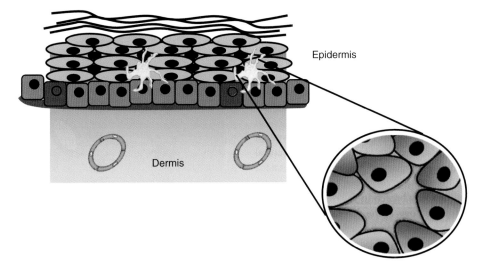

**Figure 10.2** Langerhan's cells function as immune sentinels in the skin and are located in the epidermis where they form an extensive dendritic network.

specialized structures called corneodesmosomes, which help stabilize adjoining cells and are mostly comprised of the protein corneodesmosin.

The upper layers of the epidermis are filled with the protein keratin, which not only provides the stratum corneum with its physical strength but also provides an obstacle to soluble molecules. The intracellular areas of the already densely packed epidermis also contain sphingolipids, further preventing undesirable agents entering the body and contributing to the prevention of water loss. The

dead cells of the stratum corneum are eventually sloughed off and replaced by keratinocytes that have migrated from the lower layers of the epidermis. The journey of a keratinocyte form the stratum basale to the stratum corneum takes approximately 4–6 weeks and keratinocytes that have migrated from the lower regions of the epidermis continually replace those lost from the upper layers.

The basal layer of the epidermis contains melanocytes, whose primary function is the production of melanin. This pigmented compound helps to protect the skin

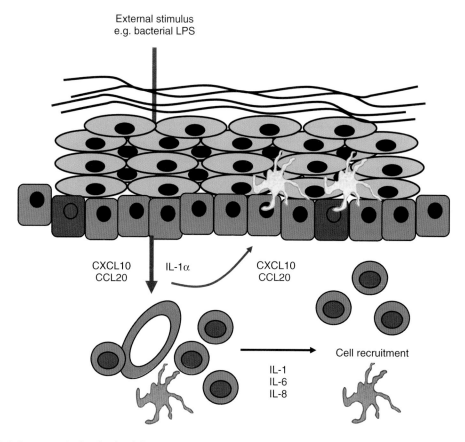

**Figure 10.6** Cells are recruited to the skin following an inflammatory stimuli. For example, the chemokines CXCL10 and CCL20 recruit leukocytes from the circulation and into sites of inflammation. Leukocytes that have been recruited add to the inflammatory microenvironment by secreting inflammatory cytokines.

LCs and dermal DCs (DDCs), thereby enhancing innate immune activation and migration of DCs into regional lymph nodes. In addition, KCs contribute to the pro-inflammatory cytokine milieu necessary for effective T and B cell responses in the dermis.

Keratinocytes are thought to play an active role in the pathogenesis of inflammatory skin disease, the most studied of which is psoriasis. Psoriasis is normally characterized by the development of highly demarcated plaques that are normally scaly, raised and red in colour and are associated with keratinocyte hyperproliferation and abnormal differentiation. Keratinocytes are specialized epithelial cells that form the protective outer barriers of the skin and play an important role in immune defence. They are an important source of antimicrobial peptides and pro-inflammatory cytokines. During psoriasis, they are able to secrete chemotactic molecules such as chemokines and antimicrobial peptides, thereby

attracting inflammatory cells into the psoriatic lesion. In turn, cytokines produced by the proliferating T cells and DCs further enhance KC activation and proliferation. For example, Th1 cells producing IFN-γ and TNF, and Th17 T cells secreting both IL-17 and IL-22, both provide the KCs with positive proliferative signals. In turn, the KCs secrete IL-1β and the chemokines CXCL8 (IL-8), CXCL10 and CCL20 among others, which further provides chemotactic signals causing the migration of more lymphocytes and DCs into the psoriatic lesion.

## 10.5 Keratinocytes secrete antimicrobial peptides

Activation of inflammatory signalling pathways in KCs can result in the release of antimicrobial peptides, which

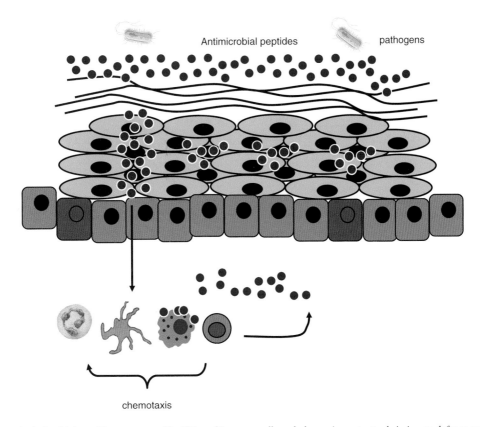

Antimicrobial peptides        pathogens

chemotaxis

**Figure 10.7**  Antimicrobial peptides are secreted by KCs and immune cells and play an important role in innate defence against microbes.

are secreted in response to inflammatory stimuli, during certain diseases such as psoriasis and in response to skin damage (Figure 10.7). Defensins contribute significantly to the milieu of antimicrobial peptides throughout the epidermis, dermis and over the skin surface. In particular, β-defensins are primarily expressed by keratinocytes and include human β-defensins 1–4 (hBD-1-4), while α-defensins are produced mainly by macrophages, granulocytes and T cells and include HD5 and HD6. A more recently discovered antimicrobial peptide, dermicidin, possesses extensive bactericidal and fungicidal properties and is secreted mainly by the eccrine sweat glands in the skin. Sweat also contains a significant amount of LL37 and defensins, and provides an ideal vehicle for the distribution of antimicrobial peptides, and other secretory factors, across the surface of the skin (Box 10.1). Indeed, patients with chronic or infected wounds tend to secrete less antimicrobial peptide, while those with enhanced antimicrobial peptide secretion have fewer skin infections, as exemplified by people who have

the inflammatory skin disease psoriasis. Keratinocytes express another family of antimicrobial peptides, the S100 proteins, which are upregulated in response to inflammation, wounding and cell stress. For example, S100A8 and S100A14 form a heterodimer that acts as an antimicrobial polypeptide, a chemoattractant and an immune stimulatory factor. The synergy between the many antimicrobial peptides therefore represents a very important first line of humoral defence across the skin. Another source of humoral immune components is breast milk, which provides neonates with important passive immunity early in life (Box 10.2).

One of the better characterized antimicrobial peptides is LL37, which is a member of the cathelicidin family. It is processed into a 37-amino acid peptide from an 18 kDa protein known as human cationic antimicrobial peptide (hCAP18), following cleavage by the enzymes elastin and proteinase 3. LL37 represents the only known cathelicidins found in humans. It is expressed at times of heightened inflammation, for example during psoriasis,

### Box 10.1. Hair Follicle and Immune Privilege.

The majority of the hair follicle lies deep within the dermal layers of the skin, while the hair follicle sheath containing the hair shaft extends out of the dermis and into the epidermis. The hair shaft eventually emerges from a pore at the skin surface, which is encased in an oily substance known as sebum (see figure accompanying this box). The hair follicles, formed predominantly of epithelial cells, act like a pocket in which the hair shaft can grow. At the base of the follicle is a structure known as the papilla that contains a capillary bed, which feeds the rapidly dividing hair stem cells. The stem cells are located in a structure known as the bulb, from which the living part of hair extends. The rest of the hair sheath is formed of dead cells and large amount of keratin fibres that are produced during the growth phase (known as anagen). One other structure associated with hair follicles is the sebaceous gland that is responsible for the production of sebum. The oily substance known as sebum keeps the hair follicle and the skin surface moisturized and consists mostly of triglycerides, lipids, wax monomers and a molecule known as squalene. The sebaceous gland in association with the pilo-erector muscle (the tiny muscle that makes hair stand on end) is known as the pilosebaceous unit.

The canal of the hair follicle is a potential entry site for pathogens and it is of no surprise that an array of lymphocytes and APCs are located along its length. The majority of LCs are to be found in the epithelium, while CD4+ and CD8+ T cells are mostly located in the subepithelial areas of the proximal hair follicle, being more frequent distal to the hair bulb or surrounding the sebaceous gland. Macrophages, B cells, NK cells and mast cells can also be detected along the hair follicle. The tissue surrounding the hair bulb and the papilla is largely devoid of immune cells and is notable for the lack of MHC class I expression. This has led to the notion that the hair follicle is an immune privileged site. The characteristics of any immune privileged site (other examples include the brain and the ocular chamber of the eye) is that allografts transplanted into these sites are not rejected by the host immune system. There are several factors which account for this immune privilege including a lack of CD8+ T cell stimulation (through the lack of MHC class I expression), the absence of draining lymphatics, the generation of immune tolerance to antigens derived from that site and the expression of immunomodulatory factors such as IL-10, TGF-β, Fas and FasL.

The special feature of immune privilege in hair follicles is that it is a cyclical phenomenon, which is intrinsically linked to the stage of cell cycle within the dividing epithelium of the hair bulb. Only during the anagen phase, when stem cells are making new epithelial cells, is immune privilege apparent. Immune responses can be effectively induced during the resting or apoptotic phases. Why only certain phases of hair shaft formation should undergo changes that invoke immune privilege remain obscure. However, it may be related to the nature of hair production, in that shaft formation involves the apoptosis of epithelial cells and their replacement with keratin fibres. In order to protect the hair follicle from immune damage elicited by apoptotic cell fragments, a switch occurs at some point that renders the follicle refractory to immune activation. Alternatively, the continuous production of hair shafts requires the maintenance of a small stem cell population. Immune privilege therefore exists to protect these stem cells and thus preserve normal hair follicle function.

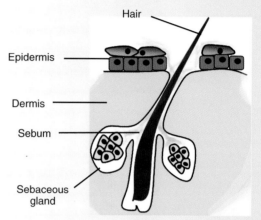

The hair follicle and sebaceous gland are responsible for secretion of sebum onto the skin surface.

wound healing or during an acute infection and possesses a broad spectrum of antimicrobial activities against bacteria, fungi and viruses. LL37 was first discovered to have antimicrobial activities due its ability to bind the bacterial product LPS in studies performed in rabbits. In addition to its antimicrobial properties, LL37 can act as an immunomodulator and a chemoattractant for a variety of cells such as DCs, neutrophils, macrophages and T cells. These migratory leukocytes express LL37 themselves; thereby augmenting the local antimicrobial environment within the skin. Several chemokines expressed by keratinocytes are also known to possess antimicrobial activities and include CXCL9 (IP-9), CXCL10 (IP-10) and CXCL11 (MIG). It is thought that some of these chemokines function in a similar way to the defensins.

### 10.6 Langerhan's cells act as immune sentinels in skin

Langerhan's cells or LCs are a subset of DC, originally derived from the myeloid lineage, that migrate to the epidermal layer of the skin during development and effectively act as immune sentinels for danger signals. In humans they are the major leukocyte found in the healthy epidermis. They are highly dendritic in nature, interact closely with keratinocytes and form a network that covers a large percentage of the skin, allowing them to trap any antigen that may enter (Figure 10.2). They are characterized by the surface expression of a C-type lectin receptor known as langerin (CD207), which is involved in the endocytosis of several pathogens through an interaction with mannose and other related sugars. More recently, the term LC has also come to cover langerin+ DCs found in other epithelial tissues, although those located in the skin remain the archetypal LC. They also express high levels of MHC class I, MHC class II and CD1 molecules. This enables them to interact and present numerous peptide antigens via MHC, or glycolipid antigens via CD1, to epidermal T cells or to T cells circulating through the regional lymph nodes (Figure 10.8).

Following phagocytosis of soluble antigen or microorganisms, activated LCs loose their dendritic appearance and migrate to skin-draining lymph nodes where they interact with T cells and initiate adaptive immunity. Migrating LCs down-regulate E-cadherin, which allows them to disengage from epidermal keratinocytes, and upregulate CCR7 facilitating their migration to lymph nodes. They also upregulate MHC class II molecules and co-stimulatory molecules such as CD40, suggesting that they lose the ability to capture antigen

and become antigen presenting cells instead. LCs continually migrate to regional lymph nodes, although the rate of migration increases during episodes of inflammation. It is now thought that LCs are able to repopulate the epidermis from a local source that is independent of the bone marrow-derived haematopoietic system. TGF-β is required for both LC development in the skin and for maintaining a functional population. More recently, studies using mice deficient in LCs suggest that these cells are dispensable for T cell activation in the skin. It may be that other DC populations are able to take over the functional role of LCs. Alternatively, this has led some to suggest that LCs are actually involved in T cell tolerance, rather than T cell activation. For example, active vitamin D, which is produced in the skin in response to UV radiation, stimulates LCs to migrate to lymph nodes and present antigen to T cells. However, rather than initiating an effector T cell response these vitamin D stimulated LCs promote the differentiation of CD4+CD25+ T reg cells, which in turn contribute to the maintenance of peripheral tolerance to antigen.

## 10.7 Dermal dendritic cells and cross-presentation of antigen

In contrast to the epidermis, the dermis contains few recognizable LCs and instead possesses specialized dermal dendritic cells (DDCs). These DDCs are located primarily in the dermal layer of the skin. They are characterized by the high expression of MHC class II and CD1 molecules and are therefore effective antigen presenting cells (Figure 10.9). They are also highly migratory, as they rapidly leave the dermis once activated and enter draining lymph nodes where they present antigen to T cells. It has been suggested that DDCs are a heterogeneous cell population comprising several sub-populations, one of which is able to cross present antigen to CD8+ T cells. Cross-presentation is an important mechanism for CD8+ T cell activation, particularly for the initiation of adaptive immune responses in the periphery. Exogenously derived antigen, which is normally presented to CD4+ T cells by MHC class II molecules, is endocytosed and processed by DDCs so that it enters the MHC class I processing pathway. This antigen is then presented to CD8+ T cells, either in the lymph nodes or in the skin, resulting in the priming of CTL responses. Cross-presentation is an important mechanism that allows DDCs (and other APCs) to monitor antigen in the skin and to initiate an appropriate immune response. The outcome of cross presentation

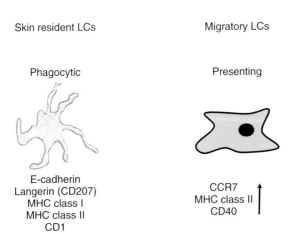

Skin resident LCs

Migratory LCs

Phagocytic

Presenting

E-cadherin
Langerin (CD207)
MHC class I
MHC class II
CD1

CCR7
MHC class II
CD40

**Figure 10.8** Langerhan's cells are a major DC population in the epidermis. Within the epidermis they are highly phagocytic and express cell adhesion molecule that help to retain them in tissue. Following activation they can migrate out of the epidermis and mature into antigen-presenting cells.

## Box 10.2. Immune System of the Mammary Gland.

Immediately after birth the newborn is placed in an environment full of microorganisms and antigens that it has never encountered before, a situation far from the relative sterility of the womb. This, together with an underdeveloped or immature immune system, means that neonates are particularly vulnerable to infectious diseases. The immunological components present in breast milk therefore represent an important aspect of very early immune defence of the young. Breast feeding is thought to be particularly important for infants living in poor sanitary conditions, although numerous benefits are also passed onto babies living in industrialized countries, including a lower rate of lower respiratory infection and maybe even decreased incidence of allergies and food intolerances. The principal immunological component of breast milk is S-IgA, which protects the infant by a process of immune exclusion, particularly in the gut. However, other factors exist in breast milk including antimicrobial peptides, complement factors, mucins, cytokines, chemokines and even leukocytes, such as macrophages and neutrophils. In addition to providing protective immune defence against infections, components that actively stimulate the infant immune system are also found in breast milk, including growth factors and hormones.

The mammary gland is an integral part of MALT and a site of significant S-IgA production. The IgA secreted by the mammary gland is mostly specific for components of those pathogens encountered by the mother; usually as a result of intestinal or respiratory infection, and should therefore be effective against pathogens that the new-born infant will contact. In effect, mammary glands are modified sweat glands, such as those found throughout conventional cutaneous tissue. Each mammary gland contains numerous alveoli that are lined with cuboidal epithelial cells, which are responsible for the secretion of breast milk. These alveoli combine to form lobules within the breast tissue that eventually open into lactiferous ducts, which eventually drain into openings within the nipple. Breast milk, and therefore all the immune components contained within, is secreted into the nipple, which has an external opening from which the infant can imbibe fluid. Breast milk that is produced during late pregnancy or as a consequence of birth is known as colostrum and is the first secretion that the newborn drinks. Due to the small capacity of the neonatal digestive tract, colostrum is a highly concentrated form of breast milk that is rich in nutrients and vitamins as well as IgA, IgM and various other immune factors described earlier. The antibodies are thought to provide the neonate with passive immunity, while the cytokines and growth factors stimulate intestinal and immune development. Considering that the nipple is a modified projection of the skin, albeit rich in nerve endings and highly vascularized, the mammary gland and breast milk immunology is best described as part of this chapter, although it could quite easily be discussed in terms of gut immunology due to the protective effects it has there.

---

can either be immune tolerance or immunity, depending on the extent of the inflammatory environment. For example, a cross-presented antigen in the absence of an infection is likely to result in immune tolerance as no danger signal is relayed to the DDC. On the other hand, a bacterial infection is accompanied by the recognition of numerous danger signals, such as the recognition of LPS by TLR4, in combination with other pro-inflammatory mediators. This then allows the DDCs to cross present antigen to CD8+ T cells and initiate an adaptive immune reaction against the bacteria.

The dermis may also play host to other DC populations, namely langerin+CD103+ DCs and plasmacytoid DCs (pDCs). These DC sub-populations are thought to be effective at processing and presenting antigen to T cells. Moreover, pDCs in particular release a large quantity of the type I interferon IFN-α in response to TLR ligation. Microbial products such as ssRNA from viruses and CpG DNA from bacteria seem especially potent inducers of IFN-α. Many of these DC populations, in particular langerin+CD103+ dermal DCs, have been discovered in the mouse and it remains to be elucidated whether equivalent populations exist in humans.

Plasmacytoid DCs have been implicated as one of the principal DCs involved in the pathogenesis of psoriasis and are one of the main producers of IFN-α in psoriatic lesions. Less certain are what signals induce the pDCs to undergo activation and set up an inflammatory reaction. One explanation is that self-DNA is detected by pDCs in association with the antimicrobial peptide LL37. This acts as a ligand for TLR9 and provides a positive inflammatory signal to the pDC. The pDCs are then thought to migrate to draining lymph nodes and induce T cells to undergo differentiation toward a Th1/Th17 phenotype. Recognition of self-DNA does not normally lead to the induction of an inflammatory response but may contribute to the suspected autoimmune nature during the development of psoriasis. Myeloid dermal DCs are also found throughout the dermis of inflamed skin and they have been demonstrated to produce several inflammatory mediators that induce and or maintain the proliferation of Th1 cells in psoriasis. For example, activated DDCs

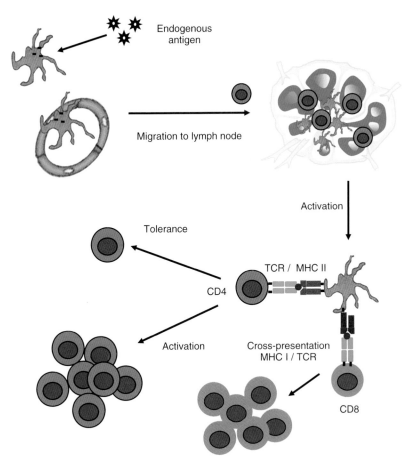

**Figure 10.9** Activated DCs migrate to regional lymph nodes and activate T cells. The skin does not possess organized lymphoid structures and therefore relies on lymph nodes to act as inductive sites. DCs prime lymphocytes within lymph nodes, followed by the recruitment of lymphocytes into the skin, where they can function as effector cells or play a role in maintaining peripheral tolerance.

are effective producers of IL-23, which in turn drives the proliferation of IL-17 producing Th cells.

## 10.8 Mast cells and NK cells in the skin

Both mast cells and NK cells contribute to the innate arm of the immune system. Although some of their roles in inflammatory diseases, such as atopic dermatitis and psoriasis, have been determined, little is known about how they function in healthy tissue or during immuno-surveillance. Mast cells can be regarded as tissue-resident granulocytes and have been primarily associated with the development of Th2-driven allergic reactions. Following activation, mast cells produce the immunomodulatory mediator histamine, which is involved in accelerating

wound healing, driving IgE production by B cells and can also acts as a vasodilator. Histamine production by mast cells is often associated with the pain and itching experience after a bee or wasp sting. More beneficially, mast cells are thought to produce compounds that may be able to neutralize certain toxins or venoms introduced into the skin by insects or other animals.

Classically, NK cells are viewed as being important for the killing of virally infected cells and preventing the spread of the pathogen. The skin will encounter numerous viruses on a daily basis, although NK cells may only come into play if the outer layers of the skin are breached. For example, an insect bite that subsequently introduces a virus into the dermis is likely to initiate an NK cell response. Virally infected cells would then undergo NK cell-mediated cytotoxicity involving perforin and granzyme release. NK cells also produce a number of proinflammatory cytokines during the early phases of an immune reaction, such as IFN-γ and TNF, and thereby direct the adaptive immune system toward a Th1-mediated response. NK cells may also play a very important role in tumour immunosurveillance, as indicated by animal studies that have shown an increased tumour burden in mice with defective NK cell activity.

## 10.9 Intraepidermal lymphocytes in the skin

In contrast to the dermis, very few lymphocytes are located within the epidermis. It has been estimated that intraepidermal lymphocytes only account for 2 per cent of the entire lymphocyte population found throughout human skin. The majority of these intraepidermal lymphocytes are CD8+ T cells, of which most are αβ+ T cells, rather than γδ+ T cells. Mice and other rodents possess a much greater number of γδ T cells in their skin compared to humans. γδ T cells can account for at least 50 per cent of all the intraepidermal T cells in mice. Furthermore, mice possess a resident γδ T cell population known as dendritic epidermal T cells (DETC) that are thought to have significant regulatory functions and are characterized by the expression of an invariant Vγ3Vδ1 TCR (or the alternative Vγ5Vδ1 TCR). These DETCs reside in the epidermal layer of the skin and take on a highly dendritic morphology, which may be as a consequence of their close interaction with tightly packed keratinocytes. In humans, the relatively small population of γδ T cells in the skin also express Vδ1. It is unclear why there is such a difference in skin T cell populations between different species, but may be a result of divergent evolutionary adaptations. It is thought, however, that Vγ3Vδ1 DETCs recognize endogenous antigens derived from damaged skin cells (DAMPs) and as a result respond to epidermal stress factors. They have also been implicated in regulating wound healing in the skin, as mice lacking γδ T cells have poor tissue repair mechanisms.

T cells found throughout human skin have a predilection to locate around the outer sheath of hair follicles and are often found in close association with LCs located within the epithelium (Figure 10.10). The migration of CD8+ T cells into the epidermis seems to be tightly regulated and may be dependent on previous encounter with antigen in the skin, as a high proportion of intraepidermal CD8+ T cells are CD45RO+ memory cells. These epidermal T cells are thought to significantly contribute to the immune surveillance network in the skin. The chemokine CCL27, produced constitutively by keratinocytes, seems to be crucial for the recruitment and maintenance of the intraepidermal T cell population.

## 10.10 Lymphocytes in the dermis

In healthy human skin, the vast majority of lymphocytes located within the dermis are αβ T cells, with CD4+ T cells normally outnumbering CD8+ T cells (Figure 10.10). These T cells are primarily located in peri-vascular areas surrounding the high endothelial venules and arterioles. These areas act as the major sites of entry for lymphocytes into the dermis. Both CD4+ and CD8+ T cells continually circulate between the skin and the draining lymph nodes and play an important role in surveying the tissue for pathogens and tumours. In addition, skin homing T cells are mostly memory cells expressing CD45RO+ (conversely they lack CD45RA). This indicates that the migratory patterns of skin homing T cells are determined by previous antigen encounter within the same tissue. Furthermore, a significant population of dermal T cells are of the Th1 lineage or possess a regulatory T cell phenotype. This predominance of Th1/Treg cells does not preclude the skin from mounting a Th2 immune response though, as evidenced by atopic dermatitis and the allergic reactions to insect bites. It may be that the epidermis is more predisposed to Th2 immune induction, while the dermis is skewed more to Th1 responses, a scenario that is reflected by the anatomical location of Th1/Treg cells in the dermis.

## 10.11 Skin homing T cells express CLA

In order for skin-specific T cells to migrate into appropriate peripheral tissues they must express the correct pattern of surface receptors, which is essential for effective immune surveillance in peripheral anatomical sites. For instance, gut-homing T cells express α4β7 integrin, in contrast to skin homing T cells that express the cutaneous lymphocyte antigen (CLA) marker (CLA is a member of the sialy Lewis-X glycoprotein family). Approximately one third of all the circulating T cells found in the blood express CLA, while very few tissue-resident T cells that express CLA are found outside of the skin. The ligand for CLA is most likely to be E-selectin, which is expressed on endothelial cells located throughout the dermis (Figure 10.11). The precise homing of T cells is particularly important for the memory T cell

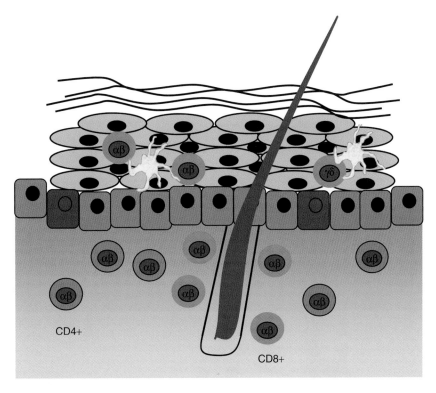

**Figure 10.10** Lymphocytes associate with hair follicles in the skin and are located throughout the dermis. The skin also has an intra-epidermal population of CD8+ αβ T cells and γδ T cells.

population, as this ensures effector cells return to the site of pathogen encounter. A large population of resident T cells are also present in the dermis, where they are ideally positioned to rapidly respond to and eliminate antigen. The resident T cell population is capable of swiftly producing pro-inflammatory cytokines, in order to attract migratory T cells and other cells of the innate and adaptive immune systems.

Inflammation in the skin causes the number of CLA+ CD4+ and CD8+ T cells to increase in the local area. Certain inflammatory diseases of the skin, such as psoriasis, are also associated with the migration of T cells, particularly CD8+ T cells, into the dermis as well as the epidermis. Other cell surface receptors are necessary for the correct homing of T cells to the skin, including the upregulation of integrins, selectins and chemokine receptors. For example, skin homing T cells express the chemokine receptors CCR4 and CCR10, the ligands for which are expressed by keratinocytes (CCL17) and endothelial cells (CCL27) respectively. These homing properties of skin-specific T cells are thought to be established within the

microenvironment of the skin, which affords some sort of molecular imprinting on memory T cells. This imprinting may be influenced by the significant amount of vitamin D that is produced predominantly in the skin in response to UV radiation. Alternatively, the correct signals for homing to the skin may be imprinted in the draining lymph nodes where circulating T cells initially encounter antigen. Either way, skin homing T cells are able to migrate back to the skin, rather than to other peripheral sites.

## 10.12 Chemokines and migration

Although the expression of CLA is important for the migration of T cells across dermal endothelia, many chemotactic mediators are involved in lymphocyte homing to the skin. One of the most studied systems is the chemokine and chemokine receptor system, which not only contributes to cell recruitment but also helps to maintain resident cell populations. Keratinocytes are thought to be a significant source of constitutively

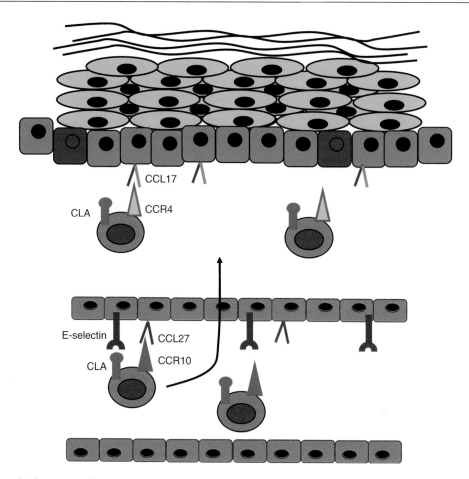

**Figure 10.11** Skin homing T cells express CLA and specific chemokine receptors.

expressed chemokines, including CCL2 and CCL20, which provide retention signals for epidermal LCs, which express and CCR2 and CCR6. LC cell retention within the epidermis is also dependent on the constitutive expression of TGF-β by keratinocytes. Migration of LCs out of the epidermis and into the dermis is thought to require the upregulation of CXCR4 and the production of CXCL12 by dermal fibroblasts. Keratinocytes are also capable of upregulating several chemokines in response to inflammatory stimuli including CXCL8 (IL-8), CCL5 (RANTES), CCL17 (TARC) and CXCL10 (IP-10). Indeed, TNF and IFN-γ released from activated T cells causes the upregulation of these chemokines. Chemokines released from keratinocytes are often accentuated in atopic dermatitis and psoriasis, and significantly contribute to the inflammatory cell infiltration observed in these skin diseases. This chemokine release

is associated with the expression of the corresponding chemokine receptors on the surface of infiltrating lymphocytes. For example, CCR4, CCR6 and CCR10 are often expressed on skin homing T cells, whereby binding to their respective chemokine ligands CCL17, CCL20 and CCL27 facilitate tissue-specific trafficking (Figure 10.12).

## 10.13 Initiation of an immune response in the skin

The series of events that lead to the induction of an antigen-specific immune response can be generalized based on observations from a number of inflammatory skin diseases and microbial infections (Figure 10.13). Initial events involve the recognition of danger signals by cells that are resident in the skin. These innate immune

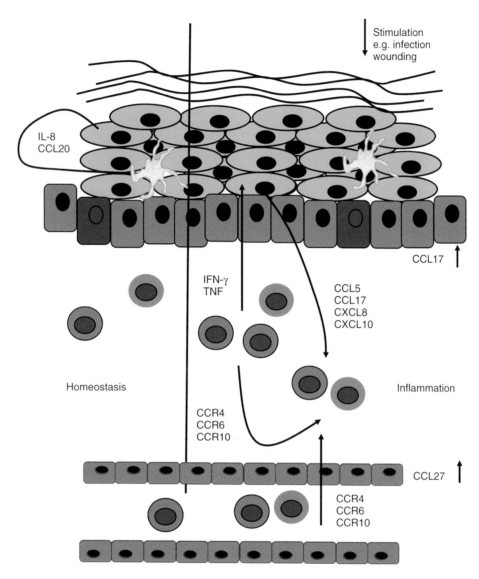

**Figure 10.12** Chemokines are necessary for cell migration n to the skin. In particular, CCR4, CCR6 and CCR10 are important chemokine receptors expressed by skin homing lymphocytes, which migrate toward the chemokine ligands CCL5, CCL17 and CCL27. Different chemokine ligands are produced by the endothelium (CCL27) compared to keratinocytes (CCL5 and CCL17.

cells, including keratinocytes and LCs that are located throughout the epidermis and DDCs, pDCs, macrophages and mast cells that are distributed throughout the dermis, respond rapidly to danger signals. Any challenge or disruption of the epithelial barrier of the skin can be dealt with quickly by these cells. For example, keratinocytes upregulate IL-1α in response to stimulation, which acts as a pro-inflammatory immune stimulus to resident immune cells and results in the recruitment of

lymphocytes, monocytes and granulocytes from the circulation. In a similar way LCs respond to various stimuli, such as microbial products that are recognized by TLRs expressed on their cell surface, and in so doing express IL-1β and other immunoregulatory cytokines and chemokines.

DCs and macrophages within the dermis respond in a similar way to pathogen associated molecular patterns (PAMPs), resulting in the release of pro-inflammatory

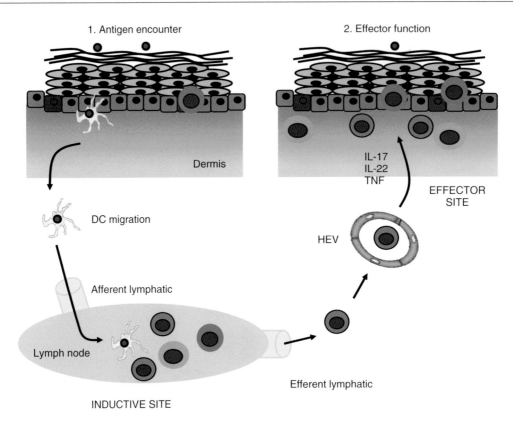

**Figure 10.13** Inductive and effector sites during skin immune responses. Skin lacks organised lymphoid follicles and therefore relies on regional lymph nodes to act as immune inductive sites.

cytokines and the upregulation MHC class II and co-receptors, which enhance their antigen presenting capabilities. Alterations in the expression of chemokines and other homing molecules allows these activated APCs to migrate out of the skin and into draining lymph nodes but at the same time has the effect of activating resident NK cells and T cells. Following activation by IL-1 and TNF APCs undergo a maturation process, whereby they become effective antigen presenters capable of inducing T cell and B cell activation in regional lymph nodes. The lack of organized lymphoid structures within the skin means that regional lymph nodes must act as immune inductor sites. Activation of the innate immune system is an effective way of activation and recruiting cells of the adaptive immune system into the skin, which then acts of the immune effector site.

T cells that are primed by mature DCs in regional lymph nodes undergo clonal expansion and differentiation into effector or memory cells that express skin homing surface receptors. The effector T cells then migrate to the site of initial antigen stimulation in the skin, where they encounter a milieu of inflammatory mediators produced by activated innate immune cells. DCs and macrophages within the inflamed tissue can then initiate T cell effector functions through the recognition of MHC molecules and the appropriate co-receptors and co-stimulatory molecules. For example, the ligation of CD80/CD86 expressed on the surface of APCs, with CD28 expressed on T cells is vital for effective activation, which is augmented by the secretion of IL-12 from DCs and IL-2 from T cells. This not only induces T cell effector functions but also results in the expression of further pro-inflammatory cytokines such as IFN-$\gamma$, IL-17 and IL-22 from CD4+ Th cells and IFN-$\gamma$ and TNF from CD8+ Tc cells. Further cytokines, expressed by NK cells (IFN-$\gamma$), macrophages (TNF) and KCs (IL-23) maintain the effector T cell population within the site of inflammation. It is also important to regulate the amount of immune stimulation in order to prevent excessive reactions and subsequent tissue damage. Regulatory T cells play a significant role in controlling excessive inflammation, through the release of immunomodulatory cytokines IL-10 and TGF-β. These cytokines have the effect of down regulating pro-inflammatory cytokines and contributes to the establishment of homeostasis, which is crucial for maintaining the normal integrity of the skin.

Of course, these series of events have been illustrated by describing the induction of a Th1-type immune response, involving the release of Th1 cytokines. It is becoming increasingly apparent that IL-17 and IL-22 production contributes significantly to inflammatory reactions in the skin. An excellent example is the immunopathogenesis associated with psoriasis, which is an inflammatory skin disease driven by Th1 and Th17 cells. In addition, the induction of a Th2 response can occur in skin and involves the production of IL-4, IL-5 and IL-13. This therefore results in a different immune phenotype and will be discussed in terms of allergy in Chapter 15. For example, atopic dermatitis is characterized by histamine release, IL-4 production by CD4+ Th2 cells and the production of IgE antibodies by B cells.

## 10.14 Cytokines

Cytokines are essential for the induction of immune responses in the skin and are involved in other biological process such as development, wound healing and tissue homeostasis. Within the skin, cytokines are produced by many cell types including keratinocytes, endothelial cells, LCs, DDCs, macrophages, NK cells, granulocytes, T cells and B cells. Depending on the context, individual cytokines have pleiotropic effects in the various cell types located throughout the epidermis and dermis (Figure 10.14). This will depend on the exact location within the skin microenvironment, the particular type of cell the cytokine is affecting, the activation status of the cell and the presence of other cytokines (and other immunomodulatory factors). For example, keratinocytes respond differently to certain cytokines depending on the inflammatory state of their environment. This is evident in their response to IL-22, which helps to maintain KC homeostasis in healthy tissue but induces KC proliferation when in combination with pro-inflammatory cytokines such as TNF and IL-17. In a similar way that other tissues respond to external insults, the various cells of the skin produce specific cytokines that reflect the status of the insult.

IL-1 is an important cytokine for the control of many inflammatory processes and is produced by a myriad of cells types including KCs, endothelial cells, APCs and most lymphocytes. IL-1α is constitutively expressed by KCs in the epidermis and is normally inactive, leading some to suggest that IL-contributes little to the induction of inflammation in the skin. However, it is upregulated in the early stages skin diseases such as psoriasis, during wound healing and during the pathogenesis of cutaneous autoimmune disease (e.g. pemphigus vulgaris). In these cases IL-1 can induce the upregulation of the adhesion molecules ICAM-1 and E-cadherin and contribute to driving Th1 T cell responses through the induction of IFN-γ, IL-6 and additional IL-1. Therefore, IL-1 can act as an instigator of inflammatory responses, particularly when acting in synergy with other pro-inflammatory cytokines such as TNF. Indeed, KCs are also active producers of TNF, as are macrophages and T cells. Ultimately, the early release of IL-1 and TNF by activated cells, results in the upregulation of chemotactic factors, adhesion molecules and pro-inflammatory cytokines and the initiation of inflammation. In fact, TNF has been viewed as the primary instigator of many inflammatory skin diseases, leading to the use of anti-TNF therapy in an attempt to treat diseases like psoriasis. Although historically, TNF has been considered the most appropriate target for therapy, other cytokines may also be useful targets in the future due to their functions in the inflammatory process (e.g. IL-2, IL-6 and IL-17).

## 10.15 Psoriasis, inflammation and autoreactive T cells

Psoriasis is described as a chronic inflammatory disease of the skin, characterized by disfiguring skin lesions. The pathogenesis of psoriasis is associated with substantial keratinocyte proliferation in the epidermis, dilation of dermal blood vessels, the generation of cutaneous plaques (appearing as dry, red and scaly skin) and is thought to be associated with an ongoing inflammatory response. It has been suggested that psoriasis is actually an autoimmune disease, although neither the auto-antigen nor the autoreactive T cell population have yet to be identified. Psoriasis may also represent a common clinical outcome resulting in the progression of one or more distinct disease pathways. For instance, it is thought that certain patients develop psoriasis as a result of a streptococcal infection, while in others psoriasis develops in the absence of detectable streptococcus. Interestingly, humans are the only known species to suffer from psoriasis.

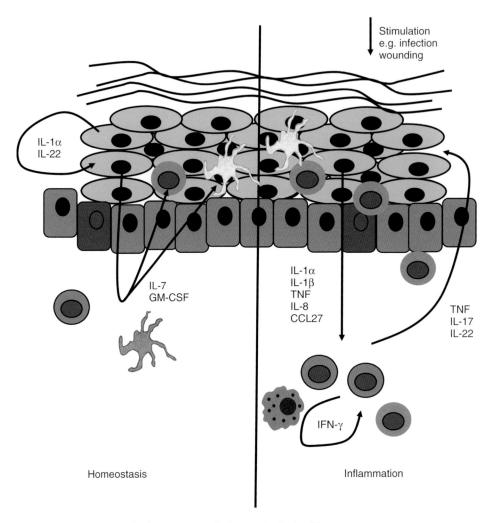

**Figure 10.14** Cytokines are important for the generation of inflammation in the skin.

One of the biggest questions relating to the pathogenesis of psoriasis is to what extent (if at all) the disease is driven by an autoimmune reaction. Central to the development of an autoimmune disease is the recognition of a self peptide by autoreactive T cells, although this peptide has never been conclusively discovered. However, it is evident that Th1 cells producing IFN-γ, and Th17 cells producing IL-17 and IL-22, seem to be the major players in the evolution of a psoriatic lesion (Figure 10.15). These helper T cells activate CD8+ T cells, which are then able to migrate into the psoriatic epidermis as well as the dermis. CD8+ T cells contribute to the production of IFN-γ, TNF and IL-17, thereby accentuating the inflammatory conditions within the psoriatic plaque. Furthermore, IL-23

produced by DCs and activated keratinocytes maintains Th1 and Th17 cells, while IL-12 from DCs is likely to provide initial activation signals to T cells in the lymph node and in the skin.

As to what the pathogenic T cells are recognizing in diseased skin is still uncertain but may be related to a common factor found throughout the skin. For example, some evidence suggests that there may be a link between streptococcal infection and the development of psoriasis. This may involve a mechanism known as molecular mimicry, whereby T cells responding to a streptococcal peptide unfortunately recognize a host molecule as a result of peptide homology. It has been proposed that T cells may respond to an auto-antigen such as keratin,

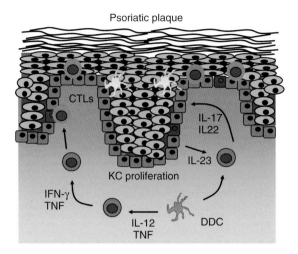

Psoriatic plaque

**Figure 10.15** Initiation of psoriatic inflammation in the skin involves several leukocytes and the expression of pro-inflammatory cytokines such as TNF, IFN-γ and IL-17.

which results in the development of an autoimmune disease. However, in the case of psoriasis it may be more appropriate to consider this a disease that is driven by a hyperactive immune system rather than a *bone fide* autoimmune disease. The central mechanism seems to be a dysregulation of the homeostatic controls that normally keep the skin immune system in check.

## 10.16 Autoimmune-mediated diseases of the skin

Of course, psoriasis is not the only immunological disease that affects the skin; many other immune-driven disorders have been characterized in humans (Table 10.1). For example, the immunobullous diseases are a group of autoimmune diseases that are characterized by

inflammation, blistering and erosions of the skin, exemplified by pemphigus vulgaris. This autoimmune disease is characterized by bullae that occur within the epidermis of the skin and the mouth as a result of a disruption in the normal adhesive properties of keratinocytes. The immune component of this disease involves a mixed infiltration of cells, notably T cells expressing IFN-γ and TNF and B cells producing large amounts of auto-reactive antibodies, particularly IgG in cases of pemphigus vulgaris. These auto antibodies are usually directed against desmosome proteins (e.g. the cadherin desmoglein-3), which are important for cell adhesion between adjacent keratinocytes (Figure 10.16). The auto-antibodies form immune complexes with their antigen throughout the epidermis and basement membrane zone. This can then lead to complement activation giving rise to immunopathology, which results in a separation of the epidermal layers from the basement membrane, a process known as acantholysis. The space between these layers is filled with tissue fluid, thereby giving rise to a blister.

Another immunobullous disease, with similar clinical manifestations to pemphigus vulgaris is bullous pemphigoid. Rather than blisters occurring in the epidermis, bullous pemphigoid blisters occur in the subepidermal layer and occur mostly on the limbs and torso. The autoimmune component of this disease are IgG autoantibodies directed against hemidesmosome, which are responsible for tethering epidermal cells to the basement membrane. These autoantibodies co-localize with the complement protein C3 along the basement membrane and are therefore thought to activate the complement cascade. The deposition of IgG and C3 also instigates the recruitment of inflammatory cells such as neutrophils, T cells and B cells. The subepithelial location of this autoimmune response results the separation of the epidermis from the dermis. Other blistering diseases, with similar manifestations in the skin, include pemphigus

**Table 10.1** Types of skin disease.

| Disease | Immune response | Characteristics |
| --- | --- | --- |
| Atopic dermatitis/eczema | Th2 | Red skin, itching, swelling, blisters |
| Psoriasis | Th1/Th17 | Raised plaques, scales |
| Lichen planus | Th1 | Papules and plaques, white scales |
| Seborrheic dermatitis | Th1? Immunodeficiency | Redness, greasy skin, Dandruff |
| Pemphigus vulgaris | Autoimmune antibodies | Blistering |
| Cutaneous lupus erythematosus | Autoimmune | Rash, erythema, psoriasiform lesions |

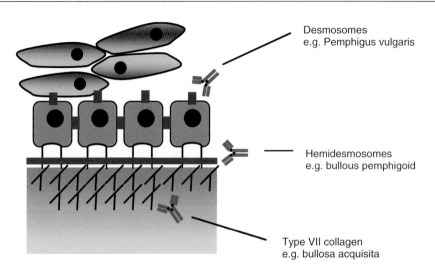

**Figure 10.16** Autoantibodies directed against various adhesion molecules in the skin are associated with several autoimmune diseases.

gestationis, which infrequently occurs in pregnant woman and is also thought to be the result of IgG directed against the hemidesmosomes of the basement membrane. Likewise, epidermolysis bullosa acquisita also involves IgG directed against the basement membrane but type VII collagen fibres seem to be the target for autoantibodies.

Dermatitis herpetiformis, otherwise known as Duhring's disease, is characterized by the formation of extremely itchy blisters on the knees, elbows, neck and shoulders. Despite its name, dermatitis herpetiformis is not caused by a herpes virus but rather the subepidermal blisters resemble those of a herpes virus infection. The lesions are characterized by an inflammatory cell infiltrate of lymphocytes, eosinophils and predominantly neutrophils. Like the other blistering diseases, dermatitis herpetiformis is associated with the deposition of autoantibodies across the basement membrane, most commonly of the IgA isotype. Individuals with dermatitis herpetiformis often have gastrointestinal disease, which is remarkably similar to coeliac disease. Indeed, the severity of dermatitis herpetiformis is often associated with dietary intake of the wheat protein gluten. A strict gluten-free diet can normally control the intestinal aspect of the disease and a decrease in the subepidermal blistering can be achieved over a period of time. It has become apparent that the major autoantigen in dermatitis herpetiformis is the enzyme epidermal transglutaminase (eTG), which co-localizes with IgA in the skin, thereby providing a link between gluten intake and disease severity. There is also a strong genetic component to the disease, whereby individuals with the HLA-B8, HLA-DR3, HLA-DQ2

haplotype have a higher risk of developing dermatitis herpetiformis and coeliac disease.

## 10.17 Systemic diseases that affect the skin

Lupus erythematosus encompasses a range of diseases that can affect several organs of the body including the joints, kidneys, brain, lung and skin. The most severe form is systemic lupus erythematosus (SLE), which is considered to be an autoimmune disease characterized by the formation of antibody-antigen complexes and a resultant type III hypersensitivity reaction that causes extensive immunopathology. In most cases of SLE autoantibodies are directed against nuclear proteins and even DNA itself, hence the potential involvement of so many different organs. Most patients with SLE experience inflammatory skin disease, which present as red rash known as malar rash (or butterfly rash). One of the known factors that trigger the dermatological manifestations of SLE is exposure to UV light. Excessive UV light causes the apoptosis of keratinocytes and results in the presentation of nuclear proteins to the immune system. It is thought that a defect in the normal apoptotic pathway may allow endogenous proteins to gain access to the antigen presentation pathway.

One form of lupus erythematosus is largely restricted to the skin, known as discoid (cutaneous) lupus erythematosus (DLE), which is characterized by red, scaly patches typically on the face and scalp. The immune factors responsible for this scarring inflammation are less clear than SLE, as not all individuals with DLE have autoantibodies directed to nuclear proteins. A related form of lupus, called subacute cutaneous lupus erythematosus (SCLE), is a clinically distinct form of the disease. SCLE is particularly photosensitive as skin lesions appear on those parts of the body exposed to sunlight. The involvement of autoantibodies in SCLE is more likely, particualry IgG directed to type VII collagen fibres located at the basement membrane of the epidermal-dermal junction. The tissue injury associated with SCLE is directed against keratinocytes and typically involves an infiltration of macrophages and T cells.

Systemic sclerosis is a chronic autoimmune disease that affects many organs of the body including the lungs, musculoskeletal system, vasculature and the skin. The characteristic pathological feature of systemic sclerosis is fibrosis and remodelling of the extracellular matrix. This is particular evident in the skin and has led to the use of the term systemic scleroderma. It is thought that a disregulated immune system drives disease pathogenesis, as autoantibodies can be detected in the blood, abnormal levels of cytokines have been demonstrated and there is a marked infiltration of leukocytes into affected areas. However, autoantibodies are rarely deposited in affected tissues, suggesting that disease progression is cell mediated. Inflammatory cells, in particular monocytes and T cells, have been shown to infiltrate the skin during systemic sclerosis pathogenesis. This cellular infiltration is associated with elevated levels of type I interferon, IL-6, IL-10 and IL-13. Macrophage phenotype is significantly affected by these cytokines, causing them to repress inflammatory cytokines and stimulating them to participate in tissue remodelling. In addition, macrophages and T cells express TGF-β under conditions favourable for tissue-remodelling. TGF-β is thought to be an important growth factor for the activation of fibroblasts and for the deposition of collagen and matrix proteins. Another growth factor, connective tissue growth factor (CTGF), has also been implicated in stimulating extracellular matrix deposition, which is secreted by activated endothelial cells. Therefore, interactions between macrophages, T cells, endothelial cells and fibroblasts appear to be central in mediating skin fibrosis.

## 10.18 Infectious diseases of the skin

The skin is continually exposed to a variety of viruses, bacteria and fungi (Table 10.2). Probably the most common virus to infect the skin is human papillomavirus (HPV). These tiny viruses infect the upper layers of the stratified epidermis, causing keratinocyte hyperproliferation and resulting in characteristic warts (otherwise known as papillomas or common warts). The immune system is actually rather adept at preventing the spread of

**Table 10.2** Infectious diseases that affect the skin.

| Infectious disease | Pathogen | Characteristics |
|---|---|---|
| VIRAL | | |
| Chicken pox | Varicella zoster | Rash, itchy pockmarks, blisters, cough, fever |
| Smallpox | Variola major/minor | Rash, fluid-filled blisters, scarring, haemorrhage |
| Herpes | Herpes simplex (type 1) | Papules, blisters, whitlow, keratosis |
| | Herpes simplex (type 2) | |
| Warts/verruca | Human papillomavirus | Raised warts (various shapes/sizes) |
| Kaposi sarcoma | Human herpes virus 8 | Nodular lesions |
| BACTERIAL | | |
| Impetigo | Staphylococcus aureus/ | Pustules, blisters, sores (ecthyma) |
| | Streptococcus pyrogenes | |
| Cellulitis | | |
| | Staphylococcus sp. | Itch, rash |
| | Streptococcus sp. | |
| FUNGAL | | |
| Athlete's foot (tinea pedis) | Trichophyton | Scaling, flaking, itch |
| Ringworm (tinea) | Trichophyton | Itching, circular growth |
| Candidiasis (thrush) | Candida albicans | Itching, burning, vaginal discharge |

the virus and is thought to involve a co-ordinated innate response involving LCs, macrophages and NK cells, while CD4+ and CD8+ T cells appear important for viral clearance. HPV viruses also cause plantar warts (verruca) on the feet and are responsible for the highly contagious sexually transmitted genital warts, that latter of which can result in alterations to the cervical epithelium leading to cancer.

Other viruses are rather more proficient at avoiding the attentions of the immune system, such as the human herpesviruses, which can persist for the entire lifetime of their host. An example is the childhood disease chicken-pox, which is caused by the varicella zosta virus (VZV) and characterized by cutaneous pustular lesions. The primary site of infection is the upper airways but the virus is capable of disseminating to other tissues, in particular the skin. It has been proposed that within the tonsils and regional lymph nodes, VZV infects CD4+ T cells that express skin homing receptors, which then deliver the virus to the skin. Virus replication within epidermal cells results in the characteristic pustules, which contain infectious viral particles that allow transmission to other hosts. Clearance of the virus is dependent on the activation of effector CD4+ and CD8+ T cells and the production of IgG and IgA antibodies. Varicella zosta is also capable of remaining dormant within the dorsal root ganglion of sensory neurons associated with the skin. Reactivation of latent virus results in the painful adult disease shingles. It is thought that memory T cells are responsible for containing latent virus and preventing reactivation. A VZV vaccine is available, which has been shown to control primary and secondary endogenous infections, although routine administered to children is not currently undertaken.

Bacterial infections, namely staphylococcal and streptococcal species, cause a variety of skin diseases such as impetigo, foliculitis and cellulitis and are characterized by a large Th1-driven inflammatory infiltrate at the site of infection. For example, staphylococcal impetigo causes a bullous disease that is easily spread from person to person and is sometimes referred to as school sores. Fungi are also common pathogens that infect the skin, particularly in immunocompromised patients following transplant surgery or in those undergoing chemotherapy. Some of the more familiar fungal infections are caused by Trichophyton species, including athlete's foot, dandruff and ringworm.

Systemic infections can also manifest themselves as lesions in the skin. For example, bacterial septicaemia caused by meningococcal infection often causes local inflammatory foci in the skin. Another good example of infectious skin manifestations occurs following prolonged HIV infection and often correspond to the development of full blown AIDS. One such secondary infection is known as molluscum contagiosum, caused by the molluscum contagiosum virus (MCV), a type of pox virus. The clinical symptoms of MCV infection are small, dome-shaped lesions that can last for years and remain infectious if touched. Another skin disease associated with HIV infection is Kaposi's sarcoma, cased by the human herpes virus 8 (HHV8), and is one of the most widely known manifestation of AIDS. The lesions associated with Kapos's sarcoma are most commonly found in skin, although they can also occur in mucosal tissues such as the gastrointestinal and respiratory tracts. These lesions are often associated with a high rate of AIDS-related mortality.

## 10.19 Summary

1. The skin is not a mucosal tissue, although it is an important body surface involved in protecting the host from infection.

2. The skin has an outer epidermis, composed of specialized epithelial cells known as keratinocytes, and an inner dermis.

3. Specialized DCs known as Langerhan's cells, form a dendritic network throughout the epidermis that act as immune sentinels.

4. Dendritic epidermal T cells (DETC) maintain homeostasis in the epidermis.

5. The majority of leukocytes are restricted to the dermis, where effector immune functions take place.

6. The skin has no organized lymphoid aggregates and therefore requires regional lymph nodes to act as immune inductive sites.

# 11 Immunity to Viruses

## 11.1 Introduction

Viral infections occur regularly across mucosal surfaces such as the respiratory and gastrointestinal tracts, and can often result in severe disease. For example, respiratory tract infection with influenza virus can cause a life-threatening form of pneumonia in naïve individuals who have never encountered the virus before. Adenoviruses and paramyxovisurses such as respiratory syncytial virus (RSV) are also common causes of respiratory diseases that often require hospitalization, especially in infants who have inexperienced immune systems. Viruses such as rotaviruses and noroviruses can cause severe gastrointestinal diseases, characterized by debilitating diarrhoea, particularly in infants and the elderly. Furthermore, the urogenital tract is susceptible to virus infection, including those viruses that are sexually transmitted. The conjunctiva of the eye and the skin are also sites of virus infection. Therefore, immunity to viruses at mucosal tissues is vital in protecting against disease.

A primary infection usually results in the development of disease due to the absence of adaptive immunity to the virus, although innate immune mechanisms are in place that help control the spread of infection and control virus replication. A secondary infection with the same virus is normally asymptomatic or far less severe than a primary infection, due to the acquisition of protective adaptive immunity. A secondary adaptive immune response to a viral infection is normally sufficient for disease prevention. Cell-mediated immune responses are induced rapidly during secondary infections, which kill infected cells and clear the virus before it has a chance to disseminate and cause disease. Antibody responses are also effective at controlling viral spread, by neutralizing viral particles before they have the opportunity to infect a host cell.

The generation of immunity at mucosal surfaces involves the combined action of both innate and adaptive immune factors. In essence, a race exists between virus replication and the development of protective immunity. The inherent immunoregulatory capacity of mucosal tissues is also a factor in driving host immune responses to viral pathogens, such that repeated exposure to the same virus is often necessary in order to maintain protective immunity. Furthermore, many viruses that infect mucosal tissues have evolved numerous immune evasion strategies that aim to subvert host immune responses. This chapter will therefore focus on antiviral innate and adaptive immune mechanisms of the host, and the strategies that viruses employ in order to evade host immunity.

## 11.2 Structure of viruses

Viruses are much smaller than human cells and even bacteria: so small that they cannot be detected under a standard light microscope. Individual viral particles are known as virions, which range in size between 10 and 300 nm in length. Each virion contains centrally located nucleic acid surrounded by an external coat, known as a capsid, which is composed of repeated protein units called capsomers. The proteins that form the capsid of viruses often form specific shapes that can be used to identify the family of viruses. Virions often have proteins that interact with the nucleic acid, known as nucleocapsid proteins. The capsid and nucleocapsid proteins can also become important antigens during an immune response and can be recognized by antibodies or become processed and presented by MHC class I molecules. Viruses that are capable of infecting humans have capsids that usually take the form of an icosahedral or sphere (Figure 11.1). Capsids can also be helical in shape, although these viruses normally only infect plants.

It is also common for viruses that infect humans to have a lipid envelope, which is derived from the plasma membrane of a host cell as a result of viral replication and the process of budding. Viruses that have a lipid envelope

*Immunology: Mucosal and Body Surface Defences*, First Edition. Andrew E. Williams.
© 2012 John Wiley & Sons, Ltd. Published 2012 by John Wiley & Sons, Ltd.

Naked capsid virus                     Enveloped virus

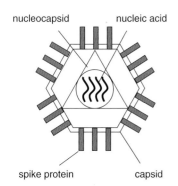

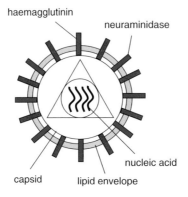

**Figure 11.1**  The structure of viruses that infect humans are usually icosahedral or spherical in shape. Nucleic acid is surrounded by a nucleocapsid. The outer protein layer is known as the capsid from which spike proteins protrude. Some enveloped viruses, such as influenza virus, also have an outer lipid membrane.

are known as enveloped viruses, while those without an envelope are known as naked capsid viruses. Often capsid proteins are capable of protruding through the envelope, known as spikes, such as the haemagglutinin and neuraminidase proteins on the surface of influenza virus. In addition, some viruses have a complex structure with an irregular shape composed of an envelope embedded with proteins, as is the case with poxviruses.

The structure of the virus also defines its physical properties. For example, enveloped viruses require water in order to survive and they are easily broken down by acids and detergents. Therefore, enveloped viruses cannot survive in the gastrointestinal tract, although they are adept at infecting other mucosal sites such as the respiratory or urogenital tracts. Enveloped viruses often require a combination of cell-mediated and humoral immunity for complete clearance. An example of an enveloped virus is herpes simplex virus, which is the cause of cold sores or genital herpes. Naked capsid viruses are relatively resistant to acids, detergents and even proteases. As a result, they can survive in the environment for a considerable length of time and are easily transmitted between individuals. They can also survive in the gastrointestinal tract. Unlike enveloped viruses, naked capsid viruses normally kill infected cells during the process of replication, although an antibody response is usually enough to prevent further infection. An example of a naked capsid virus is adenovirus, which infects the respiratory tract.

## 11.3 Classification of viruses

Viruses have traditionally been excluded from taxonomic classification systems that are used for other organisms. This is because viruses do not share the defining

characteristics of living organisms, such as respiration, reproduction, growth and sensitivity. However, virologists have now devised a taxonomic system based on the physical characteristics of a virus: namely the size, shape and symmetry of the capsid, the envelope properties and the genomic structure. Viruses are therefore classified according to order, family, subfamily and genus with the corresponding suffixes virales, virinae, viridae and virus, respectively. However, not all viruses are given an order or subfamily, as it is often sufficient to group a virus according to family and genus.

Viruses can also be classified based on the type of genome they have. This is particularly useful when describing viruses that infect humans. Viral nucleic acid can be constructed from single-stranded (ss)DNA, double-stranded (ds)DNA, single-stranded (ss)RNA or double stranded (ds)RNA. On a basic level, viruses can therefore be classified as being either a DNA virus or an RNA virus. The essential requirement of viral nucleic acid is to produce mRNA for the purpose of protein synthesis. However, viruses utilize several different mechanisms in order to synthesize mRNA depending on the nature of their genomic nucleic acid. In addition, the composition of a viral genome, whether it is ssRNA, dsRNA or DNA, is an important activator of the immune system. For instance, ssRNA and dsRNA are potent activators of inflammation, through their recognition by members of the Toll-like receptor (TLR) family.

## 11.4 Viruses replicate within host cells

The fist stage in virus replication is the attachment of the virion to the surface of a host cell. This attachment

**Table 11.1** Possible host receptors used by viruses to gain entry into cells.

| Virus family | Virus | Cell receptor |
|---|---|---|
| Adenoviridae | Adenoviruses | Integrin $\alpha_V\beta_3$, MHC class I |
| Herpesviridae | Epstein-Barr virus | Complement receptor 2 (CR2) |
| | Human cytomegalovirus | MHC class I |
| | Herpes simplex virus | Heparan sulphate |
| Myxoviridae | Influenza virus | Sialic acid residues |
| | Measles virus | CD46 |
| Picornaviridae | Rhinoviruses | ICAM-1 |
| | Poliovirus | IgG superfamily |
| Rhabdoviridae | Rabies virus | Acetylcholine receptor |
| Rheoviridae | Reoviruses | EGF receptor |
| | Rotaviruses | EGF recpetor |
| Retroviridae | HIV | CD4 |

process involves an interaction between a viral protein and a receptor expressed by the target cell (Table 11.1). For example, human immunodeficiency virus (HIV) attaches to host cells following an interaction with the virion glycoprotein gp120 and the T cell co-receptor CD4. The mechanism of entry into a host cell differs between naked capsid viruses and enveloped viruses. Naked capsid viruses are normally engulfed by receptor-mediated endocytosis so that the virion is contained within a vesicle. The vesicle is then ruptured to release the virion into the cytoplasm. Enveloped viruses enter host cells by membrane fusion between the virus envelope and the plasma membrane. Alternatively, enveloped viruses fuse with vesicle membranes following receptor-mediated endocytosis. The end result is the same, in that the nucleocapsid is deposited into the cytoplasm.

Virus replication can either occur in the cytoplasm or the nucleus, although the mechanism for both is largely the same. The genome of the virus needs to be exposed in order for replication to take place. Therefore, nucleocapsid proteins are shed to reveal the viral nucleic acid. In some cases, the viral genome acts directly as mRNA, while in other cases the viral genome is transcribed into mRNA. The translation of viral proteins occurs on cellular ribosomes, which direct the synthesis of the enzymes required for viral replication and the structural proteins that will

form new capsids. Once viral mRNA has been translated, genome replication can take place. Newly synthesized genome nucleic acid together with the translated viral structural proteins form new capsid structures. The intracellular location of viruses has important consequences for the type of immune response generated, in that cell-mediated immunity and the activation of cytotoxic effector functions are normally required for viral clearance.

The final stage in the virus replication cycle is the release from infected cells. This mechanism is different for naked capsid viruses compared to enveloped viruses. Naked capsid viruses basically induce the lysis of infected cells, which releases newly formed capsids into the extracellular environment. For enveloped viruses, the newly formed capsids, together with viral envelope glycoproteins, enter the vesicle exocytosis pathway of the cell. They are then transported to the plasma membrane where a process known as budding takes place. Envelope glycoproteins direct the budding process so that newly formed capsids are enclosed within the plasma membrane of the cell. Viral budding ultimately leads to the formation of a mature enveloped virus. These differences in viral replication have certain immunological consequences so that different defence mechanisms are required to combat enveloped viruses compared to naked capsid viruses.

## 11.5 Infections caused by viruses

Viruses infect their host primarily across mucosal membranes (Table 11.2), or following the introduction into skin as a result of an insect bite, for example. Therefore, the respiratory tract, gastrointestinal tract, urogenital tract and the conjunctiva of the eye are all susceptible to virus infection. These tissues are commonly referred to as primary sites of infection, as this is where viral and host interaction first takes place. However, during viral pathogenesis the virus may be able to spread from the primary site of infection and disseminate to other areas of the body known as secondary sites. If the infection disseminates into the blood, described as viraemia, multiple secondary sites are likely to be infected, although viral spread is usually restricted to local tissue. Viral replication is followed by viral shedding, usually within the primary site of infection, for example the respiratory tract. Viral shedding often results in the transmission of the virus to a new host.

Three forms of virus infection are recognized: acute, latent and chronic (Figure 11.2). Acute infections only last for a relatively short period of time and the disease symptoms are usually associated with a rapid increase

**Table 11.2** Viruses that commonly infect mucosal tissues.

| Mucosal tissue | Virus | Transmission |
|---|---|---|
| Respiratory tract | Influenza virus | Droplets |
| | Respiratory syncytial virus (RSV) | Droplets |
| | Rhinovirus | Droplets, contact |
| | Adenovirus | Droplets, contact |
| | Cytomegalovirus (CMV) | Droplets |
| | SARS coronavirus | Droplets |
| | Measles virus | Contact, droplets |
| Gastrointestinal tract | Rotavirus | Food, water, contact |
| | Polio virus | Food, water, contact |
| | Norovirus | Food, water, contact |
| | Coronavirus | Food, water, contact |
| Urogenital tract | Human immunodeficiency virus (HIV) | Body fluids, blood |
| | Herpes simplex virus type 2 | Body fluids |
| | Human papillomavirus | Contact |
| Skin | Varicella zoster (chicken pox) | Contact |
| | Rubella virus | Contact |
| | Smallpox virus | Contact |
| | Papillomavirus | Contact |
| | Human herpes virus 6 | Contact |
| | Parvovirus | Contact |

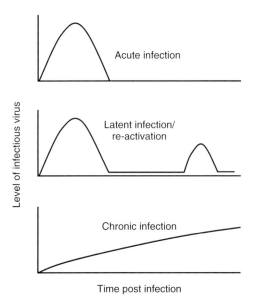

**Figure 11.2** Virus infections can take the form of acute, latent or chronic. Acute infections are short-lived and are cleared relatively quickly. Latent infection involve an initial acute phase, a period of dormancy when the virus is in a latent form and subsquent period of re-activation. Chronic infections persist in the host for a relatively long period of time.

in viral load, followed by pathogen clearance and alleviation of symptoms. An example of an acute infection is exemplified by influenza virus infection. In a latent infection, disease symptoms can reoccur following the reactivation of a virus after a period of dormancy. Virus reactivation usually occurs during periods of immunosuppression, while total viral clearance is rarely achieved. The cold sores caused by herpes simplex virus are an example of a latent infection. Chronic infection occurs over a prolonged period of time, resulting in continuous or recurrent disease symptoms and a continued elevation in viral load. The chronic liver disease caused by hepatitis B virus is an example of a chronic infection. It is often the case that the pathology observed during a viral infection is the result of the immune response, rather than the virus causing direct tissue damage. This is exemplified following infection with influenza virus or severe acute respiratory syndrome coronavirus (SARS-CoV), whereby excessive immune responses lead to immunopathology.

## 11.6 Certain viruses can infect immune cells

Several viruses are able to infect cells of the immune system (Table 11.3), which represents a form of immune evasion whereby the virus infects the very same cell that is trying to eliminate it. For example, Epstein-Barr virus (EBV) is able to infect B lymphocytes where it can remain in a latent form for many years. Although EBV primarily infects epithelial cells, viral transfer to B cells results in a persistent infection that is associated with transformation of B cells. Even though an immune response is generated against the virus, it is able to hide from the immune system in a latent form. A balance therefore exists whereby reactivation of viral replication leads to a host immune response, only for the virus to return to a latent state within B cells. Another virus that infects B cells in the chicken virus infectious bursal disease virus, which is

**Table 11.3** Viruses that are capable of infecting cells of the immune system.

| Virus | Immune cell |
|---|---|
| Epstein-Barr virus | B lymphocytes |
| Infectious bursal disease virus | B lymphocytes |
| HIV | Macrophages, T cells, DCs |
| Cytomegalovirus | Macrophages |
| Lactate dehyrdogenase virus | Macrophages |
| Measles virus | T lymphocytes |
| Human herpes virus 6 | T lymphocytes |
| Varicella zoster virus | DCs |

often fatal and has the potential to devastate whole flocks. Infection of B cells within the bursa of Fabricius causes severe immunosuppression and a necrotizing disease that can spread to other mucosal tissues such as the GALT, BALT and tonsils.

A number of viruses are known to infect T cells, including HIV, measles virus and certain herpes viruses. HIV is able to infect T cells following binding of the gp120 viral protein to the CD4 co-receptor on the surface of Th cells. Replication and the release of HIV from T cells results in elevated virus titres in the blood and a corresponding decrease in CD4+ T cell counts, which is associated with immunodeficiency. This acute phase of infection does result in a CTL response to virally infected cells, which controls virus titre in the blood but also leads to viral latency. HIV is able to remain latent in memory CD4+ T cells, particularly within lymph nodes where HIV can infect surrounding T cells more easily. Without treatment, persistent infection with HIV eventually leads to acquired immune deficiency syndrome (AIDS), which is often fatal due to secondary complications associated with an immunocompromised state. In addition, HIV infects macrophages and DCs as it has a particular tropism for CCR5. Indeed, within the mucosa of the genital tract HIV is thought to initially infect macrophages, which then transport the virus to lymph nodes where contact can be made with T cells.

## 11.7 Virus infection of epithelial cells

Epithelial cells are the first cell type to be exposed to an infectious virus. For example, reoviruses infect the gastrointestinal tract following ingestion. It is thought that they attach to intestinal M cells, which transport the virus to the basolateral surface of the epithelium. M cells may provide an easier target for viruses, as they do not possess microvolli or an extensive glycocalyx or mucous sheath. Once at the basolateral surface, reoviruses are able to infect neighbouring epithelial cells. Polio virus is also thought to be endocytosed by M cells, which results in dissemination to Peyer's patches and regional lymph nodes. The transport of poliovirus to lymphoid tissue is also thought to aid infection of secondary tissues. The combined infection of intestinal epithelial cells and migratory lymphoid cells results in the release of poliovirus into the gut lumen, where it is expelled with the faeces.

Viruses that infect the reparatory tract usually only infect the epithelial cells of the upper airways, as is the case with influenza virus, rhinovirus and RSV. However, severe influenza infections can result in dissemination to the lower airways resulting in life-threatening pneumonia. In humans, influenza primarily infects airway epithelial cells across the apical surface, which causes the destruction of the epithelia and dissemination to neighbouring cells. Influenza infections are normally restricted to the respiratory tract. This is due to sialic acid residues expressed exclusively on the surface of airway epithelial cells, which are required for haemagglutinin cleavage and viral endocytosis. Rhinoviruses also infect nasal epithelial cells across the apical surface. Replication within epithelial cells results in the shedding of virus into the nasal cavity where it is expelled and disseminated.

Epithelial cells therefore play an important role in barrier defence and in preventing virus replication at mucosal sites. Mechanical factors such as mucociliary clearance and chemical factors such as the pH of the stomach provide an inhospitable environment for many viruses. The mucous layers of the gastrointestinal and upper respiratory tracts and the secretion of IgA across the epithelial surface also prevent viral invasion. In addition, infection of epithelial cells causes the rapid upregulation of type I interferons and intracellular antiviral mechanisms (discussed later), which inhibit virus replication. In healthy individuals, these elements of innate immunity are often sufficient to prevent viral infections. If these barriers are breached then epithelial cells are also capable of secreting pro-inflammatory cytokines such as IL-1, and can recognize viral PAMPs following ligation with PRRs. These inflammatory mediators effectively stimulate innate and adaptive immune responses (Figure 11.3). Epithelial cells are also the target for NK cell and CTL effector functions, as they represent the principal cell type that hosts viral replication.

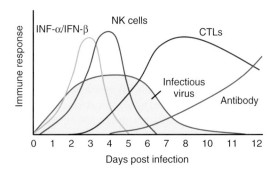

**Figure 11.3** The immune response to viruses involves several components of the immune system. Type I interferons (IFN-α/IFN-β) are produced in response to primary infection and function by inhibiting virus replication. NK cells are recruited to sites of infection where they kill infected cells. T cell responses occur 2–3 days after innate responses, which involve CTL-mediated killing of infected cells and coincides with reduction in viral load. Increases in antibody are detected 4–5 days after infection and remain elevated for months after infection.

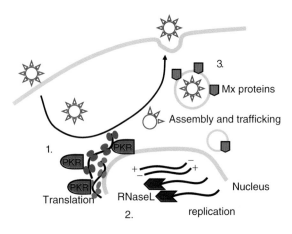

**Figure 11.4** IFN-induced expression of antiviral proteins. Type-1 IFNs cause the upregulation of 1. RNA-activated serine/threonine protein kinase (PKR), which interferes with protein translation; 2. RNaseL, which enzymatically degrades viral RNA; and 3. Mx proteins, which inhibit vesicular transport.

## 11.8 IFN-α response

Type I interferons possess potent antiviral properties and are expressed in numerous cell types that are susceptible to viral infection, such as the epithelial cells that line the respiratory and gastrointestinal tracts. Another important source of IFN-α are plasmacytoid DCs (pDCs), which tend to reside in subepithelial compartments of mucosal tissues, such as the parenchyma of the lung, and are sometimes referred to as natural interferon-producing cells. Binding of type I interferons to their receptor activates a number of transcription factors, such as interferon regulator factors (IFNs) and NF-κB, leading to the expression of genes involved in immune defence. Type I interferons have numerous antiviral properties including the expression of proteins that interfere with viral replication.

Type I interferons induce the upregulation of the double-stranded (ds) RNA-activated serine/threonine protein kinase (PKR), which results in the inhibition of protein translation and therefore inhibits viral replication. Upregulation of oligoadenylate synthase (OAS) leads to the expression of RNAseL, which is an enzyme that degrades viral RNA. In addition, the upregulation of a group of proteins, known as the Mx proteins, interfere with intracellular trafficking and therefore prevent virus particles escaping from the cell. The combination of these factors acts to prevent virus replication and dissemination (Figure 11.4).

Antiviral activity is further enhanced by the release of reactive nitrogen species, such as NO, via nitric oxide synthase upregulation. Moreover, infected cells secrete IFN-α and IFN-β, which act in a paracrine fashion so that neighbouring cells activate antiviral gene expression pathways prior to virus contact. Type I interferons also enhance presentation of viral antigens through the upregulation of MHC class I and co-stimulatory molecules and through the recruitment and activation of CTLs and NK cells. In addition, type I interferons have been demonstrated to be effective anti-tumour cytokines, in part through increased expression of tumour-specific antigens. The type III interferons, IFN-λ1, IFN-λ2 and IFN-λ3 (IL-29, IL-28A and IL-28B), have similar antiviral properties as type I interferons and utilize similar intracellular signalling cascades. They are produced by various cell types including macrophages, monocytes and pDCs, inhibiting viral replication and upregulating MHC class I expression.

## 11.9 NK cell response to viruses

NK cells are part of the innate immune system that is activated by intracellular pathogens such as viruses. They provide two important defence mechanisms: the first is to initiate cytotoxic effector functions against virally infected cells and the second is the secretion of the pro-inflammatory cytokines IFN-γ and TNF. They are activated by IL-12 and type I interferons, (IFN-α and IFN-β) that are secreted by infected epithelial cells and by tissue-resident macrophages and DCs (in particular, pDCs). As well as providing cytotoxic functionality, the effective secretion of the pro-inflammatory cytokines IFN-γ and TNF act by activating macrophages and T cells and enhancing NK cell effector functions themselves. Therefore, NK cells are important components of cell-mediated immunity against viral infection, acting to control the dissemination of viruses, and they bridge innate and adaptive immune responses through the release of cytokines.

The release of IFN-γ by activated NK cells not only acts as a pro-inflammatory mediator that enhances further cell-mediated immunity but also augments antiviral mechanisms in infected cells. For example, the production of IFN-γ by NK cells causes the upregulation of IFN-β in infected cells, which in turn stimulates intracellular antiviral processes such as those discussed in previous sections. This has the effect of preventing further virus replication within the infected cells and inhibits the spread of the virus to other cells. The interferon network is thought to be an important response to many viruses including influenza virus, RSV and human cytomegalovirus (HCMV).

The cytotoxic function of NK cells is mediated through the recognition of specific receptors on the surface of infected cells. Viral infection causes infected cells to upregulate several stress-induced molecules including the MHC family members MICA and MICB. These are recognized by the killer cell lectin-like receptor NKG2D, which activates the NK cells and initiates cytotoxic effector functions. This receptor ligation results in the formation of an immunological synapse with the infected cell and the release of perforin and granzymes onto the target cell surface (the mechanism of NK cell killing is discussed in Chapter 2). These cytotoxic factors kill the infected cell through the activation of apoptotic pathways, via activation of caspases and through pore formation and resultant cell lysis.

NK cell cytotoxicity against infected cells is enhanced by further recognition mechanisms. Importantly, certain viruses are capable of downregulating the expression of MHC class I molecules on target cell plasma membranes; an immune evasion strategy that aims to avoid CD8+ CTL cytotoxic responses. However, NK cells are able to identify the lack of MHC class I expression on the cell surface, a mechanism known as the recognition of missing self (Box 11.1). A set of molecules expressed by NK cells, known as killer-cell immunoglobulin-like receptors (KIRs), normally prevent NK cell cytotoxicity of healthy cells, which

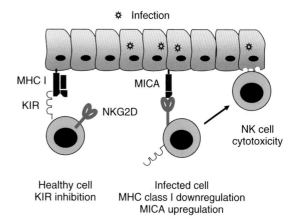

**Figure 11.5** NK cells recognize abnormal expression of receptors on the surface of infected cells. Viruses cause the down regulation of MHC class I molecules, which is identified by the lack of ligation to KIRs on NK cells (missing self). In addition, activation receptors such as NKG2D recognise MICA, which is upregulated by stressed cells. These signals activate NK cell killing mechanisms.

express high levels of MHC class I. However, the lack of MHC class I expression relieves NK cells of an inhibitory signal so that cytotoxicity can be initiated. This mechanism is thought to operate in tandem with the recognition of MICA by NKG2D, which acts as the positive signal (Figure 11.5). Other activating receptors include CD96 (T cell activation, increased late expression, TACTILE) and CD226 (DNAX accessory molecule-1, DNAM-1), which recognize CD155 expressed by target cells in response to infection. Furthermore, NK cytotoxicity is greatly enhanced by the opsonization of infected cells by antibody, which initiates antibody-dependent cellular cytotoxicity or ADCC (again discussed in depth in Chapter 2).

## 11.10 Viral evasion of NK cell responses

Many viruses have evolved immune evasion strategies that attempt to subvert or modulate NK cell effector functions, thereby demonstrating that NK cells play a central role in controlling viral infections within mucosal tissues. For example, it has been demonstrated that HCMV is able to downregulate the expression of NK cell receptor ligands on the surface of infected cells. This includes modulating the levels of the NKG2D ligand MICA and the NK cell receptor ligand CD155, thereby rendering the target cell more resistant to NK cell-mediated lysis. Moreover, polymorphisms within the human population have been

## Box 11.1. NK Cells and CTLs Collaborate during Virus Infections.

NK cells and cytotoxic T lymphocytes (CTLs) possess potent antiviral properties, through the secretion of IFN-γ and the induction of cytotoxic effector functions. The secretion of IFN-γ upregulates antiviral mechanisms in infected and neighbouring cells and activates Th cells, CTLs and macrophages, thereby augmenting cell-mediated immunity. The cytotoxic function of NK cells and CTLs, through the release of perforin and granzymes, causes the apoptosis of infected cells and prevents the spread of the virus. Due to the effectiveness of these lymphocyte populations in host defence against viral infection, many viruses have evolved mechanisms that subvert NK cell and CTL effector functions. These immune evasion strategies include the down regulation of MHC class I molecules and the inhibition of peptide loading onto MHC class I molecules. In turn the immune system has evolved strategies to combat virus evasion mechanisms. Therefore there exists an intricate interplay between the virus and the host immune system, whereby virus replication competes against immune defence.

An important example of how the immune system has evolved to fight viral evasion strategies exists in the divergent ways in which NK cells and CTLs recognize infected cells. This recognition is based on the expression of MHC class I molecules, which NK cells and CTLs both utilize during host defence. Primed CTLs recognize virus antigens presented on the surface of infected cells in conjunction with MHC class I molecules. This recognition activates CTLs and induces their cytotoxicity, thereby killing the infected cell. In turn, certain viruses modulate this immune mechanism by downregulating the expression of MHC class I molecules, thereby escaping CTL-mediated host defence. The immune system therefore turns to NK cells.

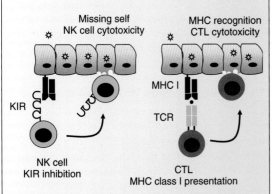

NK cells and CD8+ CTLs contribute to the killing of virus infected cells.

Rather than recognizing antigen, NK cells identify the absence of MHC class I expression on virus-infected cells.

Therefore, even though the virus has avoided the attentions of CTLs, it subsequently brings NK cells into play. NK cells express several inhibitory receptors known as killer-cell immunoglobulin-like receptors (KIRs), which recognize the presence of MHC molecules on the surface of normal cells. These KIR receptors inform NK cells that the host cell has a normal pattern of MHC molecules and therefore not to embark on a cytotoxic pathway. However, the lack of MHC molecules means that NK cells do not receive an inhibitory signal, a mechanism known as missing self. Down regulation of MHC class I molecules results in the activation of NK cells and the induction of cytotoxic processes. Therefore, although viruses aim to subvert a particular component of host defence, the immune system often compensates by utilizing a different effector mechanism. This is a prime example of host-pathogen coevolution.

found in the MICA gene, which might render some groups of people more responsive to viral infections than others. HCMV has also been shown to encode a protein that is able to inhibit another activating receptor, the natural cytotoxicity triggering receptor (NCR3) NKp30, expressed on the surface of NK cells. The pp65 HCMV protein is thought to inhibit NKp30 directly, thereby reducing NK cell activity. Similar discoveries have been made in the murine CMV, whereby a similar viral product (m157) is able to inhibit an analogous NK cell receptor in mice (Ly49H).

NK cells are thought to participate in immunity to chronic HIV infections by killing infected cells and reducing the amount of virus detectable in the blood. Indeed, individuals with chronic disease, but that do not have viraemia, have higher levels of CD56+ NK cells than those that do have viraemia. This is thought to represent a disregulation in the NK cell population, as the lack of CD56 expression on NK cells is associated with anergy, low NK cell activity and reduced expression of perforin. HIV is also known to selectively downregulate the expression of certain MHC class I molecules on the surface of infected cells, through the production of the HIV-1 negative factor (nef). It has been shown that HLA-A and HLA-B are selectively downregulated, but HLA-C and HLA-E are not. This strategy inhibits the recognition of HLA-A and HLA-B by CTLs, while NK cell inhibitory receptor ligation is maintained by HLA-C and HLA-E, which are recognized by KIR receptors. This suggests that HIV avoids vital components of an antiviral immune response by hiding from the effects of cytotoxic T cells, while at the same time inhibiting NK cell-mediated killing mechanisms (Figure 11.6).

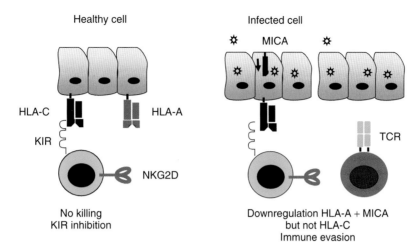

**Figure 11.6** Many viruses employ immune evasion strategies to avoid NK cell-mediated killing. For example, HIV is able to downregulate HLA-A, while retaining HLA-C expression, thereby avoiding both NK cell and CTL recognition.

## 11.11 Macrophages contribute to virus elimination

Under steady-state conditions, macrophages within mucosal tissues such as the respiratory tract function as modulatory cells that maintain tissue homeostasis. Indeed, alveolar macrophages comprise as much as 95 per cent of leukocytes within a healthy lung. However, viral infections can alter the immunoregulatory phenotype of macrophages, for example in response to type I interferons and IL-1 released from epithelial cells or the direct recognition of virally-derived PAMPs. Activated macrophages play several important roles in the early phases of host defence against viruses, including the release of IFN-β and the pro-inflammatory cytokines IL-12, IL-1 and TNF and the phagocytosis of infected cells and virus particles. The release of IFN-β by macrophages contributes to the antiviral effector mechanisms of neighbouring cells and the initiation of intracellular pathways that prevent virus replication. The secretion of IL-12 and TNF acts as a pro-inflammatory mediator for other innate and adaptive cells including neutrophils, NK cells and T cells. IL-12 and TNF production by macrophages is particularly adept at inducing NK cells to secrete IFN-γ. Furthermore, activated alveolar macrophages recruit CCR2 expressing monocytes into the lungs through the production of the chemokine CCL2 (MCP-1). These monocytes then differentiate into mDCs or phagocytic macrophages and participate

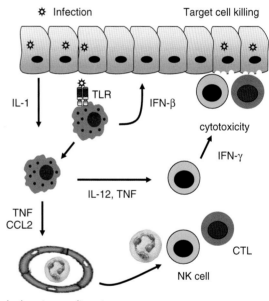

**Figure 11.7** Activated macrophages secrete several pro-inflammatory cytokines and chemokines that recruit leukocytes to the site of infection. TNF induces NK cells to secrete IFN-γ and initiate killing mechanisms. IL-12, TNF and IFN-γ promote cell-mediated immunity and CTL activity.

in antigen processing and presentation to cells of the adaptive immune system (Figure 11.7).

Macrophages are highly phagocytic cells that are thought to contribute to the inhibition of viral spread.

Phagocytosis of infected cells that have been lysed by replicating viruses, or the phagocytosis of apoptotic cells, limits viral dissemination. The phagocytosis of infected cells is aided by opsonization with antibodies, which are recognized by FcRs expressed on the surface of macrophages. Neutrophils that have migrated into the lungs in response to chemotactic stimuli also contribute to the phagocytosis of apoptotic cells.

However, heightened alveolar macrophage activation and TNF production can lead to excessive recruitment of macrophages and neutrophils into the lungs, which can cause tissue damage. This is exemplified following acute influenza virus infections, which are characterized by extensive immunopathology as a result of elevated pro-inflammatory cytokine production (e.g. TNF and IL-1β) and accumulation of activated macrophages and neutrophils.

## 11.12 TLRs and NLRs recognize virus motifs

The recognition of PAMPs derived from viruses is a key step in early host defence against infection. Numerous viral PAMPs are recognized by the immune system including ssRNA, dsRNA and cytosolic DNA. The PRRs involved in recognizing these PAMPs are the Toll-like receptors (TLRs) and the nucleotide-binding oligomerization domain (NOD)-like receptors (NLRs), although a new family of PRRs known as retinoic acid inducible gene-I (RIG-I)-like receptors (RLRs) are also known to bind viral nucleic acid. The TLRs are expressed on cell membranes and recognize extracellular or endosomal PAMPs, while the NLRs and RLRs recognize PAMPs within the cytoplasm.

The TLRs responsible for the recognition of viral PAMPs include TLR3, TLR7, TLR8 and TLR9, which are all expressed within endosomes and all recognize viral nucleic acids. TLR3 recognizes dsRNA, TLR7 and TLR8 recognize ssRNA, while TLR9 recognizes unmethylated CpG-containing DNA. These nucleic acids do not normally exist within host cells and are therefore identified as being non-self. These TLRs signal through the adaptor molecule MyD88, through an interaction with TIR domains contained in the cytoplasmic domain of the receptors (discussed in depth in Chapter 2). This results in the activation of an intracellular signalling cascade that ultimately leads to the activation of several pro-inflammatory transcription factors, including NF-κB, AP-1, interferon regulatory factor 3 (IRF3) and

IRF7. Viral RNA is particularly adept at upregulating the type I interferons IFN-α and IFN-β. Although numerous cell types, such as epithelial cells, endothelial cells and fibroblasts, are capable of expressing these TLRs and upregulating type I interferons, pDCs are particularly adept at responding to viral PAMPs. One of the defining features of pDCs is the expression and secretion of large quantities of IFN-α and IFN-β. Therefore, pDCs are important immune cells in the initiation of antiviral immune responses.

The NLR NOD2 has been demonstrated to recognize cytosolic ssRNA derived from viruses such as RSV and influenza virus. This results in the activation of the signalling molecule IRF3 and the upregulation of type I interferons, which are central to the innate immune response to viruses. This NOD2-dependent activation of type I interferons utilizes a separate pathway to the one that bacterial products, notably peptidoglycan, initiate (discussed in Chapter 12). It is thought that NOD2 requires the mitochondrial antiviral signalling protein (MAVS) that is expressed on the outer membrane of mitochondria. The interaction between activated NOD2 in response to ssRNA and MAVS on mitochondria triggers IRF3 activation and type I interferon gene transcription.

## 11.13 Activation of the inflammasome by viruses

The inflammasome is a cytoplasmic protein complex that contains numerous scaffold proteins that recruit caspase-1 in response to the presence of viral PAMPs. The recruitment and activation of caspase-1 in turn activates the pro-form of IL-1β and IL-18, both of which have potent antiviral properties. Two PRRs are particularly associated with the inflammasome, NLRP3 and RIG-1, which recognize ssRNA derived from viruses. A third PRR may also be involved in the activation of the inflammasome, known as absent in melanoma 2 (AIM2), which recognizes dsDNA within the cytosol. All three PRRs recruit the inactive pro-caspase-1 into the inflammasome (Figure 11.8). However, the interaction between PRRs and pro-caspase-1 is mediated by an adaptor molecule called apoptosis associated speck like protein containing a CARD (ASC).

The interaction between PRRs and ASC is mediated by a pyrin domain (PYD), which is associated with NLRP3 and a caspase recruitment domain (CARD) associated with ASC and pro-caspase 1. Binding of PYD and CARD cleaves pro-caspase-1 into its active form, thereby exposing its enzyme domain. Active caspase-1 is then able

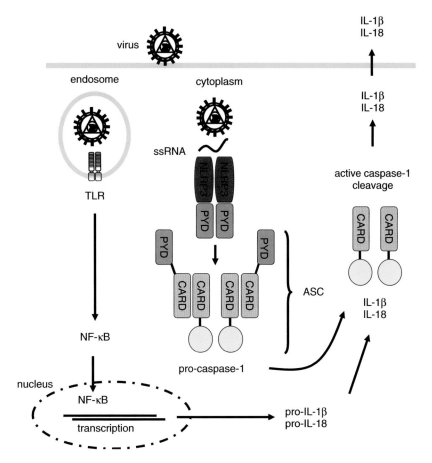

**Figure 11.8** Activation of the inflammasome is an important event during anti-viral immune responses. Transcription of pro-IL-1β and pro-IL-18 is initiated following recognition of viral PAMPs and activation of the NF-κB pathway. Using NLPR3 as an example, ssRNA is recognzsed by NLRP3, which associates with the apoptosis-associated speck-like protein containing a CARD (ASC) complex via interaction with its PYD domain and the CARD domain of pro-caspase-1. This interaction activates capsase-1, which then cleaves pro-IL-1β and pro-IL-18. Active IL-1β and IL-18 are then secreted from the cell.

to cleave pro-IL-1β and pro-IL-18, thereby releasing the active form of both cytokines. Secretion of IL-1β and IL-18 has numerous effects on the immune system, including upregulation of IFN-γ and activation of Th cells and NK cells. It is thought that two signals are actually required for IL-1β and IL-18 release. The first signal is supplied by more conventional PRR signalling such as TLR3 or TLR9, which results in the expression and accumulation of pro-IL-1β and pro-IL-18, and a second signal supplied by NLRP3, RIG1 or AIM2, which activates the inflammasome and induces caspase-mediated cleavage. The requirement for two signals is thought to ensure that the inflammasome is not activated unnecessarily by inflammatory signals.

## 11.14  Dendritic cells present virus antigens to CD8+ CTLs

The cell-mediated immune response to viral infections in an important means of killing infected cells, preventing the spread of the virus and clearing the virus from the body. Along with NK cells, CTLs are a major lymphocyte that posses cytolytic function capable of killing infected cells. In order to perform this function CTLs must recognize virus antigen presented on the surface of infected cells in the context of MHC class I molecules. However, CTLs are not primed directly at the site of infection but rather in regional lymph nodes or organized lymphoid structures associated with mucosal tissues. The principal cell type responsible for priming CTLs are DCs that have migrated from the site of infection. DCs are central to the phagocytosis and processing of viral antigens from the site of infection. Following activation, DCs migrate to regional lymph nodes where they present this antigen to CD4+ Th cells and CD8+ CTLs in the context of MHC class II and MHC class I molecules respectively.

However, viral antigens that have been phagocytosed by DCs are regarded as being exogenous antigens and therefore should be processed by the MHC class II pathway for presentation to CD4+ Th cells. How does exogenous antigen then get processed by the MHC class I pathway, which conventionally only processes endogenous antigens derived from the cytoplasm? The first explanation is that viruses directly infect DCs and therefore release proteins directly into the cytoplasm. However, most viruses infect epithelial cells at mucosal surfaces and not DCs. Therefore there has to be an alternative explanation. DCs are able to participate in a process known as cross-presentation, whereby exogenous

antigen is delivered to the components of the MHC class I processing pathway. Activation of CD8+ CTLs by this means is known as cross-priming. The mechanics of DC cross-presentation are discussed in Chapter 2.

DCs that have migrated into draining lymph nodes from the periphery also act as efficient antigen shuttles. It is thought that once inside lymph nodes, peripheral DCs transfer antigen to resident CD8+ DCs. The transfer of antigen in this manner is particularly effective at initiating the cross-presentation of antigen to naïve CD8+ CTLs. In this way, CD8+ CTLs are primed to peripheral virus antigens by DCs within lymph nodes. It may be that both migratory DCs from the periphery and resident DCs within lymph nodes are required for effective antiviral CTL responses (Figure 11.9). Antigen-primed CTLs then migrate to the site of infection where they recognize viral peptide presented by MHC class I molecules on the surface of infected cells. Activation of primed CTLs in the periphery initiates their cytotoxic effector functions, which enables them to kill infected cells.

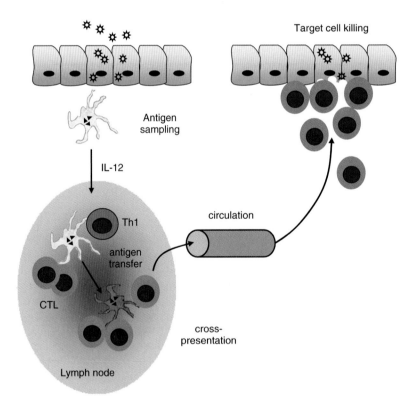

**Figure 11.9** The priming of naïve T cells occurs through a process of cross-presentation. Peripheral DCs migrate into regional lymph nodes where they prime naïve CTLs and transfer antigen to resident DCs. Primed CTLs then migrate to site of infection where they perform their effector functions.

## 11.15 T cell responses to viruses

Considering that most viruses infect epithelial cells of mucosal tissues, a T cell response directed to the epithelium is desirable for the clearance of the virus. The respiratory tract is probably the most studied mucosal tissue and will therefore be used as a model with regard to viral infection. For example, it has been demonstrated that effector T cell responses are central for the control of respiratory tract infection, including those caused by influenza virus, RSV and rhinovirus. Resident DCs within the subepithelial compartment are activated in response to infection and migrate to regional lymph nodes. Within regional lymph nodes, DCs present antigen to naïve lymphocytes and initiate a phase of T cell priming. Antigen-specific T cells then undergo proliferation and differentiation and migrate to the site of infection within the respiratory tract, where they can exert their effector functions. Significant increases in effector T cell numbers are usually observed by day 6 following infection. The cytotoxic activity of effector CTLs causes the lysis of virally infected epithelial cells, which results in the clearance of virus, usually around day 10 post infection. The numbers of T cells within the site of infection decreases as the virus is cleared from the respiratory tract, while some effector cells differentiate into long-lived memory T cells.

The type of T cell response to viral infection is influenced by the cytokine microenvironment within the infected tissues. Activated DCs in the respiratory tract upregulate IL-12, which is a key cytokine that mediates the differentiation of T cells into Th1 cells that express IFN-γ. Th1 cells secreting IFN-γ promote the proliferation and differentiation of virus-specific CTLs (Figure 11.9). The ongoing inflammatory environment of the infected respiratory tract also contain TNF-producing macrophages and IFN-γ-secreting NK cells, both of which promote the effector functions of CTLs. Infected cells express MHC class I molecules on their surface, which present peptide antigens derived from viral proteins. Primed CD8+ CTLs recognize MHC class I molecules via their TCRs and instigate the process of cytotoxicity, which ultimately results in the lysis of infected epithelial cells.

The generation of a population of memory CD8+ T cells is important for protection against secondary infections with the same virus. It has been demonstrated that a population of memory cells reside in the respiratory tract airways following primary infection. The majority of these cells are located within the epithelial compartment and show all the characteristics of effector memory cells.

However, memory T cells within the airways possess a unique characteristic in that they express very low levels of the cell adhesion molecule leukocyte function-associated antigen-1 (LFA-1), which is involved in both T cell activation and T cell recruitment into the lung. The low level of LFA-1 expression is thought to function in two ways. The first is to prevent effector memory cells from responding to inert antigens within the airways. In so doing, memory CD8+ T cells in the airways are less able to undergo proliferation or exert cytotoxic activity, thereby preventing unnecessary immune activation. Second, the low LFA-1 expression may help to retain these memory cells within the respiratory tract.

Airway memory CD8+ T cells are thought to participate in secondary recall responses, although probably more in a cytokine secreting capacity, which induces antiviral processes in subsidiary cells such as NK cells. It has also been shown that these cells may not require specific antigen stimulation, but rather respond to general inflammatory conditions. Therefore, airway memory CD8+ T cells are probably initiated during initial encounter with virus.

Memory CD8+ T cells are also found within the lung parenchyma and within lymphoid aggregates associated with BALT. These cells differ from those in the airways in that they express high levels of LFA-1 and are thought to be a mixture of effector and central memory CD8+ T cells. These memory T cells are thought to be responsible for the rapid induction of recall responses to a secondary challenge with a virus. They require antigen-stimulation, undergo proliferation and are highly cytolytic against virally infected cells. Memory CD8+ T cells are also likely to reside in secondary lymphoid organs, such as regional lymph nodes, and can be recruited into the lungs following re-infection. These antigen-specific CD8+ T cells require time to proliferate and expand and therefore are not recruited to the lungs until 3–4 days following re-infection. However, they do posses heightened cytolytic activity against virally infected cells and can rapidly eliminate virus. Therefore, recall responses in the airways and lungs involve a heterogeneous population of memory CD8+ T cells (Figure 11.10).

## 11.16 Evasion of CTL-mediated immunity by viruses

Certain viruses have evolved several immune evasion strategies that interfere with CD8+ CTL effector mechanisms, demonstrating the importance of this arm

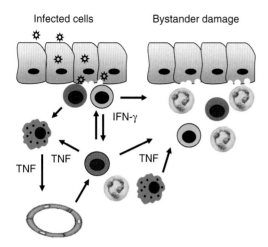

Infected cells        Bystander damage

IFN-γ

TNF    TNF        TNF

TNF

**Figure 11.12** Bystander damage and immunopathology can result following excessive cytokine production and cell recruitment. A positive feedback mechanism is largely responsible, involving cytokine production that induces leukocyte recruitment, which in turn produces more cytokines.

and hospitalization. However, rather than protecting the children from disease, those that were vaccinated developed excessive immunopathology upon infection with wild type RSV. This was associated with a more severe disease and increased hospitalization and mortality. Numerous pro-inflammatory mediators are implicated in the development of severe inflammation to viral infection, the most prominent being TNF. Excessive TNF production in the lungs can cause protracted and heightened leukocyte recruitment and bystander tissue damage. Neutrophil recruitment into the lungs and activation of the respiratory burst, which involves release of damaging reactive oxygen species (ROS), is often associated with such immunopathology.

Acute respiratory viral infections are associated with the production of numerous cytokines and chemokines, expressed by several cell types including epithelial cells, macrophages, NK cells and T cells. Excessive cytokine release in response to acute infection is sometimes referred to as a cytokine storm (hypercytokinemia) and is often the result of a positive feedback between cytokine release, cell activation and the release of more cytokines. This leads to the further recruitment of cells, activation of the endothelia, vascular leakage, activation of the coagulation cascade, release of ROS and tissue damage. The lungs are particularly susceptible to such inflammatory reactions, as vascular leak, cell recruitment and epithelial damage contribute to airway obstruction and prevent gaseous

exchange. This life-threatening condition is a hallmark of severe influenza and SARS virus infection, and is associated with immune mediated lung diseases such as acute lung injury (ALI) and acute respiratory distress syndrome (ARDS).

## 11.18 Antibody response to viruses

One of the first lines of defence against acute viral infections is the presence of natural antibody across mucosal surfaces and within the blood circulation. This form of antibody is produced by an unconventional subset of B cells, known as B1 cells or natural B cells, and predominantly occurs as pentameric IgM. Natural antibodies tend to be poly-specific, in that they can bind to a range of different antigens derived from a multitude of infectious agents. Furthermore, it is thought that acute viral infections actually induce natural antibody responses. Some natural antibodies can directly neutralize viruses and prevent dissemination by inhibiting attachment to host cells. IgM is particularly effective at viral neutralization due to its multivalent nature and its large size, which enables it to block viral surface protein from interacting with host receptors (steric hindrance). Natural antibody also forms immune complexes with virus particles. These immune complexes are effective at activating the complement cascade and initiating phagocytosis by macrophages. These virus-antibody complexes are also transported to regional lymph nodes, where viral antigens can be presented to T cell and B cells. Natural antibody is therefore able to prime adaptive immune responses and promotes the synthesis of conventional T cell-dependent antibody responses.

T cell-dependent antibody responses are also induced by more conventional means. Upon infection with a virus, DCs become activated within the mucosal tissue and transport viral antigens to regional lymph nodes, where they present this antigen to naïve T and B cells. Maturation of antigen-specific T and B cells then takes place in the lymph node, followed by migration to the mucosal effector site. IgA-producing plasma cells are rapidly recruited to sites of infection during primary immune responses and can be detected in mucosal tissues as early as three days following infection. The neutralizing effects of IgA at primary sites of infection are important in order to prevent virus dissemination and eliminate opsonized cells infected with virus. The peak of IgA production usually occurs at approximately five weeks after primary infection and tends to diminish thereafter, so that little antigen-specific IgA is detected by three months. This decrease in

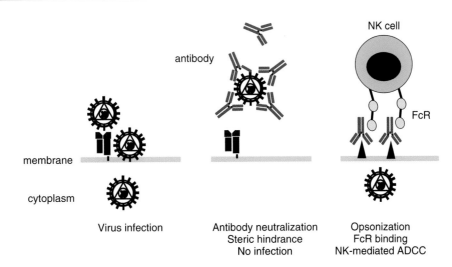

**Figure 11.13** Antibodies perform several important antiviral effector functions. In the absence of neutralizing antibodies viruses are able to attach to host cell receptors and invade the cell. Neutralizing antibodies prevent viruses from binding to cell surfaces through steric hindrance or by inducing a conformational change in viral surface proteins. Non-neutralizing antibodies can opsonise infected cells that have viral proteins on their surface as a consequence of the infectious process. Opsonization of infected cells promotes NK cell-mediated ADCC.

IgA at mucosal sites may explain why certain viruses are able to re-infect the same host. However, the longevity of IgA production increases following secondary infection.

Standard parenteral vaccinations, for example by intramuscular injection, are rather poor at inducing mucosal IgA responses in naïve individuals who have never encountered the virus. However, in individuals who have been infected with wild type virus, subsequent vaccination against that virus is more effective at inducing protective mucosal IgA responses. This may reflect the discrete homing properties of B cells to mucosal tissues, in that IgA-producing memory B cells that have been activated at a mucosal site respond more readily to subsequent antigen exposure and migrate back into the same mucosal tissue. In comparison, vaccination of naïve individuals does not induce a memory B cell population with mucosal tissue homing properties. Furthermore, the compartmentalization of mucosal B cell responses can be demonstrated following differential infection of the gut or the respiratory tract. For example, a gut infection with rotavirus induces high IgA antibody titres in the gut but not the respiratory tract, while the reverse is true for a respiratory tract infection with influenza virus. This reflects the distinct homing properties of IgA secreting B cells, which tend to migrate into the site of primary antigenic stimulation. Memory B cell responses within mucosal tissues are particularly important in protecting the host against secondary infections.

Viruses contain many antigens that can be recognized by antibody and stimulate a polyclonal B cell response (Figure 11.13). However, only a fraction of these antibodies are capable of neutralizing a virus. In order to prevent a virus from infecting host cells, a neutralizing antibody must first bind with high affinity to a viral surface protein. Moreover, the binding of antibody to a viral surface protein must interfere with cell surface receptor binding in order to prevent viral attachment or cell invasion. The mechanism of virus neutralization is thought to be the result of steric hindrance, thereby preventing viral protein–host receptor interactions, or by inducing a conformational change in the viral protein that inhibits receptor binding. Non-neutralizing antibodies are still capable of exerting important immunological effects, however, even though they tend to bind to viral epitopes that are not involved in receptor binding. For example, antibodies that bind to viral antigens deposited on the surface of infected cells are capable of stimulating the complement cascade, phagocytosis and ADCC.

## 11.19 Difference between cytopathic and non-cytopathic viruses

The different replication cycles of viruses influence the effectiveness of cytotoxic responses mediated by NK cells and CTLs. Some viruses are known as cytopathic viruses, such as poliovirus and rabies virus, which cause the lysis and killing of infected cells during the late stage of replication, which coincides with the release of many viral particles. Cytopathic viruses spread rapidly from one cell to the next, causing the destruction of cells as they replicate. Other viruses are known as non-cytopathic, such as hepatitis B and C viruses and EBV, which do not cause cell lysis during the late replication phase. Instead, non-cytopathic viruses persist within the same cell and are continually released. Therefore, the cytolytic effects of NK cells and CTLs will have differing effects on the spread of virus depending on the stage of the viral replication cycle and whether the virus is cytopathic or non-cytopathic (Figure 11.14).

Killing an infected cell as soon as it becomes infected will eliminate further spread of the virus, as it has had

no time to synthesize new virions. This is true for both cytopathic viruses and non-cytopathic viruses. Viral dissemination can be markedly reduced if the target cell is killed at the early stages of infection when only a limited number of virions have been synthesized. However, a cytotoxic response against a target cell that is infected with a cytopathic virus is only effective if the cell is killed before the virus replication cycle is complete.

Killing a cell infected with a cytopathic virus that has completed the synthesis of all new virions is equivalent to the virus lysing the cell itself. No effect on viral dissemination will have occurred and the infection is allowed to spread. This is not true for non-cytopathic viruses. Although killing an infected cell will still release non-cytopathic virus particles, the continuous secretion of the virus will have been prevented. These are

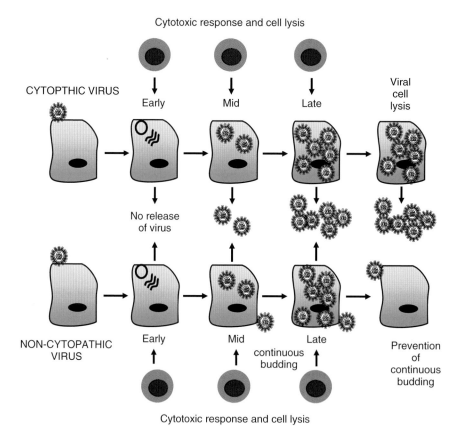

**Figure 11.14** CTL-mediated killing has differing effects on virus dissemination depending on whether the virus is cytopathic or non-cytopathic. Killing an infected cell before virion synthesis can take place prevents further dissemination of both types of virus. However, killing a cell that contains mature, late phase cytopathic viruses is equivalent to the virus lysing the cell itself and does not prevent further dissemination. However, killing a cell during the late phase of a non-cytopathic virus infection prevents further dissemination by halting the continuous budding process.

important factors when considering the kinetics of an immune response against cytopathic viruses compared to non-cytopathic viruses.

In general, cytopathic viruses cause extensive damage to infected tissue and therefore have to be eliminated as rapidly as possible; otherwise they will kill their host. Coincidentally, cytopathic viruses cause the release of more DAMPs and PAMPs compared to non-cytopathic viruses, thereby influencing the rapidity of immune reactions. Infection of a naïve host with a cytopathic virus usually requires the rapid induction of a neutralizing antibody response in order to prevent further dissemination. The first line of protection against acute infection with cytopathic viruses is afforded by natural antibody, for example T cell-independent antibody produced by B1 B cells within MALT or maternally-derived antibody. These poly-specific antibodies prevent viral dissemination and activate innate and adaptive responses and are particularly important in preventing viral infections in naïve new-born and neonatal children that have immature immune systems. When this natural immunity is breached, T cell-dependent antibody responses and CTL responses then participate in the control of viral spread and in the elimination of the virus. T-cell dependent neutralizing antibody tends to be of a higher affinity than natural antibody and has undergone isotype switching, for example to IgA, which provides effective protection at mucosal sites. Protection from secondary infections and the effectiveness of vaccinations against cytopathic viruses also rely on the generation of neutralizing antibodies produced by a population of memory B cells.

On the other hand, non-cytopathic viruses tend to cause little direct damage to the host and result in persistent infection. Non-cytopathic viruses are more adept at avoiding the effects of neutralizing antibodies; therefore CTL responses become more important in order to control persistent infection, although subsequent immune-mediated tissue damage may result. In addition, it has proven much more difficult to vaccinate against non-cytopathic viruses, due to the lack of a neutralizing antibody responses. Non-cytopathic viruses have evolved ways to avoid the effects of neutralizing antibody in order to promote chronic persistence.

## 11.20 Immune evasion by antigenic shift and drift

Immunological memory provides effective protection against subsequent encounter with the same virus. For example, neutralizing antibodies directed against viral epitopes prevent secondary infection across mucosal tissues. However, certain viruses are able to escape immune detection by antibodies by establishing mutant viruses that have altered antigenic determinants. This process in known as antigenic drift, whereby viral mutants acquire small changes to their nucleotide sequence that result in changes to surface proteins. In addition, large scale changes can also occur when whole gene segments are swapped between different strains of the same virus (Box 11.2). This is thought to occur when two separate strains of virus infect the same host cell and genetic re-assortment takes place. This process is known as antigen shift and can result in the emergence of completely new strains of virus. Influenza virus best exemplifies these processes. Variations between the same strains of seasonal influenza virus occur following the process of antigen drift, which results in slight variations in haemagglutinin and neuraminidase antigenic determinants. The emergence of new pandemic strains of influenza occur following antigenic shift, which is characterized by a completely new combination of haemagglutinin and neuraminidase proteins.

Antigenic drift and antigenic shift both have effects on the ability of pre-existing antibodies to recognize mutated influenza viruses and therefore act as important immune evasion strategies. Most neutralizing antibodies are directed against the viral haemagglutinin proteins and to a lesser extent neuraminidase. Therefore, mutations in key epitopes can alter the affinity and the avidity of pre-existing neutralizing antibodies, thereby rendering the host more susceptible to infection. These mutations are thought to accumulate over time, and have been observed in haemagglutinin residues from the 1918 H1N1 strain. Antigen shift has an even more significant impact on the neutralizing antibody response to influenza virus. The re-assortment of gene segments results in a strain of virus with a new haemagglutinin and/or neuraminidase, which the immune system has never encountered before. With regard to pandemic influenza strains, the immune system is considered naïve and does not possess any neutralizing antibodies, therefore making the host much more susceptible to infection. Similar effects due to antigenic drift and shift have also been demonstrated for CTL responses directed against peptides derived from influenza virus.

## 11.21 Vaccination and therapies against viral infections

Seasonal influenza virus infection is the leading cause of vaccine-preventable death in the United States,

## Box 11.2. Influenza Viruses and Flu Epidemics.

In the spring of 1918 a deadly infectious disease spread across America and then the rest of the world. Healthy young adults could be feeling energetic and lively in the morning and be lying dead by nightfall. The rapidity from disease onset to death was frightening, while post mortem analysis clearly showed that a dreadful pneumonia was responsible for damaging the lungs, which were full of thick liquid, blood and mucus. Today we know exactly what caused this disease: the influenza virus. Throughout the following months of 1918–1919 the deadly influenza virus spread around the globe instigating the greatest ever flu pandemic, known as Spanish flu. It has been estimated that as many as 50 million people died of Spanish flu, more than those killed during World War I. Unusually, this strain of flu affected healthy adults and not the infant and elderly populations one would expect. The Spanish flu virus was incredibly pathogenic and behaved differently to the strains of influenza in circulation today. From the body of a flu victim buried in the Alaskan permafrost, and from previously stored lung samples, the genome of the 1918 virus strain has been meticulously pieced together using modern molecular biology techniques.

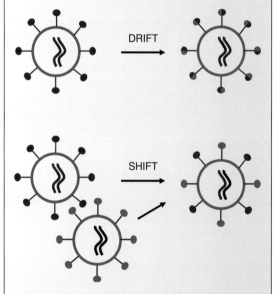

Antigenic drift and shift exhibited by influenza virus.

Infectious influenza virus requires the presence of two very important molecules on its surface. These molecules are known as haemagglutinin (HA) and neuraminidase (NA) and without these molecules the virus cannot replicate or cause disease. Variations in HA and NA can result in differences in the virulence of the virus. For example, the 1918 influenza virus has been named H1N1, being the prototype virus strain.

Other pandemic influenza strains include the H2N2 (Asian flu, 1957) and H3N2 (Hong Kong flu, 1968) strains. As one virus disappears, another strain with a slightly different haplotype emerges. Occasionally, infection with two strains can cause two viral genomes to recombine, to form H1N2 for example, and this is known as antigenic shift. The acquisition of mutations that lead to slight changes in the same virus is known as antigen drift. Due to the differences in these viral proteins, individuals that have been infected by one strain are not protected against a variant strain. For example, people who have been infected with H1N1 only have circulating antibodies specific for HA and NA from H1N1, and if they were to become infected with H2N2 they would have no protective immunity against that viral strain. This also highlights the importance of seasonal vaccinations against those strains known to be circulating throughout the human population.

The intriguing question regarding the 1918 strain is why this particular virus was so virulent. Was it due to the HA and NA proteins on its surface? The answer is not so simple. By fitting together the HA and NA genes from the 1918 H1N1 strain with the other genes of a contemporary strain, it was shown that the newly constructed virus was not hyper-virulent. Only when the entire 1918 virus was synthesized, with all its original genes, was it demonstrated to be highly pathogenic. This means that genes other than HA and NA are partly responsible for the highly pathogenic state. Furthermore, the 1918 virus does not require trypsin derived from its host for replication, as the viral NA actively cleaves HA itself. This could also explain the enhanced pathology observed with 1918 influenza as it is no longer restricted to trypsin-rich areas such as the lungs.

which can account for as many as 30,000 deaths and 200,000 hospitalizations each year. The demographic most susceptible to life-threatening infection are the very young and elderly. Influenza virus rarely causes life-threatening pneumonia in healthy people; rather severe illness is caused by underlying complications such as secondary bacterial infections, immunosuppression and health problems such as asthma. Seasonal influenza outbreaks tend to be the result of the most common circulating viral strains, which most people have some level of immunity to and therefore cause less severe disease. However, pandemic virus strains can also enter the circulation, as happened in 2009 when a new pandemic strain of H1N1 arrived on the scene. As with any new strain of influenza virus, the majority of the population will not possess protective immunity against it. Therefore, pandemic strains such as H1N1 pose a much greater risk to the general population. Seasonal influenza vaccines are therefore tailor made each year, in order to incorporate the most common strains in circulation at that moment. For example, the influenza vaccine in use at the moment is a trivalent, inactivated vaccine containing seasonal H3N2, pandemic H1N1 and an influenza virus B strain. This vaccine, which is administered intramuscularly, is suitable for any healthy person over the age of 6 months. More recently, a trivalent live-attenuated influenza virus vaccine has been developed that can be administrated intranasally and may have the added benefits of being delivered directly into the respiratory tract.

Another important vaccine directed against viruses is the measles, mumps and rubella (MMR) combined vaccine. The MMR vaccine contains three live-attenuated virus strains and is given to infants around one year of age intramuscularly. A second jab is administered before primary school age, which aims to protect those children not covered following the first shot. The MMR vaccine has drastically reduced the incidence of all three debilitating diseases in vaccinated children. Another successful vaccine is the polio vaccine, which can either be administered orally (Sabin strain) or intramuscularly (Salk strain) and has resulted in the near eradication of polio worldwide. The one vaccine that has probably had the most beneficial impact on human heath is the smallpox vaccine. The smallpox vaccine consists of a live cowpox virus (vaccinia), which is closely related to the deadly smallpox virus. Subcutaneous vaccination causes a variola at the site of injection and, due to it similarity to smallpox, induces a protective immune response against smallpox (discussed also in Chapter 17). Other vaccines have proven efficacious, including those against hepatitis A, hepatitis B, rabies virus and yellow fever virus.

Although vaccination is the most effective way of preventing disease caused by viral infection, not all vaccines are 100 per cent efficacious and not every individual receives the vaccine. Furthermore, vaccines against many viral diseases simply do not exist. Therefore, several antiviral therapies have been development. An example is antiretroviral (ARV) therapy that is commonly used to treat patients with HIV. The aim of ARV drugs is to inhibit one or more stages in the life cycle of HIV. Combination therapy with several different ARV drugs, which each inhibit a different aspect of HIV replication, has been successful in preventing disease progression and extending the lifespan of people infected with HIV. Several drugs have also been developed to combat influenza infections, including neuraminidase inhibitors such as oseltamivir. These drugs are thought to be particularly useful during outbreaks of pandemic influenza, when there is no vaccine available for the newly emerged strain.

## 11.22 Summary

1. Viruses cause a multitude of human diseases.
2. The mucosal epithelial barrier has a number of antiviral defence mechanisms.
3. NK cells play a central role in recognizing and killing virally infected cells.
4. Virus-derived PAMPs are recognized by immune cells and induce Th1 cell-mediated immunity.
5. Virus-derived peptides are presented by MHC class I molecules on the surface of infected cells and are recognized by CTLs.
6. Viruses have evolved numerous immune evasion strategies to subvert NK cell and CTL-mediated immune responses.
7. Neutralizing antibodies provide effective protection against acute cytopathic viruses.
8. Bystander tissue damage can result from excessive immune responses to viral infections.

# 12 Immunity to Bacteria

## 12.1 Introduction to bacterial immunity

Bacteria are microscopic organisms that belong to their own kingdom of classification, the Monera. They are characterized by their lack of a membrane-bound nucleus and are therefore termed prokaryotes (from the Greek meaning before nucleus). All true bacteria, and the great majority of medically important bacteria, belong to the subkingdom Eubacteria. However, not all bacteria cause disease; some are even beneficial to humans, while others are essential for our own biological processes. For instance, it has been estimated that there may be as many as 500 different species of bacteria living in the human gut, all of which are non-harmful or beneficial. Bacteria live practically everywhere on the planet from the soil in your garden to hydrothermal vents miles beneath the ocean surface. They can resist extremes of acidity, temperature, water availability and can live with or without oxygen.

Only a minority of bacterial species have evolved mechanisms that allow them to cause disease and in so doing impact on the health of humans worldwide. However, those bacteria that can cause disease are among the most deadly pathogens on Earth. The vast majority of pathogenic bacteria infect their host via a mucosal epithelium, for example the intestinal, respiratory, urogenital or conjunctival epithelium. MALT therefore plays a vital role in protecting the host from infection. This chapter details the mechanisms of host defence against pathogenic bacteria and the various ways in which bacteria try evade host defence (Box 12.1).

## 12.2 Classification of bacteria

Bacteria are such a diverse and numerous set of prokaryotic organisms, an official classification system has not been accepted, although Bergey's *Manual of Systematic Bacteriology* remains the most widely accepted. For many years, bacteria have been classified according to a simple staining procedure devised by Hans-Christian Gram in 1884, known as the Gram-reaction. This stain uses crystal violet to differentiate Gram-positive bacteria, which retain the dye, and Gram-negative bacteria, which are impervious to the dye. Whether bacteria are stained with the dye or not is a reflection of the composition of their cellular envelope (Figure 12.1), which in turn influences recognition by the immune system. Gram-positive bacterial cell walls are composed of many layers of peptidoglycan, which allow the absorption and retention of the dye. However, the Gram-negative bacterial cell wall contains only a few layers of peptidoglycan, which is surrounded by an outer cell membrane that is refractive to crystal violet. The outer membrane of certain Gram-negative bacteria is also responsible for some types of antibiotic resistance. The Gram-reaction has therefore been used as a guideline for the identification of many medically important bacteria.

Another widely used approach to identify bacteria is that based on morphology; in other words the shape and structure of the bacterium. The two main morphological types are rod-shaped bacilli and spherical-like cocci, although other shapes do occur such as spiral-shaped bacteria known as spirochetes. A list of medically important bacteria is shown in Table 12.1. Examples of medically important diseases caused by rod-shaped bacilli include cholera, salmonella, tuberculosis, typhoid fever, botulism and leprosy. Diseases caused by round cocci include meningitis and chlamydia, as well as serious hospital infections caused by Staphylococci (MRSA) that can lead to septicaemia, and respiratory tract infections (pneumonia). In terms of identifying bacteria the Gram-reaction is often used in conjunction with the organism's morphology. Of course, other morphological factors can help in the identification process such as the presence of flagellum or cilia. Other stains can also be used to differentiate types of bacteria, such as the Ziehl Neeson stain, which stains acid-alcohol fast bacilli such as mycobacteria. Although different species of bacteria may

*Immunology: Mucosal and Body Surface Defences*, First Edition. Andrew E. Williams.
© 2012 John Wiley & Sons, Ltd. Published 2012 by John Wiley & Sons, Ltd.

**Box 12.1. Bacterial Immune Evasion Strategies.**

| Mechanism | Bacterial example |
| --- | --- |
| Capsule formation | *Streptococcus pneumoniae* |
| | *Escherchia coli* |
| | *Haemophilus influenzae* |
| | *Neisseria meningitidis* |
| Avoid phagocytosis or phagolysosomal killing | *Mycobacterium tuberculosis* |
| | *Yersinia pestis* |
| | *Listeria monocytogenes* |
| | *Legionella pnumophila* |
| Inhibit antimicrobial peptides | *Salmonella sp.* |
| | *Staphylococcus aureus* |
| Kill immune cells | *Streptococcus pyogenes* |
| | *Staphylococcus aureus* |
| | *Salmonella enterica* |
| Inhibit complement | *Staphylococcus aureus* |
| | *Streptococcus pneumonia* |
| Subvert TLR/NOD recognition | *Salmonella typhi* |
| | *Porphyromonas gingivalis* |
| | *Lysteria moncytogenese* |
| Interfere with apoptosis | *Salmonella enterica* |
| | *Chlamydia trachomatis* |
| Modulate cytokines/chemokines | *Staphylococcus aureus* |
| | *Shigella flexneri* |
| Secrete toxins or enzymes | *Salmonella typhi* |
| | *Bordetella pertussis* |
| | *Escherichia coli* |
| | *Vibrio cholerae* |
| | *Haemophilus influenzae* |
| Alter cell surface | *Salmonella sp.* |
| | *Campylobacter jejuni* |
| | *Helicobacter pylori* |
| | *Yersina pestis* |
| Antigenic variation | *Neisseria meningitidis* |
| | *Neisseria gonorrhoeae* |
| Interfere with cellular pathways | *Yersinia enterocolitica* |
| | *Helicobacter pylori* |
| | *Neisseria gonorrhoeae* |

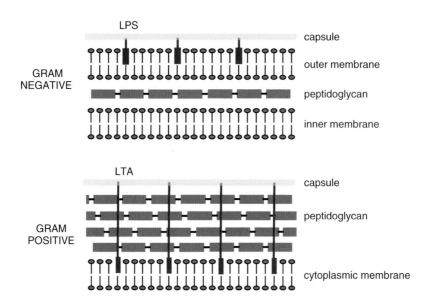

**Figure 12.1** Structure of gram-negative and gram-positive bacterial cell walls. Gram-negative bacteria possess a double membrane and have little peptidoglycan. Gram-positive bacteria have a single membrane and a thick layer of peptidoglycan.

share similar morphological or biochemical characteristics, this method of identification may not be a true representation of their evolutionary relatedness. It is becoming more common to test the genetic phenotype of bacteria, either by comparing their genomic DNA or their RNA characteristics. This highly sophisticated molecular approach allows the precise identification of bacterial species based on their unique genetic fingerprint.

## 12.3 Structure of the bacterial cell

Despite being small, single-cell microorganisms bacteria have a complex cell structure. The typical size of a bacterial cell is only 1–2 μm in length, resulting in a large surface area-to-volume ratio. This allows for the rapid uptake of molecules and expulsion of waste products and easy distribution of nutrients throughout the cell. Gram-positive bacteria have only one plasma membrane, which is surrounded by a cell wall composed of many layers of peptidoglycan. Gram-negative bacteria are slightly more complex. They too have an inner membrane surrounded by a few layers of peptidoglycan but in addition have an outer membrane, which itself is surrounded on the outside by a layer of lipopolysaccharide (LPS). The space between the inner and outer membranes is known as the periplasm, which is composed of lipoproteins and

enzymes as well as peptidoglycan (Figure 12.1). The components of the bacterial cell wall are highly immunogenic and often provide the immune system with very important signals during the induction of an immune response. However, some pathogenic bacteria have capsules that surround the entire cell, comprising mostly of carbohydrates, which often inhibit recognition by the immune system, or render them less susceptible to immune attack.

Bacteria do not possess a defined membrane-bound nucleus; rather the genetic material is carried on a single chromosome of circular DNA that is retained in a cellular region known as the nucleoid. Some bacteria also carry additional genetic material on small, circular pieces of DNA called plasmids that replicate independently of the chromosome. These plasmids are useful for transferring genetic information from one bacterium to another and often encode antibiotic resistant genes or genes that encode toxins. The cytoplasm of bacteria is full of ribosomes, which translate messenger RNA into many functional proteins. However, unlike eukaryotic cells, bacteria do not possess an endoplasmic reticulum, Golgi complex or mitochondria. In addition, certain bacteria have appendages known as pili, which are responsible for transporting plasmids and chromosomal DNA across the plasma membranes of adjacent cells. Motile bacteria can also possess one or more flagella, made of a complex of

**Table 12.1** Diseases caused by bacterial species.

| Disease | Bacterium | Incidence/year |
|---|---|---|
| Tuberculosis | *Mycobacterium tuberculosis* | 9 million |
| Leprosy | *Mycobacterium leprae* | 750,000 |
| Cholera | *Vibrio cholorae* | 5 million |
| Plague | *Yersinia pestis* | 2000 (outbreaks can occur) |
| Tetanus | *Clostridium tetani* | 1 million |
| Whooping cough | *Bordetella pertussis* | 48 million |
| Meningitis | *Neisseria meningitidis* | 200,000 (Africa alone) |
| | *Haemophilus influenzae* | |
| Syphilis | *Treponema pallidum* | 11 million |
| Gonorrhoea | *Neisseria gonorrhoeae* | 62 million |
| Typhoid fever | *Salmonella typhi* | 30 million |
| Diphtheria | *Corynebacterium diphtheriae* | 8000 (outbreaks can occur) |
| Diarrhoea | *Escherichia coli* | 200 million |
| Scarlet fever Strep throat Necrotizing fasciitis | *Streptococcus sp.* (Group A) | 600 million (low estimate) |
| Toxic shock | *Staphylococcus aureus* (MRSA) | 60 million (carriers) |
| Food poisoning | *Salmonella enteriditis* | 16 million |

structural proteins (mostly flagellin) arranged helically and powered by membrane bound proteins that act like a rotor.

## 12.4 Diseases caused by bacteria

The mere presence of a bacterium is not sufficient in itself to cause disease. We encounter millions upon millions of bacteria each day. They exist in the air that we breathe, on our skin, in our mouths and on the things that we touch; yet we rarely succumb to disease. Bacteria must first gain entry into the body before they have a chance of becoming pathogenic. Probably the most common route of infection is through the digestive tract. Undercooked meat and ill-prepared food can harbour millions of disease-causing bacteria, the ingestion of which often results in severe diarrhoea, vomiting, dehydration and stomach pain. One of the more serious bacterial infections of the digestive tract is cholera, which is caused by *Vibrio cholerae*. This highly infectious bacterium releases a potent enterotoxin (cholera toxin) that triggers ion instability in the gut and results in severe dysentery. The generic term 'food poisoning' is largely due to bacterial infections of the stomach or intestine and can result from infection by a number of differing species including *Escherichia coil, Salmonella spp.* or *Campylobacter spp.*, among others.

Diseases with huge historical significance are also attributed to bacteria. For example, leprosy and tuberculosis are caused by two species of Mycobacteria,

*Mycobacterium leprae* and *Mycobacterium tuberculosis*, respectively. These diseases have haunted mankind for thousands of years and episodes of infection can even be traced back to the ancient Egyptians. A bacterium was also responsible for the Great Plague which ravaged Europe in the seventeenth century. This devastating disease was the result of infection with *Yersinia pestis*, which was likely to be responsible for the Black Death in the fourteenth century as well. As many as 20 million people died from the Black Death across Europe in the Middle Ages, equivalent to one in every three people. Other species of bacteria are responsible for outbreaks of anthrax, botulism, typhoid fever, meningitis and syphilis among myriad other diseases (Table 12.1). The human and economic cost resulting from bacterial infections is therefore hugely significant.

## 12.5 Mucosal barriers to bacterial infection

Bacteria live on and within the human body without causing any pathology or disease. Most of these bacteria are harmless commensal species, although a minority are capable of causing disease but only if they are allowed to invade deeper into the tissue. The body surface of the skin and the mucosal surfaces of the digestive, urogenital and respiratory tracts form effective barriers to bacterial invasion. Epithelial cells have the most contact with these bacteria, a scenario most evident in the digestive tract

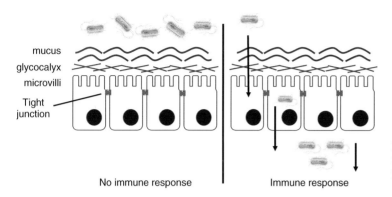

mucus
glycocalyx
microvilli

Tight
junction

No immune response

Immune response

**Figure 12.2** Barrier defence to bacterial invasion. An immune response is initiated only when barrier defence has been breached.

where billions of bacteria reside. Despite only being a single layer of cells thick, the mucosal epithelium is resistant to pathogen entry. This is mainly attributed to tight junctions, which join the cytoskeletons of adjacent cells firmly together so that any material that passes through the epithelial barrier must pass through the cell itself. As well as providing a purely physical barrier to bacterial entry, epithelial cells interact with specialized immune cells in a coordinated effort to prevent infection.

The structure of the intestinal epithelium also makes it particularly resistant to bacterial entry (Figure 12.2). The lumen of the intestine is folded in such as way that it forms numerous villi. Each villus is formed of epithelial cells known as enterocytes whose luminal membrane is coated with microvilli, which not only increase the absorptive surface area of enterocytes but produce numerous glycoproteins that form a filamentous brush border glycocalyx. This specialized brush border glycocalyx is particularly impervious to large molecules and bacteria. Epithelial cells and goblet cells of the intestine and respiratory tract

also secrete mucins, which form a mucous layer over the luminal surface adding to the protective barrier. The mucous layer interacts with bacterial appendages and helps to trap bacteria, preventing them from interacting with the epithelial surface. The respiratory tract also possesses a mucociliary elevator (discussed in Chapter 8) that carries particulate material, including microbes, away from the lower respiratory tract and towards the upper respiratory tract for expulsion.

## 12.6 Anti-microbial molecules

As well as providing a physical barrier, the epithelium of the small intestine produces a number of peptides and proteins that have anti-microbial properties (Figure 12.3). In particular, within the gut a specialized secretory epithelial cell, known as a Paneth cell, produces large quantities of anti-microbial peptides. One family of anti-microbial peptides are called defensins, which can be

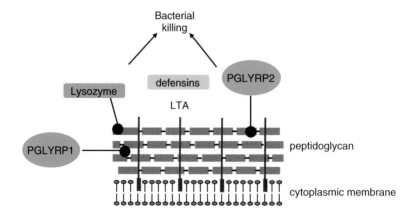

**Figure 12.3** Antimicrobial peptides target different components of the bacterial cell wall and play important roles in innate defence against infection.

divided into α-defensins and β-defensins. There may be over 50 genes that encode defensin peptides in humans and they function by disrupting the membranes of bacteria and directly killing the microbes in the gut lumen. Defensins are particularly important for the innate immune defence against enteric bacteria that infect the digestive tract.

Bactericidal/permeability-inducing protein (BPI) is a member of the LPS-binding proteins produced constitutively by epithelial cells and neutrophils. Its primary function is to directly kill bacteria by permeabilizing the cell membrane but can also opsonize bacteria for subsequent removal and has a neutralizing effect on LPS. Another family of permeabilizing proteins are the anti-microbial lectins, which are produced mainly by Paneth cells in the small intestine. Certain lectins may also coat the microvilli of enterocytes forming a protective border against bacterial invasion. Cathelicidin (otherwise known as LL37 in humans) is another anti-microbial peptide produced by epithelial cells, neutrophils and mast cells. It has a similar method of action of disrupting bacterial cell membranes. Cathelicidin may even assist in the destruction of intracellular bacteria as cathelicidin-containing vesicles fuse with the bacterial compartment.

Peptidoglycan recognition proteins (PGRP), or PGLYRPs in mammals, are an evolutionarily conserved set of small proteins with bactericidal properties. They were fist discovered as important anti-microbial peptides in the fruit fly *Drosophila melanogaster* and later discovered in mammals, of which four genes have been described in humans (PGLYRP-1-4). All PGLYRPs are soluble proteins found in a variety of tissues and secretions such as the skin, intestine, mouth, blood, sweat and tears. PGLYRP-1 is predominantly expressed by monocytes and macrophages, while PGLYRP-3 and PGLYRP-4 are expressed in tissues subjected to contact with the atmosphere (e.g. mouth and digestive tract). PGLYRP-1, -3 and -4 are able to directly kill both Gram-positive and Gram-negative bacteria by interfering with peptidoglycan biosynthesis or destabilizing their cell wall. PGLYPR-2, which is secreted from the liver and expressed by epithelial cells, has amidase activity that degrades excess peptidoglycan in order to alleviate excessive inflammation.

## 12.7 Recognition of bacterial PAMPs by Toll-like receptors

Bacteria possess numerous, evolutionarily conserved molecules such as LPS and peptidoglycan. These highly repetitive molecules are essential for the survival of the microbe. Moreover, these molecules are not expressed by the host and therefore they provide a means of discriminating between self and non-self. Such molecules, with a highly repetitive structure, are known as pathogen associated molecular patterns (PAMPs) and their expression is shared among a variety of bacterial species. Many host receptors exist that recognize PAMPs and are collectively called pattern-recognition receptors (PRRs), the most widely studied being the Toll-like receptor (TLRs) family (Figure 12.4).

In humans there are ten known TLRs (TLR1-10), four of which are very important for the recognition of bacterially-derived products (TLR2, TLR4, TLR5 and TLR9). TLR2 recognizes peptidoglycan, lipoteichoic acid and various glycoproteins, which are highly abundant in Gram-positive bacteria. TLR4 recognizes LPS, which is specific to Gram-negative bacteria, and heat shock proteins derived from stressed cells. TLR5 recognizes flagellin, which forms the protein subunits of flagella, while TLR9 recognizes bacterial DNA that has a high unmethylated CpG content. Recognition of PAMPs by one or more TLRs enables the innate immune system to respond to a set of microbial products specific to a particular pathogen. Activation of TLRs by PAMPs, and other PRRs, results in the initiation of an immune response only when a pathogen is detected. This also maintains tissue homeostasis and prevents unwanted immune reactions to harmless or commensal bacteria.

Although not a member of the TLR family, mannose receptor (MR) is also an important PRR for the detection of bacteria. MR is predominantly expressed on the

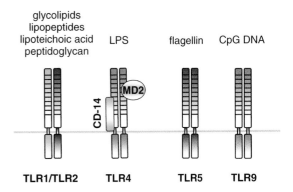

**Figure 12.4** TLRs implicated in the recognition of bacterial PAMPs.

surface of DCs and macrophages, and belongs to the C-type lectin family of receptors. It recognizes carbohydrate motifs, particularly mannose, expressed on the surface of bacteria, viruses, fungi and parasites. In addition, MR recognizes endogenous motifs such as sialy Lewis X and lysosomal proteins such as myeloperoxidase and thyroglobulin, suggesting it plays a role in homeostasis and the resolution of inflammation. MR is primarily involved in the initiation of phagocytosis by macrophages, although it requires the help of other stimulatory signals (such as TNF or IL-12) to activate cell effector mechanisms. Furthermore, MR is able to recognize endocytosed material and has been found associated with intracellular MHC class I pathways in DCs. This co-localization suggests that MR recognition of foreign antigens may assist in the cross-presentation of exogenous peptides to CD8+ T cells, via the endogenous peptide processing pathway. Some have also suggested that MR is modulatory under certain conditions, whereby IL-12 production is inhibited in DCs and a Treg phenotype induced.

Many pathogenic bacteria deploy evasion strategies to escape immune recognition by TLRs and other PRRs. For example, certain bacteria synthesize a carbohydrate capsule that envelopes the entire cell. The capsule effectively hides molecular motifs that would otherwise be recognized by PRRs. The capsule also helps to avoid opsonization by complement factors and prevent the attachment of antibodies. Furthermore, the capsule prevents the formation of a membrane attack complex by keeping complement proteins away from the cell membrane. However, certain bacterial molecules protrude through the capsule and can therefore be detected by the immune system. One such molecule is LPS, which is recognized by TLR4. Salmonella employ enzymes that modify the composition of the lipid A part of LPS, thereby rendering it far less effective at inducing an immune response. The intracellular pathogen *Lysteria monocytogenes* employs another mechanism of avoiding recognition. Intracellular Nod proteins recognize peptidoglycan, which stimulates an NF-κB-dependent immune response. However, *L. monocytogenese* produces enzymes that alter the synthesis or metabolism of peptidoglycan, which makes it far less effective at immune stimulation.

## 12.8 Complement and bacterial immunity

The complement pathway comprises approximately 30 proteins that are involved in the humoral immune response to bacterial infections. The complement cascade terminates with the insertion of a pore forming unit, known as the membrane attack complex, into the membrane of target cells and ultimately causes cell lysis of gram-negative bacteria. During bacterial infection, the classical complement cascade can be induced in response to antibody bound to the surface of opsonized bacteria. In addition, bacteria can also induce both the alternative pathway and the mannose-binding lectin pathway of complement activation (discussed fully in Chapter 2). All three complement pathways lead to the generation of a C3 convertase and the formation of the membrane attack complex. The complement system is considered to be highly protective against pathogenic bacteria and it also plays a role in keeping commensal bacterial numbers in check. However, the membrane attack complex is considerably less effective at killing gram-positive bacteria, due to the presence of a thick cell wall, compared to gram-negative bacteria. In addition to inducing the lysis of cells and causing direct bacterial death, complement proteins such as C3b opsonize bacteria for subsequent phagocytosis by macrophages and neutrophils (Figure 12.5). The importance of the complement system is clearly demonstrated in people with specific immunodeficiencies in

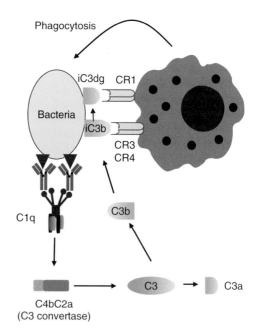

**Figure 12.5** Activation of the complement system leads to opsonization of bacteria. Recognition of opsonins such as C3b by complement receptors on macrophages results in phagocytosis of bacteria.

complement proteins, whereby they are much more susceptible to primary and especially recurring bacterial infections.

As the complement system is highly effective at providing immune protection against bacterial infection, many pathogenic bacteria have evolved strategies of escaping the effects of complement. Many stages of the complement pathway can be modulated by bacteria. With regard to the classical pathway, C1q can be prevented from binding to bacteria that have been opsonized by antibody, thereby interfering with complement initiation. For example, Staphylococcal protein A (SpA) is expressed on the bacterial cell surface and binds to the Fc region of IgG. The function of SpA is two-fold, as it interferes with C1q binding to the surface of the bacterium and it modulates macrophage function by interfering with FcR-mediated phagocytosis (Figure 12.6). Many other species of bacteria express surface proteins that interfere with the opsonization process, while the polysaccharide capsule of certain species, such as *Streptococcus pneumonia*, also interferes with opsonization and prevents phagocytosis by neutrophils and macrophages.

Bacteria are able to interfere with the generation of C3 convertase, which is one of the key stages of the complement cascade. For example, Staphylococcal complement inhibitor (SCIN) is a surface expressed protein that blocks the production of C5 by inhibiting both forms of C3 convertase (C4b2a and C3bBb) and therefore interferes with the classical, alternative and lectin pathways. SCIN effectively inhibits cell lysis and, considering C3b and its products are important for opsonizing bacteria, SCIN also inhibits complement-mediated phagocytosis. Other complement inhibitors include those bacterial products that bind to or digest components of the membrane

attack complex. For example, Streptococcal inhibitor of complement (SIC) is able to bind C5b-C7, which are essential components of the pore forming unit of the membrane attack complex. SIC is also capable of neutralizing the activity of antimicrobial peptides such as LL37 and α-defensin, demonstrating a further immune evasion strategy. One other mechanism that bacteria employ to evade the complement system is the inhibition of complement receptors (CRs). For example, group A streptococci express the M5 protein, which interferes with CR3 and therefore inhibits phagocytosis by neutrophils. These same bacteria also secrete soluble Mac-1 like protein, which binds to FcγRIIIB and prevents IgG-mediated phagocytosis and killing mechanisms of neutrophils and macrophages.

## 12.9 Neutrophils are central to bacterial immune responses

One of the characteristics of a bacterial infection is the accumulation of neutrophils within the affected tissue, demonstrating their importance in host defence against pathogenic bacteria. For example, respiratory bacterial infections are associated with neutrophil recruitment to the lungs. Neutrophils are important cells of the innate immune system that function by phagocytosing bacteria, releasing their granule contents and initiating an inflammatory response through the secretion of antimicrobial peptides, cytokines and chemokines. However, severe bacterial infections can cause the accumulation of excessive amounts of neutrophils, which can cause the bystander destruction of cells and tissue damage. In the case of bacterial infections of the lung, neutrophil accumulation and immunopathology can result in life threatening pneumonia (Chapter 8, Box 8.2), acute lung injury (ALI) or acute respiratory distress syndrome (ARDS).

Neutrophils are recruited into tissues, such as the lungs, from the bloodstream and across endothelium in response to chemotactic factors. The most effective chemoattractant for neutrophils is IL-8, which is expressed by epithelial and endothelial cells following bacterial infection. For example, airway epithelial cells produce IL-8 in response to *Streptococcus pneumoniae* infection, which recruits neutrophils to the site of infection. Other pro-inflammatory factors such as IL-12 and IFN-γ also contribute to the increase in neutrophil migration. Bacterial PAMPs such as LPS and peptidoglycan are often effective inducers of cytokines and chemokines such as IL-8, IL-12, IFN-γ and TNF. Activated epithelial cells

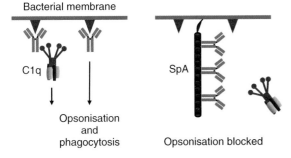

**Figure 12.6** Certain bacteria are able to evade opsonization by components of the complement system. For example, Staphylococcal protein A (SpA) is able to bind to the FC region of antibodies, thereby preventing recognition by FcRs.

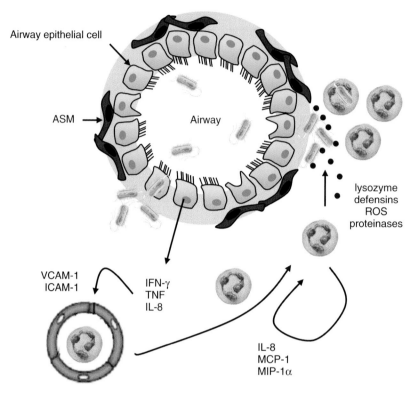

**Figure 12.7** Neutrophils are recruited from the bloodstream in response to bacterial infection. They are capable of phagocytosing bacteria and instigating killing mechanisms through the release of reactive oxygen species and enzymes. ASM, airway smooth muscle; ROS, reactive oxygen species.

and tissue resident macrophages are key producers of pro-inflammatory signals that act to recruit neutrophils from the bloodstream. Once neutrophils have migrated across the endothelium they are further attracted to foci of bacterial infection. This may involve trans-epithelial migration, as occurs during bronchial pneumonia or bacterial enterocolitis, whereby neutrophils migrate into the lumen of the airways or gastrointestinal tract respectively. Within these tissues, neutrophils are activated and initiate anti-bacterial defence mechanisms.

Neutrophils are the first cell type to be recruited into sites of infection and they are the principle cell responsible for the clearance of bacteria. They are effective at phagocytosing bacteria, which results in the uptake of bacterial cells into a phagosome (discussed in Chapter 2). Initiation of engulfment usually occurs following the recognition of bacteria opsonized with IgG or complement, through either FcγRs or complement receptors. This endocytosis does not kill the bacteria directly; rather phagosomes must mature and fuse with other vesicles known as

lysosomes, which are highly acidic and contain lytic enzymes and reactive oxygen species. The fusing of a phagosome and a lysosome produces a phagolysosome, within which bacteria are killed by the action of lytic enzymes such as neutrophil elastase, cathepsin G and several other proteinases, and reactive oxygen species such as superoxide and hydrogen peroxide. Neutrophils can also be induced to release their granule contents into the extracellular milleu. This has the effect of killing extracellular bacteria but at the same time can result in bystander damage and immunopathology. Along with epithelial cells, neutrophils contribute to the production of anti-microbial peptides, including lysozyme and human α-defensin-1, which are released form their primary granules.

In addition to bacterial killing, the release of granule proteins by neutrophils enhances monocyte and macrophage recruitment to sites of infection and activates macrophage effector functions. This is largely mediated by the release of proteinases such as proteinase-3, which attaches to endothelial cell walls and induces the expression of cell adhesion molecules. For example, proteinase-3 causes the upregulation of VCAM-1 and ICAM-1 on endothelial cells, which in turn enhances monocyte adhesion and migration across the endothelia. In addition, the antimicrobial peptide LL37 further enhances VCAM-1 expression and also acts as a chemoattractant for macrophages to migrate into inflamed tissue. Activated neutrophils are also significant inducers of pro-inflammatory cytokines and chemokines that attract more neutrophils and other immune cells to the site of infection. Neutrophilic granule components cause macrophages to secrete TNF, IL-8 and CCL3 (MIP-1α) and endothelial cells to produce CCL2 (MCP-1). Therefore, neutrophils are often central to the induction of an effective immune response against bacteria (Figure 12.7).

Neutrophil granule components also interact with G-protein coupled receptors (GPCRs) on the surface of macrophages and endothelial cells. For example, in addition to possessing antimicrobial properties, human α-defensin-1 (also known as human neutrophil peptide-1) is a ligand for purinoreceptors (e.g. P2Y6) on the surface of macrophages. The P2Y6 receptor signals through the NF-κB-dependent pathway that leads to the upregulation of IL-8. Furthermore, protease-activated receptors (PARs), expressed on the surface of macrophages, respond to proteinase-3 released by neutrophils and contribute to the signals required for IL-6 release. PARs are also thought to play a significant role in the tissue injury associated with pulmonary fibrosis, ALI and ARDS, through their expression on the surface of macrophages, endothelial and epithelial cells. Under normal circumstances, PARs participate in endothelial homeostasis and the regulation of the coagulation cascade. They normally recognize the coagulation factors thrombin and coagulation factor Xa, although they also respond to several proteases including the neutrophil enzymes cathepsin G and proteinase-3. Infection with bacteria, or stimulation with bacterial products such as LPS, and the subsequent activation of PARs on epithelial and endothelial cells, results in the upregulation of cell adhesion molecules and increases vascular permeability. Increased endothelial cell activation and resultant vascular permeability allows for the influx of more neutrophils and macrophages into the lung parenchyma. Subsequent neutrophil activation and release of granule proteins, together with pro-inflammatory cytokine production by macrophages, may then result in a positive feedback loop that causes tissue damage and neutrophilic lung disease. Interactions with components of the coagulation cascade and the innate immune system are associated with the pathogenesis of bacterial-induced sepsis, as well as pulmonary diseases.

## 12.10 Some bacteria are resistant to phagosome mediated killing

One of the strategies that bacteria use to evade the phagocytic activity of neutrophils and macrophages is to inhibit the process of opsonization, the mechanism of which has been discussed earlier. Other strategies include the direct inhibition of the biochemical activity of the enzymes and reactive oxygen species found within the phagolysosome, while some bacteria actively inhibit the metabolic pathways that are responsible for the synthesis of bactericidal factors. Some bacterial species are so adept at avoiding these killing mechanisms that they have evolved into intracellular pathogens capable of replicating inside phagocytes.

One such intracellular pathogen is the gram-positive bacteria *Listeria monocytogenes*, which infects the gastrointestinal tract following epithelial cell endocytosis and phagocytosis by resident macrophages. *L. monocytogenes* survives within its host by successfully avoiding phagolysosomal killing, which allows it to reproduce and disseminate. Phagocytosis by macrophages is induced by the recognition of lipoteichoic acid expressed on the surface of the bacteria and through ligation with complement receptors that recognize bound C1q and C3 complement proteins. The first strategy that *L. monocytogenes* employs is the secretion of several factors that cause the phagosome membrane to break down. One such factor is listeriolysin O (LLO), which is a cytolytic protein that forms pores in the phagosome membrane and results in de-acidification and loss of phagosome function (Figure 12.8). Other factors include membrane-active phospholipase C enzymes, which combine with LLO to breakdown the membrane and allow *L. monocytogenes* to escape into the cytoplasm of the macrophage. The bacteria replicate within the cytoplasm, without killing the macrophage, and allow *L. monocytogenes* to disseminate.

*Mycobacterium tuberculosis* is another intracellular pathogen that subverts macrophage killing mechanisms. *M. tuberculosis* is the cause of TB, a worldwide disease that accounts for more than 1.3 million deaths every year. It is contracted through inhalation and infects macrophages resident within the lungs, following recognition by complement receptors. It is able to escape phagocytic killing by inhibiting the maturation of the phagolysosome through the expression of an excretion system (ESX),

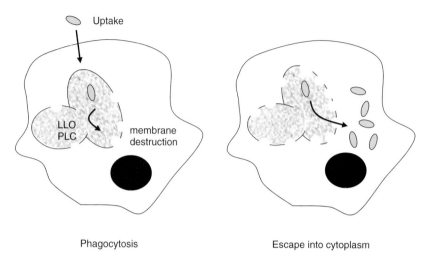

Uptake

LLO
PLC

membrane
destruction

Phagocytosis

Escape into cytoplasm

**Figure 12.8** Certain bacteria, such as *L. monocytogenes*, secrete factors that prevent normal phagolysosome function. The release of listeriolysin O (LLO) and phospholipase C (PLC) cause the breakdown of the phagolysosome membrane so that bacteria can escape into the cytoplasm.

which is thought to inhibit PI(3)P signalling. The suppression of PI(3)P signalling disrupts Ca2+ mobilization and arrests phagolysosome maturation. However, in healthy individuals the activation of macrophages following IFN-γ or IL-1β signalling is normally sufficient for phagocytosis and the killing of *M. tuberculosis*. The bacteria also attempt to respond to the macrophage by secreting a zinc metalloprotease (ZmpA) that inhibits IL-1β secretion. Therefore, a fine balance exists between the pathogen and the immune system, which often results in a chronic but non-pathogenic infection. When this bacterial persistence occurs, individuals are often referred to as carriers of latent TB. The importance of an ongoing immune response to TB is further demonstrated by the high death rate in HIV infected populations, who succumb to TB as a result of their immunodeficiency.

*Salmonella* species are also known to persist in their host for long periods of time. For example, *Salmonella enterica* serovar typhi infects its host by crossing the intestinal epithelia associated with Peyer's patches, probably by utilizing the trans-epithelial transport system of M cells. Once within the lamina propria, the bacteria replicate rapidly in the extracellular environment. They are then engulfed by macrophages (and DCs) but avoid phagocytic killing by inhibiting phagosome fusion with lysosomes. *S. typhi* is then disseminated throughout the reticuloendothelal system, for example to mesenteric lymph nodes and the spleen, by infected macrophages. Periodic lysis

of infected macrophages by *S. typhi* then causes the characteristic typhoid fever. Cytolysis of macrophages is beneficial to bacteria, as it promotes dissemination at the same as avoiding phagocytic killing. *Legionella pneumophila*, which causes Legionnaires, disease in humans, is another bacterial species known to prevent phagolysosome maturation, although it is thought that many other pathogenic bacteria subvert the phagocytic pathway as a means of immune escape.

## 12.11 NK cells and ADCC

Antibodies directed against extracellular bacteria opsonize the cell, which promotes phagocytosis and also antibody-dependent cellular cytotoxicity (ADCC). The mechanism of ADCC involves the recognition of antibodies bound to the surface of bacteria by NK cells, via cross-linking of FcRs expressed by the NK cell. The initiation of NK cell-mediated ADCC is an effective immune defence mechanism against extracellular bacteria, but is less effective at killing intracellular bacteria (Figure 12.9). Although direct lysis of bacterial cells is thought to contribute to immune protection, a more significant contribution that NK cells provide may be the production of IFN-γ. The secretion of IFN-γ by activated NK cells promotes macrophage phagocytosis and killing of endocytosed bacteria. Moreover, IFN-γ is an

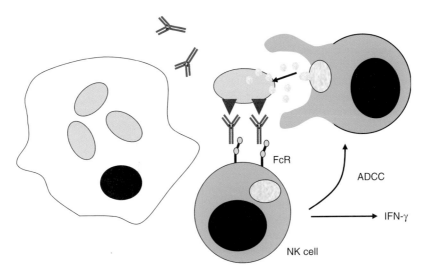

**Figure 12.9** Antibody-dependent cellular cytotoxicity, mediated by NK cells, is an effective way of killing extracellular bacteria but is less effective at killing intracellular bacteria.

important pro-inflammatory cytokine that supports the proliferation and activation of T cells. NK cells therefore provide an important link between the initiation of innate immune responses and the generation of adaptive immunity to bacteria.

Although ADCC is normally ineffectual against intra-cellular bacteria, other NK cell effector functions are capable of killing infected cells. For example, monocytes and macrophages infected with *M. tuberculosis* become targets for NK cell-mediated killing. NK cells express several receptors that recognize alterations in ligands expressed by host cells. Infected macrophages express the ligand ULBP1, which acts as a danger signal, and is most likely upregulated in response to TLR ligation by bacterial components such as peptidoglycan. The ligand ULBP1 is recognized by the NKG2D receptor on NK cells, which has a c-type lectin domain. Cross-linking of NKG2D activates NK cells and initiates perforin-mediated killing of *M. tuberculosis* infected cells. The other NKG2D lig-ands, MICA and MICB, are upregulated following cellular stress, and may also be important for the activation of NK cell cytotoxicity during bacterial infection.

## 12.12 The role of antibody in bacterial immunity

One of the key characteristics of mucosal surfaces is the high concentrations of dimeric secretory IgA (S-IgA), which is synthesized by resident plasma cells located in the subepithelial compartment. IgA performs sev-eral functions within the mucosa (Figure 12.10). Firstly, S-IgA binds to bacteria in the lumen and traps them in the mucous layer. This prevents bacteria from interacting with the epithelial surface, a process known as immune exclusion. Epithelial attachment and invasion can be fur-ther blocked by IgA binding to attachment proteins on the surface of bacteria. Transepithelial migration of bac-teria can be inhibited during pIgR-mediated transport of

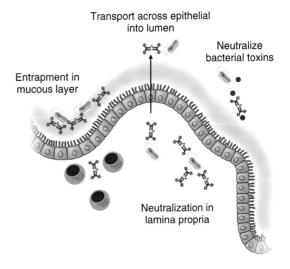

**Figure 12.10** S-IgA has a number of important functions in protecting the host against bacterial infection.

S-IgA across epithelial cells, which transfers bacteria back into the mucosal lumen. IgA is also present in the subepithelia and lamina propria where it can further inhibit bacterial attachment and cell invasion. Bacterial toxins are also neutralized by IgA, which prevents some of the pathogenic consequences of infection. Furthermore, IgA bound to the surface of bacteria is able to induce NK cell-mediated antibody-dependent cellular cytotoxicity. In addition to IgA, the urogenital tract also has significant amounts of IgG, which is thought to contribute to the protection against sexually transmitted bacteria.

Much of the S-IgA produced in mucosal tissue is polyreactive, meaning that the pool of IgA can bind multiple antigens, and is known as natural IgA. It is thought that the normal intestinal microflora is largely responsible for shaping the natural IgA response (Box 12.2). IgA is thought to be important for the transepithelial transport of bacteria and antigens from the baso-lateral surface and into the lumen, thereby limiting the potential for pathogen invasion and the accumulation of antigens. Specific antibody responses are also induced in mucosal tissue to pathogenic bacteria, although these responses are often slow to develop. It may be that pathogen-specific IgA is more beneficial for the clearance of chronic infections or for the prevention of a recurrent infection. Moreover, the generation of specific antibody responses at mucosal surfaces is an important property of a successful vaccine.

---

### Box 12.2. Intestinal Microflora Elicits Mucosal Immune Development.

The human body harbours as many as 100 trillion ($10^{14}$) microorganisms, which is ten times more than there are humans cells. Most of these microorganisms are bacteria that live in the intestines, although the skin, mouth and conjunctiva also harbour significant numbers and are collectively known as the microflora. These bacteria rarely cause disease and are therefore known as commensal bacteria. The interaction between the commensal microflora and the immune system is thought to play an important role in the development of the mucosal immune system. Without colonization of the gastrointestinal tract by commensal bacteria, the immune system of the gut does not mature properly and is far less effective at eliminating pathogens.

Most of the evidence that suggests the commensal microflora plays a pivotal role in eliciting immune development has come from experiments using germfree mice. These mice have very small Peyer's patches along the intestine, compared to mice with a normal microflora, and do not develop germinal centres. As a consequence, the numbers of mucosal B cells and the amount of IgA was significantly smaller. In addition, fewer CD4+ T cells were present in the lamina propria and fewer CD8+ T cells were found in the intraepithelial compartment. The introduction of commensal bacterial species, such as *Morganella morganii* or *Bacteroides thetaiotaomicron*, into germfree mice enhanced mucosal plasma cells and elevated levels of IgA. Similar findings have also been made in newborn babies, which are essentially germfree, whereby they only have primary B cell follicles and low levels of IgA. Both mucosal lymphoid tissue and the levels of IgA rapidly increase following birth, which is coincident with bacterial colonization.

It has been suggested that commensal bacteria induce immune cell activation, which is necessary for development of effective immune responses. For example, it has been proposed that mucosal DCs are able to survey the gut lumen and recognize PAMPs on the surface of commensal bacteria. TLR2 and TLR4 are both expressed by mucosal DCs and recognition of peptidoglycan and LPS, respectively, could induce DCs to produce cytokines such as IL-12. The production of IL-12 is required for the initiation and subsequent proliferation of CD4+ Th1 cells.

Commensal bacteria may also be required for the proper generation of immunological tolerance and the maintenance of mucosal homeostasis. Colonization of the gastrointestinal tract during the early postnatal period may be critical for developing tolerance to both commensal bacteria and dietary antigens. The mechanisms by which commensal bacteria elicit tolerance are unclear, although it is thought that the differentiation of IL-10 producing cells, such as Treg cells, is required. Non-pathogenic commensal bacteria may also deliver signals that induce Th2 responses, as well as Treg responses, both of which have immunoregulatory influence over more pathogenic Th1 responses. Disregulated mucosal tolerance can lead to debilitating diseases such as inflammatory bowel disease or Crohn's disease.

---

Several bacterial species secrete proteases that cleave and degrade IgA, demonstrating the importance of IgA in mediating protection against pathogenic bacteria. For example, *Neisseria* species, *Streprococcus* species and *Haemophilus influenzae* have all ben shown to secrete proteases that efficiently degrade IgA and prevent IgA-mediated immune exclusion. The secretion of IgA proteases into the extracellular environment is thought to promote bacterial invasion across mucosal surfaces.

## 12.13 Dendritic cells and immunity to bacteria

DCs are important members of the innate immune system that recognize danger signals delivered by pathogenic bacteria, particularly at mucosal surfaces of the gastrointestinal, respiratory and urogenital tracts. DCs express several PRRs, such as TLR2 and TLR4, which recognize molecular structures derived from bacteria (for example peptidoglycan and LPS). They also play a significant role in the uptake and processing of bacterial antigens and the presentation of those antigens to T cells, making them central for the induction of acquired immune responses. Within the lamina propria of the intestinal mucosal, DCs have been shown to extend dendrites across the epithelial cell monolayer, so that they can sample the luminal contents directly. Resident DCs become activated in response to bacterial danger signals, endocytose bacterial antigen and migrate to regional lymph nodes. The interaction between DCs and T cells within lymph nodes is important for the induction of both T cell effector responses and for the initiation of T cell-dependent antibody responses against pathogenic bacteria. Activated lymphocytes then migrate to the site of infection, such as the gut, where they perform their effector functions. DCs are so integral to the generation of acquired immune responses, that many pathogenic bacteria have evolved elaborate strategies to avoid recognition by DCs.

Several bacteria, such as *S. typhi* and *L. monocytogenes*, avoid phagocytosis by DCs by using the same evasion mechanisms as they do to avoid phagocytosis by macrophages. This prevents them from entering the antigen processing pathway and essentially avoids DC presentation of peptide to T cells. Indeed, several species secrete proteins that interfere with the MHC class II antigen processing pathway. For instance, *M. tuberculosis* actively sequesters its own antigens that would otherwise be presented to T cells. *M. tuberculosis* also downregulates the expression of MHC class II molecules on the cell surface by interfering with the transcription factor class II transactivator (CIITA) and inhibiting IFN-γ production.

*S. typhi* has an even more elegant strategy to avoid antigen presentation and the specific activation of CD4+ effector T cells. The extracellular forms of *S. typhi* are flagellated in order to promote motility. One of the components of bacterial flagellin is FliC, which is an effective ligand for TLR5 and promotes the phagocytosis of *S. typhi* by DCs. Once inside the cell, a peptide derived from FliC is efficiently processed and loaded onto MHC class II molecules for presentation to CD4+ Th cells. This induces the proliferation and expansion of FliC-specific T cell clones that are activated and have full effector function. However, these FliC-specific T cells are ineffective at protecting the host from infection. *S. typhi* avoids T cell-dependent immunity by altering its pattern

of gene expression. Once *S. typhi* is inside the intracellular compartment of a DC it no longer requires flagellin, and so it alters its pattern of gene expression so that FliC is rapidly downregulated. The immunological consequence of this is that the majority of T cell clones are FliC specific, so that the dominant immune response is directed against an antigen that the bacteria no longer express.

## 12.14 Autophagy and intracellular bacteria

Autophagy represents a process that is important in the control of intracellular pathogens, including viruses, bacteria and parasites (Figure 12.11). In addition to the proteosome pathway, endogenous material is degraded following the formation of autophagosomes, in much the same as exogenous material is degraded in lysosomes following phagocytosis. Autophagy is a natural process that occurs in most living cells and functions by recycling senescent organelles and proteins. However, autophagy also participates in the innate immune response to intracellular pathogens. Autophagosomes are composed of a double-membrane that contains material destined for degradation. The fusion of lysosomes with the autophagosomes results in the degradation of the inner membrane together with its contents. The degraded cellular material is then released back into the cytoplasm for subsequent use.

Autophagosome degradation of intracellular microbes occurs following either the formation of an autophagosome around a free bacteria or following the fusion of lysosomes with a conventional phagosome. The formation of autophagosomes is thought to result from intracellular TLR ligation, for example following the recognition of CpG motifs by TLR9 or ssRNA by TLR7. Both processes result in the fusion with lysosomes and the release of enzymes that hydrolyse and degrade the autophagosome contents. Subsequent fusion with late endosomes allows degraded bacterial antigens to enter the MHC class II pathway for presentation to CD4+ T cells (only cytosolic proteins degraded in the proteosome enter the MHC class I pathway). Autophagy therefore represents an important process in professional antigen presenting cells in particular, such as DCs and macrophages. However, the uptake of apoptotic cells by DCs, and their degradation by autophagy, may allow the cross-priming and presentation of endogenous bacterial antigens to CD8+ T cells in the context of MHC class I molecules (Figure 12.12). In addition, the combination of TLR signalling and the formation of autophagosomes

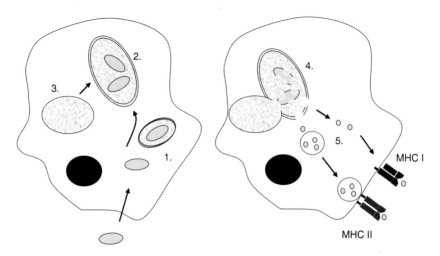

**Figure 12.11** Autophagy represents an important process in the presentation of bacterial antigens to T cells. 1. The bacteria starts as an extracellular pathogen before cell invasion into the cytoplasm. 2. Autophagosomes surround the bacteria and 3. fuse with lysosomes. 4. Autophagosome degradation releases bacterial products back into the cytoplasm. 5. Peptide is then able to presented on both MHC class I and MHC class II molecules. Cytoplasmic proteins usually only enter the MHC class I pathway.

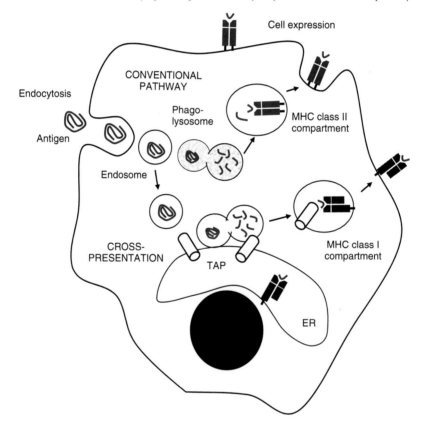

**Figure 12.12** Cross-presentation of exogenous antigen occurs in cells that are infected with bacteria or in professional phagocytic cells. Phagosomes contact ER factors, such as TAP proteins, which load bacterial peptides onto MHC class I molecules for presentation to CD8+ T cells.

upregulates several pro-inflammatory cytokines such as TNF and IFN-γ.

## 12.15 T Cells contribute to protective immunity

T cells play a significant role in maintaining tissue home-ostasis, while at the same time initiating effector responses to pathogens (Box 12.2). This is particularly evident at mucosal surfaces, especially the gut, as tolerance needs to be maintained to the commensal microbiota and harm-less antigens derived from food. The central event in driving a T cell-mediated inflammatory response is the interaction between activated DCs and T cells within regional lymph nodes. T cell responses are especially important for host defence against intracellular bacteria, which are largely impervious to the effects of antibody. The process of phagocytosis and the subsequent location of intracellular bacteria within the phagosome, results in bacterial antigens predominantly entering the MHC class II processing pathway and the presentation of peptide to CD4+ Th cells. Activated CD4+ T cells produce sig-nificant amounts of IFN-γ, which drives Th1-mediated immune responses and enhances neutrophil, macrophage and NK cell anti-bacterial effector functions. However, it is known that CD8+ T cells also participate in acquired immunity to intracellular bacteria. This involves a mech-anism of cross-presentation or cross-priming of antigen to CD8+ T cells in the context of MHC class I molecules (Figure 12.12).

The cross-presentation of antigen is thought to occur in the very same cell that the bacteria infect. For example, bacteria that are engulfed by macrophages normally enter the phagosome. It is at this early point in the process of phagocytosis that elements of the endogenous processing pathway come into contact with the phagosome. It may be that ER membranes contribute to the formation of the phagocytic cup, which allows MHC class I proteins to associate with the phagosome. For example, TAP proteins have been shown to be associated with the early phago-some. The result is the transfer of exogenously-derived bacterial antigens into the MHC classes I processing path-way and the presentation of peptide to CD8+ T cells. This process of cross-presentation is thought to function in combination with the process of cross-priming.

One suggested mechanism for the cross-priming of bacterial antigen involves the uptake of infected cells by DCs, after the infected cell has undergone apoptosis. For example, phagocytosis of bacteria by macrophages results

in the formation of a phagosome, where the bacteria replicate and cause the macrophage to undergo apoptosis. This results in the formation of apoptotic vesicles that con-tain breakdown products from the bacteria. These vesicles are then endocytosed by adjacent DCs, which shuttle the apoptotic vesicles directly into the endocytic processing pathway. In this way, bacterial antigens are processed and loaded onto MHC class I molecules for presentation to CD8+ T cells. This is thought to be an important method of cross-priming of antigen during *M. tuberculosis* and *S. enterica* infections, as both these pathogens compartmen-talize to the phagosome and prevent access to the cytosol.

## 12.16 The DTH response and granuloma in TB

Infections with *M. tuberculosis* are often associated with the formation of granulomas (otherwise known as tuber-cules) in the lungs, which can be observed as characteristic lesions in an x-ray scan from a TB sufferer. Despite causing fibrosis and local tissue damage, granulomas prevent TB bacilli from spreading, thereby containing the infection. Both CD4+ and CD8+ T cells are required for granuloma formation, while the type of immune response involved can be considered to be a delayed-type hypersensitivity reaction (discussed in chapter 15). It is often the case that immunodeficient patients, such as those with AIDS, are unable to contain TB due to the lack of T cell function and are unable to prevent reactivation. TB is a common cause of death in AIDS patients, demonstrating the importance of T cells in *M. tuberculosis* immunity.

Granuloma formation in TB follows a discrete set pat-tern (Figure 12.13). Inhaled bacilli are phagocytosed by macrophages in the lung, which release pro-inflammatory mediators and chemoattractants that recruit mononu-clear cells to the site of infection. Infected macrophages secrete the inflammatory mediators TNF, IL-1, CCL2 and CXCL10, which initially attract monocytes and neu-trophils. The upregulation of cell-adhesion molecules such as ICAM-1, within the granuloma and surround-ing epithelium, further assists with cell migration into the lung. The monocytes and neutrophils form a cen-tral zone around infected macrophages and contribute to the release of more inflammatory mediators. These heightened inflammatory signals result in the migration of CD4+ and CD8+ T cells, which form a mantle sur-rounding the central zone. The T cells also produce IFN-γ, which enhances both macrophage and neutrophil effector functions and causes them to release more inflammatory

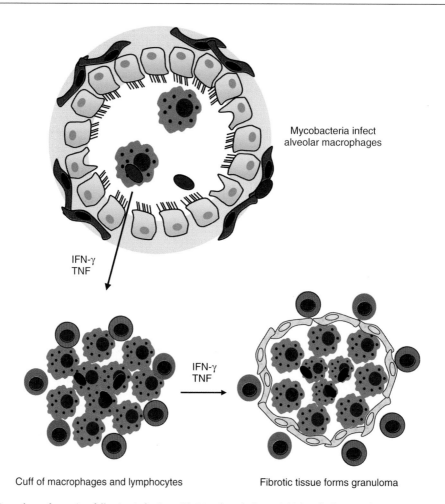

**Figure 12.13** Granuloma formation following infection with *M. tuberculosis*. An initial cuff of macrophages surrounds the phagocytosis bacteria, surrounded by T cells. This aggregate develops into a fibrotic granuloma, which prevents the spread of the bacteria.

mediators. IFN-γ produced by T cells is thought necessary to maintain the granuloma and prevent reactivation of TB. The combined production of TNF by macrophages and IFN-γ by T cells induces a self perpetuating, positive feedback loop that encourages the formation of granulomas around TB bacilli. Finally a sheath of fibrotic tissue, formed of collagen and other matrix proteins, envelopes the granuloma and effectively contains the bacteria.

## 12.17 Th17 cells in bacterial immunity

Cytokine producing, effector CD4+ Th cells can be divided into three broadly overlapping subpopulations:

Th1, Th2 and Th17. The conventional dichotomy of T cell-mediated immune responses includes Th1 cells, which secrete IFN-γ, and Th2 cells, which produce IL-4, IL-5 and IL-13. However, more recently another subpopulation of T cells has been shown to secrete IL-17, IL-21 and IL-22. These Th17 cells have been shown to be important in host defence against a number of bacterial species and in particular against extracellular bacteria (Figure 12.14). For example, *Borrelia burgdorferi* stimulates DCs to produce IL-23 and other innate cells to produce TGF-β, which cause naïve T cells to differentiate into Th17 cells. The IL-17 and IL-21 produced by Th17 cells in turn enhances bacterial clearance through several mechanisms that include the

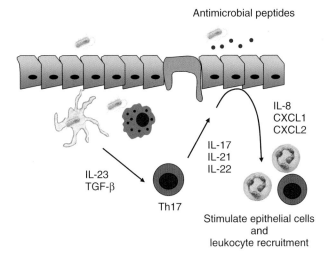

**Figure 12.14** CD4+ T cells differentiate into Th17 cells in response to IL-23 and TGF-β. Th17 cells produce IL-17, IL-21 and IL-22, which stimulate epithelial cells to release pro-inflammatory mediators that recruit further leukocytes into the site of infection.

upregulation of antimicrobial peptides (e.g. α-defensin, S100A8) by mucosal epithelial cells and increased chemokine expression (e.g. IL-8, CXCL1, CXCL2). Similarly Th17 cells have been associated with the chronic inflammation observed following *Helicobacter pylori* infection of the gastric mucosa. IL-17 stimulates the production of IL-8 within the mucosa, which is a potent chemoattractant for neutrophils. In addition, IL-17 has been demonstrated to contribute to protective immunity against several other bacteria, including *S. typhi, Klebsiella pneumoniae* and *Staphylococcal aureus.*

The potent immunostimulatory and inflammatory properties of IL-17-driven immune responses mean that excessive inflammation can sometimes lead to tissue damage and immunopathology. An example of this is the autoimmune-like reactions that occur followinga *B. burgdorferi* infection, which result in a destructive arthritis and inflamed joints. Excessive inflammation is also a characteristic of Crohn's disease, which may be the result of a Th17-mediated autoimmune response against harmless commensal bacteria or antigens derived from food. Although Th17 cells contribute to protective immunity against a number of pathogens, certain Th17 responses are associated with several immune-mediated diseases, including colitis, asthma and autoimmunity.

## 12.18 Treg cells in bacterial infection

There is an obvious requirement for the induction of a protective inflammatory response against invading bacteria, even though an unwanted consequence is bystander damage and immunopathology. Treg cells are known to provide crucial signals that prevent autoimmune reactions, maintain peripheral tolerance and help prevent tissue damage associated with protective immune responses. It has been demonstrated that pathogen-specific Treg cell populations are generated following infection with *B. pertussis* and that these cells are able to modulate Th1 responses through the production of IL-10. Treg cells are also central to the downregulation of an immune response following bacterial clearance, through the production of IL-10 and TGF-β and the subsequent inhibition of Th1 cytokine production.

Although Treg are beneficial to the host in preventing immunopathology, they are also involved in sustaining chronic bacterial infection. For example, Treg cells are observed within TB lesions and in the draining lymph nodes of TB infected lungs. It is thought that Treg cells contribute to non-pathogenic, chronic TB infections by dampening active T cell responses to *M. tuberculosis,* through the production of IL-10 and TGF-β. Even during the active phase of an immune response to the bacillus, Treg cells secrete the immunoregulatory cytokines IL-10 and TGF-β. Although this acts to prevent tissue damage within the lung microenvironment, it does provide the bacteria with a better opportunity to replicate and disseminate. This immune regulation may also contribute to disease progression in advanced TB, where a switch from Th1-type to Th2-type immunity is sometimes observed. This is often associated with diminished protective immunity and severe disease. Other chronic bacterial diseases, such as the persistent gastric inflammation found in *H. pylori* infections, are also attributed to the effects of Treg cells.

## 12.19 Unconventional T cells

Unconventional T cells, including NKT cells and γδ T cells, posses a limited TCR repertoire and are not restricted by conventional MHC molecules. Rather, they are thought to recognize lipid motifs presented in association with MHC-like molecules such as CD1. NKT cells and γδ T cells therefore share properties of both innate and adaptive immune cells. NKT cells specifically recognize CD1d, which is expressed on epithelial cells of the gastrointestinal tract, macrophages and DCs. NKT cells also have an invariant TCR that is restricted to Vα14Jα18, which recognizes glycolipid antigens presented by CD1d. They are activated early on in an immune response to bacteria and, depending on extrinsic signals, can produce either IFN-γ or IL-4. They are therefore capable of inducing either Th1 or Th2-mediated immune responses. NKT cells have been shown to participate in the protective immune response to several bacterial species, including *B. burgdorferi*, *S. pneumonia* and *M. tuberculosis*. They seem to be involved in initiating inflammatory responses to invading microbes by upregulating pro-inflammatory cytokines such as IFN-γ and attracting other leukocytes, particularly neutrophils, to the site of infection. NKT cells also have pleiotropic functions within the immune system, as they may contribute to the immunoregulation observed in chronic infections, such as in *H. pylori* infection, where they are thought to function in conjunction with Treg cells.

Compared to αβ T cells, the circulating population of γδ T cells in the blood is relatively small. However, γδ T cells comprise a significant population within mucosal sites such as the gastrointestinal, respiratory and urogenital tract, as well as the skin, particularly within the intraepithelial compartment. The anatomical location of γδ T cells allows them to act as early responders to infection and cell stress. γδ T cells are referred to as unconventional T cells as they either recognize phospho-antigens derived from bacteria or stress ligands derived from host cells (these host factors are also known as danger associated molecular patterns or DAMPs). Cells respond to bacterial infection by upregulating several DAMPs on their cell surface. These include the MHC-related molecules CD1c, MICA and MICB, which are recognized by γδ TCRs, and the ULBP ligands, which are recognized by the NKG2D receptor expressed by γδ T cells and NK cells. These signals induce the proliferation of γδ T cells and the expression of pro-inflammatory cytokines. In response to intracellular bacteria γδ T cell produce IFN-γ and TNF, while extracellular bacteria induce IL-17 production, thereby influencing the type of immune response depending on where the bacteria resides. γδ T cells are also capable of directly participating in bacterial clearance through the release of perforin and granzymes, and the secretion of antimicrobial peptides. Therefore, γδ T cells provide early recognition of cellular stress and provide key pro-inflammatory signals to other innate immune cells and cells of the acquired immune response.

## 12.20 Vaccination against bacterial diseases

Vaccination provides a powerful tool in the fight against bacterial infections (discussed at length in Chapter 17). Immunization induces a specific immune response that provides immunological protection against infection when the real bacteria are encountered. Vaccines against a number of bacterial diseases have been developed including diphtheria, tetanus, pertussis, TB and meningitis. Probably the most common vaccine, which is given to young infants, is the combination jab that protects against five pathogens: *Corynebacterium diphtheriae* (diphtheria), *Clostridium tetani* (tetanus), *Bordetella pertussis* (pertussis or whooping cough), *Haemophilus influenzae* type b (meningitis) and polio virus. Another common vaccine given to teenage children is the Bacille Calmette-Guérin (BCG), which is essentially *Mycobacterium bovis*, a bovine form of *M. tuberculosis*, which protects against human TB. A vaccine against childhood pneumococcal disease, which is caused by *S. pneumoniae*, is also in common use. These vaccines have provided protection against some of the most infectious diseases in the world and have proven invaluable for the improvement of public health. The mechanisms by which these vaccines induce protective immune responses are discussed in detail in Chapter 17.

## 12.21 Summary

1. Most bacteria infect their host across a mucosal epithelium or the skin.

2. The immune system is able to differentiate between commensal bacteria and pathogenic bacteria.

3. The commensal microflora contributes to normal immune system development.

4. Several mechanisms exist to protect the host against pathogenic bacteria, including the synthesis of antimicrobial peptides, complement proteins and antibodies.

5. Neutrophils are the first leukocytes to be recruited to sites of bacterial invasion.

6. Phagocytosis is an important process in the killing and removal of bacteria.

7. Bacteria have evolved several immune evasion strategies that attempt to subvert host immune defence.

# 13 Immunity to Fungi

## 13.1 Introduction

Fungi are a diverse group of eukaryotic organisms that belong to their own Kingdom of classification. As eukaryotes they possess a membrane bound nucleus containing multiple chromosomes and a genetic organization similar to that of animals. Importantly, certain fungi are capable of causing disease in many animal species including humans. Individuals who are immunocompromised as a result of HIV infection, cancer chemotherapy or immunotherapy following organ transplantation are particularly susceptible to infection. Fungi can affect various parts of the body and, depending on the route of infection, cause a particular form of disease. Fungal disease can be the result of superficial (infection of the skin), subcutaneous (infection of the subcutaneous tissue, dermis and adjacent tissue) or systemic infection (infection of deep organ tissues, particularly the lungs). A list of relevant fungal diseases that affect humans is depicted in Table 13.1.

Clinically relevant fungi occur as either yeasts or moulds. Yeasts are single celled microscopic structures that reproduce asexually, while moulds are multicellular structures comprising long filaments known as hyphae. Rather than being separate organisms, yeast and mould represent a distinct stage in the life cycle of a fungus, so that any one species may be capable of taking the form of either stage depending on various environmental signals. To become fully infectious fungi normally require a morphological change from one life cycle stage to another. Fungi that are able to change morphologically are called dimorphic fungi and represent the most widespread pathogenic species. Although less well studied than viruses, bacteria and eukaryotic parasites, fungi are very important veterinary and human pathogens. Fungal immunity represents a distinct area of immunobiology, due to the complex interaction between the host and the fungi.

## 13.2 Morphology of fungi

One of the most characteristic features of fungi is the composition of their outer cell wall. The rigid cell wall provides support for the cytoplasm, protection from the external environment and is responsible for the overall shape of the fungi. The cell wall of fungi comprises both structural and non-structural elements (Figure 13.1). The principal structural component is the polysaccharide chitin, which is a polymer of N-acetyl-D-glucosamine (GlcNAc) produced in the cytoplasm and subsequently integrated into the cell wall during synthesis. Chitin, in combination with various glucans, forms microfibrils within a larger matrix structure that give the cell wall its rigidity. Importantly chitin is recognized by receptors on various innate and adaptive immune cells, including mannose receptor, dectin-1 and TLR2. The non-structural elements are composed of other key polysaccharide and protein macromolecules. The main polysaccharides include mannans, galactans and rhamnomannans, the latter being responsible for the immunogenicity of many pathogenic fungi. Proteins are less abundant but include enzymes such as proteases, lipases and those responsible for breaking down polysaccharides. Other minor components include lipids, inorganic salts and various pigment molecules.

The fungal plasma membrane is different to that of mammalian membranes; however, its function as a semipermeable barrier remains the same. It is composed mostly of lipids, although it contains the non-polar ergosterol instead of cholesterol as its main sterol. Most anti-fungal agents are directed toward the ergosterol component of the plasma membrane or the disruption of its synthesis pathways. The lipid bilayer consists mostly of phospholipids and sphingolipids with proteins integrated throughout the membrane structure. Movement of molecules across the lipid bilayer occurs either by diffusion or active transport.

*Immunology: Mucosal and Body Surface Defences*, First Edition. Andrew E. Williams.
© 2012 John Wiley & Sons, Ltd. Published 2012 by John Wiley & Sons, Ltd.

**Table 13.1** Medically important fungal diseases.

| Pathogen | Disease | Route of infection |
| --- | --- | --- |
| Candida albicans | Candidiasis (thrush) | Vagina or mouth |
| Cryptococcus neoformans | Cryptoccosis (pneominia, meningitis) | Lung |
| Histoplasma capsulatum | Histoplasmosis (resembles tuberculosis) | Lung |
| Aspergillus fumigatus | Aspergillosis (pneumonia, aspergillomas) | Lung |
| Blastomyces dermatitidis | Blastomycosis | Lung |
| Coccidiodes immitis | Coccidiomycosis | Lung |
| Paracoccidiodes brasiliensis | Paracoccidiomycosis | Lung |
| Sporothrix schenckii | Sporotrichosis | Dermis and subcutaneous tissue |
| Pneumocystis carinii | Pneomocystosis (pulmonary cyst formation) | Lung |
| Microsporum sp. | Tinea manus (ringworm of hand) | Skin |
| Trichophyton sp. | Tinea capitis (ringworm of scalp) | Skin |
| Trichosporum beigelii | White peidra | Hair shaft |

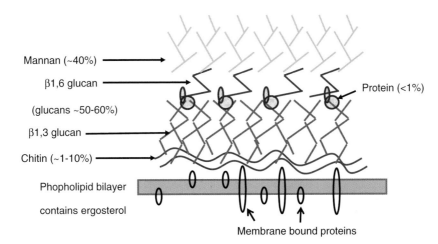

Mannan (~40%)

β1,6 glucan

Protein (<1%)

(glucans ~50-60%)

β1,3 glucan

Chitin (~1-10%)

Phopholipid bilayer

contains ergosterol

Membrane bound proteins

**Figure 13.1** The structure of the fungal cell wall. The fungal cell wall is a complex structure composed mostly of structural carbohydrates. The principle component is a molecule known as chitin, which forms microfibrils that give the fungal cell wall its rigidity and form its inner-most layer. Glucans are polysaccharides that can either be branched (β1,6 glucan) or non-branched (β1,3 glucan) and form the largest constituent ($\sim 50 - 60\%$) of the cell wall. The outer most layers contain mostly mannan, which comprise a range of polymers derived from mannose and form the basis of much of the cell wall's immunogenicity. Minor components of the cell wall include proteins and inorganic salts.

Fungi also possess a membrane-bound nucleus, which consists of a double nuclear envelope surrounding chromatin and a nucleolus. The number, size and shape of chromatin varies between species, although chromatin structure is relatively homogenous and consists of DNA contained in long filaments similar to mammalian chromosomes. Most fungal DNA is contained within the nucleus, although extra-nuclear DNA is also present such as plasmid DNA contained within mitochondria. In addition, fungi possess other sub-cellular structures characteristic of eukaryotic cells such as an endoplasmic reticulum, golgi apparatus, mitochondria and vesicles (Figure 13.2). Movement of cytoplasmic organelles (or chromosomes during cell division) is often controlled by microtubules composed of the protein tubulin. These are important structures for the control and maintenance of many cellular processes.

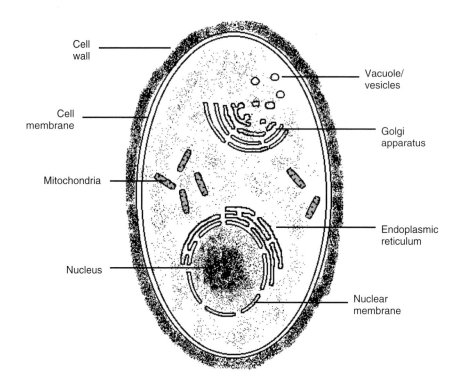

**Figure 13.2** The morphology of a fungal cell. The structure of fungal cells is similar to that of most eukaryotic cells. The cell itself is surrounded by a cell wall, which is complex and often immunogenic in many species. The plasma membrane and the membrane of sub-cellular organelles comprising of a lipid bilayer and contains ergosterol instead of cholesterol. Fungi possess a membrane bound nucleus containing a number of chromosomal structures and a variety of organelles such as an endoplasmic reticulum, golgi apparatus, numerous vacuoles and mitochondria.

## 13.3 Yeasts

Yeasts represent a particular single-celled stage in the life cycle of a fungus, rather than a specific phylogenetic classification. However, many species of fungi are solely unicellular and the term yeast generically refers to these organisms. Yeasts reproduce by a process known as budding, whereby the parent cell divides to form a daughter cell called a blastoconidia. The dividing nucleus and the cytoplasm of the parent cell are initially continuous with the daughter cell. New cell wall and plasma membrane synthesis occurs at the site of budding as the daughter cell grows. Blastoconidium separation occurs when a septum is formed between the two cells as a result of chitin deposition and eventually the daughter cell buds off (Figure 13.3). Certain medically important yeasts, such as *Candida albicans* and *Cryptococcus neoformans*, reproduce in this way. In some circumstances, blastoconidia are produced without the final separation of cells and the formation of elongated filaments called pseudohyphae results. Some yeast, including *C. albicans*, can also form true hyphae.

The majority of diseases caused by fungi are the result of infection with yeast cell stage organisms. *C. neoformans* occurs as a yeast in the environment. Infection is thought to occur when yeast are inhaled in conjunction with other material such as bird droppings or vegetable matter (these infectious boluses are known as formites). Upon contact with lung tissue *C. neoformans* retains its yeast morphology but develops a large, protective capsule around the cell wall, which is composed mostly of glucuronoxylomannan (GXM).

## 13.4 Moulds

The characteristic feature of moulds is the development of filamentous structures known as hyphae (Figure 13.3).

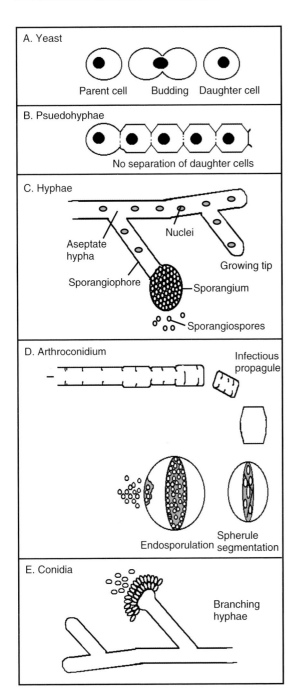

**Figure 13.3** Structural variation of fungal species.

Moulds grow by a process known as apical elongation, whereby new cellular units remain attached to the parent cell. This results in the typical filamentous structure. Occasionally cell elongation occurs at an angle to the original direction of growth and these individual branches are referred to as hyphae or mycelia. An interwoven network of different branching mycelia is called a mycelium. In addition to apical elongation some moulds can reproduce by the production of conidia or spores (Figure 13.3). These spores are produced in sac-like cells called sporangia, which form on specialized hyphae called sporangiophores. The sporangia consist of a large protoplasm, which is cleaved to form the individual sporangiospores. Infection with the hyphal forms of fungi is less common than yeast cell infection, although notable examples include *Microsporum canis* and *Trichophyton tonsurans*, which cause infections of the skin known as ringworm or tinea.

## 13.5 Fungal dimorphism

The morphology of a fungus is often intimately associated with the pathogenicity of the organism and its ability to cause disease. Many medically important fungi have the ability to undergo morphological differentiation, from their non-pathogenic environmental phenotype to a more pathogenic phenotype in the host. The propensity for morphological change is known as fungal dimorphism (Figure 13.4). Dimorphic fungi account for most fungal infections in humans and include *C. albicans*, *Coccidioides immitis*, *Blastomyces dermatitidis* and *Histoplasma capsulatum*. Why do some fungi require a morphological change to become pathogenic?

Most dimorphic fungi live as saprophytic mould in the environment (the mould is more suited to growth and survival). The initial infectious agent is usually a fragment of filamentous hyphae (for example, during *B. dermititidis* infection) or a conidia or spore derived from a specialized reproductive structure on hyphae (for example *C. immitis*). These are generally inhaled into the lung where they rapidly undergo morphogenesis into a more pathogenic yeast form (the yeast is more suited to growth and survival within the host). The yeast can then invade the surrounding tissue, disseminate throughout the host and progress to a systemic disease. An alternative to this process is morphological differentiation from a yeast form to a hyphal form, as exemplified by the commensal fungus *C. albicans*. As a yeast *C. albicans* is non-pathogenic until it changes into a more invasive hyphal or psuedohyphal

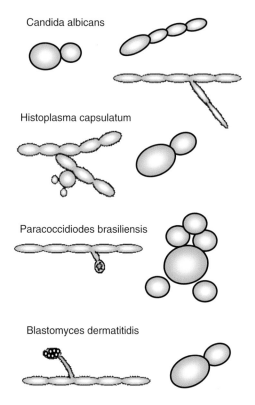

Candida albicans

Histoplasma capsulatum

Paracoccidiodes brasiliensis

Blastomyces dermatitidis

**Figure 13.4** Dimorphism in fungal species. Many fungal species possess the ability to change from one morphology to another. For example *C. albicans* changes from a single-celled yeast into a more pathogenic hyphal form during host tissue invasion. However, most dimorphic fungi change from a hyphal form (usually the morphology present in the outside environment) into a yeast form during the infection process.

form. The yeast stage of *C. albicans* is normally found in the mucosa of the vagina. If the host becomes immunocompromised then morphological differentiation into a hyphal form occurs and *C. albicans* is able to invade the surrounding tissue and cause disease.

*C. immitis* undergoes a unique morphological differentiation when infecting a host. The environmental stage comprises filamentous hyphae from which an infectious, unicellular structure called a propagule develops. The propagule grows when it comes into contact with host tissue into a large multinucleated spherule. The spherule undergoes segmentation, producing membranous compartments in which mitosis occurs, resulting in the development of many endospores. As the endospores mature, the spherule eventually swells and bursts, releasing the endospores and initiating the infection process.

## 13.6 Diseases caused by fungi

In order for a fungus to cause disease it must actively invade host tissue. Fungal infections are also known as mycoses and can be classified according to the route by which they infect the host and the degree of tissue involvement associated with the disease (Table 13.1). Superficial infections are localized to external sites such as the skin, nails and hair. The most common superficial mycosis is ringworm or tinea caused by one of several species of dermatophyte, such as a member of the Trycophyton, Microsporum or Epidermophyton genera. The disease thrush or candidiasis, caused by the commensal fungus *C. albicans*, is also classified as a superficial mycosis as it infects the mouth or vagina.

Subcutaneous mycoses infect the dermis or subcutaneous tissue of the skin and usually arise following damage or wounding to the skin. These infections are rare and mostly confined to tropical areas, are slow to develop and often result in chronic infection. An example is sporotrichosis caused by *Sporothrix schenckii*, which enters the dermis following wounding such as an insect bite.

The more widespread mycoses are the result of a systemic infection, affecting the internal organs of the host such as the lung or gastrointestinal tract. Inhalation of fungi or fungal fragments is an everyday occurrence, which rarely results in the development of disease in a healthy individual. However, immunocompromised people including AIDS patients, those undergoing chemotherapy, or transplant recipients are much more susceptible to fungal-related diseases. Systemic mycoses can be further classified into primary pathogenic fungi and opportunistic fungi (Table 13.2).

Primary pathogenic fungi can infect previously healthy individuals via the respiratory tract. An example is histoplasmosis caused by *H. capsulatum*, which differentiates from an environmental mould into pathogenic yeast when

**Table 13.2** Pathogenic and opportunistic fungi.

| Pathogenic fungi | Opportunistic fungi |
| --- | --- |
| *Histoplasma capsulatum* | *Candida albicans* |
| *Trichophyton tonsurans* | *Cryptococcus neoformans* |
| *Blastomyces dermatitidis* | *Aspergillus fumigatus* |
| *Coccidioides immitis* | *Mucor mucorales* |
| *Paracoccidioides brasiliensis* | *Pneumocystis carinii* |

inhaled. It causes a disease similar to tuberculosis, but dissemination to other parts of the body can occur in the immunocompromised. Opportunistic fungi are ubiquitous and only infect those with defective immune systems or other physiological defects. Aspergillosis is the most common opportunistic fungal infection, caused by the ubiquitous environmental mould *Aspergillus fumigatus*. This fungus can infect the lungs, inner ear and sometimes disseminate throughout the body. *A. fumigatus* releases unicellular conidia, which develop from a branching hypha. Individual conidia are inhaled and subsequently transform into branching hyphae that invade the surrounding host tissue. The infection results in pneumonia and the formation of small tumours called aspergillomas that possess a 'fuzzy-ball' appearance. *A. fumigatus* is a fine example of a mould with non-pathogenic and pathogenic hyphal forms. *C. neoformans* is another opportunistic pathogen that is becoming increasingly more widespread mainly due to the HIV/AIDS pandemic (Box 13.1: Fungal infections of the immunocompromised). Inhalation of yeast cells results in a chronic pulmonary infection that can develop into a severe infection of the central nervous system, culminating in life-threatening meningitis. Immunocompromised patients can also develop pneumonia as a result of *Pneumocystis carinii* infection. This intracellular pathogen was originally thought to be a protozoa but molecular analysis revealed it to be a red yeast fungi. It can cause fatal pneumonia through the formation of multinucleated cysts within lung tissue.

## 13.7 Immune response to fungi

Humans are constantly exposed to fungal microorganisms but why are fungal diseases so uncommon? The immune system in a healthy individual is more than capable of preventing infection and fungal disease. However, an intact immune response to fungal infection is critical in preventing an individual succumbing to disease. This is exemplified by the high incidence of fungal-related diseases associated with the immunocompromised. A defective immune system cannot adequately protect the host from fungal invasion and results in the progression of disease, pathology and tissue damage. Many mycoses are the result of opportunistic fungal species that take advantage of a defective immune system. Although not natural pathogens, opportunistic fungal infections often result in serious diseases. Most healthy individuals, with intact immune systems, are able to eradicate fungal microorganisms before disease can progress.

The mechanisms of fungal immunity are diverse and often depend on the particular species in question. A combination of innate immune and adaptive immune mechanisms is responsible for protection (Figure 13.5). Furthermore, the type of adaptive response (Th1 versus Th2, for example) can also influence the severity of disease.

## 13.8 Innate immunity

Innate immunity forms the first line of defence against fungal infection. Many aspects of innate immunity exist and include the barrier function of mucosal surfaces (epithelium of the airways, lung, intestine and vaginal tract), the effect of anti-microbial products (defensins and collectins), recognition of pathogen-associated molecular patterns (PAMPs), phagocytosis by macrophages and neutrophils, the antigen processing and presenting capacity of DCs and the function of innate lymphocytes (NK cells, NKT cells and γδ T cells).

The innate immune response has a number of important functions. Firstly, it provides a physical barrier to fungal invasion and actively expels microorganisms and particulate matter. Secondly, the secretion of microbicidal products and phagocytic processes directly kill fungi. Thirdly, cells of the innate immune system provide an instructive utility for the rest of the immune system. The production of cytokines and chemokines initiates a pro-inflammatory environment that attracts and activates other immune cells. The ligation of specific PAMPs supplies the immune system with information that enables an appropriate response to develop. In addition, the processing and presentation of antigen and the upregulation of co-stimulatory molecules activates and directs the adaptive immune response.

## 13.9 Mucosal barriers to fungal infection

Most pathogenic fungi infect via the mucosal surfaces of the respiratory tract. However, such mucosal surfaces provide both a structural and non-structural barrier to fungal invasion, which physically separates the non-sterile external environment from the sterile internal tissue. In order for disease to progress, fungi must first adhere to the epithelial surface of the mucosa. Various mechanical forces in the respiratory tract prevent this from occurring, including the coughing reflex, the turbulent flow of air during respiration and the mucociliary clearance system.

## Box 13.1. Fungal Infections of the Immunocompromised.

Despite the continuous exposure to ubiquitous fungi within the environment most people never succumb to infection. Only in recent times have the incidence of mycoses intensified, principally due the increase in immunocompromised individuals. By far the most significant contributing factor is the HIV/AIDS pandemic, although cancer chemotherapy patients, those undergoing transplantation or patients with leukaemia or other haematological defects are also at higher risk of developing mycoses. A non-pathogenic species becomes a primary opportunistic pathogen in those people with immune defects. The three most prevalent mycoses of the immunocompromised host are cadidiasis, aspergillosis and cryptococcosis.

The majority of HIV infected individuals develop oropharyngeal candidosis (OPD), which may spread further into the oesophagus and bronchioles in advanced AIDS. In HIV-infected females candidal vulvovaginitis (CVV) is also common. These diseases are related to a markedly reduced CD4+ T cell count. A sizeable defect in CD4+ T helper cells diminishes the effectiveness of both cell-mediated and antibody-mediated immunity. However, HIV-infected individuals rarely succumb to disseminated infection. This suggests that immune mechanisms other than cell-mediated immunity are responsible for preventing dissemination, although an intact cell-mediated immune response is required for the control of superficial infection.

OPD is usually treated with azole-based therapeutics such as fluconazole. However, in recent years azole-resistant strains of *C. albicans* have emerged, accentuating the problems associated with treating immunodeficient patients. Furthermore, non-albicans Candida species have been recognized as new emerging species, such as *C. glabrata*, *C. parapsilosis* and *C. krusei*, which can penetrate deep body tissues. These species may also be resistant to common anti-fungals.

Bone marrow transplant, lung and heart-lung transplant patients are particularly susceptible to pneumonia caused by *A. fumigatus*. Aspergillosis is less common in people infected with HIV and is primarily a cutaneous infection. Pulmonary infection following transplantation follows a biphasic pattern, which correlates with the initial neutropenia and later with the development of graft-versus host disease. Immuno-suppressive drug treatment following transplantation exacerbates fungal dissemination. For instance, macrophages loose their ability to kill *A. fumigatus* following corticosteroid treatment.

*C. neoformans* is the second most common opportunistic pathogen in HIV-infected patients (only *C. albicans* is more prevalent). It primarily infects the lung causing chronic pneumonitis in individuals with defective cellular immunity. Fungal dissemination can lead to meningitis or meningoencephalitis, which is often lethal, and represents the leading mycosis of the central nervous system. It has been demonstrated that CD4+ T cells protect the brain against *C. neoformans* infection. In the immunocompetent host an

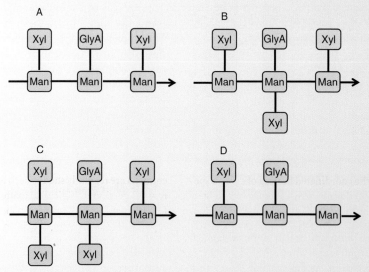

Cryptococcus GXM serotypes. Cryptococcus species can be divided into four main serotypes based on the composition of the GXM (glucuronoxylomannan) capsule. Each serotype differs in its incorporation of xylose (Xyl) and glucuronic acid (GlyA) onto the standard mannose (Man) backbone. Serotypes A and D are ubiquitous opportunistic fungi of *C. neoformans* variety *neoformans* that only cause disease in the immunocompromised, while serotypes C and B are *C. neoformans* variety *gattii* and are restricted to tropical area where they can cause disease in immunocompetent individuals.

inflammatory response develops in the brain, characterized by an infiltration of mononuclear cells, granulocytes, CD4+, and CD8+ T cells. A granulomatous lesion develops that contains the yeast, which is absent in AIDS patients lacking effective T cell immunity. Dissemination of *C. neoformans* to the brain is life-threatening if left untreated and causes nausea, vomiting, headaches and loss of memory. Cryptococcosis is normally treated with amphotericin B and fluconazole but treatment is often not completely effective without a fully intact immune system.

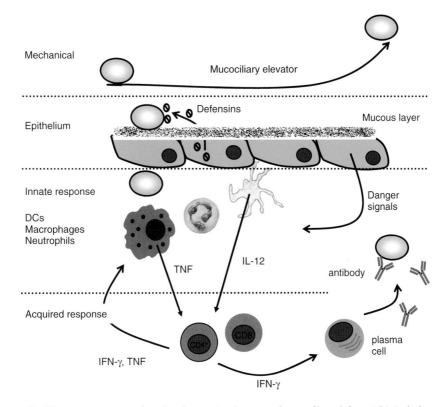

**Figure 13.5** Generalized immune response to fungal pathogens involves many layers of host defence. This includes mechanical and physical barrier defence at the epithelium, innate immune responses, cell-mediated T cell responses and antibody-mediated B cell responses.

The latter involves the coordinated beating of cilia on specialized epithelial cells that acts to move particulate matter toward the upper respiratory tract and mouth. Coughing or swallowing then removes particles from the respiratory tract. This process is aided by the continuous secretion of mucus, which forms a gelatinous layer that traps and protects the underlying epithelium, effectively preventing fungal adherence. Mucus principally consists of large glycoproteins called mucins, which are secreted from specialized goblet cells lining the mucosa. Goblet cells can increase in number and accelerate production of mucins upon encounter with fungi. Furthermore, cells from the epithelia are regularly shed so as to maintain function and remove persistent particulate matter. The general barrier function of the mucosa and associated mechanisms of expulsion is known as non-immune exclusion.

## 13.10 Anti-fungal molecules

Invasive fungi that evade the physical and mechanical barriers are able to cause damage to underlying host tissue such as the epithelial lining of the respiratory tract. A number of small, anti-fungal molecules are expressed in response to cell damage and include defensins, collectins and surfactant proteins. These have both direct killing activity and indirect activity toward fungi. Many of these molecules are constitutively expressed and are always present in the respiratory lumen but they are also upregulated during fungal infection. For example, α-defensins are contained within the azurophilic granules of neutrophils, which are discharged at the site of infection in response to inflammatory mediators such as IL-1 and IL-8. Similarly the production of β-defensins by epithelial cells is enhanced in response to TNF and IL-1 production. Defensins are highly positively charged peptides that bind to negatively charged repeated-motifs on fungal cells, causing membrane disruption by forming permeabilizing pores. Defensins also act as chemoattractants for DCs, neutrophils and T cells and they can activate the alternative complement cascade.

Collectins are C-type lectins that bind to carbohydrate structures on the surface of fungi including *A. fumigatus* conidia and *C. neoformans* acapsulated yeast. Mannose-binding lectin (MBL) is an example of a collectin that binds to glycosylated proteins and structural polysaccharides such as mannan and GXM. Other C-type lectins include the surfactant proteins (SP) A and D, which contain conserved C-type lectin carbohydrate recognition domains (CRDs) that bind to PAMPs. SP-A and SP-D inhibit fungal invasion by forming cross-linked bridges between carbohydrate structures on fungal surfaces by a process of agglutination. SPs also enhance macrophage and neutrophil phagocytosis through opsonization, stimulate the production of the inflammatory cytokines TNF, IL-1 and IL-6, increase ROS and NO release and act as chemoattractants for macrophages and PMNs.

## 13.11 Recognition of fungal PAMPs by Toll-like receptors

How does the innate immune system provide an effective line of defence against fungal infections? An important aspect of fungal immunity is the molecular interaction between pattern recognition receptors (PRRs) on host cells and evolutionary conserved molecules on fungal surfaces known as pathogen-associated molecular patterns (PAMPs). One group of PRRs are the Toll-like receptors (TLRs), which are highly conserved cellular receptors that recognise various cell-wall components of fungi. Ligation of TLRs results in an intracellular signalling cascade mediated by the adaptor protein MyD88 and leading to the production of pro-inflammatory cytokines such as TNF and IL-1β. Furthermore, certain TLRs distinguish

between different fungal components so that a variety of fungal pathogens can be recognized and an appropriate immune response activated (Figure 13.6).

Zymosan is a cell wall component of many fungi that contains various polysaccharide components including β-glucan and mannan. TLR2 is responsible for the recognition of zymosan and the subsequent production of inflammatory cytokines. The activation of innate immune cells, such as DCs and macrophages, by TLR2 ligation with zymosan is assisted by the lectin-binding receptor dectin-1. The β-glucan component of zymosan is specifically recognized by dectin-1, resulting in the production of TNF, IL-1 and IL-8. This illustrates the importance of receptor cooperation in the effective recognition of specific pathogenic fungi.

TLR4 is important for the recognition of mannan, a cell-wall component of many fungi including *C. albicans*. Ligation of TLR4 with mannan requires the presence of CD14, which is crucial for TLR4 signalling and the production of inflammatory cytokines. Immunological resistance to *C. albicans* is reduced in mice that are deficient in TLR4, although release of TNF and IL-1β occurs in a TLR4-independent manner, suggesting other receptors recognize different fungal components. The *C. neoformans* cell-wall component GXM is also recognized by TLR4 and CD14, although GXM only partially activates TLR signalling pathways also suggesting that other receptors (such as TLR2) are involved in fungal recognition. Other fungal cell-wall components that activate TLR-dependent signalling pathways include glucans, chitin, phospholipomannan (PLM), lipoteicheioc acid and conidial and hyphal components from *A. fumigatus*. In addition, *C. albicans* and *A. fumigatus* are able to signal through TLR9, the PRR for unmethylated CpG-rich DNA. It is likely that many PRR act in concert to initiate downstream signalling events that eventually result in immune activation and pathogen clearance.

## 13.12 Complement and fungal immunity

The complement system comprises a cascade of complement factors that are sequentially activated leading to the formation of a protein complex called the membrane attack complex (MAC). There are three separate pathways that can activate the complement cascade (refer to Chapter 2 for a more detailed description of the complement cascade), the classical pathway, alternative pathway and a pathway initiated via mannose binding lectin (MBL) opsonization. All three pathways involve the generation

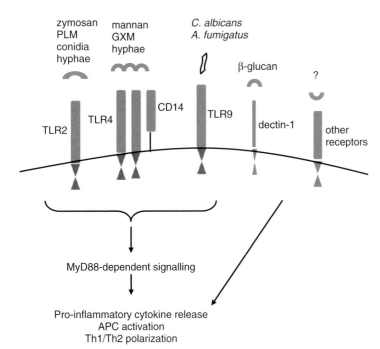

**Figure 13.6** TLRs recognize fungal components. TLRs and other PRRs recognize fungal PAMPs and initiate protective Th1-mediated immune responses. Zymosam, PLM and conidia or hyphal components from certain fungi are recognized by TLR2. Mannan and GXM are recognized by TLR4 (in conjunction with the CD14 receptor), while DNA from *C. albicans* and *A. fumigatus* is recognized by TLR9. Other receptors such as dectin-1 are able to recognize various fungal components such as β-glucan. TLRs signal via the MyD88 pathway, which leads to transcription factor activation, gene expression and Th1-dominated immune response.

and cleavage of C3 convertase into C3a and C3b. While C3a is involved in inflammatory cell recruitment and acts as an anaphylatoxin, C3b continues the complement cascade that eventually forms the membrane attack complex (MAC). The MAC forms a pore in the fungal cell membrane, or in an infected host cell membrane, causing osmotic lysis and direct killing of the pathogen (Figure 13.7).

The complement system has other functions other than MAC formation. Both myeloid and lymphoid cells express several receptors that recognize certain factors of the complement cascade. An important mechanism of pathogen clearance is the opsonization of fungal cells by complement factors, particularly C3b fragments (e.g. iC3b) that enhance macrophage phagocytosis by binding via complement receptors (CR). For example, efficient clearance of *C. neoformans* requires the recognition of C3 fragments bound to the GXM-rich capsule via CR3 (CD11b/CD18) and CR4 (CD11c/CD18). In addition, complement factors are important for the generation of a protective Th1-mediated immune response involving the production of pro-inflammatory mediators such as IL-8 and TNF. The complement factors C3a, C5a and C3b also possess chemotactic properties by enhancing the migration of immune cells into the site of inflammation.

MBL is an evolutionarily conserved pathogen recognition molecule and a member of the collectin family of proteins. It is involved in the recognition of a range repetitive carbohydrate structures, allowing it to bind to a variety of components in fungal cell walls. MBL is thought to provide three important anti-fungal mechanisms: (1) the activation of the complement cascade; (2) opsonization of pathogens leading to phagocytosis; and (3) the initiation of inflammation. MBL activation of the complement pathway is distinct from the classical and alternative pathways. Once bound to carbohydrate moieties on the fungal cell surface MBL, in association with MBL-associated serine proteases (MASPs), becomes activated and cleaves the complement proteins C4 and C2 (in a similar way to C1 esterase). This generates a C4b2a complex and acts as a C3 convertase, generates opsonic C3b fragments and begins the complement amplification loop (Figure 13.7). Although less well studied, MBL also opsonizes fungal cells, independently of the complement system, and thus assists in the clearance of pathogen. Furthermore, MBL is thought to directly induce the release of inflammatory cytokines such as TNF, IL-1β and IL-6 from macrophages, possibly through receptor binding on these cells.

Mannose receptor (MR) is a member of a group of glycoproteins known as lectins (CR3 is also a type of lectin).

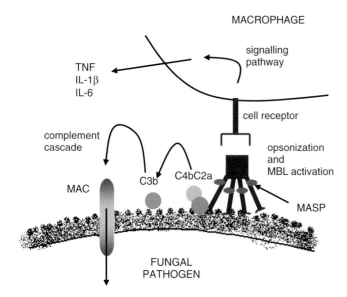

MACROPHAGE

Figure 13.7 Mannose-binding lectin attaches to fungal cells. Mannose-binding lectin (MBL) is able to activate the complement pathway that is involved in the lysis of fungal cells. MBL recognizes polysaccharide (mannose) motifs on the fungal cell surface. The complement components C4aC2b binds to MBL and in association with MBL associated serine proteases (MASPs) activates its C3 convertase properties, which initiates the complement cascade eventually resulting in membrane-attack complex formation. In addition, MBL is able to directly bind the MBL receptor on the surface of macrophages leading to pro-inflammatory cytokine production.

Macrophages and certain endothelial cells constitutively express MR. It is the best characterized lectin (Figure 13.8) and is involved in the recognition of both host and fungal factors, having the highest affinity for oligosaccharides such as mannose, fructose and galactose. MR is able to bind a range of pathogen-derived molecules that possess rich mannose content, such as mannan from *C. albicans*. Ligation of the MR primarily mediates macrophage phagocytosis and may also induce the production of IFN-γ, TNF and IL-12 and activate the respiratory burst.

## 13.13 Dendritic cells link innate and adaptive fungal immunity

The initiation of an appropriate immune response to fungal pathogens is a critical event in the generation of an effective immune response. Dendritic cells (DCs) have the unique ability to activate naïve $CD4^+$ T cells and provide the adaptive immune system with important instructive signals. Most fungal pathogens induce a protective Th1 immune response, largely due to the effective recognition of PAMPs by TLRs and other PRRs on DCs that lead to Th1 cytokine production. DCs are also able to differentiate between fungal morphotypes, so that one set of PRRs detects fungal yeasts, while another set recognize hyphae. Differential receptor ligations result in specific down-stream signalling events that translate into cytokine production, upregulation of co-stimulatory

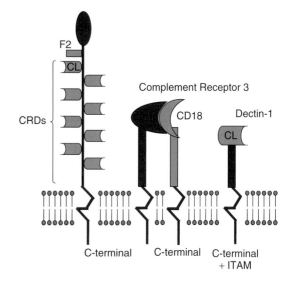

Figure 13.8 Structure of MR, CR3 and dectin-1. (CDR, carbohydrate-rich domain; CL, C-type lectin-like domain.)

molecules and antigen peptide/MHC complexes, and polarization toward a Th1 or Th2 immune response.

Activation of DCs requires the recognition and phagocytosis of fungi. In particular the ligation of MR and CR3 are important for internalization of fungi, while the involvement of TLR2, TLR4 and TLR9 appear less crucial. DCs have a unique ability to phagocytose different fungal morphotypes, exemplified by internalization of yeast or hyphal forms of *C. albicans* and conidial or hyphal forms of *A. fumigatus*. The phagocytosis of these morphotypes occurs through different mechanisms. The yeast of *C. albicans* and the conidia of *A. fumigatus* are phagocytosed by coiling phagocytosis, while the hyphal forms of these fungi are phagocytosed by zipper-type phagocytosis. Coiling phagocytosis involves the formation of an extended, finger-like structure called a pseudopod, which wraps around yeast or conidia (Figure 13.9). Before the fungal structure is endocytosed the unilateral pseudopod coils around the central fungal structure to form a pseudopod whorl. This structure is also flanked by lateral pseudopods. A phagosome is eventually formed by the fusion of the pseudopod whorl and the yeast or conidia is internalized. This form of phagocytosis usually depends on PRRs such as MR and CR3, which detect carbohydrate structures and complement factors on the fungal cell surface respectively.

Zipper-like phagocytosis is the more conventional form of phagocytosing large exogenous structures (Figure 13.9). Extending pseudopods wrap around the hyphae in a symmetrical manner, following the contour of the particle. Internalization of hyphae occurs by pseudopod fusion and phagosome formation. This form of phagocytosis depends on the recognition of opsonized hyphae and involves FcγR and CR3. Both mechanisms of phagocytosis result in the entry of fungal structures into the endocytic pathway, where various phagolysosomal enzymes degrade them. Although DCs

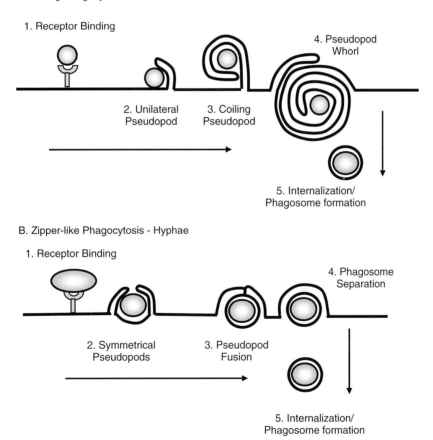

**A. Coiling Phagocytosis - Yeast and conidia**

1. Receptor Binding
2. Unilateral Pseudopod
3. Coiling Pseudopod
4. Pseudopod Whorl
5. Internalization/ Phagosome formation

**B. Zipper-like Phagocytosis - Hyphae**

1. Receptor Binding
2. Symmetrical Pseudopods
3. Pseudopod Fusion
4. Phagosome Separation
5. Internalization/ Phagosome formation

**Figure 13.9** Phagocytosis of fungal cells. Phagocytosis is necessary for the clearance of invading fungal cells. Recognition of fungal components by receptors on the surface of phagocytes (macrophages) initiates the process of phagocytosis. Two important mechanisms exist, coiling phagocytosis (A) and zipper-like phagocytosis (B). Coiling phagocytosis utilizes a membrane extension called a pseudopod to wrap around the fungal cell, while zipper-like phagocytosis uses two symmetrical pseudopods that close around the fungal cell. Membrane fusion events then occur during phagosome formation and the fungi is internalized.

are the most adept at recognizing different fungal morphotypes other phagocytes, such as macrophages and neutrophils, are able to internalize fungal structures by similar mechanisms.

## 13.14 DCs provide the adaptive immune response with instructive signals

How does the innate immune system and the adaptive immune system interact during fungal immunity? An essential link between innate and adaptive immunity is provided by DCs via the production of molecular signals, such as cytokines, chemokines and ROS. The ligation of TLRs by fungal PAMPs, recognition of complement factors by CRs and cross-linking of FcRs by opsonizing antibodies is central for the production of these instructive signals. A cascade of intracellular signalling events is initiated in response to different fungal factors. For instance, IL-12 production by DCs is dependent on the MyD88 signalling pathway and its production is upregulated by TLR2, TLR4 or TLR9 ligation. Specific receptor-ligand interactions greatly modify the subsequent immune response to infection. For example the yeast form of *C. albicans* activates TLRs and MR induce a protective Th1-type immune response via the production of IL-12. The induction of Th1 immunity involves the enhanced uptake and killing of canidial yeast and the development of a cell-mediated immune response. Conversely the hyphal form of *C. albicans* activates CR3, which inhibits IL-12 production, promotes IL-4 and induces a non-protective Th2 immune response. Opsonins such as the complement factors C3b and C5b, and antibody are able to modify the phenotype of DCs through their respective receptors. The variety of receptor-ligand interactions may explain the complicated and diverse immune responses induced by fungal pathogens (Figure 13.10).

A similar situation can occur during *A. fumigatus* infection, whereby conidia are phagocytosed through MR binding resulting in protective Th1 immunity. However, hyphal forms of *A. fumigatus* are phagocytosed following CR3 and FcγR binding, which induce a Th2 immune response that leads to fungal-induced pathology. However, most fungal infections are controlled by Th1 immune responses, characterized by the production of one or more cytokines such as IL-12, IL-18, IFN-γ and TNF. The phenotype of DCs will determine the nature of the subsequent adaptive immune response by controlling Th1/Th2 cell development.

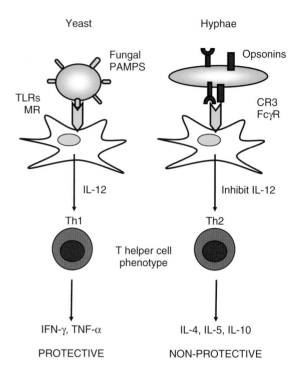

**Figure 13.10** DC polarization of the immune response. The phenotype of the immune response to fungi can dictate whether the response will be protective or non-protective. The generation of Th1 immunity is protective and involves the recognition of fungal PAMPs by TLRs and MR on the surface of DCs (associated with IL-12 production). Th2 immune responses are unprotective and involve the activation of CR3 and FcRs. Generally yeast induce Th1 responses while hyphae induce Th2 responses.

## 13.15 Macrophages are important APCs during fungal infection

Macrophages are professional antigen-presenting cells that phagocytose fungal pathogens and present processed antigen to T cells. They occur in peripheral tissues at a much higher frequency than DCs. Although their capacity to present antigen to T cells is less than DCs, they play an important role in fungal inhibition. They also provide various inflammatory signals by way of cytokine, chemokine and oxygen and nitrogen free radical production, which activates the cellular immune response. In addition macrophages directly kill invading fungi through phagocytic killing mechanisms and the initiation of the respiratory burst. In turn, cytokine production by activated T cells and NK cells enhances

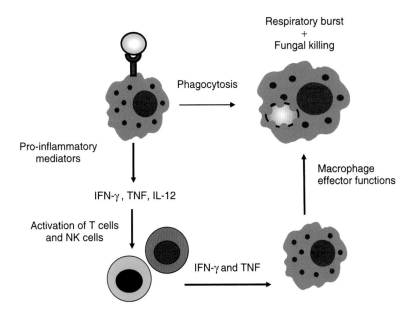

**Figure 13.11** Macrophage effector functions against fungi. Macrophages are essential for controlling fungal infection and clearing pathogens from the host. Firstly activation of macrophages induces the production of pro-inflammatory cytokines that attract and activate other immune cells. Secondly, macrophages possess various killing mechanisms including phagocytosis and the generation of the respiratory burst. The respiratory burst comprises various reactive oxygen and nitrogen free radicals that cause direct damage to fungal cells.

macrophage effector mechanisms and fungicidal activity (Figure 13.11).

Macrophages express various cell surface receptors, such as MRs, FcRs, CRs, TLRs and integrin receptors (CD18). Binding of opsonized or un-opsonized fungi via one or more of these receptors initiates rapid phagocytosis of the pathogen. The internalization process activates the respiratory burst and the production of cytokines such as IL-12, IFN-γ and TNF. An important receptor on macrophages is the lectin-like receptor dectin-1, which recognizes β-1,3 and β-1,6 linked glucan polymers on the fungal cell surface. Therefore dectin-1 acts as a β-glucan receptor that participates in the phagocytosis process independent of opsonization.

## 13.16 Neutrophils participate in the inflammatory response to fungi

The early phase of the immune response to fungi involves the recognition, internalization and processing of antigenic components by professional APCs. The next phase is the inflammatory response, characterized by leukocyte migration into the area of fungal invasion, driven by the release of pro-inflammatory mediators. Neutrophils form the most abundant part of this early cellular infiltrate and possess a number of anti-fungal effector functions that act to contain the invading pathogen and prevent dissemination. Neutrophils are able to phagocytose various fungi, although not as effectively as DCs or macrophages and they generally lack the ability to present antigen to T cells. They directly kill fungi by participating in the respiratory burst and through the release of non-oxidative defensins. They also act to maintain the inflammatory environment by sustaining macrophage reactivity and attract cells of the adaptive immune response.

## 13.17 NK cells provide inflammatory signals to macrophages

The importance of NK cells in certain infections, such as viral infections, has been well documented (see Chapter 11), but their role in fungal immunity is less clear. Rather than directly killing fungi, NK cells provide stimulatory signals to macrophages that enhance phagocytosis, particularly through the production of IFN-γ. NK cells are thought to be necessary for the continued fungicidal activity and killing mechanism of macrophages in certain fungal infections. This undoubtedly involves the maintenance of Th1 cytokine production.

NKT cells and γδ T cells are other innate lymphocytes that may be important in controlling fungal infections. NKT cells express a restricted T cell receptor repertoire, usually Vα14 in mice and Vα24 in humans. Like γδ T cells, NKT cells recognize the non-classical MHC molecule CD1, ligation of which leads to cell activation. The primary function of these two cell types is the rapid production of cytokines following stimulation. NKT cells in particular produce large quantities of both IFN-γ and IL-4 and therefore play a role in directing the Th1/Th2 phenotype of the immune response, which is a critical determining factor in the control of fungal infection.

## 13.18 Adaptive immunity to fungi

Infections with fungal pathogens are a common occurrence but the development of clinical disease is rare. Most immunocompetent individuals are found to possess circulating anti-fungal antibodies and are able to mount an effective proliferative response to fungal components indicating effective B and T cell memory. These factors suggest that exposure to fungal pathogens is initially controlled with the subsequent generation of life-long acquired immunity. Furthermore, a familiar characteristic of fungal immunity is the generation of a granuloma at the site of infection, for example in the lung, indicative of a protective Th1-driven delayed-type hypersensitivity (DTH) response. In addition, the susceptibility to fungal infections in patients with T cell deficiencies demonstrates the importance of acquired immunity (Box 13.1: Fungal infections of the immunocompromised).

What aspects of the cell-mediated immune response determine disease susceptibility or resistance? A Th1 immune response characterized by the production of IL-12 by APCs and IFN-γ and TNF by various leukocytes leads to protective immunity and disease resistance. On the other hand, a Th2 immune response, characterized by the production of IL-4 and IL-5, results in non-protective immunity and disease susceptibility (Figure 13.10). This concept has been demonstrated both experimentally and clinically. Neutralization or ablation of IL-12 or IFN-γ leads to enhanced disease, while patients with defective IFN-γ production or increased levels of Th2 cytokines often present with disseminated infections. Elevated levels of IL-4 also predispose toward enhanced fungal disease. IL-4 diminishes phagocytic activity and provides vital

signals for Th2 cell commitment and the generation of fungal allergy. Therefore, early signals derived from DCs and macrophages, particularly IL-12, are important determining factors in the generation of Th1 polarized T cell responses. Consequently, the maintenance of protective immunity involves the continued production of IFN-γ and TNF by CD4+ and CD8+ T cells.

## 13.19 The DTH response and granuloma formation inhibit fungal dissemination

The establishment of Th1 cytokine production and the generation of fungal-specific CD4+ T helper cells and CD8+ cytotoxic T cells contribute to the inhibition of fungal replication and disease. Although cytokine production by T cells reinforces the effector functions of phagocytes, they also contribute to fungal killing and containment. For example, during *C. neoformans* infection of the lung macrophages and CD4+ T cells are responsible for containing yeast in the alveoli and preventing the pathogen from gaining access to the bloodstream (Figure 13.12). Macrophages form a cuff around the invading yeast and eventually fuse to form multi-nucleated giant cells. This

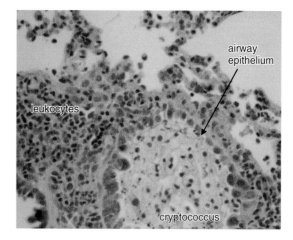

**Figure 13.12** Crytococcus restricted to the airways. An effective Th1 immune response involves the activation of DCs and macrophages and the production Th1 cytokines such as IL-12, TNF-α and IFN-γ. This activates CD4+ and CD8+ T cells which restrict pathogen growth and spread and enables the pathogen to be killed and cleared from the host. For example, *C. neoformans* causes serious meningitis if allowed to disseminate to the brain. An infiltration of immune cells around the site of infection effectively restricts it to the airways in the lungs. It is not able to penetrate host tissue and therefore can not cause damage.

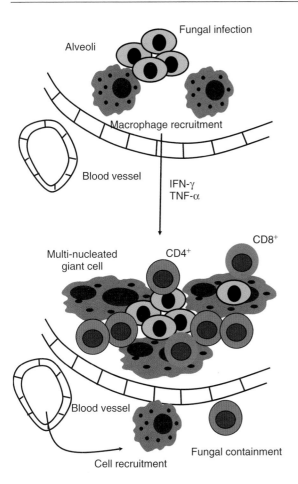

**Figure 13.13** Granuloma formation during fungal infection.

process is dependent on CD4+ T cells and forms the basis of granuloma formation (Figure 13.13). Cytokine release by T cells, particularly IFN-γ, helps to recruit inflammatory cells into the site of granuloma formation where the yeast in contained.

Another aspect of cell recruitment is the development of the DTH response. Cytokines produced by macrophages and CD4+ T cells, such as IFN-γ and TNF, attracts CD8+ CTLs to the area of the granuloma. CTLs are important for restricting many fungal infections including *C. neoformans* and *C. albicans*. Firstly they contribute to the production of protective cytokines and secondly they are able to kill infected macrophages and possibly fungal cells directly. The killing mechanism of CTLs involves the release and activity of perforin and granulysin, present in CTL granules. Perforin punctures holes in cells inducing cell lysis, while granulysin induces cell death pathways.

The mechanism is dependent on IL-15 and the presence of CD4+ T cells.

The production of IL-10 and the generation of Treg cells have paradoxical roles to play during fungal infection. On the one hand Treg cells, through the production of IL-10, are responsible for the downregulation of inflammatory responses to fungi and hence prevent excessive tissue damage. CD4+CD25+ Treg cells prevent immunopathology during pulmonary *P. carinii* infection and allow for *C. albicans* persistence and memory T cell formation. Conversely, IL-10 has a negative effect on IFN-γ production and can lead to chronic fungal disease (as exemplified during persistent candidal diseases). Although IL-10 and Treg cells are essential for immuno-regulation, they may also allow fungal replication and dissemination. Such pleiotropic effects depend on the state of the microenvironment, the cytokine milieu and the timing of IL-10 production or Treg cell generation.

## 13.20 The role of antibody in fungal resistance

Healthy individuals possess serum antibodies specific for fungal pathogens, although the precise role that antibodies play during pathogenesis and immunological protection are unclear. However, antibodies against *C. albicans* and *C. neoformans* are thought to be significant during protective immune responses to these pathogens. Antibodies have a number of effector functions during fungal infection. They can neutralize fungal cells by preventing them from attaching to host cell surfaces and inhibiting fungal invasion. They can neutralize fungal virulence factors and they are important opsonizing molecules for receptor-mediated phagocytosis (Figure 13.14).

The most significant contribution that antibodies make during fungal immune responses is their ability to neutralize invading fungi, although they also contribute to the opsonization of infected host cells and fungal cells. A key stage in the production of protective antibody is the development of Th1 cell-mediated immunity and the production of IFN-γ. Isotype switching to IgG2a antibody enhances neutralization and complement fixation and requires the active presence of IFN-γ. This is exemplified by GXM-specific antibody production during experimental *C. neoformans* infections in mice, whereby IgG2a (and to a lesser extent IgG2b) is protective but IgG3 is non-protective. While IgG2a is a poor opsonizer, IgG3 is highly effective at cross-linking FcRs and inducing phagocytosis-mediated Th2-cytokine production, which may account

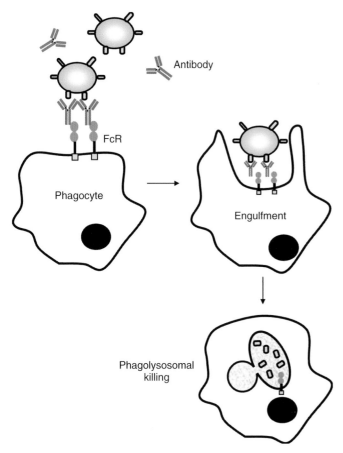

**Figure 13.14** Antibody neutralization and opsonization of fungi. Antibodies play an important role in protecting the host against certain fungal species. They are able to neutralize fungal toxins and prevent fungi from infecting cells. Antibodies bind to antigens on the fungal cell surface and prevent the fungus from attaching to cells and hence stop the infection process. Furthermore, antibodies are effective opsonizers that bind to the outside of fungal cells and tag the pathogen for receptor-mediated phagocytosis.

for IgG3 being non-protective (Figure 13.15). The IgA isotype, found predominantly at mucosal surfaces and sites of fungal entry, may also be important in neutralization of fungal cells and fungal-derived toxins and in initiating antibody-dependent cellular cytotoxicity (ADCC).

Not all antibodies directed to fungal components are protective, even though the isotype may be appropriate. It is now clear that the epitope specificity of the antibody is crucial in determining whether the antibody will be protective or non-protective. This has been demonstrated for antibodies directed to both the *C. neoformans* capsule component GXM and the *C. albicans* phosphomannoprotein complex. Furthermore, some antibodies actually enhance pathology by interfering with protective antibodies or by subverting the immune response to a non-protective Th2 phenotype. This complicated situation may have given rise to the difficulty in determining the significance of antibody during many fungal infections. Therefore, protective antibodies rely on being

of the appropriate isotype, directed against a suitable epitope and of sufficiently high titre as not to be masked by non-protective antibodies (Figure 13.15).

Some individuals develop allergic fungal disease and obstructive pulmonary disease due to repeated exposure to fungal spores. These atopic individuals possess high levels of IgE and a Th2 cytokine pattern dominated by IL-4 and IL-5. Such a situation is observed during allergic *A. fumigatus* infection, whereby tissue invasion by the pathogen leads to excessive eosinophilia, IgE and CD4+ Th2 cell infiltration. These factors can result in poor protective immunity and may lead to chronic lung disease and fungal dissemination.

## 13.21 Vaccination and immunotherapies

A Th1 response is critical for protection against a number of human fungal pathogens and efforts to develop an effective vaccine have focused on stimulating such a

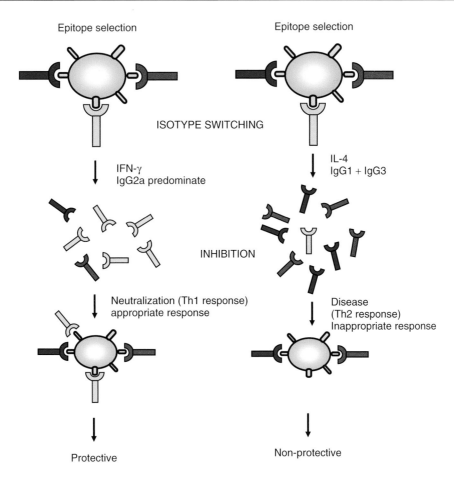

**Figure 13.15** Antibody selection during fungal infection. The generation of particular antibody isotypes may play a role in deciding whether an immune response is protective or non-protective. Th1-mediated immunity induces the production of the IgG2a antibody isotype, while Th2-mediated immunity leads to IgG1 and IgG3 antibody production. This antibody isotype selection is dependent on the Th1 cytokine IFN-γ. IgG2a is effective at neutralizing fungi and leads to protective immunity. On the other hand, IgG1 and IgG3 are effective at opsonization and activation of FcRs but these events lead to non-protective Th2 immunity. During a Th1 response the predominant IgG2a inhibits the activity of IgG1 and IgG3, while during a Th2 response IgG1 and IgG3 inhibit the activity of IgG2a.

response. However, no vaccines against fungal diseases are available at the current time. The problems faced by vaccinologists are compounded by the prevalence of fungal diseases in the immunocompromised who respond poorly to many vaccines. Why does this pose a serious challenge? Immunocompromised patients require an effective vaccine at a time when their immune system is dysfunctional. The timing of any such vaccination is therefore crucial, as individuals would benefit most from a vaccine prior to immunosuppressive drug treatment or when they are least immunosupressed.

Considering the problems faced in developing effective anti-fungal vaccines an alternative approach is to manipulate the immune system so as to improve the immune response in the immunocompromised. One way that this may be achieved is to enhance Th1 immunity or reduce Th2 immunity. For example, treatment with Th1-type cytokines, such as IFN-γ or IL-12, may promote anti-fungal responses. Alternatively inhibiting Th2-type cytokines may favour the development of a more protective Th1 immune response. A second approach is to specifically target receptors on DCs in order to stimulate the appropriate response. The use of TLR ligands such as CpG DNA could be used to enhance DC function and provide the necessary signals for the development of protective Th1 immunity. Such immunotherapeutic interventions are in the early stages of development but could provide potential treatments in the future.

## 13.22 Fungal immune evasion strategies

For a fungus to cause disease in a human host it must actively invade subepithelial tissue and disseminate throughout the body. Diseases that are caused by dermatophytes, such as ringworm, only invade the non-living cells of the outermost layers of skin, the disease state being caused by the immune reaction and not the fungus itself. Fungal pathogens that are able to cause disease in deep tissues possess multiple mechanisms that account for their virulent state.

Many pathological states induced by fungal invasion are associated with a morphological change. Although correlations have been made between fungal dimorphism, or a particular stage in the life cycle of a fungus, morphogenesis is not necessarily the critical determinant for pathogenesis. For example, yeast cells of *C. albicans* morphologically differentiate into hyphae, which are then able to penetrate host tissue by way of the pressure exerted at the hyphal tip. However, morphogenesis does not necessarily have to involve large changes in cellular structure. For example, *C. neoformans* remains as a yeast throughout the infection process, although it does increase the GXM-rich capsule coat (unencapsulated yeast are not pathogenic).

Despite the predominance of hyphal forms during invasion in some species of fungi such as *C. albicans*, most dimorphic fungi utilize the yeast form during invasion. Important pathogenic species including *H. capsulatum*, *B. dermatitidis*, *C. neoformans*, *Paracoccidiodes brasiliensis* and *S. schenckii* all form a yeast during the infection process. The transition from a multinucleated form to a yeast often involves a shift in gene expression that is associated with the production of virulence factors. These may include enzymes that favour penetration of host tissue or factors that are able to modify the immune system. Certain fungal products interact with the immune system to either divert the response away from a protective one or to manipulate it in such a way as to reduce its effectiveness. Such immuno-modulatory products form part of fungal immune evasion strategies.

## 13.23 Immuno-modulatory fungal products

Fungi can evade host defences at various stages of infection or at various times during an immune response. These include avoiding endocytosis, escaping phagolysosome killing, neutralizing the effects of damaging free radicals, hiding from an antibody response and manipulating receptor usage during phagocytosis. Fungal pathogens have a complex cell wall structure that is often surrounded by a polysaccharide coat. For example, the GXM-rich polysaccharide coat of *C. neoformans* possesses anti-phagocytic properties, which are mainly due its high negative charge that causes electrostatic repulsion between the cryptococcal cell and the host macrophage. The GXM capsule coat of *C. neoformans* is an important virulence factor. As well as participating in the avoidance of phagocytosis it is also known to downregulate co-stimulatory receptors on APCs, reduce inflammatory cell recruitment, interfere with the DTH response and decrease the effectiveness of the antibody response. GXM also forms the basis of a *C. neoformans* classification system based on its structure and in particular on the number of xylose residues.

Other *C. neoformans* factors that act to modulate the immune response include melanin and mannitol. Melanin is thought to be an effective free radical scavenger, hence reducing the effectiveness of the respiratory burst generated by macrophages and neutrophils. Antifungal molecules such as defensins are also neutralized by melanin. In addition, melanin may contribute toward the negative charge of *C. neoformans* rendering endocytosis less effective. Mannitol is produced during infection and may also act as a free radical scavenger, in addition to being associated with central nervous system pathology following *C. neoformans* infection. Likewise, *A. fumigatuts* produces a range of superoxide dimutases and catalases, which are responsible for protection against free radicals. *H. capsulatum* is also highly resistant to the effects of the respiratory bust and is able to replicate within macrophages, although the factors responsible for such resistance remain unknown. In addition, *H. capsulatum* is able to avoid effective opsonization and so reduce receptor-mediated phagocytosis and activation of the respiratory burst.

## 13.24 Evasion of phagolysosomal killing

The evasion of phagocyte killing and survival/replication inside macrophages are important virulence traits of both *H. capsulatum* and *C. neoformans*. *H. capsulatum* seems to be resistant to the effects of the respiratory bust but does not actively disregulate it. Instead this intracellular yeast alters the pH of the phagolysosome. Maintenance of a pH between 6.5 and 7.0 inactivates lysozymal enzymes such as hydrolases, thus decreasing the capacity of macrophage killing. *C. neoformans* is another fungus that possesses mechanisms to avoid phagolysosomal degradation within macrophages. Both the capsular GXM polysaccharide and melanin play important roles in survival and intracellular replication. Once yeast is internalized, melanin protects *C. neoformans* from free radical damage, enzymes and antifungal molecules, while the polysaccharide disregulates lysosome fusion and controls the phagolysosome contents. The yeast is able to replicate and eventually the macrophage is lysed releasing *C. neoformans* into the extracellular space.

## 13.25 Modifying the cytokine response

It is important to note that a protective immune response requires Th1 polarization and production of IL-12, IFN-γ and TNF. Modulation of the immune response toward a Th2 phenotype will render the host susceptible to clinical disease. The cytokine balance is often affected by the receptor usage on DCs or macrophages. For example, the ligation of complement receptor 3 (CR3), without the associated ligation of FcRs by antibody leads to an attenuation of phagocyte activation and a reduction in IL-12 production. This may ultimately enable intracellular survival. *H. capsulatum* uses CR3 to gain entry into macrophages and the lack of antibody-mediated opsonization may be an important factor in fungal pathogenesis.

## 13.26 Summary

1. Most diseases caused by fungal species are the result of an infection with an opportunistic pathogen and only infect immunocompromised individuals
2. Fungal morphogenesis plays an important role in pathogenesis.
3. Pathogen associated molecular patterns (PAMPs) are recognized by pattern recognition receptors (PRRs) on host cells.
4. Opsonization of fungi by complement proteins and antibodies initiates both innate and adaptive immune responses.
5. Delayed-type hypersensitivity reactions control fungal dissemination and prevent further infection.
6. Fungi have evolved many immune evasion strategies in order to subvert host defence mechanisms.

# 14 Immunity to Parasites

## 14.1 Introduction

In the context of this chapter the term parasite refers to a pathogenic eukaryotic organism. These include microscopic single-celled eukaryotes from the kingdom protozoa (protista) and the macroscopic, multicellular helminth and nematode worms. Such organisms are classified as parasites if they cause disease or harm to a host organism. The tissues most frequently affected are internal organs such as the intestine and lungs, or cells within the blood stream, and the affecting parasites are therefore known as endoparasites. Parasites infect their host following injection into the bloodstream, penetration into the skin or directly into a mucosal tissue. Therefore, body surface and mucosal immune responses are vital for host protection. Parasite infections exhaust a great many human and veterinary recourses worldwide, particularly in tropical and sub-tropical regions. The economic loss and medical encumbrance due to parasitic burdens is immense and therefore represents an important aspect of immunology and infectious disease research. Intimate host–parasite relationships have evolved over millions of years and have created unique mechanisms of host immune defence but also numerous parasite immune evasion strategies. This chapter identifies the medically important parasites and relates their structure and life cycles to the various immune strategies employed by the host to combat parasitic infection. Furthermore, the mechanisms that these complex parasitic organisms employ to evade host immune defence will be examined.

## 14.2 Protozoa are diverse unicellular eukaryotes

The protozoa comprise a diverse kingdom of single-celled eukaryotic organisms with over sixty thousand species described. Protozoa used to be classified within the kingdom animalia but notable biochemical, structural and physiological differences led to the protozoa being placed within a broad taxon known as the kingdom protista (unicellular eukaryotes that are difficult to place in the animal, plant or fungi kingdoms). Most protozoa are animal-like, microscopic organisms that range in size ($10-200\,\mu$m), shape and their method of motility. The 10,000 or so parasitic species of protozoa can be further classified into different groups depending on their genetic relationship into more defined phyla (Table 14.1). However, much effort has been made to classify parasitic protozoa based on morphological, biochemical and behavioural properties. Therefore, four groups of protozoa exist and include the flagellates, amoebae, sporozoa and the ciliates. The flagellates, so called because the primary method of locomotion is by flagella, include the parasites that cause leishmaniasis, giardiasis and trypanosomiasis. The amoebae, which move via extendible pseudopodia, include parasites that cause amoebic dysentery. The sporozoa, which have no single means of locomotion, include the parasites that cause malaria and coccidiosis and are characterized by an apical complex (they are also referred to as apicomplexans). The ciliates, which are distinguished by numerous cilia on the outer cell surface, do not frequently cause human disease and the only species that infects humans is *Balantidium coli*, which infects the intestines.

## 14.3 Structure of the protozoan cell

The diversity within the protozoa prevents a detailed description of the cellular structure of every form of protozoa, although many generalizations can be made. As animal-like eukaryotes, the protozoa possess a plasma membrane, a nucleus (often with a nucleolus), endoplasmic reticulum with ribosomes, sometimes a Golgi body

*Immunology: Mucosal and Body Surface Defences*, First Edition. Andrew E. Williams.
© 2012 John Wiley & Sons, Ltd. Published 2012 by John Wiley & Sons, Ltd.

**Table 14.1** Classification of parasitic protozoa. Much effort has been made to classify parasitic protozoa based on morphological, biochemical and behavioural properties, as well as genetic relationship.

| Phylum | Characteristics | Species |
| --- | --- | --- |
| Apicomplexa | Apical organelle complex<br>Obligate intracellular parasites<br>Complex life cycles | *Plasmodium falciparum*<br>*Eimeria tenella*<br>*Toxoplasma gondii* |
| Sarcomastigophora | Contains the amoebas and flagellates<br>Amoebas – pseudopodia<br>Flagellates – eukaryotic flagellum | Amoebas – *Entamoeba histolytica*<br>Flagellates –<br>*Trypanosoma brucei Giardia duodenalis* |
| Ciliophora | Also called ciliates<br>Locomotion by means of cilia | *Balantidium coli* |
| Microspora | Unsegmented filamentous algae<br>Produce infectious spores | *Enterocytozoon cuniculi*<br>*Nosema corneum* |
| Ascetospora | Spore forming | *Haplosporidium Iusitanicum* |
| Myxozoa | Spore forming<br>Infectious amoebula stage | *Tetracapsuloides bryosalmonae*<br>*Myxobolus cerebralis* |

or mitochondria, vacuoles and they often have specialized structures particular to certain groups or species. For purposes of clarity, we shall look at the ultrastructure of the flagellates, which include the trypanosomes, and those species that cause leishmaniasis and giardiasis. Although most protozoa reside in cells within the bloodstream and therefore do not directly infect mucosal sites, they represent an important group of parasites responsible for causing devastating diseases worldwide. Most of these parasites are also transmitted via the bite of an insect vector and often induce an immune reaction within the skin. Therefore, due to their significant effect on human health they will be discussed in this chapter. Leishmania species, which directly infect the skin, and giardia, which infects the intestines, shall also be discussed in this chapter. Furthermore, the apicomplexan malaria parasites are also responsible for millions of deaths in tropical and subtropic regions and, due to their impact on human health, will also be discussed.

Trypanosomes are flagellate parasites that are responsible for many human and veterinary diseases such as sleeping sickness and Chagas's disease (human) and nagana and surra (animals). Trypanosomes are typically small and elongate being 15–30 µm in length and 2–3 µm in width (Figure 14.1). They are characterized by a cytoplasmic structure known as the kinetoplast, which contains its own nucleus and kinetoplast DNA.

Trypanosomes are characterized by the flagella apparatus, which consists of a basal body where the structure originates and a flagellum that is continuous with an undulating membrane running along the length of the body. Trypanosomes also possess a nucleus, mitochondria, Golgi body, endoplasmic reticulum and various vacuoles.

The plasmodium parasite that is the causative agent of malaria has a number of life cycle stages characterized by slightly different body structures. The basic body structure is similar therefore we shall focus on the stage that invades human red blood cells (erythrocytes) called a merozoite. The shape of a merozoite is less elongated than the sporozoite stages in the mosquito, and much less so than the trypanosome body shape (Figure 14.2). It is approximately 1.5 µm long and 1 µm in diameter and possesses an outer and two inner membranes. The outer membrane retains a surface coat that plays a role in immune evasion. At one end of the merozoite is the apical complex, which consists of three polar rings from which microtubules are formed. Plasmodium parasites also have a cytosome, which is involved in the uptake of material from the external environment including host cell cytoplasm. In addition, plasmodium parasites have an endoplasmic reticulum, various vacuoles and mitochondria that are closely associated with a spherical body thought to store energy.

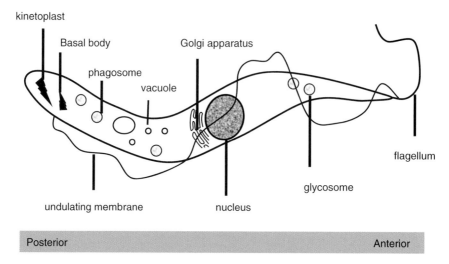

**Figure 14.1** Structure of the trypanosome cell. Trypanosomes are unicellular protozoan parasites with a cellular structure similar to other eukaryotes.

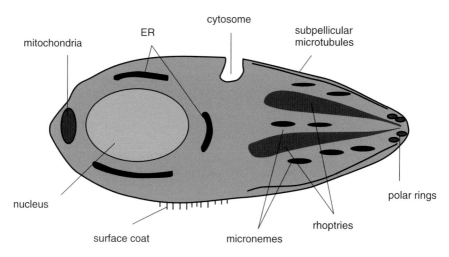

**Figure 14.2** Structure of the plasmodium merozoite. The merozoite is the form taken by Plasmodium species that is responsible for infected host red blood cells. It possesses a number of specialized structures that enable it to effectively infect cells.

## 14.4 Life cycle of protozoan parasites

The complexity of protozoan parasites is reflected in their varied and multifaceted life cycles. Most protozoan parasites have a separate set of life cycle stages in a different host species other than humans. The secondary host species is usually an insect that acts as a vector for the transmission of disease. The parasite is transferred from one human host to another by the vector. For example, a mosquito from the Anopheles family transfers the malaria parasite from human to human. Malaria is transmitted from one infected human individual to an uninfected human as a result of blood meals taken by the mosquito. The spread of disease is also exaggerated by the existence of reservoir species. These are often mammalian species, such as horses and cattle, which harbour the parasite and augment the spread of disease to the human population: they literally act as a reservoir for the parasite. A disease caused by a pathogen that replicates in more than one species is called a zoonosis. It is therefore important to appreciate the life cycle of protozoan parasites in order to fully understand the type of immunity generated against them.

## 14.5 The life cycle of *Trypanosoma brucei*

*Trypanosoma brucei* infection of humans causes African sleeping sickness and is transmitted from host to host by the tsetse fly Glossina. *T. brucie* undergoes a number of morphological changes in the tsetse fly and its human host. When the tsetse fly bites, it transfers metacyclic slender forms of the parasite into the host. At first trypanosomes are located in the dermal connective tissue where they induce an inflammatory reaction around the bite site. From there they migrate to the lymphatics and then into the blood where they undergo extensive rounds of multiplication by a process of binary fission. A large parasite load in the blood initiates a transformation from a slender form into an intermediate and then a stumpy form. Trypanosomes then invade the choroid plexus, cross the blood–brain barrier and enter the brain and cerebrospinal fluid, causing the characteristic neuropathy that characterizes sleeping sickness. Another bite from a tsetse fly transfers the metabolically active stumpy forms from the human host back to the tsetse fly midgut where the parasites undergo further rounds of division. In the tsetse fly gut, a 'slender' procyclic form is adopted, and they enter the proventriculus and eventually the salivary glands. In the salivary glands they adopt an epimastigote form and divide again, eventually transforming into the human infective metacyclic form, so that subsequent transmission to a human host can proceed.

## 14.6 Life cycle of *Leishmania species*

*Leishmania* species cause serious disease in humans, although the life cycle is much simpler than that of trypanosomes. Many species exist across Africa, Asia and South/Central America causing one of three forms of disease known as visceral, cutaneous or mucocutaneous leishmaniasis. The most serious disease is known as kala-azar, which is caused by *Lieshmania donovani* resulting in deep-tissue (visceral) pathologies of organs such as the liver. The less serious cutaneous disease of the skin is caused by a multitude of species such as *L. tropicana, L. major, L. mexicana* or *L. amazonensis*. The morphology

of all the *Leishmania* species is very similar, making identification of different species difficult. *Leishmania* species have only two morphotypes, the amastigote form that infects human macrophages and the promastigote form found in the insect vector sandfly (*Phlebotomus* species in Africa/Asia and *Lutzomyia* species in the Americas). The intimate relationship between the parasite and the host macrophage makes Leishmaniasis an intriguing pathogen to study from an immunology perspective.

The life cycle of *Leishmania* species (Figure 14.3) begins with a bite from a female sandfly transmitting the promastigote form into the dermal tissue of the skin. Macrophages readily phagocytose the parasites where transformation into the amastigotes form occurs. The amastigotes are highly resistant to the lysosomal and oxidative killing mechanisms of macrophages. During cutaneous leishmaniasis the amastigotes remain in the skin, and during visceral leishmaniasis the parasites are carried by macrophages to other organs of the body. Mucocutaneous leishmaniasis only occurs after chromic infection, maybe 30 years or more, whereby the parasite migrates to the oropharyngeal mucosa and results in disfiguring disease. Replication within macrophages (or other cells of the reticulo-endothelial system) results in the characteristic lesions and the release of some amastigotes into the blood stream. Another blood meal from a sandfly transmits the parasite back to the vector where it transforms into the promastigote form in the gut, divides and migrates to the pharynx eventually invading the proboscis of the insect. A new round of human transmission can begin.

## 14.7 The life cycle of *Plasmodium falciparum*

Malaria is endemic in more than 100 countries across the globe, meaning that nearly half of the world population could be exposed. The World Health Organization estimates that 300–500 million people are infected with malaria each year, making this one of the worlds leading causes of disease. There are four main species that infect humans, *Plasmodium falciparum, P. vivax, P. ovale* and *P. malariae*. By far the most severe and common form of malaria is malignant tertian malaria that causes cerebral damage. This is the result of infection with *P. falciparum*. However, all species of malaria parasite that infect humans have almost identical life cycles, with the most characteristic phase being the invasion and lysis of human red blood cells. The high fevers associated

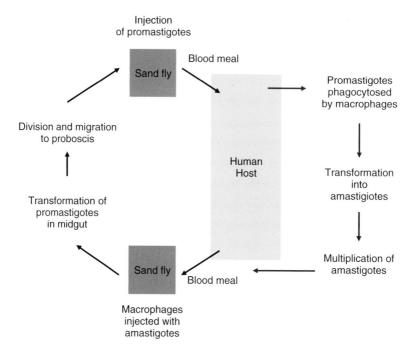

**Figure 14.3** Life cycle of *Leishmania species*. The sand fly injects the leishmania parasite into the skin of a human host where it is taken up by a macrophage. *Leishmania* amastigotes reside within macrophages where they replicate and are taken to other sites in the body.

with malaria are the result of synchronous and periodic rupture of infected red blood cells.

The *Anopheles* mosquito transmits all plasmodium species that infect humans. The life cycle begins when sporozoites are introduced directly into the bloodstream when the mosquito bites (Figure 14.4). They migrate to the liver in only 30 minutes where they enter a round of multiplication called pre-erythrocytic schizogony. Following division merozoites exit the liver and enter the blood stream where they invade erythrocytes (red blood cells). Once inside the erythrocyte the parasite undergoes erythrocytic schizogeny and transforms into a ring-like form called a trophozoite. When schizogeny is complete the erythrocyte ruptures and many merozoites are released to begin the erythrocytic schyzogeny cycle again. The rupture of all infected erythrocytes takes place within 7 hours, a remarkable feat of timing, which results in massive parasite load in the blood (parasitaemia). During *P. falciparum* infection the trophozoites become much more adherent and can block capillaries in the brain, thus causing the cerebral form of malaria. However, some merozoites do not enter the erythrocytic cycle but differentiate into gametocytes to begin a stage of sexual reproduction.

Two types of gamete are formed, the male microgamete and the female marcogemate, which arise from separate merozoites. The mosquito takes these up during a blood meal and fertilization of the macrogamete takes place in the insect gut. A zygote develops into a motile form called an ookinete, which invades the gut wall of the mosquito and develops into an oocyst. Cell division takes place in the oocyst and after approximately 15 days many sporozoites are released. The sporozoites then migrate to the salivary gland where they are injected into a human host to begin the life cycle again. The complexity of the malaria life cycle reflects an evolutionary game of cat and mouse played out between the parasite and its host, whereby the host has evolved particular immune components to combat malaria while the parasite has evolved many immune evasion mechanisms.

## 14.8 Helminths are multicellular, macroscopic parasites

Helminth is a rather inexact term used to describe the parasitic members of the platyhelminth and nematoda phyla.

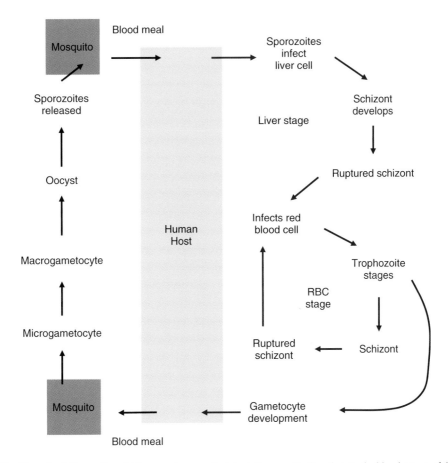

**Figure 14.4** The life cycle of *Plasmodium falciparum*. A mosquito injects the parasite directly into the bloodstream of the human host. It first migrates to the liver and then back into the bloodstream where it infects red blood cells. The rupture of red blood cells by dividing parasites causes the relapsing fevers that are associated with malaria.

In fact, only two classes within the platyhelminth family contain parasitic species, the cestodes and the digenea (trematodes). Helminths are able to infect a range of mammalian species causing a diverse array of debilitating and even life-threatening diseases. Helminths comprise many species from several taxonomic groups so they vary in size, shape, and structure, have differing reproductive requirements, utilize varying life cycles and cause a number of different diseases. Due to space limitations a restricted account of helminth diversity can be given and attention will be paid to *Schistosoma mansoni*, which is a digenean helminth that causes bilharzia and *Ascaris lumbricoides*, which is a nematode parasite that infects the gut. Comparisons with other helminth parasites will be

given and a list of diseases caused by helminths can be found in Table 14.2.

## 14.9 Structure of the trematode *Schistosoma mansoni*

*S. mansoni* is a digenean platyhelminth, otherwise known as a trematode or fluke. It is the causative agent of schistosomiasis or bilharzia that affects up to 200 million people worldwide, making this disease second in importance only to malaria. These flukes have several morphological types, each associated with different stages in its life cycle. The typical digenean body structure comprises a

**Table 14.2** Diseases caused by helminth infections. Helminths comprise a large group of invertebrate parasites and include cestodes (tapeworms), trematodes (flukes) and nematodes (roundworms).

|  | Species | Disease | Tissue |
|---|---|---|---|
| Cestodes | *Taenia saginata* | Beef tapeworm | Intestine |
|  | *Taenia solium* | Pork tapeworm | Intestine |
|  | *Echinoccocus granulosus* | Hydatid disease | Liver and lungs |
|  | *Diphyllobothrium latum* | Diphyllobothriasis | Intestine |
|  | *Multiceps multiceps* | Cestodiasis | Brain |
| Tematodes (Digenea) | *Fasciola hepatica* | Liver fluke | Intestine and liver |
|  | *Schistosoma mansoni* | Schistosomiasis | Mesenteric veins and liver |
|  | *Schistosoma haematobium* | Schistosomiasis | Vesical veins and bladder |
|  | *Schistosoma japonicum* | Schistosomiasis | Mesenteric veins and gut |
|  | *Schistosoma intercalatum* | Schistosomiasis | Mesenteric veins and gut |
|  | *Heterophyes heterophyes* | Fluke | Intetsine |
| Nematodes | *Ascaris lumbricoides* | Ascariasis | Intestine |
|  | *Strongyloides stercoralis* | Hookworm-like | Intestine |
|  | *Necator americanus* | Hookworm | Intestine |
|  | *Trichinella spiralis* | Trichinellosis | Intestine |
|  | *Toxocara spp.* | Roundworm | Intestine |
|  | *Wuchereria bancrofti* | Elephantiasis or Bancroftian filariasis | Lymph and Blood |
|  | *Onchocerca volvulus* | Onchocerciasis or River blindness | Skin and eye |
|  | Loa loa | Eye worm | Eye |

distinct anterior to posterior axis, with a mouth at the anterior end and an excretory opening at the posterior. Digeneans possess an oral sucker, a ventral sucker, a pharynx, an oesophagus, a well-developed closed intestine and numerous internal organs associated with sexual purposes. Schistosome species are rather different to the majority of digeneans in that they possess very elongated bodies as an adaptation to living in the blood vessels of their hosts.

The outer body covering of *S. mansoni* changes during the transformation from a larval form (cercaria) to an adult form (schistosomulum) when the parasite penetrates a human host. The outer layer is called a tegument and consists of syncytial epithelium that is bounded apically and basally by a trilaminate plasma membrane and covered apically by a glycocalyx. On transformation from a cercaria to a schistosomulum, a number of important changes in the glycocalyx take place, most notably the glycocalyx is lost and replaced by a heptalaminate (seven layered) membrane. This helps to protect the parasite within the host and is also involved in nutrient uptake.

## 14.10 Life cycle of *Schistosoma mansoni*

The vast majority of helminth trematodes infect a member of the mollusca phylum as an intermediate host. In the case of *S. mansoni*, eggs are released in the faeces of infected humans into an open water source. In the water the eggs hatch and release miracidia, which then penetrate and infect a species of water snail from the genus Biomphalaria (Figure 14.5). This snail can be found across Africa, Asia, South America and parts of the Arabian Peninsula. Within the intermediate snail host, the miracidia undergo morphological changes to form a sporocyst, which eventually develop into cercariae. The reproductive efficiency of a sporocyst is so great that a

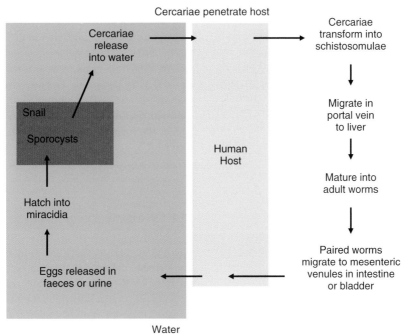

Water

Cercariae penetrate host

**Figure 14.5** Life cycle of *Schistosoma mansoni*. Cercariae penetrate the skin of a human host and migrate into the bloodstream, undergoing great morphological change as they do. Paired adult worms mate within venules and produce eggs that pass through the intestine wall and hatch out in the environment. Miracidia infect water snails where numerous multiplication events produces many cercariae.

single structure can give rise to a million cercariae. These cercariae are responsible for penetrating the human host and resemble an embryonic version of the adult form (fluke). The transmission of the disease therefore occurs when humans enter schistosome-infected waters.

Cercariae produce a number of lytic enzymes that assist the parasite gain entry through human skin and into the blood system. During penetration, the cercariae lose their tails and become schistosomulae. These then migrate through the circulatory system to the portal veins surrounding the liver. It is hear that sexual reproduction takes place when a male and female fluke come together. The male and female schistosomes remain in a paired state of permanent copula and migrate to the mesenteric veins of the large intestine where they lay eggs. When multiple eggs are laid, the delicate structure of the venules or capillaries rupture and the eggs are released into the sub-mucosal tissue and eventually into the intestinal lumen where they can be expelled with the faeces. The eggs can then hatch into miracidium in the water and continue the life cycle by infecting another snail host.

Adult flukes remain within the blood vessels of the gut for many years, causing chronic intestinal schistosomiasis. Other species of schistosome have very similar life cycles although they may differ in the location within the human circulatory system. For example, *S. haemotobium* causes bladder schistosomiasis as it inhabits the veins of the bladder wall and the eggs are passed out with the urine.

## 14.11 Structure of the nematode *Ascaris lumbricoides*

*A. lumbricoides* is one of the most prevalent parasites that infects man (up to 1 billion people may be infected). It occurs throughout the globe and in particular in areas where human faecal matter is present around dwellings. It is also the largest parasitic nematode (can reach a length of 40 cm) and causes the disease ascariais, which is associated with large numbers of parasites in the intestine. A high parasite load in the intestine can cause blockage and in more severe cases worms can migrate to the bile duct, pancreatic duct, oesophagus and even the mouth. In previously infected individuals, ascaris worms can cause allergic bronchitis and urticaria.

All nematodes possess the same basic structural body organization. They are also known as round worms and as such have an unsegmented, elongated, cylindrical body (Figure 14.6). They have an anterior to posterior axis with an intestine running the length of the body. Nematodes retain their shape through hydrostatic pressure exerted by the protein-rich, fluid pseudocoel and actively pump food through the intestine by the action of the pharynx. The contents of the intestine pass out of a rectum located in a posterior position. The development of nematodes entails four larval stages (L1–L4), involving size increases and moulting.

## 14.12  The life cycle of *A. lumbricoides*

The life cycle of *A. lumbricoides* (Figure 14.7) begins when viable eggs are ingested (note that there is only one definitive host). The eggs are extremely hardly and can persist in cold, dry conditions for years. When environmental conditions are more favourable the egg develops into a first stage larva (L1) but does not hatch. The hatching process usually begins in the duodenum of the host and the released larvae enter the second stage (L2). These larvae then penetrate the mucosal wall of the intestine and migrate via the blood to the liver where they develop

into third stage larvae (L3). From the liver the larvae then migrate to the heart and then the lungs. This migration process takes approximately 5–6 days after which the larvae rupture the small capillaries in the lungs and migrate up the trachea. The larvae are then swallowed and enter the intestine once again, now appearing as fourth stage larvae (L4) by about nine days post infection. Full sexual maturity is not reached until after the final L4 moult. Reproduction results in the formation of an egg that is laid in the intestine of the host and expelled with the faeces.

## 14.13  Immune responses to parasites

The type of parasite species and its location within the body largely determines the type of immune response mounted by the host. For example, microscopic protozoan parasites that reside within the blood circulation will activate a different immune response to a large nematode worm living within the gut. Similarly, intracellular parasites will elicit distinctive immune responses compared to extracellular parasites. Generally each immune response will utilize one or more components of both the innate and adaptive immune systems. Usually immunological protection is defined as the complete expulsion from the host.

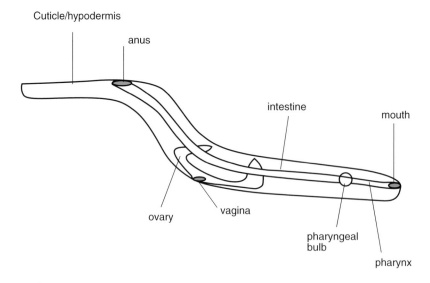

**Figure 14.6**  Structure of *Ascaris lumbricoides*. Most nematodes have a similar structure to ascarids.

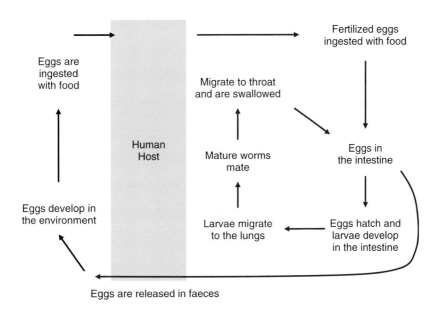

**Figure 14.7** Life cycle of *Ascaris lumbricoides*. The life cycle of this nematode does not rely on a vector or intermediate host. Unhatched eggs are ingested with environmental products. They hatch in the intestine, migrate to the lungs and are ingested again as egg-producing adults in the intestine.

## 14.14 Innate immunity to trypanosomes

Trypanosomes directly enter the bloodstream following a bite from a tsetse fly. The host is exposed to many molecules derived from the invading parasite. Of particular importance are the variable surface glycoprotein (VSG) molecules expressed by the parasite, which coat the entire organism. These molecules are important in both host immunity and in immune evasion by the parasite (refer to section on immune evasion strategies). Macrophage activation by parasite components is a characteristic of the immune response to African trypanosomes. For example, the GPI-anchor domain of VSG, which tethers these proteins to the surface of the parasite, may be responsible for macrophage activation. This results in the production of IL-12 and TNF and initiates a Th1-dominated T cell response. TNF production has also been associated with cachexia and fever characterized by African sleeping sickness. Macrophages also produce reactive nitrogen species and are able to control the first round of parasitaemia before adaptive immunity can be established.

*T. brucei* is essentially an extracellular parasite that lives and feeds within the bloodstream. Humans are resistant to bovine *T. brucie* that infect cattle. One of the major factors responsible for this resistance is trypanosome lysis factor (TLF), which is highly cytotoxic to the parasite. However,

*T. brucei gambiense* and *T. brucei rhodesiense* are resistant to TLF cytotoxicity as they prevent endocytosis of this factor and thereby inhibit its function. The action of TLF can be used to distinguish trypanosomes that are potentially infectious to humans from those that are not.

## 14.15 Adaptive immunity to trypanosomes

The adaptive immune response to trypanosomes involves both a T cell-mediated responses and antibody producing B cells and is intimately related to the various immune evasion strategies employed by African trypanosomes (discussed in section 14.28). A general immunosuppressive state is induced by the parasite that affects effector T cell and B cell functions and is mediated by IL-10. Expression of IL-10 downregulates the pro-inflammatory immune response even though continuous exposure to the parasite induces TNF production. The lack of an inflammatory response due to IL-10 together with constant TNF production may be responsible for the meningoencephalatic sleeping sickness observed in the late stages of the disease. In addition, *T. brucei* employs a sophisticated method of varying its surface coat so as to evade specific immunity. Therefore, a protective immune response can only be achieved by a decreased expression of the immunosuppressive cytokine IL-10 and enhancement of IFN-γ and TNF. Furthermore, both CD8+ CTLs and antibodies that recognize non-variable components of parasite-derived molecules can circumvent problems of immune evasion. However, without anti-trypanosome drug treatment the disease is thought to be fatal in most cases.

## 14.16 Innate immunity to plasmodium

The development of a protective immune response and resolution of disease is associated with the control of parasite replication in the first two weeks after infection. Innate immune mechanisms can restrict the parasite load in the blood, reduce erythrocyte infection and therefore prevent death. However, total parasite clearance requires help from the adaptive immune system. Furthermore, a successful immune response relies on the production of type-1 cytokines such as IL-12, IFN-γ and TNF that drive CD4+ Th1 cells, cell-mediated immunity and long-term antibody responses (Figure 14.8).

DCs play an important role in the initiation of immunity to malarial parasites. It has recently been discovered that TLR9 recognizes a by-product of plasmodium haemoglobin digestion called haemozoin. TLR9 is usually associated with the recognition of unmethylated DNA

from bacteria. Following ligation of TLR9 by haemozoin DCs become activated, produce IL-12 and upregulate a number of co-stimulatory molecules such as CD40 and CD86. IL-12 production is essential for the development of Th1 responses to the early rounds of parasitaemia. In addition, the increased expression of the chemokine receptor CCR7 allows DCs to migrate to lymph nodes where they interact with naïve T cells.

Macrophages have the ability to phagocytose infected erythrocytes, a process which may be independent of antibody and FcR-mediated phagocytosis. Rather, CD36 expressed on macrophages may bind to a protein expressed by the parasite such as PfEMP1 (*P. falciparum*-encoded erythrocyte membrane protein 1). Phagocytosis has the effect of reducing parasite load in the blood and controlling parasitaemia before the adaptive immune system is activated. The production of reactive oxygen and nitrogen species are also involved in controlling parasitaemia during both the liver and blood stages of infection. In a similar way NK cells are able to control the early rounds of parasitaemia mainly through the production of IFN-γ. High IFN-γ levels then further activate macrophages and amplify the adaptive immune response. In addition, NKT cells and γδ T cells augment this early response by producing more IFN-γ, which in turn stimulates antigen-specific CD4+ and CD8+ T cells.

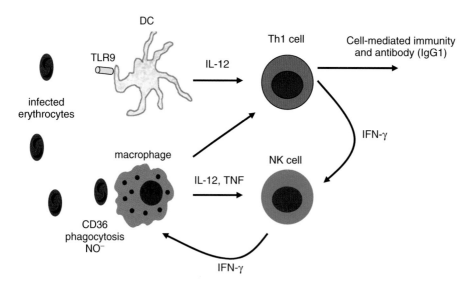

**Figure 14.8** Innate immunity to plasmodium. Infected erythrocytes are phagocytosed by macrophages and DCs. PRRs and CD36 ligation induce the production of IL-12 and initiates the development of Th1 immunity. The Th1 phenotype is reinforced through IFN-γ from NK cells and Th1 cells.

## 14.17 Adaptive immunity to plasmodium

A protective adaptive immune response involves the development of a Th1 phenotype, whereby IL-12, IFN-γ and TNF, produced by APCs and NK cells, activate CD4+ Th1 T cells. These CD4+ T cells recognize parasite-derived peptides in conjunction with MHC class II molecules expressed on APCs. Consequently plasmodium-specific CD8+ CTLs recognize antigen presented by MHC class I expressed on infected hepatocytes and parasite-specific antibodies are produced by activated B cells. These adaptive mechanisms are essential for both pre-erythrocytic (i.e. liver stage) and erythrocytic (i.e. blood stage) immunity to malarial infection (Figure 14.9).

Protective immunity against pre-erythrocytic stages of malaria is principally mediated by CD8+ CTLs that kill infected hepatocytes and is dependent on CD4+ T cell help. Hepatocytes present parasite-derived peptides on their surface in conjunction with MHC class I molecules. This process is also dependent on IFN-γ production; the main source of production being NK cells and CD8+ T cells. IFN-γ consequently induces the up-regulation of iNOS and enhances NO⁻ production further restricting the parasite. Antibodies may be less important for protecting the host against the pre-erythrocytic stages of malaria. This is due to the short time (as little as 30 minutes) that the parasite takes to migrate to the liver, providing antibodies only modest opportunity to neutralize the parasite.

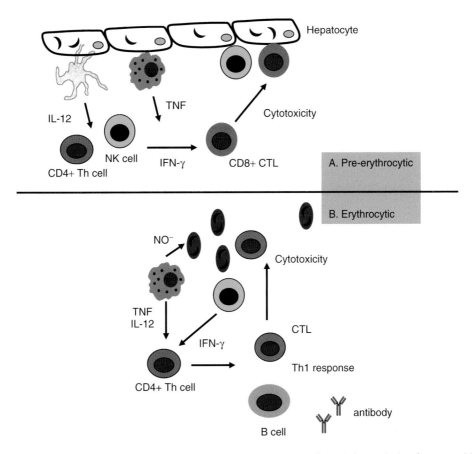

**Figure 14.9** Pre-erythrocytic and erythrocytic immunity to plasmodium. A. Pre-erythrocytic immunity involves recognition of infected liver cells by DCs and macrophages and the resultant production of IL-12 and TNF. This activates NK cells, CD4+ T cells and in turn CD8+ CTLs, which ultimately kill infected hepatocytes. B. Erythrocytic immunity relies on the recognition and phagocytosis of infected red blood cells and the production of IL-12 and TNF. This activates CD4+ and CD8+ T cells and ultimately enhances antibody production by B cells.

Immunity to erythrocytic stage malaria involves both cell-mediated and antibody-dependent mechanisms. CD4+ T cells elicit the help of both CD8+ CTLs and provide help to B cells in order to produce antibodies, mainly through the release of pro-inflammatory cytokines such as IFN-γ and TNF. In addition, CD4+ T cells may provide protection through effector mechanisms that are independent of CD8+ T cells and antibody. The generation of Th1 cells in the early stages of infection lead to a cell-mediated response, CTL activity and activation of macrophages. This is dependent on IFN-γ production and controls primary parasitaemias. However, during persistent or chronic infection IL-4 and IL-10 are upregulated and provide more of a Th2 environment. A Th2-dominated response is necessary for elimination of the parasite, although sterile-protective immunity usually does not develop for 10–15 years of constant exposure.

Antibody plays a vital role in the generation of protective immunity and is essential for the complete clearance of the malarial parasite. Th1-driven antibody production of human IgG1 and IgG3 antibody subclasses are important for opsonization of the parasite and subsequent macrophage and cell mediated effector functions. Antibody-dependent elimination of parasites or infected cells involves macrophage phagocytosis and production of reactive nitrogen and oxygen species through the binding of bound antibody to FcRs. Similarly the development of antibody-dependent cellular cytotoxicity involves cross-linking of FcRs on the surface of NK cells and CD8+ T cells. During chronic malaria IL-4 and IL-10 upregulation results in IgG2 production and is associated with resistance to the parasite at the end of the infective season.

## 14.18 Immunity to *Leishmania* – Th1 versus Th2

*Leishmania* species are responsible for a variety of diseases, most notably visceral leishmaniasis or kala azar (meaning black fever) caused by *L. donovani* and cutaneous leishmaniasis caused by numerous species (e.g. *L. major*). Major advances in the understanding of the immunity to *Leishmania* species have been achieved using mouse models of the disease. Two genetically distinct strains of mouse are routinely used; the largely resistant C57BL/6 mouse and the susceptible BALB/c mouse. The main difference

in disease outcome is associated with the type of immune response that these mice develop to *Leishmania*. Resistant C57BL/7 mice develop a Th1-mediated response, which is dependent on IFN-γ production, while BALB/c mice develop a Th2-mediated response and are therefore susceptible to disease. A Th1-type response is essential for the prevention of parasite replication and restriction of *Leishmania*-induced lesions.

The first cells to enter the site of infection are neutrophils, which start to phagocytose *Leishmania* parasites and secrete chemokines such as IL-8. This stimulates the migration of monocytes or macrophages into the site of infection. However, macrophage migration to the site of infection is an essential part of the parasitism process as *Leishmania* promastigotes effectively bind to surface molecules on macrophages and elicit internalization. Once inside the macrophage they transform into amastigote forms, undergo replication by binary fission and eventually rupture the macrophage and infect other cells. Conversely, interaction of macrophages with infected cells induces macrophage activation and Th1 cytokine production (IL-12, TNF and IFN-γ), a necessary progression to protective immunity and a resistant phenotype. In addition, the production of ROS and RNS, such as $H_2O_2$ and $NO^-$, control replication and aid killing of the parasite. However, arguably the most important series of events is the initiation of Th1-mediated immune responses.

Macrophages together with DCs also present parasite-derived peptides on their surface in conjunction with MHC class I and MHC class II molecules. CD4+ T cells will become activated following TCR:MHC class II molecule ligation in association with co-stimulatory receptor ligation involving CD28 on the T cell and CD80/CD86 on the APC. Recognition of MHC class II on the surface of macrophages that have ingested dead amastigotes, or other infected macrophages, is a key event in the generation of parasite-specific CD4+ T cells (Figure 14.10). In addition, CD4+ T cells will differentiate into Th1 cells as a result of IL-12 production from macrophages or DCs. Th1 cells then begin to express IFN-γ, which reinforces the type-1 cytokine phenotype. This Th1 polarization is necessary for protective immunity and is the case in the C57BL/6 mouse model of infection. The reverse occurs in the BALB/c mouse model as little IL-12 is expressed by APCs and CD4+ T cells differentiate into Th2 cells that produce IL-4 resulting in a non-protective phenotype. CD4+ Th1 cells are the dominant cells that

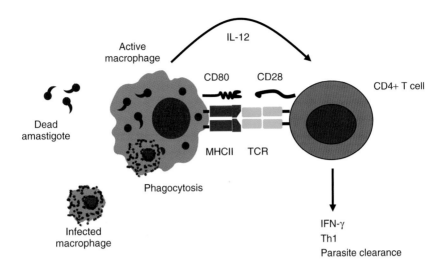

**Figure 14.10** Immunity to Leishmania. Dead amastigotes and infected macrophages are phagocytosed by viable macrophages. Activation leads to IL-12 production and upregulation of MHC class II and co-stimulatory molecules. This in turn activates CD4+ T cells to produce IFN-γ and initiate Th1 immunity.

provide resistance to experimental leishmaniasis. Along with NK cells, CD8+ T cells are potent producers of IFN-γ during infection and therefore contribute to disease resistance. However, the exact role that CD8+ T cells play in protective immunity is less clear. They may be more important during secondary infections by providing a pool of parasite-specific memory cells and they are also associated with protective immune responses to dermal *Leishmania* infection and resolution of disease.

## 14.19 Immunity to Giardia

*Giardia* species are protozoan parasites that infect the small intestine of humans, the most common species of which is *Giardia intestinalis* (otherwise known as *Giardia duodenalis*). The disease caused by these parasites is known as giardiasis, which is characterized by disruption of the intestinal epithelium that, in severe cases, can lead to malabsorbtion of nutrients. *Giardia* have a relatively simple life cycle and do not require a secondary host. Humans become infected following the ingestion of water or food items that are contaminated with parasite cysts. Once inside the small intestine, they undergo a morphological change into the trophozoite form and rapidly replicate by binary fission. Replication can be so extensive that the parasite forms a barrier between the intestinal epithelium and the lumen. This causes damage

to the structure of the villi and is thought to prevent normal uptake of nutrients. Transformation back into the cyst form facilitates expulsion from the host and into the environment, thereby completing the life cycle.

Infection with *G. intestinalis* is characterized by an acute phase and a chronic phase. In healthy individuals the immune response is normally sufficient to clear the parasite during the acute phase, which is coincident with the first signs of infection. Chronic infections usually only occur in immunocompromised patients. Several innate mechanisms are thought to contribute to host defence against the parasite and include the acid environment of the stomach and small intestine, antimicrobial factors such as defensins and enzymes produced in the gut. Complement proteins are also thought to participate in innate defence, by directly forming membrane attack complexes in the surface membrane of trophozoites and opsonizing the parasite for subsequent phagocytosis and ADCC. Opsonization of the parasite is also related to the antibody response of the host. For example, IgM and IgG are thought to provide better protection against *Giardia* than secretory-IgA, due to their more effective activation of the complement cascade.

A T cell-mediated cellular immune response is thought to be required in order to achieve complete clearance of *G. intestinalis*. This invoves activation of CD4+ Th cells and the recruitment of CD8+ T cells into the intestinal mucosa. However, *Giardia* employs an immune evasion strategy in order to avoid host adaptive immunity. Trophozoites express variable surface proteins (VSPs) that act as the main antigen for T cells. However, trophozoite cell division is associated with alterations in the VSP coat, which helps the parasite avoid cell-mediated immune responses and prolongs survival in the host. Clearance of chronic infections is characterized by the production of antibodies specific to the VSPs and it is therefore assumed that a combination of cell-mediated and antibody-mediated immunity is required for pathogen clearance.

## 14.20 Immunity to schistosomes

Schistosome worms are large multicellular parasites that undergo extensive morphological changes during the course of infection. They can persist inside a human host for many years even though they induce a complex and vigorous immune response, which is dominated by a CD4+ Th2 response. The pathology associated with schistosome infection is also directly related to the immune response. Furthermore, the immune response differs between acute schistosomiasis (3–5 weeks after infection), which induces a TNF-driven Th1 response, and chronic schistosomiasis (8 weeks onwards) characterized by a Th2 response with associated granuloma formation in the liver (Figure 14.11). Th1 immunity during acute schistosomiasis is directed to the migrating immature worms

in the first weeks after infection and is characterized by the release of pro-inflammatory cytokines such as TNF, IL-1 and IL-12. A persistent Th1 response during primary schistosomiasis results in elevated immunopathology and decreased survival. The Th2-mediated granuloma formation during chronic schistosomiasis is a result of an immune response to schistosome eggs trapped in tissues such as the liver and is a type IV delayed-type hypersensitivity (DTH) reaction. The main cytokines involved in granuloma formation are IL-4, IL-10 and IL-13 and Th2 immune responses are necessary for enhanced survival and reduced immunopathology. Immuno-regulation of the initial Th1-mediated inflammatory response, through the production of IL-10, contributes to reduced pathology and the development of Th2-dominated immunity. After approximately three months after infection, the Th2 immune response is also downmodulated and adult schistosomes are allowed to persist inside their human host.

## 14.21 Innate immunity to schistosomes

Adult helminth worms are poor inducers of DC activation, although egg-derived products are strong Th2 inducers. This may contribute to an effective shift from Th1 immunity in the acute phase to a highly polarized Th2 immune response in the chronic phase following the laying of eggs. Antigens derived from the egg downregulate the expression of CD80, CD86 and CD40, although DCs

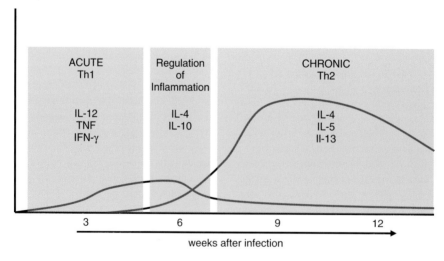

**Figure 14.11** Th1 and Th2 immune response to schistosomes. Immunity to schistosomes is complicated and bi-phasal. Initially a Th1 response is generated to immature flukes and is called the acute phase. This is followed by a period of immuno-regulation and a shift to a long-lasting Th2 response during the chronic phase.

still migrate to lymph nodes where they activate naïve CD4+ T cells. The development of Th2 cells requires the influence of IL-4, however. It has been suggested that IL-10 production by DCs blocks the expression of IL-12 therefore allowing the production of IL-4 from other sources and allows a Th2 immune response to develop.

Cells of the innate immune system contribute significantly to the killing of immature worms in the first days after infection. The more mature a schistosome worm becomes, the less susceptible it is to immune attack. The principal effector mechanism against immature worms is antibody-dependent cellular cytotoxicity (ADCC). Antibodies specific to schistosome antigens need to be present in order for this killing mechanism to be effective and therefore ADCC is particularly important during secondary infections, rather than in primary infections. Antibodies bound to the surface of schistosomes are recognized via FcRs on macrophages and eosinophils (Figure 14.12). The main antibody isotype involved in ADCC is IgE and it is particularly effective at activating eosinophils and causing degranulation. Cross-linking of FcεRs on eosinophils by IgE bound to the surface of schistosomes causes the release of toxic components contained in eosinophilic granules. This killing mechanism is aided by macrophage activation and release of ROS and RNS in the vicinity of the schistosome.

Non-specific immunity to schistosomes may involve the activation of the complement cascade and contribute to parasite killing. There is potential for both the classical and alternative complement pathways (see Chapter 2)

to be activated by the parasite, although schistosomes are known to counter the activity of these proteins by expressing various complement factor-binding proteins. Firstly, the classical complement pathway can be initiated by C1 binding to IgM or IgG on the surface of schistosomes. However, some reports indicate that schistosomes have a receptor for the Fc portion of immunoglobulin on their surface, which is able to bind the Fc portion of immunoglobulins. Therefore no free Fc region on IgM or IgG would be available for C1 binding, rendering the classical complement cascade ineffective. Mannose-binding lectin (MBL) can also potentially activate the alternative complement pathway and C4c can be detected on the parasite surface, thus indicating a role for complement in anti-schistosome immunity. However, the parasite is known to express C2, C3, C8 and C9 binding proteins that all may interfere with the alternative complement cascade.

## 14.22  Adaptive immunity to schistosomes

A strong antibody response is mounted against schistosomes and primarily involves IgE secretion. These antibodies are recognized by FcεRs on the surface of macrophages and eosinophils where cross-linking activates ADCC killing mechanisms. IgE production becomes elevated at the same time there is a shift from Th1 to Th2 immunity and is coincident with an increase in IL-4, IL-5 and IL-13 production. IL-5 is also a survival factor

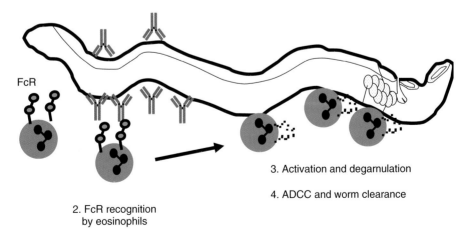

1. Antibodies bind surface of schistosome

FcR

2. FcR recognition
   by eosinophils

3. Activation and degarnulation

4. ADCC and worm clearance

**Figure 14.12** Antibody-dependent cellular cytotoxicity (ADCC) against schistosomes. Antibodies (mostly IgE) bound to the surface of the parasite are recognized by FcεRs on eosinophils. Cross-linking of antibody leads to activation, degranulation and ADCC.

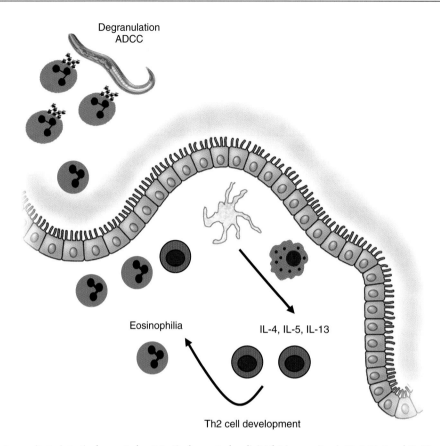

Degranulation
ADCC

Eosinophilia

IL-4, IL-5, IL-13

Th2 cell development

**Figure 14.15** Immunity to intestinal nematodes. Intestinal nematodes elicit Th2 immunity via IL-4, IL-5 and IL-13 production. This attracts and activates eosinophils to the sites of infection and initiates ADCC.

through the production of the regulatory cytokines IL-10 and TGF-β. Furthermore, IL-10 production prevents non-protective Th1 cell development. These cytokines have also been implicated in reducing the incidence of allergic immune reactions, a phenomenon known as the 'hygiene hypothesis'. The hygiene hypothesis basically states that a higher incidence of infections, such as intestinal worm infections, in early life prevents the development of allergic reactions in older life. However, in the short term, immunoregulatory cytokines prevent excessive immunopathology as a result of nematode infection.

## 14.27 Immune evasion strategies of parasites

Parasites come into direct contact with immune cells circulating within the blood stream or resident immune cells at sites of infection. For the parasite, this poses a great risk of being eliminated from the host. Therefore, many parasites have evolved key strategies that aim to counteract the actions of the immune system. These range from continuous alterations of the surface coat, as is the case of trypanosomes, to secreting products that interfere with the normal immune response, exemplified by certain helminths. Extracellular parasites possess mechanisms that evade humoral responses, such as complement-mediated lysis, while intracellular parasites evade phago-lysosomal killing mechanisms. In addition, some parasites interfere with adaptive immunity by disregulating antigen processing and presentation or cytokine production.

Humans can also be infected with parasites that normally infect and cause disease in other host species (Box 14.1). The term zoonosis is used to describe a particular parasite that infects another animal species but can

## Box 14.1. Rare Parasite Infections.

Hundreds of millions of people are infected with one or more parasites at any one time. The most serious parasitic diseases that infect man are considered the most prevalent, such as malaria, schistosomiasis and trypanosomiasis. However, there are many parasitic organisms that are rarely mentioned because the frequency of infection is very low. These infections nevertheless cause serious diseases in man, often resulting in unusual deaths. For example, *Naegleria fowleri* is a type of amoeba found throughout the world in soil and water. The amoeba infects man by entering the nose when swimming in warm fresh water. It then migrates to the brain where it causes meningoencephalitis, leading to vomiting, hallucinations,

dizziness, seizures, loss of bodily control and eventually death in only three days.

*Lagochilascaris minor* is a nematode worm that infrequently infects man causing skin lesions within neck tissues and visceral larva migrans (inflammation due to migration of nematode larvae). Such infections are usually restricted to South America as the normal hosts for the parasite are wild cats such as leopards. Many rare nematode (and other helminth) infections have been reported (see table accompanying this box). Humans often ingest nematode larvae from uncooked meat and fish and inadvertently infect themselves with a parasite. In most cases, no symptoms develop and the worm is expelled; only in rare cases do such infections cause severe illness or result in mortalities.

Rare parasitic infections. Most individuals are unaware of the many parasites that are able to infect and cause disease in man.

| Species | Taxa | Natural Host/Niche | Disease |
|---|---|---|---|
| *Balantidium coli* | Protozoan | Water | Intestinal cysts and diarrhoea |
| *Blastocystis hominis* | Protozoa | Water or food | Diarrhoea, abdominal pain, itch |
| *Cyclospora cayetanensis* | Protozoan | Water or food | Intestinal cysts and diarrhoea |
| *Isospora belli* | Protozoa | Environment | Diarrhoea and abdominal pain |
| *Lagochilascaris minor* | Nematode | Wild cats | Visceral larva migrans |
| *Angiostrongylus cantonensis* | Nematode | Rats | Eosinophilic meningitis |
| *Anisakis simplex* | Nematode | Fish and squid | Acute abdominal pain |
| *Capillaria hepatica* | Nematode | Fish eating birds | Liver inflammation and eosinophilia |
| *Gnathostoma spinigerum* | Nematode | Domestic animals | Ocular or visceral larva migrans |
| *Opisthorchis felineus* | Fluke | Molluscs | Biliary duct pain and stones |
| *Paragonimus westermani* | Fluke | Crustaceans | Bronchitis and pneumonia |
| *Hymenolepis dimunita* | Tapeworm | Rodents | No symptoms (rare abdominal pain) |

be transmitted to humans. These parasites can cause opportunistic infections and often result in debilitating diseases in susceptible individuals.

## 14.28 Trypanosome variant surface glycoproteins (VSGs)

Trypanosomes are entirely covered with an outer later of protein called variant surface glycoprotein (VSG), which is highly immunogenic and an important parasite antigen that induces a potent antibody response. Specific immunity toward parasites expressing a particular VSG ultimately destroys the parasite through a combination of antibody-mediated humoral and cellular mechanisms.

However, within any given population of trypanosomes there are variant parasites that express a different VSG. These parasites are not recognized by the immune response and can therefore go undetected and produce subsequent parasitaemias (Figure 14.16). The number of possible VSGs that a population of trypanosomes can produce is almost limitless. Each trypanosome possesses hundreds of VSG genes, which are linked to telomere expression sites so that new VSG transcripts are produced during replication of metacyclic forms. This provides an essential immune evasion strategy for trypanosomes so that a chronic infection ensues whereby the immune system is always trying to catch up with the ever-changing parasites. This also makes it extremely difficult to develop a vaccine against African trypanosomiasis.

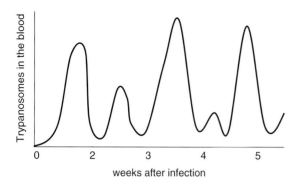

**Figure 14.16** Immune evasion by African trypanosomes. The variable surface glycoprotein (VSG) coat on the trypanosome cell surface is highly polymorphic but also highly immunogenic. Antibodies directed to one VSG quickly clear those parasites expressing that particular VSG. However, parasites expressing a different VSG can escape specific antibody-mediated immunity. This produces waves of different trypanosome populations in the blood.

## 14.29 Plasmodium life cycle contributes to immune evasion

The life cycle of the malaria parasite involves rapid changes in location within the host (Figure 14.17). The parasite is firstly deposited in the bloodstream where it rapidly migrates to the liver and invades hepatocytes. There is therefore a shift from an extracellular to an intracellular existence and is thus an excellent means to avoid humoral immunity such as antibody and complement. Large numbers of parasites are then released back into the blood where they quickly invade red blood cells and become intracellular once again. Further rounds of replication and cell lysis of host cells result in transient extracellular forms and dominant invasive intracellular forms.

Plasmodium species possess a number of other immune evasion strategies. These include polymorphisms within the major surface proteins. For example, differences in surface protein expression occur between the different life cycle stages. Sprozoites express the CSP (circumsporozoite protein), merozites express MSP-1 (merozoite surface protein-1), while trophozoites express PfEMP-1 on infected erythrocytes. Changes in surface coat expression act to confuse and outrun host defences. Furthermore, CSP is composed of several tandem repeats that are thought to prevent antibody affinity maturation, and MSP-1 may bind 'blocking' antibodies that act to

downregulate immune responses. In addition, PfEMP-1 enhances erythrocytes binding to endothelia via CD36 and ICAM-1, which prevents the infected cells entering the spleen and other organized immune sites and thus reducing the capacity of the immune system to develop adaptive responses and clear the parasite.

## 14.30 *Leishmania* evade phagocytic killing

*Leishmania* species reside in host macrophages where they are able to evade phagocyte killing mechanisms and therefore effectively hide from the immune system inside one of its own cells. One important molecule expressed by *Leishmania* during the infection process is a lipophosphoglycan (LPG), which actively blocks phagosome and lysosome fusion, so that the parasite is able to develop into the amastigote form in the absence of toxic lysosomal products. In addition, a gp36 surface protease inhibitor blocks the activity of phago-lysosomal enzymes should lysosome fusion take place. The amastigote then adapts to life within the phagosome of the macrophage. LPG and gp36 also assist promastigote survival by scavenging and preventing the action of reactive oxygen species. Inhibiting apoptosis and antigen presentation on MHC class II molecules further enhances survival inside the macrophage. *Leishmania* endocytose MHC class II molecules and target them for degradation. In addition they inhibit peptide loading and therefore interfere with T cell activation and the generation of adaptive immunity.

## 14.31 Immune evasion strategies of helminths

Helminth worms are large multicellular organisms that reside in sites of the host, such as the blood or intestines, which makes them susceptible to immune attack. The main mechanism of immunity is ADCC, involving the effector activities of eosinophils and CTLs. The large size of many helminths renders them relatively resistant to the effects of antibodies and complement, an evasion of immunity intrinsically related to their morphology. Their life cycle also assists in evading specific immune responses. An immune response is generated to a particular stage in the life cycle, although helminths move around the body

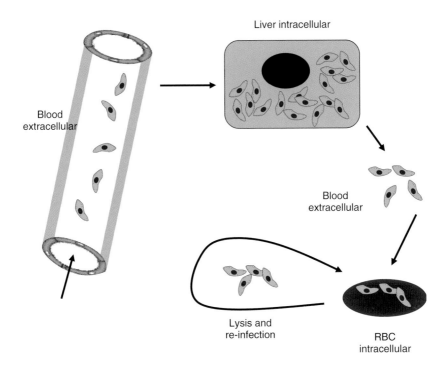

Liver intracellular

Blood extracellular

Blood extracellular

Lysis and re-infection

RBC intracellular

**Figure 14.17** Immune evasion strategies of Plasmodium. The nature of the plasmodium life cycle makes it difficult for the immune system to keep track of the parasite. Moving From the blood, to the liver, to the blood and in and out of erythrocytes is problematic for generating effective immunity that clears the parasite.

and therefore evade specific immune defences. However, the immune response often results in expulsion of the invading parasite. Therefore helminths have evolved several sophisticated immune evasion strategies to avoid the immune response of the host. These include shedding of the surface coat, incorporating host-derived products in order to hide from the immune response, and the expression of a number of factors that inhibit or misdirect the immune system.

C-type lectins are a diverse group of animal receptors that recognize carbohydrate structures. One example is mannose-binding lectin expressed on the surface of macrophages, which recognizes mannose motifs on the surface of microbes and activates the complement cascade and initiates phagocytosis. C-type lectins contribute to immune defence in both vertebrates and invertebrates. A C-type lectin from *Toxocara canus* called TES-32 is thought to interfere with several aspects of host immune defence. Firstly it may interfere with leukocyte migration into sites of infection by competitively inhibiting receptors such as L-selectin (Figure 14.18) and preventing cell adhesion. Surface expressed C-type lectins, on

*Schistosoma mansoni* for example, may also bind host glycoproteins such as IgG, C3 and MHC molecules. The absorption of host products onto the surface of parasites may hide these pathogens from immune cell attack.

Helminths express numerous proteins that alter the phenotype of the immune response. C-type lectins expressed by helminths may selectively bind to Th1 cells and skew the immune response away from a protective Th2-mediated response. However, the majority of worm infections result in the generation of Th2-mediated responses (Figure 14.19). The initiation of Th2 responses is the result of specific activation of APCs such as DCs and macrophages. Helminth glycans are thought to drive immunity toward a Th2 phenotype by acting like a PAMP. For example, glycans found in conjunction with the schistosome egg antigen (SEA) are largely responsible for the Th2-driven granuloma formation observed following schistosome infection. In fact, if the SEA is de-glycosylated it loses its ability to induce a Th2-type response. Such helminth glycans also enhance the production of the regulatory cytokines IL-10 and TGF-β. This firstly acts to reduce the intensity of the immune

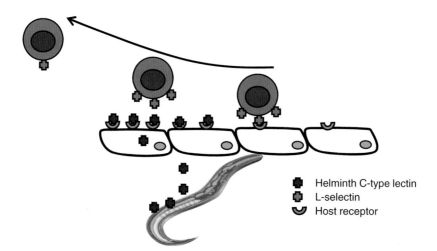

**Figure 14.18** C-type lectins from helminths inhibit host cell adhesion. These C-type lectins bind to the same cell adhesion molecules as L-selectin on lymphocytes. This competition reduces the ability of the lymphocyte to bind to epithelia and inhibits migration into sites of infection.

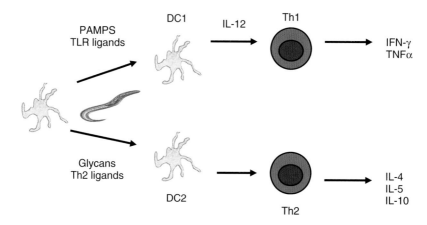

**Figure 14.19** Generation of Th2-mediated immunity. Certain molecules expressed by helminths divert the immune system towards a Th2 immune response and away from a Th1 inflammatory response. This may be an immune evasion strategy used by these parasites to avoid cell-mediated killing mechanisms.

response against the worm and secondly IL-10 and TGF-β production has been directly associated with reduced levels of allergy in individuals with high worm burdens.

A further way in which helminths can downregulate immunity is through the expression of cytokine homologues. For example *Brugia malayi* expresses homologues of TGF-β and a chemokine involved in macrophage migration called macrophage migration inhibitory factor (MIF). Several aspects of host immunity are downregulated by TGF-β, including APC function, phagocytic killing and T cell activation. Expression of MIF is rather

more specific in that it will act to inhibit the migration of macrophages into the site of inflammation.

## 14.32 Summary

1. Parasites are eukaryotic organisms that can infect mucosal tissues and the skin.
2. Immunity to single celled protozoa is largely governed by the life cycle of the individual parasite and its intracellular location.

**3.** Th1-driven, cell-mediated immunity is normally required for the clearance of protozoan parasites.

**4.** Multicellular parasites, such as helminth worms, live in the extracellular environment and induce a Th2 mediated immune response.

**5.** The production of IgE and the activation of eosinophils are characteristic features of worm infections.

**6.** Parasites have evolved complex immune evasion strategies to combat host immunity.

# 15 Disorders of the Immune System

## 15.1 Introduction to immune disorders

The immune system provides effective defence against invading microorganisms and potentially harmful substances. However, aberrations in the normal functioning of the immune system can often lead to disease. Immune disorders fall into three broad categories. The first category includes those diseases that result from an excessive or overactive immune response, such as allergies and asthma. The second are due to the generation of immune responses directed against self antigens and are known as autoimmune diseases. The third are characterized by an abnormal immune system resulting from an inherited genetic defect or mutation and are referred to as immunodeficiency diseases. All three groups of immune-mediated diseases can affect the mucosal immune system in some way.

The immune mechanisms responsible for mediating allergic responses and autoimmune reactions often share a common feature, which is a lack of proper immune regulation (immunodeficiencies are different, as they are caused by genetic defects). Without appropriate regulatory signals, immune cells have the potential to proliferate and exert excessive immunological effector functions. Excessive inflammation generally leads to tissue damage and disease. Moreover, the immune-mediated tissue damage is caused by the inappropriate induction of a specific adaptive immune response to an antigen that would normally be considered harmless. At the heart of immunoregulation, and therefore immunological disease, is a mechanism known as tolerance. Central tolerance occurs in the thymus and bone marrow, where T cells and B cells that are reactive to self antigens are deleted from the immune system. Peripheral tolerance occurs

in secondary lymphoid tissues or the periphery and acts to maintain central tolerance and ensure that T cells and B cells do not respond to harmless antigens. When this mechanism of tolerance is disregulated, lymphocytes respond inappropriately to harmless antigens. In the case of allergies, lymphocytes respond to environmental antigens, while autoimmune reactions are associated with lymphocyte responses to self antigens. This chapter will therefore focus on the various disorders of the immune system that affect mucosal tissues and the immunological mechanisms responsible for the generation of disease.

## 15.2 Types of allergy

Allergy is the most common immune-mediated disorder and is the result of an immune response to an environmental antigen (Table 15.1). Any substance that causes an allergic reaction is known as an allergen. Most people do not mount an immune response against allergens but individuals who are susceptible to allergies often mount an allergic immune response that can cause disease. Allergens are found throughout the environment and can be inhaled, ingested or come into contact with the skin. Common examples of allergens include grass pollen, house dust mites and animal fur. Depending on the nature of the allergen, allergies tend to develop either on a seasonal basis or are a constant problem. For example, grass or tree pollen is released only for a short period in the springtime and therefore people develop seasonal allergies. On the other hand, house dust mites live along side us all of the time and therefore present a constant problem to allergic individuals. Although most allergens are airborne particles, such as grass pollen, others include

*Immunology: Mucosal and Body Surface Defences*, First Edition. Andrew E. Williams.
© 2012 John Wiley & Sons, Ltd. Published 2012 by John Wiley & Sons, Ltd.

**Table 15.1** Types of allergy that affect mucosal tissue and body surfaces.

| Disease | Affected tissue | Symptoms |
|---|---|---|
| Allergic rhinitis | Nose, eyes | Runny nose, itchy eyes, swelling |
| Asthma | Airways, respiratory tract | Shortness of breath, bronchoconstriction, coughing, wheezing |
| Anaphylaxis | Systemic, skin | Vasodilatation, bronchoconstriction, swelling, itchy skin |
| Eczema | Skin | Itchy skin, redness, rash |
| Contact dermatitis | Skin | Itchy skin, redness, rash |
| Insect venom | Skin, systemic | Itchy skin, redness, rash |
| Drug allergy | Gastrointestinal tract, skin, systemic | Vomiting, diarrhoea, abdominal pain, itchy skin |
| Food allergy | Gastrointestinal tract, skin, systemic | Vomiting, diarrhoea, abdominal pain |

food allergens, venom or toxins from insects, contact allergens such as heavy metals and some drugs or medications. Most people are not born with an allergy but rather develop an allergy to a particular substance over time as a result of repeated contact with that substance. The development of an allergic immune response involves an initial sensitization event (first contact). Subsequent exposures to the same allergen then stimulate an allergic immune reaction. Repeated exposure to an allergen can often lead to a more chronic, long-lasting disease such as occurs in people who develop asthma.

Allergic asthma is one of the most prevalent forms of allergy that affects the respiratory tract, which results from contact with an airborne allergen. Asthma primarily affects the airways and is characterized by coughing, wheezing, increased mucus production and the narrowing of the airways (known as bronchoconstriction). The most extreme cases of asthma are associated with severe bronchoconstriction, which can result in a fatal inhibition of gaseous exchange, as occurs during an asthma attack. Airborne allergens also account for episodes of allergic rhinitis (otherwise known as hay fever), which affects the nose and upper respiratory tract, and allergic conjunctivitis, which is characterized by itching and redness of the eyes.

Allergies of the gastrointestinal tract include those resulting from contact with an allergen derived from food and are characterized by vomiting, diarrhoea and stomach pain. An allergy to peanuts is one of the most common forms of gastrointestinal allergy and can be so severe that it sometimes leads to anaphylactic shock and occasionally death. Anaphylaxis is the most severe form of allergy, which is a whole body reaction to an allergen and involves vasodilatation of blood vessels, very low blood pressure and bronchoconstriction. Food allergies also involve reactions to other kinds of seeds and nuts, to

proteins contained within egg and diary products and to shellfish. Adverse allergic reactions to certain medications, such as aspirin and penicillin, are often mediated through the gastrointestinal tract.

The skin is also susceptible to allergic reactions due to its constant interface with the environment, examples of which include contact dermatitis and eczema. The most common cause of contact dermatitis is contact with allergens derived from plants, for example poison ivy, and results in intense itchy rashes. Heavy metals such as gold and nickel, and the use of latex gloves, are also known to cause allergic reactions in the skin, although these immune responses are slower to develop and often involve slightly different immune processes known as delayed hypersensitivity (Box 15.1, hypersensitivity reactions). In addition, toxins or venoms introduced into the skin by biting and stinging insects can cause acute reactions at the site of the puncture wound. Some people respond so vigorously to insect bites that they develop a potentially lethal anaphylactic reaction, which is intensified by direct injection of allergen into the bloodstream.

An allergic immune response is essentially an immediate hypersensitivity reaction (type I hypersensitivity), which is associated with the activation of mast cells and basophils. Type I hypersensitivity reactions are also associated with the secretion of IgE, which is recognized by the high affinity Ig$\varepsilon$RI expressed on the surface of various cells including mast cells, basophils and eosinophils. The cross-linking of Ig$\varepsilon$RI by IgE causes the degranulation of mast cells and basophils and the release of granule contents into the immediate environment. This induces a rapid and often severe allergic immune response that is sometimes referred to as atopy. Individuals who are particularly susceptible to developing IgE-mediated allergies are regarded as being atopic.

## Box 15.1 Hypersensitivity Reactions.

Hypersensitivity reactions are often damaging, sometimes even fatal, and can be considered to be the result of an excessive immune reaction. A prerequisite for developing a hypersensitivity reaction is a previous exposure to an antigen that renders a person sensitized to that antigen. Hypersensitivity reactions occur following re-exposure to the same antigen and can be classified into four types based on the immune components that are involved.

Type I hypersensitivity reactions are commonly known as allergies (discussed in detail within this chapter). An allergic response is caused by the pre-sensitization to a particular antigen known as an allergen. Examples of allergens include animal dander, house dust mites and pollen. The induction of type I hypersensitivity reactions involves the cross-linking of IgE molecules on the surface of mast cells and basophils, which causes the release of allergic mediators such as histamine, prostaglandins and type 2 cytokines. The release of allergic mediators results in the symptoms associated with allergies such as sneezing, coughing, itching and constriction of the airways. Chronic exposure to an allergen also involves activation of CD4+ Th2 cells, release of IL-4, IL-5 and IL-3 and the activation of eosinophils (as occurs during allergic asthma).

Type II hypersensitivity is also known as antibody-dependent cytotoxicity and occurs following the binding of antibodies to the surface of one's own cells. Under normal circumstances antibodies are prevented from binding to the surface of healthy cells. However, following certain infections, foreign antigens can be deposited on the cell membrane. Alternatively, self antigens can also be recognized by antibodies during certain autoimmune reactions. Either way, this results in antibody binding to antigen bound to the surface of host cells, which results in the recognition by NK cells and macrophages through binding to their FcRs. More specifically, IgG bound to the surface of cells is recognized by FcγRIII, which induces NK cells and macrophages to initiate antibody-dependent cellular cytotoxicity (ADCC). Another form of type II hypersensitivity results from the binding of complement components to surface bound antibodies (IgG or IgM). The activation of the complement cascade leads to cell lysis and ultimately tissue damage. Examples of type II hypersensitivity reactive include acute rheumatic fever, myasthenia gravis and Grave's disease.

Type III hypersensitivity is also known as immune complex disease and occurs when antibodies cross link soluble antigen and form antibody-antigen complexes. Under normal circumstances, large antibody-antigen complexes are recognized and phagocytosed by macrophages, which results in the removal from the system. However, small antibody-antigen complexes evade detection and become deposited into the walls of blood vessels, the glomeruli of the kidneys or within the joints. This deposition within solid tissue results in the activation of inflammatory reactions

through opsonization and release of inflammatory mediators. Examples of type III hypersensitivity reactions include reactive rheumatoid arthritis, various forms of vascularitis and are a prominent feature in systemic lupus erythematosus.

Type IV hypersensitivity reactions are also known as delayed type hypersensitivity (DTH) responses and involve the activation of memory CD4+ Th cells and CD8+ CTLs. Unlike the other types of hypersensitivity reaction, DTH responses take longer to develop (2–14 days) and are independent of antibody-mediated mechanisms. Circulating memory Th cells recognize peptides presented in conjunction with MHC class II, which activates and causes them to initiate an inflammatory reaction involving the release of IFN-γ and TNF. This in turn recruits inflammatory cells, including activated CTLs, which are responsible for killing cells expressing the relevant antigen. Type IV hypersensitivity can involve either Th1 or Th2-mediated immune responses. The main example of type IV hypersensitivity is allergic contact-dependent dermatitis, which results in exposure to chemicals such as nickel or the active agent in poison ivy. An example of Th1-driven type IV hypersensitivity is the Tuberculin reaction in response to Mycobacterium tuberculosis antigen injected into the skin, otherwise known as the Mantoux test. As type IV hypersensitivity reactions rely on antigen-specific T cell responses, they are often confused with genuine autoimmune responses, which have been unconventionally classified as type V hypersensitivity reactions (discussed in detail in this chapter).

## 15.3 Sensitization and the acute phase response

A type I hypersensitivity response can be divided into two phases, the acute phase and the late phase and is a consequence of prior sensitization to an allergen (Figure 15.1). One of the major characteristics of allergic reactions is the differentiation and expansion of Th2 cells and the production of Th2-dependent cytokines, which are central for sensitization to allergen. In particular, the Th2 cytokines IL-4 and IL-13 are necessary for the secretion of IgE by plasma cells. IL-4 and IL-13 induce B cells to undergo antibody class-switching, resulting in the secretion of significant quantities of allergen-specific IgE. Activated Th2 cells also upregulate CD40L, which stimulates CD40 on B cells and provides a co-stimulatory signal that triggers B cell differentiation and antibody production. The development of a Th2-mediated immune response is indicative of all allergic reactions.

Following sensitization, the acute phase occurs immediately after re-exposure to an allergen, while the late

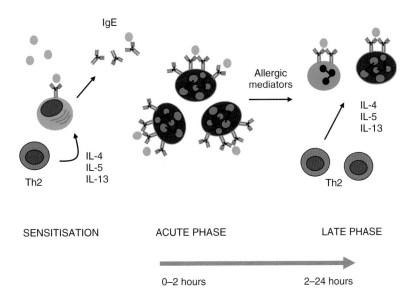

**Figure 15.1** An allergic immune response requires initial sensitization to an allergen. Subsequent exposure to that allergen induces an immediate-type hypersensitivity reaction, the acute phase response, followed by a delayed-type hypersensitivity reaction, the late phase response.

phase tends to develop 2–24 hours after the acute phase. The central event during the acute phase is the activation of mast cells and the rapid release of allergic mediators. This response is also mediated by IgE, which is bound to the mast cell surface by high affinity IgεRI molecules. The recognition of an allergen by surface bound IgE causes the cross-linking of IgεRI molecules, which triggers mast cell degranulation. Atopic individuals, who are more susceptible to allergic diseases, have a much denser coating of IgE on the surface of their mast cells. This means that exposure to an allergen that is specifically recognized by IgE results in the rapid initiation of a type I hypersensitivity response. Even a tiny amount of allergen is enough to trigger an allergic reaction in atopic individuals. In non-atopic people, even though some allergen-specific IgE may be present, it is not enough to trigger an allergic reaction. The presence of receptor bound IgE on the surface of mast cells is a consequence of a previous event known as sensitization, which is dependent on the activation of Th2 cells and the secretion of IgE by plasma cells. Sensitization to an allergen can be considered to be the first stage in the development of a specific allergy.

## 15.4 Mast cell degranulation

The binding of IgE to the high affinity IgεRI on the surface of mast cells is a critical event for the activation,

degranulation and the release of pro-inflammatory mediators (Figure 15.2). The activation of mast cells through IgεRI signalling also requires the cross-linking of IgE by specific antigen and the aggregation of receptor complexes. IgεRI is a member of the immunoglobulin receptor superfamily and is a tetramer composed of an α-chain, which is responsible for binding to IgE, and a β-chain plus two γ-chains, which are responsible for the initiation of intracellular signalling events. Receptor cross-linking and the formation of IgεRI aggregates effectively bring together the necessary signalling molecules and accessory molecules that are required for an effective signalling cascade. The β-chain and γ-chains both contain immunoreceptor tyrosine-based activation motifs (ITAMs) that, when phosphorylated following receptor cross-linking, provide docking sites for the tyrosine receptor kinsases Lyn and Syk (Figure 15.3). The kinases Lyn and Syk possess SH2-domains that provide high affinity binding sites for the ITAM motifs on the β-chain and γ-chains.

The docking of Lyn and Syk onto the ITAM motifs causes the phosphorylation and activation of these kinases, which subsequently causes the phosphorylation of another membrane bound adaptor molecule called LAT (linker for activation of T cells). The activation of LAT induces the recruitment of several intracellular signalling molecules, including the phospholipase PLCγ, and the eventual activation of protein kinase C (PKC)

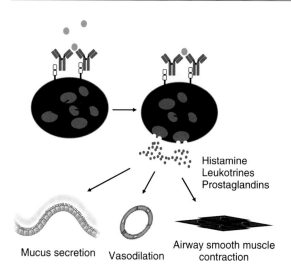

**Figure 15.2** Mast cell degranulation is an important mediator of the symptoms associated with the cute phase response. The release of histamine and other allergic mediators affects various tissues and causes airway smooth muscle contraction, vasodilation and mucus secretion.

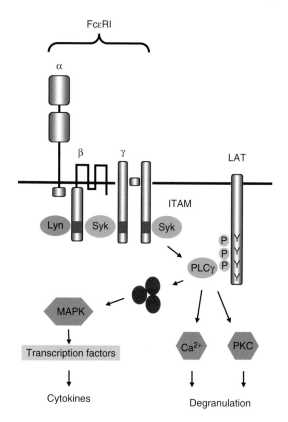

**Figure 15.3** The high affinity IgεRI and mast cell signalling. The tyrosine receptor kinases Syk and Lyn are recruited to IgεRI following IgE cross-linking. This induces downstream signalling events involving activation of MAPK, PKC and $Ca^{2+}$ mobilization.

and the mobilization of intracellular $Ca^{2+}$ reserves. Various signalling cascades, involving a number of different protein kinases, have been implicated in the complex signalling events that lead to mast cell activation. These downstream signalling events initiate the process of degranulation and the activation of MAPK-dependent transcription factors and NF-κB, among others. Collectively, receptor signalling results in the release of granule contents from mast cells and the production of pro-inflammatory mediators such as TNF, IL-1, IL-4 and IL-13.

Mast cell granules already contain preformed mediators, including histamine, serine proteases, proteoglycans (such as heparin) and leukotrienes. Therefore, activation of IgεRI signalling events leads to the rapid fusion of granules with the plasma membrane and the almost immediate release of granule contents. Histamine is an important allergic mediator as it causes the contraction of smooth muscle cells, which in turn causes bronchoconstriction and mucus production (Figure 15.2). Leukotrienes can also stimulate bronchoconstriction and mucus secretion, as well as acting as a potent chemotactic factor and vasodilator. In addition, the serine protease tryptase is thought to contribute to the life-threatening consequences of analphylactic shock. Mast cell degranulation therefore plays a significant role in allergic diseases such as asthma and anaphylaxis.

Mast cell degranulation and cytokine release is also mediated by other cell surface receptors. For example, KIT (CD117) dimerization, as a result of stem cell factor (SCF) binding, is able to synergize with FcεRI cross-linking and enhance MAPK signalling as well as activate the JAK-STAT pathway. Furthermore, the cytokine receptors for IL-4 and IL-5 are able to enhance FcεRI-mediated cytokine release, while ligation of the chemokine receptors CCR1 and CCR3 enhance degranulation and IL-13 release, respectively. Other receptors that can activate mast cells include the complement component receptors C3aR and C5aR and various TLRs including TLR3 and TLR4.

## 15.5 Late phase response

The inflammatory mediators released during the acute phase response, result in the migration of leukocytes into the original site of allergen exposure. From the point of antigen exposure, the late phase reaction can have a delay of 6–12 hours, and is therefore considered to be a delayed-type hypersensitivity reaction (type IV hypersensitivity). The late phase response often subsides within a couple of days but in certain circumstances can lead to the development of chronic allergy that persists for a lengthy period of time. The characteristic features of a late phase reaction include the production of Th2 cytokines (e.g. IL-4, IL-5 and IL-13) and chemokines (e.g. IL-8 and CCL2). Several cell types are responsible for the release of these mediators, including activated mast cells. These inflammatory mediators work in conjunction with the contents of mast cell granules, such as histamine and leukotrines, so as to recruit an inflammatory cell infiltrate (Figure 15.4). Furthermore, TNF produced by mast cells and activated T cells functions by enhancing the inflammatory potential of other Th2 cytokines. Even though TNF has traditionally been associated with Th1 responses, it contributes to the migration and stimulation of immune cells and is significantly elevated during allergy and anaphylaxis.

The inflammatory infiltrate during a late phase reaction includes neutrophils, eosinophils, macrophages and T cells. Neutrophils and macrophages are effective phagocytic cells but they also produce inflammatory cytokines, such as TNF, which augments any inflammatory reaction taking place. Eosinophils are a significant contributor to the allergic response. Like mast cells, they express the high affinity IgεRI on their cell surface, which initiates degranulation and the release of further mediators, including protaglandins, reactive oxygen species and cytokines (e.g. TNF, IL-1 and IL-5), which augment the allergic response and cause tissue damage. The T cell population mostly comprises Th2 cells, which are effective at maintaining an allergic environment through the release of IL-4, IL-5 and IL-13. As well as providing signals that support eosinophil function and longevity, they also signal to B cells to produce more IgE. Importantly, a proportion of T cells and B cells differentiate into a memory cell population, which is more easily triggered following a subsequent encounter with the same allergen.

## 15.6 Allergic asthma

Asthma is a common disease of the airways that is characterized by chronic inflammation, airway hyper-responsiveness, mucus production, bronchoconstriction and reversible airway obstruction. In general terms, there are two forms of asthma, one of which affects atopic individuals (extrinsic asthma) and the other affects non-atopic individuals (intrinsic asthma). Extrinsic asthma involves an allergic immune reaction to a specific allergen, is seasonal in occurrence and is therefore referred to as allergic asthma. Intrinsic asthma does not involve an allergic immune response but instead is associated with acute episodes of severe bronchoconstriction in response to exercise, stress or environmental irritants. Many of the symptoms of allergic and non-allergic asthma are the same, such as coughing, wheezing and shortness of breath, although the initial trigger may be different.

Typical allergens that induce an episode of allergic asthma include molds, house dust mite allergen, pollen and animal dander. The allergic immune response generated following sensitization and exposure to these allergens causes inflammation and bronchoconstriction, which is often reversible and treatable with medications (this is often not the case with intrinsic asthma). The ease with which allergens induce allergy and narrowing of the bronchi is what make airway hyper-responsiveness a distinctive feature of asthma. Allergic asthma sufferers often take drugs to relieve the symptoms, such as the beta$_2$-adrenoceptor agonist salbutamol, which acts as a bronchodilator that helps to relax the muscles within

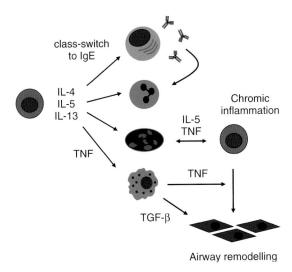

class-switch to IgE

IL-4
IL-5
IL-13

TNF

Chromic inflammation

IL-5
TNF

TNF

TGF-β

Airway remodelling

**Figure 15.4** Th2 cells secrete IL-4, IL-5 and IL-13, which influence several other leukocytes during the generation of an allergic immune response. This includes class-switching to IgE, eosinophil recruitment, mast cell activation and remodelling of the airways. Further recruitment and activation of Th2 cells establishes a late phase response, while persistent inflammation can lead to tissue remodelling.

the airways. Preventative drugs are also used, such as budesonide, which is a corticosteroid that reduces the sensitivity of the immune response to an inhaled allergen. These drugs are often dispensed using an inhaler.

## 15.7  Mast cells and the early phase allergic asthma

Allergic asthma is triggered when inhaled allergens are recognized by IgE on the surface of mast cells, which cause the cross-linking of IgεRI molecules and subsequent activation and degranulation of mast cells within the airways. It is thought that tissue resident mast cells and mast cells recruited from the blood in response to chemokines such as CCL11 and IL-8, participate in an asthmatic response. In addition, airway epithelial cells release stem cell factor (SCF, otherwise known as kit-ligand), which specifically binds to the receptor KIT (c-Kit) expressed on cell surface of mast cells. This causes mast cells to migrate to the airway epithelium and also brings them into close proximity to smooth muscle cells. SCF, IL-4 and IL-5 prime mast cells so that they become more responsive to FcεRI cross-linking by allergen. This priming mechanism is thought to be disregulated in asthmatics, so that mast cells are undergo degranulation more easily when in the presence of allergen.

The degranulation of mast cells results in the release of numerous bronchoconstrictors such as histamine, which is preformed and stored in mast cell granules, and leukotriene-$D_4$ and prostaglandin-$D_2$, which are transcribed following activation. The activation and release of allergic mediators from mast cells is associated with the early phase of allergic asthma and is equivalent to an immediate type hypersensitivity reaction. Histamine acts as an immediate mediator of bronchoconstriction, while leukotrienes and prostaglandin $D_2$ act on smooth muscle cells over a longer period of time and contribute to the chronic symptoms of airway contraction observed in asthma. Mast cells also express the cytokines IL-4, IL-5, IL-13 and TNF, which are all involved in the development and continuation of allergic immune responses (Figure 15.5). These cytokines act in an autocrine manner, by perpetuating the activation of mast cells, and also function by attracting and activating other leukocytes, including eosinophils and Th2 cells. The subsequent recruitment of these leukocytes is associated with the late phase response.

## 15.8  Epithelial cells can trigger allergic asthma

Epithelial cells are important mediators of allergic asthma, particularly as they release several chemokines

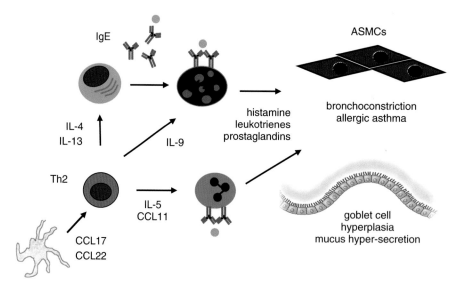

**Figure 15.5** Generation of an allergic immune response characteristic of asthma. An initial episode of sensitization to allergen is associated with T cell-dependent B cell activation and IgE antibody secretion. Repeated exposure to allergen results in mast cell degranulation and eosinophil recruitment, leading to downstream effects that result in an asthmatic phenotype.

and cytokines in response to tissue damage. Indeed, epithelial cell injury and airway hyper-responsiveness are thought to be important triggers for allergic responses associated with asthma. Although the factors that trigger airway hyper-responsiveness are not entirely clear, several likely candidates include viruses, bacteria or bacterial products such as LPS, toxins and allergens. For example, virus infections are known to be significant inducers of asthmatic exacerbations. Not only do viruses cause direct damage to epithelial cells, they also induce robust inflammatory responses. Epithelial cells upregulate and secrete a number of inflammatory mediators in response to infection, including CCL5, CCL11, IL-1β and type I interferons. Similarly, microbial PAMPs such as viral ssRNA or LPS derived from bacteria can activate PRRs expressed by epithelial cells, including TLR4. Furthermore, damaged cells release several endogenous DAMPs, which are also recognized by PRRs and result in the expression pro-inflammatory genes.

The release of pro-inflammatory mediators from epithelial cells contributes to the activation of mast cells and the recruitment of eosinophils and Th2 cells into the airways. Recent findings have demonstrated that IL-25, IL-33 and TSLP (thymic stromal lymphopoietin) secreted by epithelial cells can activate and promote cytokine release from mast cells. IL-25, IL-33 and TSLP can also promote the differentiation of Th2 cells, either directly or through the activation of airway dendritic cells. TSLP in particular can stimulate DCs to promote Th2 responses. It may be that these cytokines, which are produced by epithelial cells and other innate immune cells, act in conjunction with IL-4 during the promotion of Th2 responses in allergic asthma (Figure 15.6). Therefore, epithelial cells represent a key cell type that can trigger asthmatic immune responses, following the release of allergic mediators and the activation of downstream effector cells.

One of the key characteristics observed in allergic asthma is airway remodelling, associated with epithelial cell hyperplasia, fibroblast proliferation and general thickening of the airways. The wound repair mechanisms normally present in the airway are also affected by asthma, in part through the enhanced production of fibroblast growth factors (FGF-1 and FGF-2) and TGF-β by epithelial cells, which can result in fibroblast and smooth muscle cell proliferation. Furthermore, it seems likely that airway smooth muscle cells also contribute to the production of inflammatory mediators (e.g. IL-1β and CXCL10), as well as implementing the constriction of the airways and undergoing hyperplasia in the airway wall.

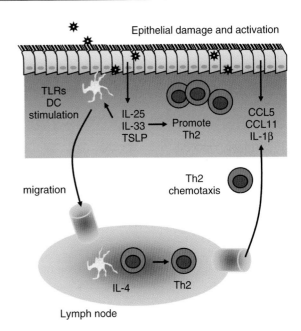

**Figure 15.6** Epithelial cells contribute to the promotion of allergic immune responses in asthma. Asthma exacerbation often results from epithelial damage, following a virus infection, for example. This results in production of several mediators such as the chemokines CCL5 and CCL11, which enhance leukocyte recruitment into the airway. In addition, epithelial cells secrete IL-25, IL-33 and TSLP, which promote Th2 cell differentiation and proliferation. These cytokines also stimulate DCs further promoting Th2 cell differentiation and may function in conjunction with IL-4.

Furthermore, IL-25, IL-33 and TSLP, which are expressed by epithelial cells, are thought to stimulate a recently discovered population of natural helper cells. These lymphoid-like cells are thought to be located throughout the airways, where they provide an innate source of Th2 cytokines. As natural helper cells are capable of Th2-cytokine secretion, including IL-5 and IL-13, they are thought to participate in immune responses to parasites and have been implicated in allergic diseases such as asthma. Indeed, stimulation of natural helper cells with IL-25, IL-33 or TSLP causes the upregulation of IL-5 and IL-13 and therefore contributes to the Th2 cytokine environment, goblet cell hyperplasia and eosinophilia. Similar Th2 cytokine-producing innate cells have also been found in GALT (multi-potent progenitor type 2 (MPPtype2) cells), the spleen (nuocytes), mesenteric lymph nodes (innate type-2 helper cells) and fat-associated lymphoid clusters (FALCs) in the mesentery (natural helper cells). All these innate cells respond to IL-25 and IL-33 and are capable of expressing

IL-5 and IL-13. Therefore, natural helper cells may be important for the establishment of Th2-mediated immune responses.

## 15.9 T cells and the late phase of allergic asthma

T cells play a central role in the development of immune responses in asthma, during both the sensitization phase and during the progression of the late phase. DCs are thought to play a significant role in the activation and recruitment of Th2 cells in asthma, not only in the cytokines they produce but also in presenting antigen to Th2 cells. DCs are often activated in response to an allergen, whereby they downregulate IL-12 and upregulate IL-10, which promote the differentiation of Th2 cells. In addition, DCs upregulate the chemokines CCL17 and CCL22, which help to recruit Th2 cells into the airways. Th2 cells that have been recruited to the airways express IL-4 and IL-13, which act on B cells to undergo class-switching to IgE production, and secrete IL-5, which is an important cytokine for the differentiation of eosinophils in the bone marrow and for the maintenance of eosinophil function in the airways. During allergic asthma, Th2 cells also secrete the cytokine IL-9, which stimulates the proliferation of mast cells and therefore acts to accentuate and reinforce the development of allergic asthma. The cytokines IL-4, IL-9 and IL-13 also induce goblet cell hyperplasia, and elevated mucus production, which is a characteristic of allergic asthma. The activation of Th2 cells is thought to be crucial for the progression of the late phase and for the persistence of chronic inflammation and airway remodelling (Figure 15.5). Therefore, an ongoing asthmatic immune response can be considered a delayed-type hypersensitivity reaction (Box 15.1).

More recently Th17 cells, which secrete IL-17 and IL-22, have been implicated in enhancing the inflammation associated with allergic asthma. These cytokines may exacerbate the asthmatic phenotype and synergize with the Th2 cytokines IL-4, IL-5 and IL-13 to increase airway hyper-responsiveness. Indeed, the levels of IL-17 and Th17 cells increase in patients with asthma and are related to the severity of disease. However, contradictory experimental observations suggest that IL-17 has a more complex role in asthma, which may be related to the differing activities of the various IL-17 isoforms, namely IL-17A and IL-17F. Although IL-17A and IL-17F are elevated in patients with severe asthma, they may have more of role in recruiting neutrophils,

rather than influencing eosinophilia or allergic responses. The importance of Th17 cells in asthma pathogenesis therefore remains unclear.

The immunopathology associated with asthma is characterized by an eosinophil infiltration into the airways, which is thought to contribute to the allergic state and may be central to the chronic nature of the disease. In some asthmatic patients, eosinophils are the predominant cell type located throughout the conducting airways. Persistent asthma is often characterized by the prolonged presence of eosinophils in the airways, which are activated and persist in the presence of the Th2 cytokine IL-5. Eosinophils are thought to cause direct damage to airway epithelial cells following IgE-mediated degranulation and the release of the granule proteins eosinophil cationic protein (ECP), eosinophil peroxidase (EPO) and major basic protein (MBP). As well as causing direct tissue damage, these granule proteins are thought to mediate bronchoconstriction, along with histamine. In addition, eosinophils secrete several cytokines such as IL-4 and IL-5, which contribute to the chronic inflammation observed in allergic asthma.

## 15.10 Allergic rhinitis

Allergic rhinitis is the result of an immune response to an environmental allergen that leads to inflammation of the nasal passages and ocular conjunctiva. The pathogenesis of allergic rhinitis shares many parallels with asthma, in that it is characterized by a type I allergic hypersensitivity reaction involving early and late phase responses. The principal trigger for allergic rhinitis is the cross-linking of IgE on the surface of mast cells by a specific allergen. Mast cell activation and subsequent degranulation causes inflammation of the nasal mucosa, the overproduction of mucus, dilation of blood vessels and the associated itching and congestion. Histamine, leukotiene-$D_4$ and prostaglandin-$D_2$ all play a significant role in effecting the immune response during allergic rhinitis. Mast cells also release pro-inflammatory mediators such as IL-5, which attracts Th2 cells and eosinophils. Th2 cells in turn secrete IL-4, IL-5 and IL-13, which has the combined affects of stimulating B cells to produce IgE and recruiting both eosinophils and more mast cells. As with asthma, an initial sensitization phase is necessary for heightened allergic reactions to the same allergen, and involves the activation and differentiation of Th2 cells, which is dependent on antigen presentation by DCs, and the production of allergen-specific IgE by plasma cells.

Subsequent encounter with allergen then results in mast cell degranulation, due to IgE that is already bound to the cell surface, and the resultant manifestations of an allergic immune reaction. Due to the shared immunological mechanisms, atopic individuals often develop both asthma and allergic rhinitis. Moreover, the trigger of those responses is often due to the same allergen, such as tree pollen or animal dander.

## 15.11 Skin allergy and atopic dermatitis

Atopic dermatitis is the most common form of chronic inflammatory skin disease and is characterized by epidermal hyperplasia, epidermal oedema, extended rete ridges and an inflammatory infiltrate into both the dermis and epidermis, which consists mainly of memory CD4+ T lymphocytes. Atopic dermatitis is also associated with elevated levels of IgE and with an increase in the number of dermal mast cells and eosinophils. The acute phase of the inflammatory response in atopic dermatitis involves the production of the cytokines IL-4, IL-5 and IL-13, predominantly by Th2 cells. However, as the disease progresses, Th2-type cytokines decrease, while some Th1-type cytokines increase. The chronic phase of the disease therefore requires the involvement of the Th1 cytokines IL-12 and IFN-γ. The chronic disease also seems dependent on the continued recruitment of eosinophils into the skin, which are dependent on IL-5 production. Therefore, chronic atopic dermatitis appears to be a complex disease involving the interplay between several Th1 and Th2 cytokines and the ongoing recruitment of inflammatory cells (Figure 15.7).

## 15.12 Food allergies

The immune system of the gastrointestinal tract must balance the need for immune defence against invading pathogens, with the induction of tolerance to harmless agents such as food. A breakdown in tolerance induction to food items may lead to the development of gastrointestinal disease. Allergic reactions against food items can cause debilitating illness, which in some situations can be fatal. The principal cause of food allergies is thought to be the induction of excessive Th2-mediated immune reactions. Sensitization to a food allergen results in an immediate hypersensitivity reaction upon subsequent exposure to the same allergen. In much the same way as allergic reactions are mediated in the airways, most food

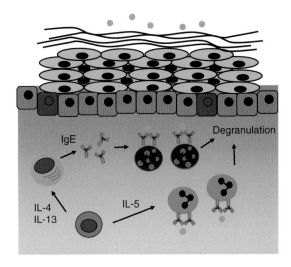

**Figure 15.7** Atopic dermatitis is characterized by Th2 cells secreting IL-4. IL-5 and IL-13, which activate mast cells in the dermis and result in the recruitment of eosinophils. Mast cell and eosinophil degranulation results in damage to the epidermis.

allergies in the gastrointestinal tract involve IgE, which is specific to the antigen, and the activation of mast cells, basophils and Th2 cells. Common examples of food items that cause allergies include peanuts, milk and diary products, fish and shellfish.

Under normal circumstance the epithelium provides an effective barrier to the absorption of proteins, as well as commensal bacteria. Nutrient uptake is a tightly regulated process that usually avoids stimulatory contact with immune cells resident in the gut. However, food allergies have been associated with damage or disruption to the epithelial layer, which is likely to increase contact with immune cells such as APCs and T cells. Although epithelial barrier function is important in maintaining intestinal homeostasis, disregulation of oral tolerance is thought to be the primary cause of allergic immune reactions to food. Importantly, Treg cells are essential for the maintenance of oral tolerance, and function in a similar way as Treg cells do during peripheral tolerance (Figure 15.8).

The process of maintaining tolerance to food items involves the production of regulatory cytokines such as IL-10 and TGF-β. Individuals with food allergies have been shown to express much lower levels of IL-10 and TGF-β and they have corresponding higher levels of inflammatory cytokines, including IFN-γ and TNF, in addition to elevated type 2 cytokines IL-4, IL-5 and IL-13. Treg cells therefore provide the necessary signals that prevent activation of Th2 cells. Under healthy

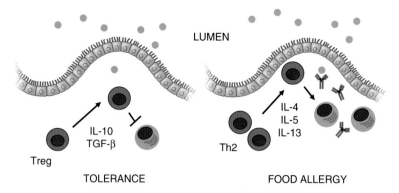

**Figure 15.8** The development of allergic reactions to food is associated with a lack of immunoregulation afforded by Treg cells. Food allergy occurs in the absence of tolerance.

circumstances, immunoregulation causes T cells to undergo anergy (a state of unresponsiveness) or cell death, so that any allergen-reactive T cells are deleted. This is the basis for oral tolerance, as Th2 cells are necessary for the induction of active immune responses, including the production of allergen-specific IgE by B cells. Neonates are though to be particularly susceptible to food allergen sensitization due to their immature immune system and the relative lack of Treg cells. One of the approaches used to treat food allergies is the targeted induction of tolerance within the gastrointestinal tract. This involves feeding allergic patients small quantities of the allergen and then increasing the dose over a period of time. The aim of this therapy is to induce oral tolerance by decreasing sensitization to the allergen and increasing the maximum amount of allergen the patient can tolerate.

## 15.13 T cell subsets in allergy

The dominant T cell subset present in allergic reactions is a Th2 cell population. These cells are driven by the expression of the transcription factor GATA3, which is essential for the differentiation into a Th2 phenotype. This is not the case with CD4+ T cells that differentiate into Th1 cells, which are characterized by the expression of the transcription factor T-bet (Figure 15.9). GATA3 is therefore a typical marker of CD4+ T cells found within asthmatic airways and in skin with atopic dermatitis. The cytokines that drive the differentiation of Th2 cells are IL-4 and IL-33, which are probably initially secreted by mast cells, basophils or NKT cells. This is in contrast to Th1 development, which is driven by IL-12 expressing DCs and the promotion of IFN-γ expression. The

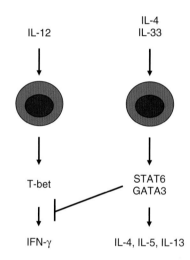

**Figure 15.9** Th2 cell differentiation is mediated by the transcription factors STAT6 and GATA3, which are upregulated in response to IL-4 and IL-33.

upregulation of GATA3 in Th2 cells is regulated by IL-4 signalling and the activation of the signalling molecule STAT6, through the phosphorylation of Jak1 and Jak3. Phosphorylation of GATA3 enables it to translocate into the nucleus and initiate gene transcription. The *IL-4-IL-5-IL-13* locus is directly affected by GATA3, causing the expression of all three of these Th2-type cytokines. In addition, GATA3 inhibits the expression of IFN-γ (through the down regulation of the IL-12R) and it augments its own expression, acting to further reinforce Th2 cell differentiation. GATA3 is also thought to cooperative with another transcription factor, NFAT, during the activation of Th2 cells and the upregulation of IL-4, IL-5 and IL-13.

Another characteristic of individuals who suffer from allergies is the lack of immune regulation afforded by Treg cells. Atopic people often have a smaller Treg cell population than non-atopic individuals. For example, there are less TGF-β-producing or FoxP3-expressing T cells in the lungs of asthma patients. The decrease in Treg cell numbers may also be linked to the hygiene hypothesis (Box 15.2), which suggests that increases in atopy are related to a reduced parasitic burden. In other words, the lack of immune stimulation results in a lack of normal Treg cell numbers.

## 15.14 Mechanisms of autoimmune disease

Autoimmune diseases are a family of diseases characterized by an inappropriate immune response that is directed towards self constituents, resulting in an excessive response against its own cells, which leads to immunopathology and tissue damage. For a disease to be classified as an autoimmune disease it must fulfil one or more criteria. There must be evidence of the presence of autoreactive antibodies directed against a self antigen, that these antibodies are able to induce immunopathology, that T cells are present that react to autoantigens and that these autoreactive cells can transfer an autoimmune phenotype in an experimental model. The distinction between a bone fide autoimmune disease and an immune-mediated disease with no autoimmune component is often blurred. For example, psoriasis is sometimes considered to be an autoimmune disease of the skin, even though no autoimmune antigen has been discovered. In a similar way, chronic obstructive pulmonary disease is associated with the chronic activation of the immune response, although the involvement of an autoimmune antigen remains debatable. These diseases are not typical autoimmune diseases, but rather diseases that are mediated by excessive immune reactions, and will therefore be discussed elsewhere in this text.

Central to the development of autoimmune disease is the lack of immune regulation afforded by immunological tolerance to self molecules. Under normal circumstances, T cells that are reactive to self antigens are deleted in the thymus during negative selection (discussed in chapter 3), a process that is known as the clonal deletion theory. This forms the basis of central tolerance, which is further supported by peripheral tolerance, which is mediated by Treg cells and involves the induction of T cell anergy (known as the clonal anergy theory). B cells also undergo negative

selection in the bone marrow, whereby self reactive B cells are deleted. Indeed, the majority of autoimmune diseases are thought to be mediated by self-reactive antibodies, rather than directly mediated by the T cell system. The question remains as to how autoimmune diseases are initiated in the first place, especially considering the mechanisms of tolerance that are in place.

## 15.15 Disregulation of tolerance and autoimmunity

The overriding mechanism of autoimmune induction is the disregulation of immunological tolerance and the breakdown in normal immunoregulatory processes. Several mechanisms have been proposed that may lead to the activation of self reactive T cells and B cells. Most individuals with autoimmune disease possess a deficiency in Treg cells. It is thought that Tregs play an important role in preventing the activation of self-reactive T cells and in maintaining a state of self-tolerance in the periphery. The immunoregulatory cytokines IL-10 and TGF-β are central to the mechanism by which Treg cells induce T cell anergy. A lack of proper Treg cell function and a reduction in IL-10 and TGF-β could therefore lead to the activation of self-reactive CD4+ Th cells. These helper T cells are also necessary for the activation of B cells, the production of self-reactive antibodies and for the activation of potentially immunopathogenic CD8+ CTLs. In addition, it is often the case that the antigen autoreactive antibodies recognize is not the same antigen to which CD4+ T cells are reactive against. This is known as T cell/B cell discordance. It may be that any antigen-responsive T cell is capable of providing the necessary signals for B cell activation during autoimmunity. This means that T cells responding to a pathogenic insult may be all that is required for the activation of B cells and the production of self-reactive antibodies. This mechanism is assisted by the direct presentation of antigen by a B cell to a CD4+ Th cell (Figure 15.10).

Some autoimmune reactions are thought to be driven by a mechanism of BCR-mediated positive feedback, which may contribute to T cell/B cell discordance during the initiation of autoimmunity. It is interesting to note that several autoimmune diseases are characterized by antibodies that are directed to B cell surface receptors. Examples include BCRs themselves, the complement receptor CD21, TLR9 and FcγRs. The binding of an antibody to a cell surface receptor is often enough to activate the signalling pathway for that receptor. Antibody activation of cell surface receptors would then provide the B

## Box 15.2 The Hygiene Hypothesis.

The hygiene hypothesis was introduced to explain the disparity between the high incidence of allergic disease in Western societies and its low incidence in under-developed societies. Asthma is by far the most serious of these allergic diseases, causing more than 100,000 hospitalizations in the UK alone, and clinical signs of asthma in children have increased by 5 per cent every year for the past 20 years. The increased incidence of allergy is thought to be related to the increased sensitivity of children to environmental allergens. Moreover, there appears to be a causal relationship between the lower incidence of childhood infections in Western societies and a higher rate of allergic disease. Most allergic individuals have elevated levels of IgE, which is responsible for triggering allergic responses, and are therefore considered to be atopic. What then is causing this allergic epidemic?

An association has been made between the incidence of allergy and the frequency of infectious episodes. One observation links the incidence of hay fever to family size and the incidence of infection early in life. A small family size means that children are exposed to fewer infections compared to large families. The lack of infectious episodes in childhood has therefore been associated with a higher incidence of allergy in adulthood. The hygiene hypothesis basically states that a low pathogen burden results in the development of allergy, while a high pathogen burden somehow protects individuals from developing allergy. Furthermore, sanitized Western societies have seen major advances in vaccination, antibiotic treatments and better food hygiene, which all lead to fewer infections compared to those less developed areas.

The mechanism responsible for the association between allergic disease and low pathogen burden is thought to be an immune-mediated phenomenon. One explanation relates to the division of the immune system into Th1- and Th2-mediated immunity. Importantly, allergic reactions are characterized by highly polarized Th2 responses, which are associated with IL-4 and IL-5 and the production of IgE. In addition, neonatal immune systems seem to be skewed more toward Th2-type immunity. Therefore, it has been proposed that viral and bacterial infections of infants are required to induce Th1 responses that can inhibit the development of allergic Th2 responses. However, epidemiologic evidence suggests that children with Schistosome worm infections have a lower incidence of allergic disease, even though Schistosome worms induce Th2 responses themselves. How can one Th2 response inhibit the generation of another Th2 response?

The answer may be found in how the immune system is regulated. The cytokines IL-10 and TGF-β are two immunoregulatory cytokines that dampen immune responses and are known to increase during both Th2-mediated worm infections and Th1-mediated viral/bacterial infections. Therefore, the more infections that are encountered, the stronger the immunoregulatory network and the less likely an individual is to develop allergic disease. Take away those infections, however, and the lack of immune regulation may allow allergic responses to develop. Indeed, disregulated production of IL-10 has been demonstrated in patients with allergic asthma, while polymorphisms in the *IL-10* gene have been associated with asthma susceptibility. In addition, the acquisition of a normal gut microflora is thought to enhance IL-10 production in order to maintain tolerance; a balance that may be lost in more sanitized countries. It may be that a combination of Th1-mediated immune education and the generation of a strong immunoregulatory network both contribute to the mechanism underlying the hygiene hypothesis.

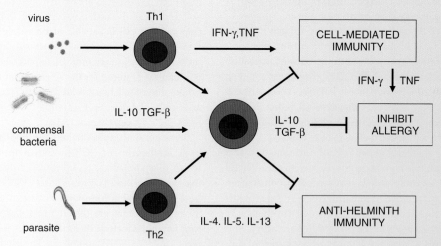

The hygiene hypothesis proposes that an immune-mediated mechanism is responsible for the aberrant incidence of allergies in Western societies.

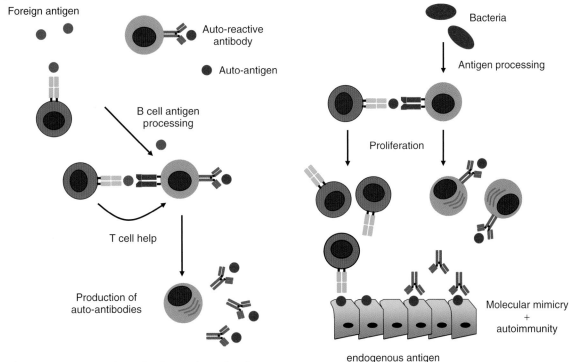

**Figure 15.10** Auto-reactive antibodies may be produced as a result of T cell/B cell discordance. The antigen that the B cell is specific for may not be the same antigen that the T cell recognizes.

**Figure 15.11** Autoimmune responses maybe the result of molecular mimicry. Activated lymphocytes that recognise pathogen-derived antigens cross react with similar endogenous antigens.

cells with stimulatory signals and result in proliferation and further auto-antibody production. This mechanism may be the reason why so many autoimmune diseases are self perpetuating. Positive feedback of self-reactive B cells may also be the result of antibodies binding to hormone or cytokine receptors, which induce the production of mediators that favour the survival of B cells.

Another mechanism that has been implicated in the induction of autoimmune responses is known as molecular mimicry (Figure 15.11). This involves a close structural similarity between a foreign antigen, derived from a bacterium for example, and a self antigen. An antibody response against that bacterium would result in the synthesis of antibodies directed to the foreign antigen but also to the self antigen. In principle, this mechanism could also apply to antigenic peptides recognized by both CD4+ and CD8+ T cells. Such cross reactivity could therefore lead to the immunopathology observed in autoimmune disease. Molecular mimicry has been implicated in driving several autoimmune diseases including multiple sclerosis, myasthenia gravis and psoriasis. A further mechanism that may

mediate autoimmunity involves the polyclonal activation of T cells or B cells following the release of a super-antigen by a pathogen. Super-antigens are capable of providing T cells or B cells with activation signals, following binding to TCRs or BCRs outside of the antigen binding site. As they are independent of antigen specificity, super-antigens are able to activate a polyclonal population of lymphocytes, some of which could recognize antigens derived from their own cells.

Several autoimmune diseases target mucosal tissues such as the gastrointestinal tract, lungs and the eye. However, most tissues or organs throughout the entire body are potential targets for autoimmune reactions. Probably the most common autoimmune diseases are rheumatoid arthritis (which affects the joints), diabetes melitis (otherwise known as type 1 diabetes) that causes destruction of insulin-secreting pancreatic islet cells and inflammatory bowel disease (IBD) (which affects the intestines). The term IBD is actually a general description of a set of more specific diseases that include Crohn's disease and ulcerative colitis (this will be discussed in more detail later). Another common autoimmune disease of the intestines is coeliac disease, which is caused by an excessive immune reaction to gluten proteins found in wheat products.

## 15.16 Inflammatory bowel disease

Inflammatory bowel disease (IBD) is a term used to describe a collection of autoimmune diseases that affect the small intestine and the colon, two of the most common of which are Crohn's disease and ulcerative colitis. The symptoms associated with IBD include abdominal pain, diarrhoea, vomiting and rectal bleeding, which can usually be treated with anti-inflammatory drugs, although severe cases may require surgical removal of the worst affected areas. The major differences between Crohn's disease and ulcerative colitis are the location of the gastrointestinal tract that each affects and the immunopathology that they cause. Crohn's disease can actually affect the entire length of the gastrointestinal tract, although the most frequently affected area is the ileum (small intestine), especially around Peyer's patches where inflammatory foci originate. The immunopathology associated with Crohn's disease affects the entire intestinal wall, including the epithelium, underlying mucosa and much of the sub-mucosal tissue. Crohn's disease is considered to be mediated by CD4+ Th1 cells producing the cytokines IFN-$\gamma$, TNF, IL-17 and IL-23. Ulcerative colitis is mainly restricted to the colon, usually only affects the epithelial mucosal lining and is thought to be mediated by an unconventional Th2-like immune response, which involves the production of IL-5 and IL-13 and the activation of NKT cells. IBD can also result in the development of autoimmune reactions outside of the gastrointestinal tract. For example, patients with IBD can also develop arthritis-like symptoms, eye and skin inflammation.

Most autoimmune diseases occur in tissues that are regarded as being sterile environments, such as the joints, thyroid gland or nervous system. IBD is slightly different as it occurs in a tissue that is far from sterile, due to the presence of enormous numbers of commensal bacteria along the entire length of the gut. The immune components that are involved in IBD therefore reflect the tissue environment. For example, one of the major factors that lead to IBD is thought to be the breakdown in tolerance to commensal bacteria. Within the intestine there is always a balance between defending the mucosa against infection, while at the same time maintaining tissue homeostasis and tolerance to commensal microorganisms. Treg cells are widely considered to provide the immunoregulatory signals that maintain this tolerance and any aberration in their normal function can lead to excessive cell activation. Gastrointestinal tolerance is an important concept, while commensal bacteria can actually be considered to be part of the self antigen repertoire. Although the commensal microflora seems to be essential for the initiation of chronic IBD, a strong genetic predisposition to the disease has been determined, whereby mutations in various immune genes make individuals more susceptible to disease. One example is a mutation in the peptidoglycan receptor NOD2, which is expressed by mucosal epithelial cells, DCs and macrophages. A decrease in the function of NOD2 is thought to render people more susceptible to bacterial invasion, for example due to a decrease in antimicrobial peptide secretion. The result is bacterial invasion, an increase in tissue damage, loss of immune tolerance and the development of excessive inflammation in the gut. Damage to the integrity of the mucosal lining and subsequent inflammatory signals is thought to be another key event in IBD.

With regard to Crohn's disease, excessive immune reactions in the intestine are thought to be mediated by Th1 cells and the release of IFN-$\gamma$ and TNF. Ulcerative colitis, on the other hand, has long been considered a Th2-like disease, with an increase in IL-5 and IL-13. However, recent investigations have revealed that the Th17 sub-population of T cells may be responsible for driving immunopathology in both Crohn's disease and ulcerative colitis. Therefore, rather than the conventional Th1 or Th2 cytokines driving the excessive inflammation, IL-17 may be more important in mediating the immunopathology observed in IBD. The effects if IL-17 are largely dependent on the upregulation of IL-23, which acts to maintain Th17 cells and causes the release of proinflammatory mediators from epithelial cells, macrophages and T cells. In addition, the inflammatory phenotype associated with IBD is accentuated by the lack of immunoregulation afforded by Treg cells. Individuals who have active IBD have been shown to express lower amounts of the immunoregulatory cytokines IL-10 and TGF-$\beta$. These cytokines have a profound effect on the activation of Th1, Th2 and Th17 cells, through the inhibition of inflammatory cytokine production and the inhibition of antigen presentation by DCs. Indeed, mice that are deficient in IL-10 develop intestinal inflammation that has a similar phenotype to IBD pathogenesis. Treg cells therefore have important functions during the maintenance of mucosal tolerance, the inhibition of T cell activation and the down regulation of ongoing immune responses.

There is evidence that patients with IBD have autoreactive antibodies specific for components of the intestinal epithelium, cytoskeleton or lymphoid system and that these antibodies contribute to the immunopathology observed in IBD. This is evident for both Crohn's disease and ulcerative colitis. For example, epithelial cell-associated components expressed on the surface of goblet cells are thought to be the target of some autoantibodies in IBD, while anti-neutrophil cytoplasmic antibodies have been detected in Crohn's disease. Furthermore, the link between a breakdown in tolerance to commensal bacteria and the development of IBD is supported by the presence of autoreactive antigens that are also specific to *Eschericia coli* or *Saccharomyces cerevisiae*. This may involve a mechanism of molecular mimicry between bacterial structures and autoantigens.

## 15.17 Coeliac disease

Coeliac disease is an autoimmune disease triggered by the ingestion of food items containing the wheat storage protein gluten. Coeliac disease is characterized by chronic diarrhoea, abdominal pain, bloating, fever and weight loss. Individuals with ongoing coeliac disease possess circulating antibodies against gluten proteins and the autoantigen tissue transglutaminase II (Ttgase), demonstrating the importance of the immune system in mediating disease pathogenesis. Genetic studies have also shown that people with certain polymorphisms in their MHC class II locus are predisposed to developing coeliac disease. In particular, a polymorphism in the HLA-DQ2 heterodimer is the major predisposing factor, suggesting that CD4+ Th1 cells are important in mediating the immunological aspects of the disease.

The wheat storage protein gluten is composed of proteins that are rich in the amino acids proline and glutamine and are therefore known as prolamines. The most immunogenic of these prolamines are the gliadins. Gluten proteins are also resistant to enzymatic digestion and are absorbed across the mucosa in a relatively intact form. It is thought the CD4+ T cell epitopes responsible for driving immunopathogenesis are derived from gliadins. Furthermore, gliadins acts as a substrate for the Ttgase enzyme, which may form a complex with gliadins, which then functions as a hapten for the stimulation of Ttgase-specific antibodies. The enzymatic activity of Ttgase may also enhance the binding properties of gliadin peptides to

MHC class II molecules, thereby augmenting CD4+ Th1 responses and the production of the pro-inflammatory cytokine IFN-γ. Therefore, a lack of tolerance to gluten proteins is responsible for driving coeliac disease.

The pathogenesis of coeliac disease is associated with episodes of inflammation within the intestinal mucosa. It has been suggested that intra-epithelial lymphocytes (IELs) contribute significantly to disease progression. Under healthy conditions Treg cells produce IL-10 and γδ T cells within the IEL compartment produce TGF-β, which are sufficient for the suppression of gluten-specific T cells and for the maintenance of tolerance. However, in coeliac disease, lesions within the intestinal mucosa are full of CD8+αβTCR+ CTLs that express the cytotoxic molecules FasL, perforin and granzyme. Therefore, during disease the regulatory populations of Treg and γδ T cells are overwhelmed by pathogenic CD8+ CTLs. This shift in phenotype seems to be the result of the presentation of gluten peptides to CD8+ CTLs and the production of IL-15 by epithelial cells in the gut. IL-15 functions by inducing the proliferation and differentiation of CTLs and their activation of cytotoxic effector functions and the induction of IFN-γ release. In summary, the breakdown of tolerance to gluten, and the resultant development of coeliac disease, is the result of a combination of immunological factors involving the activation of gluten-specific CD4+ T cells, the production of anti-Ttgase auto antibodies and a shift from a regulatory to a cytotoxic IEL population.

## 15.18 Systemic lupus erythematosus

Systemic lupus erythematosus (SLE or lupus) is a systemic autoimmune disease that can affect any part of the body but frequently the skin, lungs, eyes, muscoskeletal system, liver, kidney, heart and blood vessels. One of the key features of SLE pathogenesis is the production of antibodies directed against self antigens. The formation of antibody/antigen complexes results in tissue damage wherever these complexes precipitate. For example, immune complexes become lodged in the blood vessels of the kidneys, liver and lungs where they induce an inflammatory response and cause cellular damage. In SLE, a principal self antigen recognized by auto-reactive antibodies is actually DNA, or the proteins that bind to DNA in the nucleus such as histone proteins, which are thought to be derived from apoptotic cells. Due to the ubiquitous presence of DNA in all cells of the body, immune complexes can form in practically any tissue, which partly explains why SLE affects multiple organs.

The production of autoantibodies by B cells requires an interaction with CD4+ Th1 cells and the production of pro-inflammatory cytokines such as IFN-$\gamma$ and TNF. This interaction also involves ligation of CD40 on B cells with CD40L on T cells, a receptor ligation that delivers important signals for B cell proliferation and antibody production. Like many other autoimmune diseases, SLE is a consequence of a lack of immunological tolerance. Defects in both Treg cells and NKT cells are thought to be important in allowing SLE autoimmunity to develop. Treg cells and NKT cells produce TGF-$\beta$ and IL-10, which are effective at inhibiting T cell activation and maintaining tolerance to self antigens. By suppressing CD4+ T cell help, B cells do not receive the stimulatory signals needed for autoantibody production. A decrease in the Treg cell population is a common feature of individuals with SLE. Enhancing the Treg cell population, following treatment with corticosteroids for example, may alleviate SLE by restoring peripheral tolerance to self antigens.

One of the triggers for SLE is a defect in the apoptotic pathway. Apoptosis, or programmed cell death, is a tightly controlled process that enables the efficient removal of dead or dying cells from the body. The process of apoptosis also ensures that cellular proteins and DNA do not leak out into the tissue microenvironment. During apoptosis, DNA, cytoplasmic and nuclear proteins are normally prevented from entering the extracellular space by the phagocytic activity of tissue resident macrophages. However, some SLE patients have a defect in a specialized subset of phagocytes known as tingible body macrophages that reside in the germinal centres of secondary lymph nodes. These macrophages are responsible for the endocytosis of apoptotic B cells, which have undergone unsuccessful somatic hypermutation. It is thought that this defect in phagocytosis enables cellular proteins and DNA to escape into the extracellular environment and allows DCs to present B cell-derived antigens to T cells. The consequence of this defect may result in a breakdown in tolerance to self proteins. It is often observed that a deficient phagocyte network and an increase in apoptotic lymphocytes lead to the accumulation of apoptotic bodies in lymph nodes, where there is an increased opportunity for the presentation of autoantigens to T cells and B cells (Figure 15.12). Furthermore, DNA derived from apoptotic B cell nuclei may be able to activate both DCs and B cells, through the ligation with TLRs such as TLR9, resulting in further B cell activation and the generation of auto-reactive antibodies.

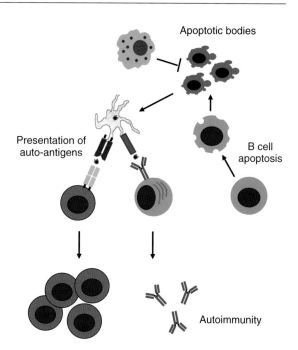

**Figure 15.12** The generation of auto-reactive T cells and B cells in SLE may be associated with defective clearance of apoptotic bodies and the presentation of self antigens.

## 15.19 Other autoimmune diseases

Many other autoimmune diseases affect organs and tissues in the body that do not belong to the common mucosal immune system (Table 15.2). Two of the most widespread of these diseases are rheumatoid arthritis and type 1 diabetes. Rheumatoid arthritis primarily affects the synovium of joints and is thought to be the result of abnormal B cell/T cell interactions and the production of autoantibodies. Type 1 diabetes mellitus is characterized by the autoimmune destruction of insulin-secreting pancreatic islet cells, which results in decreased insulin production and defective glucose metabolism. Considering that they affect a significant percentage of the human population, the immunological mechanism that drives their pathogenesis will be briefly discussed.

Rheumatoid arthritis (RA) is associated with chronic inflammation of the synovium of joints and destruction of cartilage, although it can also have systemic effects including inflammation in the lungs, pericardium and skin. The presence of autoantibodies in the serum and synovial fluid are characteristic of the disease. A common family of autoantibodies found in RA sufferers are

**Table 15.2** Autoimmune diseases.

| Disease | Tissues affected | Immune mechanism |
|---------|------------------|------------------|
| Alopecia areata | Skin, hair | T cells |
| Autoimmune hemolytic anemia | Red blood cells | Antibody |
| Autoimmune lymphoproliferative syndrome | Lymphocytes | Fas/FasL |
| Autoimmune thrombocytopenic purpura | Platelets | Antibody |
| Coeliac disease | Gastrointestinal tract | T cells |
| Crohn's disease | Gastrointestinal tract | T cells |
| Diabetes mellitus type 1 | Pancreas | T cells |
| Graves' disease | Thyroid gland | Antibodies |
| Guillain-Barré syndrome | Nervous system | Antibodies |
| Hashimoto's thyroiditis | Thyroid gland | T cells |
| Lupus erythematosus | Various | Antibodies |
| Myasthenia gravis | Muscles | Antibodies |
| Psoriasis | Skin | T cells |
| Rheumatoid arthritis | Joints | Antibodies |
| Sjögren's syndrome | Exocrine glands | T cells |
| Ulcerative colitis | Gastrointestinal tract | T cells |
| Vasculitis | Vasculature | Antibodies |
| Wegener's granulomatosis | Granulocytes | Antibodies |

known as rheumatoid factors, which are directed to the Fc region of IgG molecules. These rheumatoid factors are usually self reactive IgM molecules, although they can switch to IgA, and are therefore effective at forming immune complexes, particularly within synovial joints. Another family of autoantibodies frequently found in RA target citrullinated proteins such as type I and type II collagens, fibrinogen and fibrin. Although corresponding T cell autoantigens have not been discovered in RA, CD4+ T cells are required for antibody production and individuals with certain polymorphisms in MHC class II molecules (for example HLA-DR4) are more susceptible to developing RA.

TNF also appears to contribute to immunopathogenesis, as some of the symptoms of RA can be blocked with anti-TNF therapy. More recently, Th17 cells have been demonstrated to play a central role in driving pathology, through the production of IL-17. The deposition of immune complexes within synovial joints is thought to activate macrophages, and to a lesser extend DCs, through binding to Fc receptors. Rheumatoid immune complexes also activate complement proteins, further activating macrophages and DCs, which in turn express MHC class II molecules and provide immunostimulatory signals (TNF and IL-12 for example) that activate CD4+ T cells. This acts to further establish an inflammatory reaction that eventually leads to tissue damage, tissue

remodelling (including proliferation of synovial stromal cells) and joint destruction.

Type 1 diabetes is sometimes referred to as juvenile diabetes and is an autoimmune disease, with a significant genetic basis, that ultimately results in the destruction of insulin-producing pancreatic β-cells. It is thought that an environmental factor is required to trigger autoimmunity against β-cells and, although the exact trigger is not certain, some have suggested that this is a virus. The immunopathogenesis of type 1 diabetes is dependent on CD4+ Th cells and CD8+ CTLs. However, the production of autoantibodies may contribute to the early stages of autoimmunity and act to enhance T cell function. The targets of autoantibodies include insulin itself, although these might only reflect β-cell destruction rather than directly contributing to pathogenesis. The destruction of pancreatic β-cells is primarily mediated by CTLs that have been activated by Th1 cells. In order to induce apoptosis, CTLs utilize both the Fas-mediated killing pathway and the cytolytic pathway involving the release of perforin and granzyme. Macrophages also contribute to β-cell apoptosis through the release of reactive oxygen species.

The main genetic risk factor for the development of type 1 diabetes lies in the MHC loci within the insulin-dependent diabetes mellitus-linked gene 1 (IDDM1). During thymic development, IDDM1 peptides are loaded onto MHC class II molecules and are presented to immature T cells so as to induce central tolerance. Polymorphisms within IDDM1 result in differences in peptide presentation and therefore variations in the level of negative selection and subsequent T cell tolerance. It is thought that those individuals, who are more susceptible to developing type 1 diabetes inherit an IDMM1 polymorphism that is detrimental to T cell tolerance. Activation of these autoreactive T cells in the periphery, by an environmental factor such as a virus, then results in an adaptive immune response against pancreatic β-cells. This level of autoimmunity would also require a lack of peripheral immunoregulation, which is normally afforded by Treg cells. Furthermore, the lack of immunological suppression is partly responsible for the ongoing destruction of β-cells and the chronic nature of type 1 diabetes.

## 15.20 Immunodeficiencies

Our understanding of how the immune system works has been significantly advanced by studying human immunodeficiencies. It is often the case that the function of a complex system can only be resolved when a vital piece of the machinery is removed. Immune deficiencies can either be genetically inherited, known as a primary immunodeficiency, or can be acquired throughout life, known as a secondary immunodeficiency. Inherited primary immunodeficiencies are rare genetic diseases resulting in a defect in the immune system that renders the individual prone to infectious diseases. Examples of primary immunodeficiencies include severe combined immunodeficiency (SCID), Wiskott-Aldrich syndrome, DiGeorge syndrome, common variable immunodeficiency (CVID) and X-Linked agammaglobulinemia (XLA), although there are approximately 150 known primary diseases. Acquired immunodeficiencies are usually the result of a disease that perturbs one or more aspects of the immune system, again making the person more susceptible to infection. Examples include acquired immune deficiency syndrome (AIDS) caused by HIV infection, lymphomas, myelomas and even age.

The most life-threatening form of immunodeficiency is SCID, which is caused by inheriting one of several genes that lead to an immune defect. The consequences of inheriting SCID are a complete lack of functional T cells and B cells, and in some cases NK cells and NKT cells as well. The most frequently inherited form of SICD is caused by a mutation in the common γ-chain (Cγ), which is a vital component of numerous cytokine receptors, including the IL-2R. This mutation is inherited from the maternal X chromosome. Therefore only males have this form of SCID, although females can be carriers for the mutation. Males who have this form of SCID have exceptionally low T cell and NK cell numbers. Although B cell numbers are normal, their activation is severely impaired due to the lack of T cell help. The second most common form of SCID is caused by a mutation in the gene that encodes adenosine deaminase (ADA). A defect in ADA function results in the toxic build-up of metabolites, particularly in lymphocytes, leading to reduced T cell, B cell and NK numbers, which are even lower than in Cγ-type SCID. Other causes of SCID include mutations in the janus kinase Jak3, the cytokine receptor IL-7R and the TCR complex components CD3δ, CD3ε or CD3ζ. A very rare form of SCID is caused by a defective recombinase activating gene (RAG1 or RAG2), which are essential for TCR and BCR generation. Without function TCRs or BCRs, neither T cells nor B cell can differentiate from immature lymphocytes, although people with this form of SCID do have NK cells. Rag1 and Rag2 knockout mice have been used extensively for the investigation of immunodeficiency diseases.

Wiskott-Aldrich syndrome is an X-linked recessive disease characterized by defective T cell and B cell responses, repeated infections, eczema and defective platelets, which cause the patient to bleed easily. This immunodeficiency is caused by a mutation in the Wiskott-Aldrich syndrome protein (WASP), which is expressed during haematopoiesis, although its function remains unknown. The main cause of the immunodeficiency is a reduction in the amount of immunoglobulin, in particular IgM and IgG. Abnormalities in the B cell system are further emphasized by Wiskott-Aldrich syndrome sufferers having a higher then normal susceptibility to B cell lymphomas. XLA is another immunodeficiency disease caused by the defective production of antibodies. The mutation responsible for XLA lies in the gene that encodes the protein Bruton's tyrosine kinase (BTK), which is necessary for the proper expression of immunoglobulin. Individuals with XLA therefore possess immature B cells but no mature B cells, as they are unable to pass through the necessary checkpoints during differentiation. XLA is therefore characterized by the complete absence of any antibody at all.

DiGeorge syndrome is caused by a deletion in part of chromosome 22 that affects normal embryogenesis. Infants with DiGeorge syndrome therefore have various defects in one or more organs, although the severity of the defect and the organ involved can differ widely. The thymus is one organ that is often affected in

DiGeoge syndrome, which leads to a diminished T cell differentiation and a subsequent immunodeficiency. DiGeoge syndrome sufferers are more susceptible to infections, although most do develop immunity by adulthood. More common fatalities are due to defects in other organ systems such as the heart.

Examples of other primary immunodeficiencies include common variable immunodeficiency and CD40 deficiency, which are similar to XLA; ZAP-70 and MHC antigen deficiencies, which result in T cell defects; and various complement deficiencies, which render individuals more susceptible to bacterial infection. Other examples include immunodeficiency disorders that cause an over expression of an immune molecule or a hyperactive immune system. For example, hyper-IgM syndrome is characterized by heightened levels of IgM and the inability to generate other immunoglobulin isotypes, resulting in a higher susceptibility to recurrent infections. X-linked lymphoproliferative disorder (XLP) manifests itself following infection with Epstein-Barr virus and results in a mononucleosis that leads to extensive tissue damage. XLP can also cause B cell lymphomas and various autoimmune complications such as vasculitis. Similarly, autoimmune lymphoproliferative syndrome (ALPS) is characterized by autoimmunity, as a result of a defect in lymphocyte apoptosis caused by a mutation in the gene encoding Fas. In summary, although autoimmune diseases are rare, the consequences of inheriting one are that there is a significant impact on the health of an individual.

## 15.21 Summary

1. The importance of a properly functioning immune system is clearly demonstrated when normal immune function is lost and immune-mediated disease develops.
2. Allergies are the result of excessive and unnecessary Th2-mediated immune responses to an allergen, and are examples of hypersensitivity reactions.
3. Allergic reactions involve a sensitization phase, an IgE-mediated acute phase and a cell-mediated chronic phase.
4. Autoimmunity occurs when the immune system reacts against a self antigen and causes tissue damage.
5. Most autoimmune diseases are mediated by the deposition of autoreactive antibodies.
6. Immunodeficiencies are the result of a genetic mutation inherited from one's parents.

## With contributions from Prof. Clare Lloyd

Head of Leukocyte Biology Section, National Heart and Lung Institute, Imperial College, London

# 16 Mucosal Tumour Immunology

## 16.1 Introduction

Many cancerous tumours develop at the epithelial surfaces of the mucosa, such as the gastrointestinal tract and respiratory tract. These epithelial sites are rich in mucosal associated lymphoid tissue. Indeed, the mucosal tissue is the portal of entry for the vast majority of pathogens; it houses an array of commensal microorganisms and provides a barrier between the environment and the delicate internal organs of the body. The immune system plays a vital role in maintaining mucosal homeostasis, providing protection against infectious agents and preserving immunological tolerance to innocuous external substances such as food and commensal microorganisms. In addition, the immune system continuously surveys mucosal (and other) sites for cellular abnormalities, including tumour cells.

The importance of the immune system in combating tumour development was first demonstrated in 1891 by William B. Coley at Memorial Hospital in New York (Figure 16.1). He noticed that patients undergoing tumour regression often presented with a coincident fever and infectious episode. The regression was then attributed to the actions of the immune system resulting from the infection. Coley then attempted to imitate these fevers by injecting cancer patients with a concoction of bacterial toxins from *Streptococcus pyogenes* and *Seratia marcescens*. On his first attempt he injected the toxins directly into a large abdominal tumour of a 16-year-old man. The patient reacted with chills and fever as if infected and Coley had a remarkable success in that the tumour completely disappeared over a three-month period. Unfortunately, Coley's toxins, or mixed bacterial vaccine (MBV), is no longer in full use and toxin production ceased in 1953. However, interest in the anti-tumour properties of Coley's toxin has returned in recent years and still represents an essential insight into the importance of the immune system in killing tumour cells.

Cancer is predominantly a disease of old age, an observation attributed to the gradual attrition of the immune system and the inability to detect or control tumour cells. This is related to the important concept of the 'immune surveillance of cancer', which was tentatively introduced to explain the low incidence of tumours in early life. It basically states that immune cells circulate throughout the body and detect and destroy abnormal cells. However, this model has been cast in doubt as some immunodeficient mice do not have an increased susceptibility to tumours. It may be more likely that the number of subtle mutations required for a cell to become cancerous takes a certain length of time coincident with the age of the individual.

The development of tumours can also be viewed from a slightly different, evolutionary perspective. Tumour cells are selected for survival on the basis of their capacity to proliferate indefinitely but also on their lack of immunogenicity. In effect, tumour cells hide from the immune system because they look just the same as normal somatic cells. It now seems that the natural immune system can slow the progression of tumours but not completely prevent tumour development, further increasing the selection pressure on tumours to survive. This chapter therefore describes how the immune system detects tumour cells, prevents the development of cancer and how immunity can be utilized for anti-tumour vaccination or immunotherapy. Furthermore, the various immune evasion mechanisms, employed by tumour cells to avoid cell death, shall be discussed. Due to the vast array of different cancers (Table 16.1) such discussion will be restricted to solid tumours of the mucosal epithelia.

## 16.2 Transformation into cancer cells

For a normal cell to turn into a tumour cell a series of genetic and epigenetic changes must occur. This is a gradual and accumulative process which may take many

*Immunology: Mucosal and Body Surface Defences*, First Edition. Andrew E. Williams.
© 2012 John Wiley & Sons, Ltd. Published 2012 by John Wiley & Sons, Ltd.

**Figure 16.1** Dr William Coley at work, in front of medical students. Source: http://www.cancerdecisions.com/022003_page.html#top.

**Table 16.1** Cancer affects many different mucosal tissues.

| Type of cancer | Affected Tissue |
| --- | --- |
| Adenocarcinoma | Oesophagus |
| | Bladder |
| | Colon |
| | Lung |
| | Vagina |
| | Cervix |
| Cholangiocarcinoma | Liver |
| | Gallbladder |
| Squamous cell carcinoma | Lung |
| | Oesophagus |
| | Bladder |
| | Colon |
| | Cervix |
| Large cell carcinoma | Lung |
| Hepatocellular carcinoma | Liver |
| Small cell lung cancer | Lung |
| Pleural mesothelioma | Lung |
| Peritoneal mesothelioma | Stomach |
| | Intestines |
| Superficial spreading melanoma | Skin |
| Nodular melanoma | Skin |
| Lentigo meligna melanoma | Skin |
| Basal cell skin cancer | Skin |

years to culminate. The accumulation of mutations in many alleles of genes that govern cell proliferation is required in order for transformation to take place. The process of making tumour cells is often termed neoplastic transformation, which literally means new growth, with regard to abnormal cell division. Mutations in a number of important genes that control cell proliferation are usually involved in neoplastic transformation, such as proto-oncogenes, apoptotic genes, growth factors and tumour-suppressor genes. The additive effect of mutations in several of theses genes produces a cancerous cell, which is able to proliferate without the normal cell cycle control mechanisms in place, eventually generating a tumour mass and then a malignant cancer (Figure 16.2).

## 16.3 Proto-oncogene activation

A proto-oncogene is essentially a gene involved in cell signal transduction that provides a mitogenic stimulus when activated. For a proto-oncogene to become an oncogene an alteration in its normal activity must occur, such as an increase in concentration or enzymatic activity, a lack of regulation or over-expression, or enhanced stability that extends its functional half-life. These changes may be the result of a mutation in the coding sequence of the proto-oncogene, a genetic duplication or an alteration in one or more of its regulating proteins. Examples of proto-oncogenes include signalling molecules such as protein kinases (e.g. Ras, c-myc) and growth factors such as epidermal growth factor (EGF) or vascular-endothelial growth factor (VEGF). The cumulative changes in several proto-oncogenes can be traced throughout the development of cancer (Figure 16.3). During the progression of colorectal cancer a series of genetic alterations occurs that leads from early adenomatous polyp formation to late stage colon carcinoma. In nearly all early adenomatous polyps, a mutation in either the adenomatous polyposis coli (APC) or β-catenin tumour-suppressor gene can be detected; in intermediate adenomas a mutant Ras oncogene can be detected; then an alteration in another tumour-suppressor gene on chromosome 18; and finally half of advanced colon carcinomas carry a mutation in the tumour-suppressor gene p53. However, it may be impossible to detect all the genetic changes that occur in any given cancer and perhaps it may be that each individual cancer has unique mutant alleles.

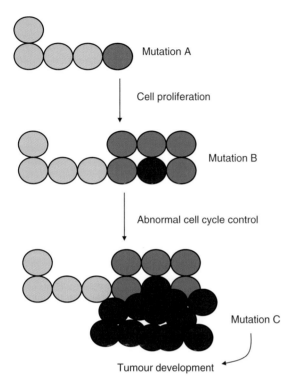

**Figure 16.2** Accumulation of mutations is required for neoplastic transformation.

## 16.4 Mutation in the p53 protein

Neoplastic transformation requires the accumulation of a number of allelic mutations over a period of time, perhaps decades. Tumour cells also possess a higher rate of mutation than normal cells and they seem poor at repairing damaged DNA. This is partly due to a breakdown in the DNA repair mechanism and a fault in the normal pathways that lead to the apoptosis of cells possessing damaged DNA. One important protein that is mutated in the vast majority of human cancer cells is the transcription factor protein p53. This is responsible for halting cell cycle progression until DNA repair has taken place and it can trigger apoptosis so as to eliminate the damaged cell. The primary function of p53 is to activate pro-apoptotic genes such as Bax and downregulate anti-apoptotic genes such as Bcl-2, following cellular stress that initiates DNA damage. A mutation in the p53 gene renders pro-apoptotic gene activation ineffective, cell cycle arrest and apoptosis no longer occur, and therefore the tumour cell can continue proliferating.

## 16.5 Mutant Ras proteins enhance proliferation

Ras proteins are retroviral-associated DNA sequences originally isolated from murine sarcoma viruses. They are now known as proto-oncogenes that positively regulate the cell cycle. For example, mutations in Ras-family genes have been associated with progression of nearly a third of all human cancers including lung adenocarcinomas. Ras is a signal transduction molecule that interacts with many downstream targets such as Raf, itself a proto-oncogene, and functions to drive cells through the cell cycle. Mutations in Ras result in over-expression of the molecule, causing cells to divide uncontrollably.

## 16.6 Aneuploidy and colorectal cancer

Numerous cancers are characterized by changes in the structure and/or number of chromosomes. A situation whereby a cell possesses an abnormal number of chromosome pairs is called aneuploidy and is often a distinctive feature of colorectal cancer, in particular during the pathogenesis of familial colon cancer (Figure 16.4). This chromosomal instability may be a result of the very high rate of cell division that cells in colon crypts undergo as a normal process of cell turnover. This may be combined with the mutation of genes that would normally induce apoptosis of the cell with aneuploidy, allowing the cancer cell to escape and continue further cell divisions. Moreover, deletion of a chromosome, or at least part of a chromosome, can result in a loss of heterozygosity. This causes a deletion in tumour suppressor genes and altered gene expression that gives the cancer cell a proliferative advantage.

## 16.7 Tumourigenesis

Solid tumours form as a result of unregulated cell division and, in the case of colorectal cancer, arise through epithelial crypt cell dysplasia thus forming an adenomatous polyp. Cancer of the epithelium is called a carcinoma and the tumour is considered malignant when it invades the underlying sub-mucosal tissue. Change from an adenomatous polyp to a carcinoma is indicative of such disease progression but is not a prerequisite of disease, as colon carcinomas cover a range of tumour types and not just malignancies. Nevertheless, the general

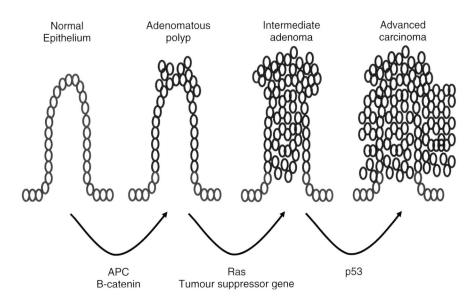

Figure 16.3 Development of colorectal cancer from normal colon epithelium is a multi-stage process. Mutations in a number of key tumour suppressor genes are associated with the different stages.

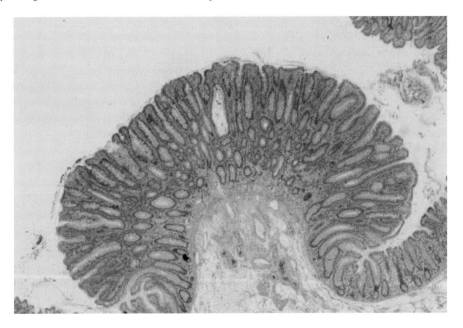

**Figure 16.4** Adenomatous polyp of the colon.

principle governing most epithelial cancers is a breakdown in normal cell growth coupled with an increase in tumourigenic potential. The collapse in normal tissue homeostasis may be the result of injury, infection or inflammation. A corresponding increase in cell division may be due to tissue repair or in response to inflammatory mediators or growth factors. These factors are then complemented by genetic changes in the epithelial cells, usually as a result of contact with environmental carcinogens (such as tobacco smoke). Once a tumour is established, it becomes governed by its own cellular responses and can be viewed as its own separate organ.

The most common form of cancer is lung cancer and denotes carcinomas arising from the epithelial cells of the lung. The epithelial cells of a normal adult lung have a low replication rate and therefore are not inherently susceptible to tumour development. However, the proliferation rate of epithelial cells in both the alveoli and bronchi is extensive following lung injury. Exposure to carcinogens enhances the accumulation of genetic mutations in oncogenes or tumour-suppressor genes resulting in epithelial cell hyperplasia and eventually in carcinoma. Progression of a cell to a cancerous state can be divided into a number of phases. Firstly, tumour cells undergo extensive proliferation through the overexpression of growth factors such as VEGF or FGF and overexpression of growth factor receptors. This presents the cell with a mitogenic signal that results in an increase in cell signalling events and transcriptional activation. This then allows the tumour cell to enter a phase of cell cycle progression. For example, cell signalling events involving oncogenic Ras or MAP kinase

culminates in active cell division and further mutations in the cell cycle control protein p53 accentuate cell cycle progression. Finally, tumour cells are adept at evading apoptosis, by either enhancing pro-apoptotic protein expression (e.g. Bax) or blocking anti-apoptotic protein function (e.g. p53, Bcl-2).

## 16.8 Angiogenesis

As tumours grow in size the requirement for metabolites, waste removal and in particular oxygen increases as cells are placed outside the limit of gaseous diffusion. One way in which tumours resolve this problem is to induce angiogenesis, the formation of new blood vessels (Figure 16.5). Angiogenesis is often initiated as a result of hypoxia within the enlarging tumour mass, although the precise factors that regulate the 'angiogenic switch' remain poorly understood. Nevertheless, specific

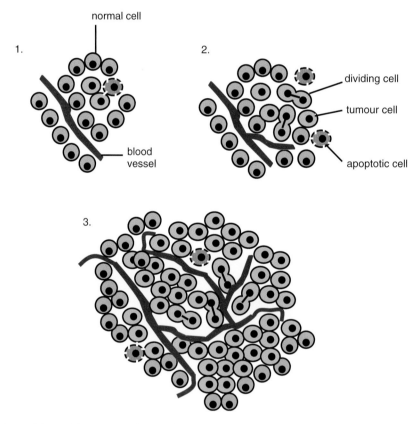

**Figure 16.5** Angiogenesis in a solid tumour. Without a blood supply a solid tumour is unable to gain enough nutrients to survive. Therefore, tumours stimulate the process of angiogenesis so that they can survive and grow.

pro-angiogenic genes are switched on in tumour cells, which encourage the growth of new blood vessels and allow the tumour access to the vasculature. However, the blood vessels that form during tumour angiogenesis are different from the normal vasculature, mainly due to a lack of appropriate control mechanisms. Tumour-induced angiogenesis often produces blood vessels that continue to grow and display haemorrhagia due to excessive permeability. Angiogenesis is an important process during tumour progression and one that has been targeted for cancer drug therapy. A cessation in angiogenesis often leads to a cessation in tumour growth. Dormant tumours do not display angiogenesis and the rate of apoptosis is higher than proliferation, therefore keeping the tumour in check. Following the angiogenic switch, this balance is lost and the rate of proliferation far exceeds that of apoptosis and the tumour can progress.

## 16.9 Metastasis

Another characteristic of many malignant tumours is the ability of individual cells to migrate out of the primary site of tumour growth, into the bloodstream or lymph and disseminate to other tissues, causing secondary tumour growth. This process is known as metastasis and tumour development in distant tissues is referred to as metastatic growth. The mechanism of cell migration is essentially identical to that used by lymphocytes in normal inflammatory situations (see Chapter 5). This requires the same receptor-ligand interactions involving integrins, cell-adhesion molecules and chemotaxins. In addition to single cell migration, clusters of cells or multicellular sheets can also metastasize from primary epithelial tumours.

## 16.10 The immune system and cancer

Coley is generally regarded as the first person to realize the importance of the immune system in combating tumour progression. However, it has been much more difficult to experimentally demonstrate the importance of an immune response against tumours. This was eventually achieved by transplanting tumours into syngeneic mice. Following tumour transplantation, the recipient mice experience tumour growth without the induction of an immune response. If mice are injected with tumours

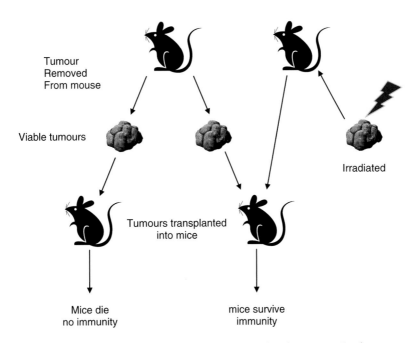

**Figure 16.6** An immune response to tumour cells was demonstrated using irradiated tumours. Mice that were transplanted with viable tumours from recipient mice went on to die. However, mice that were previously immuniszed with irradiated tumours survived. This shows that mice are able to mount an effective immune response against tumour cells, which provides immunity and protection.

that have been irradiated before transplantation, so that they lose their ability to proliferate, then they are protected from subsequent viable tumour transplantation (Figure 16.6). Furthermore, these effects can not be reproduced in T cell-deficient mice indicating the importance of T cells during the protective immune response. These experiments also revealed the existence of tumour associated antigens (TAAs), which are presented to T cells in the context of MHC molecules. Mice immunized with irradiated tumour cells but transplanted with a different tumour are not protected from tumour growth and neither are mice with a T cell deficiency (Figure 16.7). The TAAs from the separate tumours are completely different; therefore no protective memory response is generated.

Both innate and adaptive immune systems play a role during anti-tumour immunity. There are several key issues in the field of cancer immunology, namely, 1) does the immune system recognize tumour cells as self or non-self; 2) are tumour cells recognized by the immune system as dangerous; 3) is the immune surveillance hypothesis a real immunological event; 4) how significant are tumour-specific antigens; and 5) how do tumours evade the host immune response? Tumour immunity involves the same principal cell types as anti-pathogen immunity. APCs

process and present peptide on their surface in conjunction with MHC molecules and are provided with secondary 'danger signals'. T cells recognize the peptide-MHC complex and interact with co-stimulatory molecules on the APC, which provide the appropriate signals for activation and proliferation. T helper cells can elicit both CTL responses and activate B cells to produce antibody. These immune cells then initiate their effector functions, provide the correct inflammatory mediators and specifically destroy tumour cells. These essential steps toward effective immunity rely on a number of important events. APCs and T cells must recognize the presence of tumour antigens, this recognition must be in an appropriate context (non-self/danger) so as to provide a secondary signal for activation, and the tumour must not overcome immune surveillance, thereby escaping immune defence. Other immune cells, in particular NK cells and γδ T cells, also provide essential anti-tumour effector functions.

## 16.11 Immune surveillance

The Erhlich-Thomas-Burnet hypothesis of immune surveillance states that the immune system is capable of recognizing and destroying nascent tumour cells, as a part of its natural process of detection through circulation around the body. Paul Erhlich first proposed this theory in 1909, envisaging that immune surveillance could suppress a high frequency of carcinomas. Burnet and Thomas furthered this concept 50 years later by stating that thymic-derived cells constantly surveyed tissue for neoplastic cells. The immune surveillance hypothesis has been expanded again in recent times. Mice deficient in the recombinase activating gene 2 (Rag2) cannot express functional T or B cell receptors and therefore lack αβ T, B, NKT and γδ T cells. These mice demonstrate a significantly reduced capacity to eliminate chemically-induced sarcomas and they develop spontaneous epithelial tumours. In human cases of immunodeficiency, the frequency of tumours is higher, exemplified by the development of Kaposi's sarcoma in HIV infected patients. Furthermore, tumour suppression by immune surveillance was demonstrated to be dependent on IFN-γ production by lymphocytes and the importance of tumour cell killing by CTLs has also been highlighted. On the other hand, tumours that develop within immune privileged sites do not occur at any higher frequency than tumours at other sites do. In addition, the most common epithelial cancers occur at similar

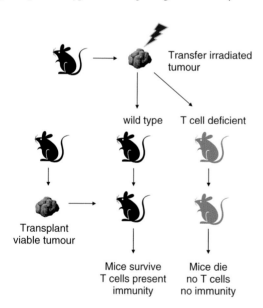

Transfer irradiated tumour

wild type    T cell deficient

Transplant viable tumour

Mice survive
T cells present
immunity

Mice die
no T cells
no immunity

**Figure 16.7** Tumour immunity requires the presence of T cells. Mice transplanted with viable tumours develop immunity to the tumour if they have been previously immunized with irradiated tumours. However, mice that have a T cell deficiency go on to die, even though they have been immunized. This demonstrates that T cells are required for the development of anti-tumour immunity.

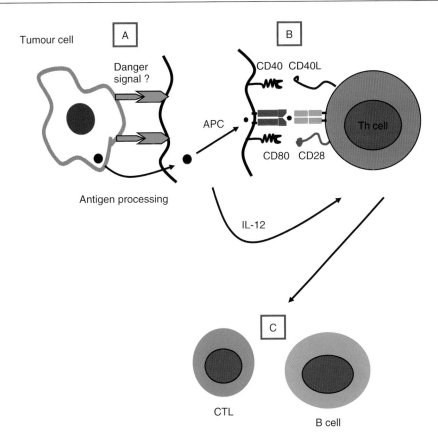

**Figure 16.8** Immunogenicity of tumour cells is influenced by the recognition of danger signals by APCs (A) and the presentation of tumour antigens to CD4+ Th cells (B). Tumour antigens are presented to T cell for antigen processing by APCs. T cell help is also required for the generation of CTL responses and antibody responses to tumour cells (C).

rates in immunosuppressed individuals and those that do occur at a higher rate are associated with viral infections. In general, immune surveillance may be important for some cancers but less so for others and may depend on the clinical setting or immune status of the individual.

The effectiveness of immune surveillance is limited by the immunogenicity of the tumour cell (Figure 16.8). It has been proposed that the vast majority of transformed cells are detected by the immune system and eliminated. However, tumours still develop and grow. Therefore, it is thought that tumour cells are selected to become less immunogenic as a result of pressure from immune surveillance. In effect, these tumour cells hide from the immune system as they are not recognized as being non-self/dangerous and therefore immune cell activation does not occur. Furthermore, specific elimination of tumour cells often requires the expression of tumour associated antigens, which are produced only in the context of a

tumour cell and presented on the surface with MHC class I molecules. The various ways in which tumours evade the immune system shall be discussed in a later section but for now we shall focus on the positive aspects of immunity to tumours.

## 16.12 Immunogenicity of tumour cells

Most tumours abnormally express one or more proteins. As such, the vast majority of tumour associated antigens are derived from self proteins that are over-expressed, expressed in the wrong tissue, or the protein is altered by mutation such that it appears foreign (Box 16.1). However, the generation of T cell immunity and subsequent T cell help to B cells and therefore antibody production, requires the expression of MHC and co-stimulatory molecules (Figure 16.8). This is a

**Box 16.1 Tumour Associated Antigens (TAA).**

| Classification | antigen | Cancer/tissue distribution |
| --- | --- | --- |
| Mutated genes | p53 | Ubiquitous |
| | alpha-actinin-4 | lung carcinoma |
| | Bax | ubiquitous |
| | Bcl-2 | ubiquitous |
| | B— catenin | melanoma |
| | cdc27 | melanoma |
| | Hsp-70 | renal carcinoma |
| | k-ras | pancreatic adenocarcinoma |
| | n-ras | melanoma |
| | myosin class I | melanoma |
| Over-expressed genes | p53 | ubiquitous |
| | neu/Her-2 | ubiquitous |
| | cyclin-D1 | ubiquitous |
| | M-CSF | liver, kidney |
| | mdm-2 | ubiquitous |
| | Muc-1 | glandular epithelia |
| | RAGE-1 | retina |
| | telomerase | testes, haematopoetic |
| Differentiation antigens | carcinoembryonic antigen (CEA) | gut carcinoma |
| | Kallekrein-4 | prostate |
| | mammaglobin-A | breast cancer |
| | tyrosinase | melanoma |
| | Melan-A/MART-1 | melanoma |
| | gp-100/Pmel-17 | melanoma |
| Cancer testes antigens | MAGE-related proteins | testes |
| Viral Antigens | HTLV-1 (Tax) | T cell leukemia |
| | Hepatitis-B (HBsAg) | hepatic carcinoma |
| | EBV (membrane antigen) | Burkitt's lymphoma |
| | Papilloma virus (E6 and E7) | cervical carcinoma |

prerequisite for immunogenicity without which the tumour can avoid significant immunity. A major characteristic of tumour cells is their low expression of MHC class II and co-stimulatory molecules CD80 and CD40. In addition, tumour cells do not provide APCs with danger signals, making immune cell activation more difficult. This lack of immunogenicity is further augmented by the presentation of weak self-peptides. Tumours are often more immunogenic during the early stages of growth and become less immunogenic as they mature. This may be related to the selection pressure placed on the tumour from the immune system to escape effective immunity. The immunogenicity of tumour cells can be enhanced, however. The addition of IFN-γ or IL-2 to the tumour environment increases the expression of MHC and co-stimulatory molecules,

thereby making them more immunogenic. Certain cancer immunotherapies target the sensitivities of tumour cells to cytokine treatment, while others attempt to increase co-stimulatory molecule expression. Increasing the immunogenicity of tumours may be a realistic way of treating human cancer and facilitating immune responses to weak tumour antigens.

## 16.13 Recognition of transformed cells

The whole process of immune surveillance involves both innate and adaptive immune cells. One of the central questions asked of immune surveillance is how transformed tumour cells are recognized from normal tissue cells? Firstly, CD4+ and CD8+ T cells recognize tumour antigens presented on MHC class II and MHC class I molecules respectively. Tumour antigens can be categorized into distinct types based on their cell distribution and origin of expression (these will be discussed in detail later). Secondly, tumour cells abnormally express other antigens that can be recognized by immune cells during tissue surveillance. These include stress-related proteins such as heat-shock proteins or cell matrix proteins that act in the same way as TLR ligands during pathogenic insults. Alternatively, some tumour-derived molecules can initiate the cytolytic activities of NK cells or CD8+ T cells. One example is the over-expression of ligands for the NKG2D receptor, allowing NK cell-mediated killing of tumour cells (discussed in more detail later). Conversely, the down regulation of MHC class I increases the capacity of NK cells to kill tumour cells. Subsequent cellular stress and tumour cell killing resulting in the release of inflammatory mediators such as IFN-α, TNF, TNF-related apoptosis-inducing ligand (TRAIL) and heat-shock proteins, ultimately leads to the activation of other arms of the immune response.

## 16.14 Tumour associated antigens

Tumour associated antigens (TAAs) can be categorized into five groups, defined by their ability to induce CTL responses in a MHC class I restricted manner:

1. self proteins transcribed from mutated genes (e.g. mutated forms of p53;
2. over-expressed antigens (e.g. neu);
3. differentiation antigens such as melanocyte differentiation antigens (e.g. Melan-A/MART-1, gp-100/Pmel17);
4. cancer testes antigens expressed on normal testes and transformed cells (e.g. MAGE related proteins); and
5. viral antigens (e.g. EBV antigens).

TAAs are an essential component of tumour-specific T cell responses and provide a target for CTL-mediated tumour cell killing. However, some tumour associated antigens do not fall into any of these categories as they do not induce CTL responses but rather they induce CD4 or humoral responses. Furthermore, the phase tumour rejection antigen (TRA) is often used to describe a TAA that activates an immune response to a tumour and assertively kills tumour cells, thereby reducing its mass.

TAAs can be broadly divided into universal tumour antigens and patient-specific tumour antigens. Universal tumour antigens are the products of normal gene expression, for example, the MAGE family of proteins. As these antigens are expressed in normal physiological conditions,

they will have triggered some degree of T cell tolerance and are therefore poor inducers of tumour immunity. However, universal tumour antigens that are expressed only at an early stage of embryonic development or within certain tissues, such as immune privileged sites, will have the capacity to induce significant anti-tumour immune responses. Patient-specific tumour antigens are the products of mutated genes that are particular to an individual tumour, such as a mutated p53. Immunity can therefore be generated against the mutated peptide as tolerance to this 'new' antigen has never taken place. Patient-specific tumour antigens should theoretically be more effective inducers of T cell immunity than universal tumour antigens. Moreover, immune responses generated to mutated proteins that are expressed only in the tumour cell are less likely to instigate autoimmunity. Universal tumour antigens are expressed in tissues other than the tumour and could therefore be the target of tumour reactive T cells.

## 16.15 Carcinoembryonic antigen in colorectal cancer

Colorectal cancer cells are poorly immunogenic as T cells obtained from inflammatory infiltrates within the tumour are not cytotoxic and do not proliferate in response to isolated tumour cells. However, substantial efforts have been made to ascertain the TAAs involved in the infiltration of immune cells and for their potential as vaccine targets. Carcinoembryonic antigen (CEA) is widely expressed on normal intestinal tissue cells but is abnormally over-expressed on colorectal cancer cells. It is therefore classified as a universal tumour antigen, although human CEA belongs to a large family of 29 immunoglobulin-like proteins, all of which are expressed on the cell surface. The function of CEA in healthy tissue is unclear, although it may be involved in cell adhesion or possess anti-microbial properties. CEA is being extensively tested as a potential TAA in the development of a vaccine for colorectal cancer. T cells derived from the blood of colorectal cancer patients have been shown to be reactive to CEA-derived peptide and produced IFN-γ *ex vivo*. However, natural immunity to CEA is lacking due to its expression in healthy, post-embryonic tissue. T cells are likely to be rendered tolerant to CEA and T cell unresponsiveness to CEA results in the progression to malignant carcinoma. Antibodies to CEA can be detected in colorectal cancer patients but are only transient and of comparatively low affinity. Ways in which peptide epitopes from CEA can be made more immunogenic are currently being sought.

## 16.16 Melanoma differentiation antigens

The immune response to melanomas has been extensively studied using a technique known as the mixed lymphocyte tumour cell culture. CTLs reactive against specific TAAs kill the melanoma tumour target cell in an MHC-dependent manner. This has allowed the isolation of the specific CTLs responsible for tumour cell killing and the establishment of melanoma associated antigens. Unfortunately these antigens are not immunogenic *in vivo* and do not elicit the proliferation of CTLs or curtail tumour progression. For example, melanoma differentiation antigens include a TAA referred to as melanA/MART-1 and represent probably the most widely studied TAA. MelanA/MART-1 is expressed in both normal melanocytes and transformed tumour cells. MHC

class I restricted CTLs specific for MelanA/MART-1 have been found in as many as 50 per cent of patients with melanoma but the presence of the TAA-specific cells does not directly correlate with improved clinical outcome. It does suggest, however, that natural immunity does exist against TAAs. Further to this concept, CTLs have been demonstrated against other melanoma differentiation antigens, including gp100 and tyrosinase.

## 16.17 Viral tumour associated antigens

It may be the case that immune surveillance of tumours evolved to detect virus-induced malignancies (Figure 16.9). It has become clear that many human cancers are associated with viral infection. For example, Epstein-Barr virus (EBV) is associated with Burkitt's lymphoma and Hodgkin's disease, hepatitis B and C viruses (HBV/HCV) with hepatic carcinoma, human papilloma virus (HPV) with cervical cancer, herpes

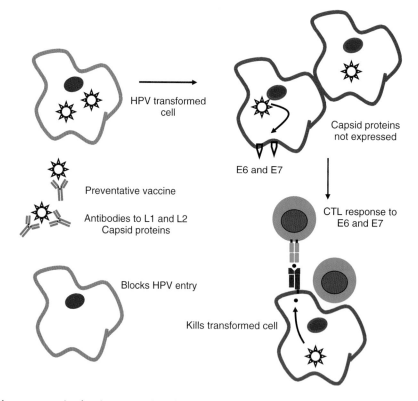

**Figure 16.9** Viral tumour associated antigens may trigger immune responses to transformed cells and to the virus itself. A cell infected with human papilloma virus (HPV) causes transformation into a tumour cell. Expression of HPV antigens and processing by the endogenous MHC class I pathway allows presentation of peptide to CTLs, which in turn kill the transformed cell. In addition, antibodies directed against HPV capsid proteins neutralize the virus and prevent further infection.

virus (HHV-8) with Kaposi's sarcoma and human T lymphotropic virus (HTLV) with T cell leukaemia. Proteins encoded by the virus can be presented on MHC molecules by the transformed cell and are therefore classified as TAAs. However, it is still unclear whether these viral products can protect against tumour growth or the development of cancer. The importance of virus-derived TAAs may be illustrated by the increased incidence of virus-induced malignancies following T cell deficiencies due to HIV infection.

HPV is the leading etiological agent responsible for the development of cervical cancer. The HPV virus expresses two proteins found in nearly all human cervical cancers, the early proteins E6 and E7. These proteins are responsible for initiating cell transformation and maintaining tumour cell growth. It may be possible to vaccinate individuals with existing HPV infections and/or cervical cancer with E6/E7 products. Alternatively, prophylactic vaccines have been proposed, which contain the capsid proteins L1 or L2. Although, these HPV capsid proteins are poorly expressed on transformed cells they may form the basis of a preventative vaccine that inhibits viral infection, rather than acting as a therapeutic vaccine for established HPV-induced transformation.

HPV infects the stratified epithelium of the genital and intestinal tract. In severe cases lesions develop that can lead to high-grade squamous intraepithelial lesions and eventually cancer. HPV-induced cancers are poorly immunogenic and only express weak TAAs E6 and E7. The lack of antigen presentation is further accentuated by the general immunosuppressive tissue microenvironment of the genital tract. There are few professional antigen-presenting cells and inflammatory cytokine production is low: an evolutionary strategy that avoids potentially damaging immune reactions around the delicate cervical and uterine areas of the body. In addition, HPV is thought to directly suppress the immune response of the host. For example, HPV-induced tumours may express extremely low levels of GM-CSF and thus prevent migration of DCs into the area. However, E6 and E7-specific CD4+ and CD8+ T cells have been reported at the tumour site and in peripheral blood, suggesting that vaccination against these HPV antigens may be viable in the future.

## 16.18 Effector molecules during tumour immune surveillance

The type of immune response generated to tumours is largely determined by the availability of MHC/peptide complexes on the surface of tumour cells, the activation of professional APCs and the proliferation of tumour reactive T cells. Of further importance are the types of cytokines produced by both the tumour and the prevailing immune response. Although CTL responses provide the most effective killing mechanisms, both cell-mediated Th1 responses and Th2-dominated humoral responses can be elicited by tumours. On the one hand IL-12, IFN-γ and TNF production enhances Th1 CTL responses, whereas IL-4 and IL-5 provide stimulation for anti-tumour antibody responses. The phenotype of the immune response is dependent on the cytokine expression of APCs, in particular DCs. APCs direct both the cytokine production of a given response and augment T cell activation through efficient antigen presentation and co-stimulation.

IFN-γ directly and indirectly affects tumour growth. Firstly, IFN-γ has direct cytotoxic activity toward some tumour cells but its most important role is its secondary activity of promoting NK cell activation and cellular killing mechanisms. IFN-γ upregulates MHC class I and class II molecules on tumour cells, thereby making tumours more immunogenic, and IFN-γ deficient mice have a higher rate of spontaneous tumours. In a similar manner IL-12 significantly enhances the activity of CD8+ CTLs, which in turn can target tumour cells for destruction. The most widely used cytokine during cancer immunotherapy is IL-2. This cytokine is known to enhance the anti-tumour activity of NK cells, probably through the enhanced production of IFN-γ and TNF. In addition, TNF has direct cytotoxic activity on tumour cells by upregulating pro-apoptotic signalling pathways. However, despite its moniker as a necrosis factor, TNF can actually promote the growth of some tumours and may even encourage angiogenesis. Even though many cytokines have been successfully used to treat tumour growth, including TNF, some have also been associated with tumour survival. For example, long-term exposure to IL-15 may enhance lymphoma formation, while MIF may interfere with the function of p53.

## 16.19 Dendritic cells modulate anti-tumour immune responses

DCs are the most specialized of all the professional antigen-presenting cells. Although T cells and NK cells provide the most effective means of anti-tumour immunity, DCs are essential for T cell activation (Figure 16.10). Infiltration of DCs into tumours occurs most readily

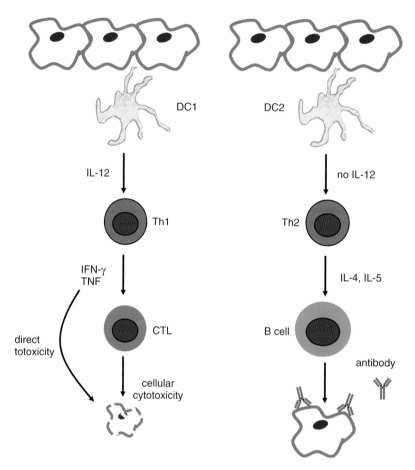

**Figure 16.10** DCs are important for directing anti-tumour immune response. DCs can process and present tumour antigens to T cells and elicit either cell-mediated or antibody-mediated immunity to tumour cells. For example, Th1-mediated cellular cytotoxicity is initiated by DC1 cells, while B cell maturation is initiated by DC2 cells.

when cell damage has taken place and when the tumour is unable to prevent migration. Tumours are differentially susceptible to DC infiltration depending on their stage, which significantly affects their disposition to immune cell attack. For example, immature tumours tend to be resistant to DC infiltration and are poorly immunogenic. Increased proliferation, cell damage and tumour mass increases immunogenicity (more 'danger signals' are released) and hence DCs both infiltrate the tumour more readily and more effectively migrate to regional lymph nodes. Immunogenicity is therefore directly related to DC activation. The more immunogenic a tumour is, the higher the expression of MHC and co-stimulatory molecules on DCs and therefore the greater the potency of T cell activation. However, even though larger tumours

are more immunogenic the tumour mass may be too great for effective immunity to take place.

The potency of DCs to induce T cell activation has been utilized with the advent of DC-based immunotherapy against tumours (Figure 16.11). Modification of DCs *in vivo*, to make them more immunogenic, has been successful in generating anti-tumour immune responses. DCs taken directly from the tumours of patients and then expanded *in vivo* have largely been unsuccessful due to the immunosuppressive microenvironment of the tumour. However, DCs isolated elsewhere from a patient, expanded/activated *in vivo*, loaded with tumour antigens and then reintroduced into the patient may provide a method for successful DC vaccination. A necessity for any anti-tumour vaccine is the appropriate activation of

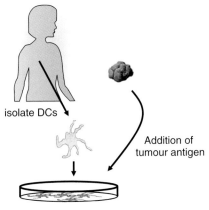

isolate DCs

Addition of tumour antigen

Expand and active DCs in vitro

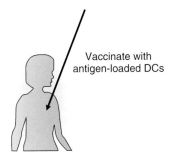

Vaccinate with antigen-loaded DCs

**Figure 16.11** DC-based tumour vaccines. DCs are isolated directly from the tumours of patients and grown and expanded in the laboratory. In addition, antigens derived from the tumours are added to the culture, which activates the DCs. These expanded and activated DCs are then injected back into the patient with the hope they will invade the tumour and initiate an immune response against the tumour.

tumour-specific T cells. The potential danger of such an approach is that the immune response may attack healthy tissue that expresses the same antigens as the tumour: a situation referred to as Ehrlich's horror autotoxicus, a severe autoimmune reaction.

## 16.20 Tumour reactive T cells are activated in lymph nodes

The migration of tumour-derived products, tumour cells or DCs loaded with TAAs to regional lymph nodes has both beneficial and detrimental outcomes regarding the cessation of tumour development. On the one hand, lymph nodes are the sites where DCs and T cells interact in order to initiate a programmed immune response. DCs that are sufficiently activated and loaded with TAAs

are able to present peptide to T cells trapped in the matrix of the lymph node. These T cells are then able to migrate out of the lymphatics, to the site of tumour growth and initiate effector functions. On the other hand, tumour cells that migrate to regional lymph nodes are able to proliferate in these lymphatic tissues and disrupt the tissue architecture. The tumour cells can also migrate out of the lymphatics and disseminate to other sites of the body, thus forming metastases.

## 16.21 NK cell recognition – missing self

NK cells can kill abnormal cells using a mechanism that is not reliant on prior sensitization or activation and is independent of specific antigen presentation. This process of natural killing is mediated by a series of activation and inhibitory receptors. NK cells express a number of killing receptors on their surface, which recognize ligands on tissue cells (Figure 16.12). Killing of healthy cells is prevented by the expression of killer-inhibitory receptors (KIRs or killer-immunoglobulin-like receptors) on NK cells, such as Ly49 or CD158 in mice. KIRs recognize MHC class I molecules on target cells and prevent the initiation of killing (Figure 16.13). On cells that downregulate MHC class I molecules, such as certain tumour cells, the NK cell killing can continue because the inhibitory signal is removed. This is known as recognition of missing self. The same principle applies to cells transplanted from a donor patient into a recipient. As these two individuals possess slightly different MHC class I molecules the NK cells recognize the transplanted cells as missing self. In other words the NK cells do not recognize these unrelated MHC molecules, interpret the transplanted cells as possessing no self MHC molecules and initiate killing of the 'abnormal' cells. As tumour cells often downregulate MHC class I as a means of escaping adaptive immune responses, the NK cells recognize this as missing self and initiate killing.

## 16.22 NKG2D receptor on NK cells

Some cells that possess no MHC class I molecules are resistant to NK cell mediated killing and some cells that have normal expression levels of MHC class I are killed by NK cells. These findings demonstrate that NK cells express stimulatory receptors, which are important during cell-mediated killing of target cells. One such receptor is NKG2D, which possesses several cellular ligands that are only upregulated on abnormal cells

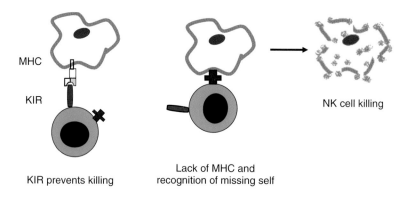

**Figure 16.12** NK cells recognize missing self from tumour cells. Certain tumour cells downregulate the expression of MHC class I molecules, which enables them to escape CD8+ CTL-mediated killing. However, this down regulation is identified by NK cells, due to the lack of an inhibitory signal provided by KIRs. NK cells are then able to initiate target cell killing mechanisms.

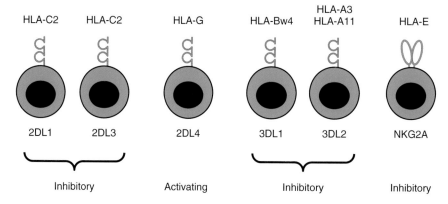

**Figure 16.13** NK cell express several KIR molecules, which recognize a variety of HLA (MHC class I) molecules. Most KIRs are inhibitory receptors, although some can activate NK cells.

undergoing stress (Figure 16.14). Such ligands include the unconventional MHC class I chain-related proteins MICA and MICB. Tumour cells expressing MICA or MICB are killed by NK cells irrespective of their MHC class I expression. Upon NKG2D receptor ligation NK cells express the pro-inflammatory cytokines IFN-γ and TNF, which are important molecular signals to activate innate and adaptive immune responses. NKG2D receptors are also expressed on αβTCR+CD8+ T cells and γδ T cells, further extending the potential importance of these types of receptor in identifying transformed cells.

## 16.23 Macrophages and neutrophils phagocytose tumour cells but support tumour growth

Macrophages and neutrophils can phagocytose and kill tumour cells by antibody-dependent cellular cytotoxicity (ADCC). Antibody bound to TAAs on the surface of tumour cells is recognized by the FcRs on macrophages and neutrophils. The cross-linking of antigen complexes on FcRs results in the activation of phagocyte killing mechanisms and the controlled release of reactive oxygen species by macrophages and neutrophils. Macrophages and neutrophils also release a number of cytokines such as TNF and IL-12 that have direct and indirect anti-tumour activities respectively. However, the infiltration of macrophages and granulocytes into the tumour mass may contribute to the malignancy and in certain tumour models the cytokines released by these cells aid tumour growth. Macrophages in particular have been associated with poor prognosis. Certain tumours over-express macrophage chemoattractants and in turn macrophages produce several cytokines or growth factors (e.g. colony stimulating factor-1, CSF-1) that contribute to tumour progression and angiogenesis. Like many aspects of the

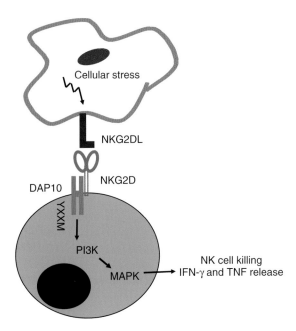

**Figure 16.14** NKG2D receptor killing of tumour cells is mediated by NKG2D, which recognises its ligand expressed on stressed tumour cells. Signalling requires the co-receptor DAP10, which is phosphorylated following ligation and causes the activation of PI3K an MAPK, which triggers NK cell-mediated killing and cytokine release.

immune system leukocytes possess both advantageous and detrimental qualities regarding tumour immunity.

## 16.24 Immune cells can augment tumour growth

A major problem for the generation of immunity to cancers is that the tumour is often recognized as self, the TAAs are weakly immunogenic and the tumour is selected to prevent immune cell damage. However, macrophages, dendritic cells, neutrophils and lymphocytes comprise a large percentage of the tumour mass, even though immune rejection does not occur. It has long been postulated that leukocytes are involved in the neoplastic process, in the maintenance of tumour mass and in angiogenesis. Indeed, tumourogenesis has been associated with an infectious episode and resultant inflammatory immune reaction.

During a normal response to infection and tissue damage leukocytes are recruited to the area of distress where they initiate an inflammatory reaction. Migration of cells into the site of infection involves cytokines, chemokines and danger signals released by damaged cells. Once the pathogen is cleared the inflammatory reaction is resolved through the release of regulatory cytokines so that tissue remodelling can take place. Further cytokines and growth hormones are needed to direct both leukocytes and other somatic cells to engage in a process of wound healing. This often involves cell proliferation, for example epithelial cell proliferation during intestinal or lung remodelling, and angiogenesis. Tumours are thought to employ similar chemotactic signals and growth factors that recruit leukocytes and support tumour growth. For example, epidermal growth factor (EGF) and vascular endothelial growth factor (VEGF) produced by macrophages may promote angiogenesis and therefore enhance tumour growth. This concept has led to the notion that tumours are in effect 'wounds that never heal'.

## 16.25 Immune evasion strategies

It is clear that the immune system can not prevent the growth of tumours in all clinical situations. In fact, natural immunity rarely develops to the vast majority of malignancies. The principal reason is that tumours effectively escape immune effector mechanisms. Furthermore, tumour cells are actively selected on their ability to avoid the immune system. Tumour cells employ many strategies to avoid cell death, enhance proliferation and escape an immune response (discussed below).

One of the key questions concerning cancer immunology is how do immune cells recognize a tumour cell, even though the tumour is derived from 'self' tissue? The default mechanism of T cells, for example, is to recognize one's own cells as self and therefore become tolerant to those cells. Tumour cells often fit into this category of being recognized as self and therefore escape immune recognition. However, T cell tolerance to self antigen is not the only issue, as DCs must also become activated in order to initiate a T cell response. This is referred to as the 'danger model', in that tumour cells may be able to hide from the immune system because they do not appear dangerous to professional APCs (Figure 16.15). Danger signals include antagonists to TLRs (PAMPs), heat-shock proteins released from stressed cells, other stress-related factors (DAMPs) and apoptosis signals. Without any danger signals, DCs do not upregulate MHC or co-stimulatory molecules, they do not express pro-inflammatory cytokines and they do not migrate to regional lymph nodes. The tumour is classified as being

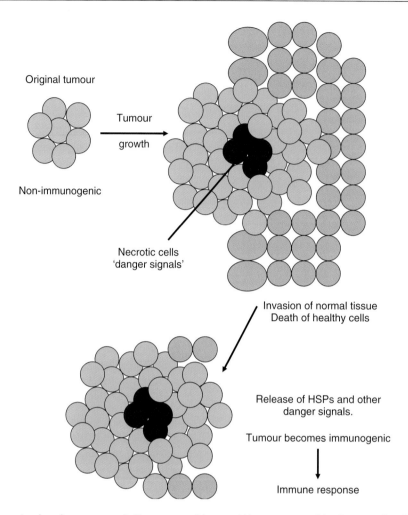

**Figure 16.15** Danger signals and tumour growth. Tumours are able to avoid immune recognition by concealing danger signals and in effect hiding from immune cells such as DCs. Only when danger signals are detected does the tumour become immunogenic, for example following cellular stress or apoptosis.

non-immunogenic, as it is does not induce an immune response.

## 16.26 Darwinian selection and tumour cell escape

Like all other biological systems, tumour cells are governed by Darwinian selection over time. One of the hallmarks of neoplastic cell growth is genetic instability, an advantageous tool upon which natural selection can work. Tumour cells receive significant selective pressure from their surroundings, be it from the need for nutritional components in order to survive or from the

immune system. A tumour cell clone will give rise to a number of tumour cell variants as a result of genetic instability, a situation known as tumour heterogeneity (Figure 16.16). These variants will comprise the tumour mass and all the clones will potentially be targeted for immune cell destruction. Those clones are therefore under selection pressure from the immune system and those better equipped to evade the intentions of the immune response will survive, while those without a selective advantage will be destroyed. The surviving clones will then form the basis of the progressive tumour mass. The same process of natural selection will result in survival of clones following immunotherapy. Those better able to evade the immunotherapy will survive and proliferate

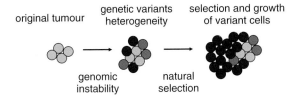

original tumour  genetic variants  selection and growth
heterogeneity  of variant cells

genomic  natural
instability  selection

**Figure 16.16** Darwinian selection of variant tumour cells. Genomic instability of tumour cells can lead to genetic variants within a population. Some of these variants will have a selective advantage over their neighbour and will eventually dominate the tumour. This process enables tumour cells to be selected based on their enhanced survival and subsequent growth.

in further generations. This process of immunological selection, whereby tumour cells survive, is also known as immuno-editing or tumour sculpting and can be broadly divided into three stages; elimination, equilibrium and escape. Elimination of original tumour cell clones by the immune system selects for tumour variants among the heterogeneic population. Equilibrium then ensues between tumour cell death and proliferation of tumour cells, resulting in a stable tumour mass. Escape of tumour cell variants eventually occurs and the tumour can progress and growth ensues as a result of immune escape (Figure 16.17).

## 16.27 Cytokine environment and tumour escape

The tumour microenvironment is host to a complex array of both tumour and immune cells, which express several cytokines and chemokines responsible for suppressing the immune response. The immunosuppressive cytokines IL-10 and TGF-β have been associated with tumour progression and elevated levels of these molecules are frequently observed in cancer patients. IL-10 blocks both the differentiation of Th1 T cells and inhibits effector functions. It also inhibits the production of IL-12 by DCs and stops antigen presentation. TGF-β also suppresses the immune system on a number of levels. Firstly it enhances the production of IL-10, therefore accentuating the immunosuppressive effects. TGF-β inhibits the effector function of T cells and macrophages and also reduces the ability of T cells to differentiate into effector cells. T cell proliferation, cytotoxicity and cytokine production are all inhibited by TGF-β.

Vascular endothelial growth factor (VEGF) is another factor produced by tumour cells that contributes to immune escape. VEGF plays a role in the vascularization process that is so crucial in tumour growth. In addition VEGF has an immuno-suppressive effect as it inhibits NF-κB activity in DCs and therefore prevents DC maturation. A lack of DC maturation is associated with poor T cell activation and hence a lack of tumour immunogenicity. DCs are basically prevented from migrating to the lymph node; they fail to express elevated levels of MHC or co-stimulatory molecules and therefore T cell activation is significantly diminished.

## 16.28 Tumours have disregulated MHC expression and antigen presentation

Down regulation of MHC class I molecules on the surface of tumour cells is a long known phenomena. Altered

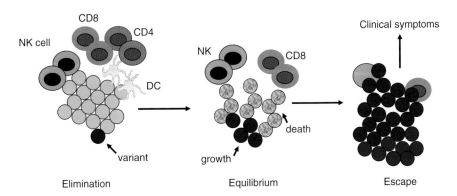

**Figure 16.17** Selection of tumour variants and tumour escape. Competition exists between the immune responses, which is trying to eliminate the tumour, and the tumour itself, which is striving to survive and grow. Genetic variants increase the opportunity for tumour cells to escape the immune system and tip the balance away from immunity and toward tumour development.

**Table 16.2** HLA phenotypes in tumour cells.

| Phenotype | Causes | Effect | Cancer Type |
|---|---|---|---|
| I Total loss of HLA class I | Impaired transcription, $B_2m$ mutation, TAP mutation | No antigen presentation to any $CD8^+$ T cells | Burkitt's lymphoma Colorectal, Lung Melanoma, Prostatic |
| II HLA haplotype loss | Loss of haplotype on chromosome 6. Loss of heterozygosity | No presentation by that haplotype to $CD8^+$ T cells | Cervical, melanoma pancreatic |
| III Locus downregulation | HLA-A, HLA-B and/or HLA-C locus downregulation | Diminished presentation to $CD8^+$ T cells | Cerivical, melanoma |
| IV HLA allelic loss | HLA-B44 allele loss | Diminished presentation to $CD8^+$ T cells | Cervical, Colorectal melanoma |
| V Compound phenotype | Combination of phenotypes | Additive effect of more than one phenotype | Cervical, melanoma |
| VI Unresponsiveness to IFNs | Unable to respond to IFN-α/β | No upregulation of HLA class I | Gastric, Renal |
| VII Abnormal Ia and Ib expression | Abnormal expression of HLA molecules | Increase in HLA-E expression. Inhibition of NK cells | Melanoma, Gastric |

expression of MHC can be either irreversible, due to a genetic mutation, or reversible, due to abnormal regulation by the tumour cell. A prescient example of a genetic mutation that impairs MHC class I molecule expression is a mutation in β2-maroglubulin, a protein that is essential for the correct folding of MHC class I complexes and expression on the cell surface. Another example is a mutation in one of the TAP proteins, which function in the processing of peptide antigens for their presentation on the cell surface in the context of a MHC molecule. Without the processing of peptides, the MHC class I complex can not assemble correctly and transport to the cell membrane is aborted. Disregulation of MHC molecules by tumour cells may also be the result of a complete loss of expression of an entire locus. An important HLA locus (human MHC locus) is located on chromosome 6 and contains HLA-A, HLA-B, HLA-C, the non-classical MHC molecule MICA and the MHC class II members HLA-DR. HLA-DQ and HLA-DP (Table 16.2). Therefore, the lack of expression of this entire locus will have a fundamental influence on the effectiveness of antigen presentation on the surface of tumour cells. However, the immunological consequence of a lack of MHC expression may result in enhanced NK cell function, through the activation of KIR receptors.

## 16.29 Tumour escape through Fas/FasL

The interaction between Fas (CD95) and its ligand FasL (CD95L) is an important system that triggers apoptotic pathways in numerous cells throughout the body and is a key element during development and in the maintenance of tissue homeostasis. Fas is a TNF receptor super-family member that, upon stimulation, activates an intracellular signalling cascade that culminates in apoptosis, which is dependent on the interaction between caspase 8 and the death domains located within the FADD domains of the Fas receptor (Figure 16.18). There are two mechanisms that tumour cells utilize in order to avoid killing by the immune system. The first is to express FasL on their cell surface, while the second is through one or more mutations in the Fas signalling cascade. In a similar way to immune privileged sites such as the eye, tumour cells express FasL, which interacts with Fas expressed on the surface of immune cells such as NK cells or CTLs. The activation of Fas on NK cells and CTLs then induces the intracellular signalling cascade and the cells undergo apoptosis, a process sometimes referred to fratricide. The second mechanism involves a mutation in the tumour cell that renders Fas signalling ineffective. One of the ways in

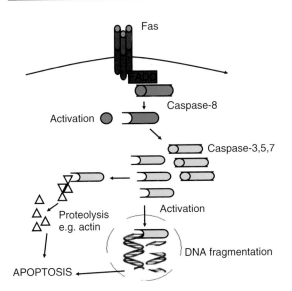

**Figure 16.18** Death receptor signalling pathways. Fas is an example of a receptor which is capable of inducing the apoptosis pathway. Activation of Fas with its ligand FasL recruits several downstream pro-caspases, which become activated and participate in proteolysis of cytoplasmic proteins and fragmentation of DNA.

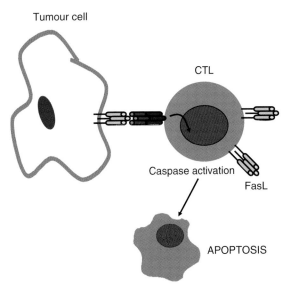

**Figure 16.19** FasL expression on tumour cells induces fratricide. Most immune cells express the Fas receptor, which contribute to normal immunoregulation. However, tumour cells are capable of upregulating FasL, which causes apoptosis of immune cells that would otherwise target the tumour cell. This process of killing one's own cells is known as fratricide.

which NK cells and CTLs kill their target cell is through the expression of FasL. Therefore, defective Fas signalling pathway in tumour cells inhibits this killing mechanism and promotes survival (Figure 16.19).

One such cancer that exploits FasL is colon cancer. FasL is expressed on the surface of colon tumour cells, which significantly decreases the effectiveness of NK cell and T cell killing mechanisms by inducing apoptosis in these cells. Furthermore, it is thought that this immune escape facilitates the development of tumours and also the ability of colon tumour cells to metastasize to secondary tissues such as the liver. In addition, colon cancer cells have defective Fas signalling and are therefore insensitive to FsaL-mediated killing themselves. This is related to a decrease in the expression of caspases and the pro-apoptotic protein Bak, and an upregulation in the anti-apoptotic Bcl-2. Similar findings have also been attributed to the pathogenesis of oesophageal cancers. The development of melanomas may also involve a defective Fas:FasL interaction. In normal skin, epidermal keratinocytes and melanocytes express FasL, which acts as an immune surveillance molecule that normally removes abnormal or transformed cells. However, exposure to excessive amounts of UV light downregulates the expression of FasL within the epidermis, thereby allowing transformed cells to survive and melanomas to develop.

## 16.30 Summary

**1.** Mucosal tissues are host to a number of cancers, for example, colon cancer or lung cancer.

**2.** The immune system plays an important role in killing tumour cells, preventing the development of cancer.

**3.** The immunosurveillance hypothesis states that immune cells are constantly inspecting host tissues for the presence of cancerous cells.

**4.** Tumour associated antigens can be presented to T cells, ensuring immunosurveillance.

**5.** NK cells are able to recognize and kill tumour cells that are lacking MHC class I molecules.

**6.** Tumour cells undergo Darwinian selection and evolve strategies to escape or subvert the host immune system.

# 17 Vaccination

## 17.1 Introduction

Vaccination is the most effective way of preventing illness caused by infectious diseases. It has been estimated that more than 11 million people worldwide die as a direct result of an infectious pathogen each year – the leading cause of mortality and morbidity know to man. Vaccination therefore has had a remarkable impact on the incidence of infectious diseases throughout history and will continue to have an effect against many vaccine-preventable diseases in the future. The biggest success story for vaccination has been the global eradication of smallpox, thanks largely to the efforts of Edward Jenner (Figure 17.1) in the late eighteenth century. Along with Louis Pasteur, he was one of the pioneers of modern day vaccination, due to their efforts to develop a vaccine against cholera and smallpox respectively. Jenner was the first person to empirically show that protection to the smallpox virus could be acquired through the immunization of a less deadly but related virus called cowpox. He noted that many farm hands who worked closely with diary cattle were often immune to the effects of smallpox. He attributed this to a mild infection of the cowpox, which causes small lesions (pocks or varioles) on the skin of infected individuals. Jenner determined that he could protect against smallpox by immunizing another person, in this case his gardener's eight-year old son, with the puss from the pocks he collected from the farm workers. The boy was then infected (variolated) with smallpox virus and assuredly he was protected from the dreaded disease.

The term vaccination actually refers to immunization with the cowpox (*Variola vaccinae*) and its protection against smallpox (*Variola major*) but is now a universal term ascribed to the method of inducing a protective immune response to any microorganism. Although Jenner is widely acknowledged to be the father of vaccination, this method was being practised throughout Asia and Europe long before. The ancient Chinese inhaled dried pocks as a means of protecting themselves from smallpox, while Lady Mary Wortley Montagu brought the variolation method from Turkey to the aristocracy of England in 1721. However, Jenner laid the foundations for the first principles of vaccination that immunity to a pathogen can be acquired through immunization with a related organism.

## 17.2 The principles of vaccination

Vaccination relies on the generation of acquired immunity. More specifically, vaccination induces an immune response to the antigenic products contained in the vaccine, which results in the development of memory immune cells. When a live pathogen is then encountered the memory cells recognize the same antigenic components (or close variants of) and generate an immune response against that pathogen. The memory response is more rapid and ardent than the primary response to the vaccine; hence the pathogen is contained and destroyed effectively with the minimum amount of damage to the host.

The ideal vaccine needs to fulfil a number of criteria to be effective (Table 17.1). Historically vaccinologists endeavoured to develop live-attenuated vaccines derived from either the disease-causing microorganism or from a closely related form of the pathogen. The vaccinia virus used to eradicate smallpox is an example of a live-attenuated vaccine, which is a tissue culture adapted strain of cowpox. Attenuated vaccine strains should cause no or little disease but induce an effective immune response to the intended pathogen. Inactivated vaccines employ whole bacteria or viruses that are rendered non-pathogenic by inactivating them, usually by chemical, UV or heat treatment. The pathogen can no longer replicate inside its host and therefore cannot cause disease but still induce an immune response. The inactivated cholera

*Immunology: Mucosal and Body Surface Defences*, First Edition. Andrew E. Williams.
© 2012 John Wiley & Sons, Ltd. Published 2012 by John Wiley & Sons, Ltd.

Doctor Jenner about to Vaccinate a Child.

**Figure 17.1** Doctor Jenner vaccinating a small child. This cartoon depicts Edward Jenner (1749–1823) vaccinating a small child with the cowpox in order to induce a protective immune response to the smallpox virus. Thanks to the efforts of Jenner, the effectiveness of the modern vaccinia virus vaccine, and the work done by the WHO, smallpox has been eradicated from the world. The last case of naturally occurring smallpox was in 1978 and it remains to be the only disease eradicated by vaccination to date. Picture courtesy of The Edward Jenner Museum, Berkley, Gloucestershire, UK.

**Table 17.1** Criteria for an ideal vaccine. The ideal vaccine should contain a number of properties that make it as effective as possible. However, most vaccine formulations do not meet all the criteria listed in the table. The development of a successful vaccine strives to meet most of them.

| CRITERIA FOR VACCINE |
| --- |
| Long-lasting immunity (without causing autoimmunity) |
| High level of stability at the immunization site |
| Safe for human use (non-toxic) |
| Protect against all variants of the pathogen |
| Reduce transmission of pathogen and length of infection |
| Transfer immunity across the placenta and protect the newborn |
| Be efficacious worldwide in all age groups |
| Inexpensive to manufacture |
| Not require multiple immunizations |
| Non-invasive administration |

vaccine is an example of a killed-bacterial vaccine. Acellular vaccines consist of fractions of bacteria (with the toxic components inactivated) or empty virions (virus capsules) that retain their immunogenic components. For example, the diptheria vaccine contains an inactivated toxin that has been treated with formaldehyde to form a toxoid. The formaldehyde cross-links the amino acids in the diptheria toxin rendering it non-toxic and safe for use as a vaccine. Other vaccines include polysaccharide or protein-polysaccharide conjugates such as the *Haemophilus influenzae* b conjugate vaccine that protects against a certain form of bacterial meningitis. A recent development in vaccine design uses plasmid DNA, containing a pathogenic gene that is expressed in the host, and is called DNA or nucleic acid vaccination. Considering that the vast majority of infections occur across mucosal barriers such as the gut and the respiratory tract, much effort is also being made on the development of a mucosal vaccine. Delivery of a vaccine directly to a mucosal tissue would have the added benefit of inducing an immune response that more closely resembles natural immune defence and should induce memory cells that are capable of homing directly to the site of infection.

Successful vaccines must take into account the type of pathogen, the tissue which it infections and its mode of cell entry. Pathogens that reside in the extracellular environment are more susceptible to antibody-mediated immune responses than those that reside inside cells. Therefore, in order to eradicate extracellular pathogens, or prevent intracellular pathogens invading cells, an antibody-mediated immune response is desirable. Inactivated bacteria or viruses that do not invade or replicate in host cells are ideal vaccines to induce strong B cell-mediated antibody responses. On the other hand, pathogens that replicate in host cells are more susceptible to cytotoxic T cell responses that kill infected cells and prevent the pathogen from spreading. Live-attenuated vaccines that retain a low level of replication potential, or DNA vaccines, are more effective at inducing cell-mediated immune responses. The location of pathogen entry and replication is also an important consideration for vaccine design. Most pathogens enter the body across mucosal surfaces such as the lung or intestine so an effective mucosal immune response is desirable. The ideal vaccine would induce both antibody and cell-mediated immune responses at the appropriate site of pathogen entry.

## 17.3 Passive immunization

The method of passive immunization was used long before the advent of active vaccination. There are two basic forms of passive immunization, natural passive immunization and artificial passive immunization. The transfer of pre-formed antibodies from the mother to the foetus during pregnancy constitutes the majority of naturally acquired immunity. Only IgG is transferred across the placenta from mother to foetus and is called maternally-derived antibody. For example, maternally-derived antibody is able to provide some protection against tetanus, diptheria, rubella and polio virus. Another source of maternal antibody is the colustrum of milk, which contains IgG as well as IgM and IgA, which is ingested by the neonate and therefore enters the intestinal tract. Passive immunization therefore provides protection at an early age before acquired immunity can fully develop. However, maternally-derived antibody can sometimes be a problem for neonatal or infant vaccination regimes. Any pre-existing antibodies specific for the vaccine strain can bind and neutralize the vaccine, lowering its efficacy and impairing the development of acquired immunity. The timing of certain vaccinations is a very important clinical consideration for the development of acquired immunity.

Artificial passive immunization involves the transfer of serum components from an immune individual to a naïve recipient. Von Behring was awarded the first Nobel prize for medicine in 1901 for the discovery that serum from tetanus toxoid immunized horses could be transferred into naïve mice in order to protect them from tetanus challenge (Figure 17.2). Horse serum is commonly used as a source for anti-venom antibodies against black widow spider bites and for anti-toxin antibodies against botulism. However, the potential dangers of using horse serum include anaphylactic shock to horse IgE contained in the serum and hypersensitivity type III reactions to the IgG and IgM. However, human serum IgG has been used against tetanus, rabies and more recently against Ebola and severe acute respiratory syndrome (SARS) virus infections. Human serum is often difficult and expensive to obtain and there is also a risk that the serum contains other human pathogens. Passive immunization is useful when no other alternative is available, when there is little time for active immunity to develop or when immuno-compromised individuals are at risk. No acquired immunity or memory is generated following passive immunization and it only has a transient effect.

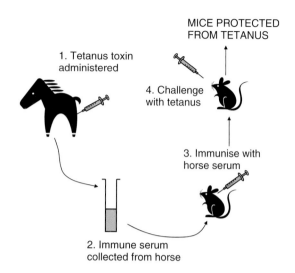

**Figure 17.2** Artificial passive immunization. The serum from horses that have been immunized with tetanus toxin can be used to passively immunize mice against tetanus challenge. Von Behring demonstrated that immune serum collected from immunized horses could be administered to mice in order to protect them against tetanus challenge.

There is currently no passive antibody therapy for most infections therefore active immunization remains the best possible option to prevent disease.

## 17.4 Active immunization

Active immunity is generated either through natural infection or by vaccination. Ideally this leads to acquired immunity whereby memory lymphocytes are generated that induce a heightened immune response upon secondary exposure to the same pathogen. There are a series of immunological events that take place from the initial administration of the vaccine to the development of acquired immunity to a pathogen and protection against disease (Figure 17.3).

## 17.5 Processing of the vaccine for immune recognition

Antigen processing and presentation is the first stage in the generation of an immune response to an antigen, such as a vaccine. Antigen-presenting cells (APCs) such as dendritic cells (DCs), macrophages and B cells provide an essential link between the innate and adaptive immune

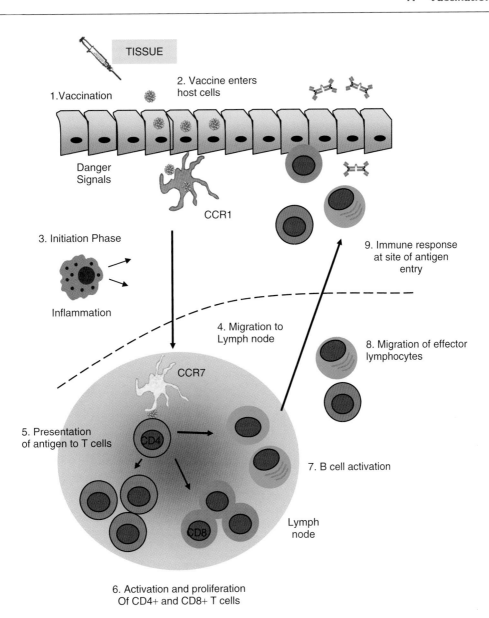

**Figure 17.3** Mechanism of acquired immunity following vaccination. Vaccine is administered (1) and enters the site of vaccination (2). The initiation phase (3) involves the uptake of vaccine antigen, (4), upregulation of CCR7 and the downregulation of CCR1, CCR2 and CCR5. In the regional lymph node APCs present antigen to CD4+ and CD8+ T cells (5). This induces the activation of T cells (6) and the activation of B cells (7). The activated lymphocytes migrate out of the lymph node (8) and enter the site of vaccine administration in order to induce an immune response (9).

responses by processing antigen and presenting peptides on their cell surface to CD4+ T-helper (Th) cells and CD8+ cytotoxic T lymphocytes (CTLs). Immature APCs migrate into the site of antigen entry in response to 'stress' signals derived from the infected cell such as epithelial cells and tissue resident macrophages. The production of cytokines and chemokines, for example, is an important first step in the migration of immature DCs to the site of vaccination. Immature DCs express the chemokine receptors CCR1, CCR5 and CXCR1, which enables them to respond to the chemokines CCL3 (MIP-1α), CCL5 (RANTES) and CXCL8 (IL-8) produced by tissue cells. In order for an immune response to be initiated APCs must first recognize the antigen as foreign. They achieve this by recognizing pathogen-associated molecular patterns (PAMPs) via Toll-like receptors (TLRs) and other scavenger receptors in addition to internal danger signals derived from damaged cells (DAMPs). Activation of APCs in this way is crucial as only they are able to initiate a primary immune response to a previously unseen antigen.

Antigen enters APCs through one of two pathways, the endogenous pathway or the exogenous pathway (previously described in Chapter 3). The processing and presentation of a vaccine will depend on the characteristics of that vaccine, which is an important consideration during effective vaccine design. A live-attenuated vaccine will induce a different type of immune response than a particulate vaccine. For example, a live virus vaccine will replicate in the host cell cytoplasm and provide antigen for both MHC class I and MHC class II processing. However, a particulate protein antigen will be engulfed by the APC and therefore predominantly provide antigen for MHC class II processing (Figure 17.4).

Following antigen uptake APCs migrate from the site of antigen encounter to a regional lymph node where they interact with cells of the adaptive immune system (Figure 17.3). APCs undergo many changes in the course of maturation. DCs, for example, lose the ability to endocytose and instead upregulate MHC expression in order to effectively present peptide to T cells. The chemokine receptors CCR1, CCR5 and CXCR1 are downregulated and replaced by CCR7, which binds to the lymphoid chemokine CCL19 (MIP-3β). CCR7 expression allows DCs, loaded with peptide, to migrate to regional lymph nodes. In order to generate an effective primary T cell

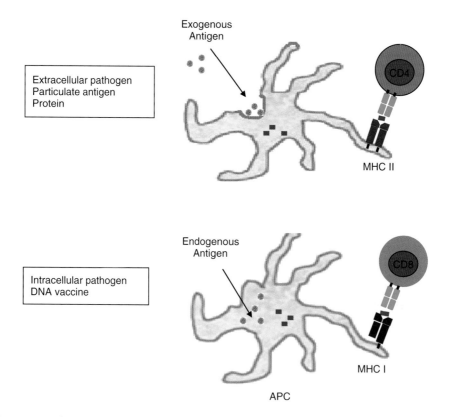

**Figure 17.4** Exogenous and endogenous antigen processing. Exogenous antigen is endocytosed by APCs and enters the MHC class II processing pathway. Peptide is then presented to CD4+ Th cells in conjunction with MHC class II molecules. Endogenous antigen (from infectious virus for example) enters the MHC class I processing pathway. Peptide is then presented to CD8+ T cells in conjunction with MHC class I molecules. Endogenous antigen can also enter the MHC class II pathway by means of cross-presentation so that it is presented to both CD4+ and CD8+ T cells.

response, all APCs upregulate certain co-stimulatory ligands such as CD80/CD86, CD40 and OX40L (refer to Chapter 3). Within the lymph node DCs present peptide in the context of MHC to T cells and provide co-stimulatory signals to initiate the adaptive immune response. When activated both CD4+ and CD8+ T cells start to proliferate and migrate to the site of antigen encounter, assisted by the cytokines and chemokines produced by activated APCs. APCs in conjunction with T helper cells also activate B cells in the lymph node, signalling them to proliferate and begin antibody production. B cells then migrate to peripheral sites where antibody is able to neutralize and opsonize exogenous antigen.

## 17.6 Adaptive Immune response following vaccination

After vaccination DCs take up antigen in the periphery and migrate to regional lymph nodes where they present processed antigen (peptide) to T cells and activate the adaptive immune response. Activated naïve T cells differentiate into primary effector T cells and upregulate surface molecules that allow them to migrate into peripheral tissue. Pathogen-specific CD8+ CTLs proliferate rapidly and expand 1000-fold in only a few days, while CD4+ helper T cells have a more limited expansion phase. During the effector phase, CTLs produce components of the perforin and granzyme systems and express FasL-Fas, which provide the CD8+ T cell with its cytotoxicity. This CTL response is responsible for killing infected cells and preventing pathogen spread. Following the downregulation of the effector phase, a subset of both T cells and B cells enter the memory pool where they form long-lasting, antigen-specific memory cells. Upon re-exposure to the pathogen memory cells respond more quickly and with a larger number of antigen-specific cells (Figure 17.5). These memory cells do not prevent the pathogen from infecting but rather prevent the pathogen from causing disease. The success of many vaccines is reliant on the ability to induce an effective memory recall response, which is why some vaccines require regular booster immunizations.

The most important qualities of a vaccine are efficient uptake by APCs and immunogenicity. If the vaccine is not processed efficiently or lacks immunogenicity it will not stimulate long-term immunological memory. To enhance the immunogenicity and increase the longevity

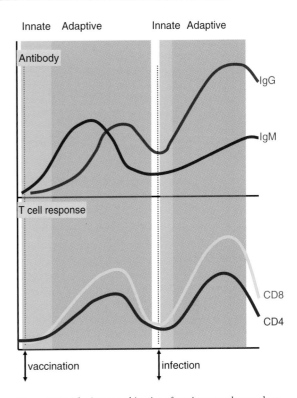

**Figure 17.5** The immune kinetics of a primary and secondary immune response. A secondary response is more rapid and larger in magnitude than a primary response. A secondary response generates a larger amount of high affinity IgG compared to lower affinity IgM. The magnitude of both the CD4+ and CD8+ T cell populations is enhanced during a secondary response compared to a primary response.

of a vaccine, a number of vaccine formulations often contain an adjuvant.

## 17.7 Vaccine adjuvants

Adjuvants have the desired effect of enhancing the immunogenicity of a co-administered vaccine to promote a stronger secondary response to the pathogen than would occur with vaccine alone. The mechanisms responsible for the increased immunogenicity vary according to the type of adjuvant used. Adjuvants can increase the magnitude, speed and duration of the immune response to the vaccine; they can influence the antibody isotype produced; enhance cell mediated immune responses; increase the effectiveness of the immune response at the site of pathogen entry, for example, at mucosal sites;

**Table 17.2** Types of adjuvant.

| Microbial Products | Oil Emulsions | Gel-based | Particulate | Synthetic | Cytokines |
|---|---|---|---|---|---|
| Cholera toxin (CT) | Freund's complete adjuvant | Potassium aluminium sulphate (alum) | Liposome | Polyphazene | IFNγ |
| Labile toxin (LT) | Freund's incomplete adjuvant | Aluminium phosphate | ISCOMS | Copolymers | IL-2 |
| Pertussis toxin (PT) | MF59 | Calcium phosphate | Microparticles | MPL-analogues | IL-12 |
| CpG motifs | SAF | Aluminium hydroxide | Virosomes | | IFNα/β |
| MPL | | | Saponins | | GM-CSF |

enhance the development of immunity in immature individuals; and reduce the dose of vaccine necessary to induce an effective immune response.

There are several different forms of adjuvant (Table 17.2) that can be broadly categorised into separate groups. The first are microbial derived adjuvants and include cholera toxin (CT), *E. coli* heat-labile toxin (LT) and CpG motifs, which rely on the natural receptors and pathways of the innate immune system for their adjuvanticity. The second group are oil emulsions and include MF59, which is licensed in Europe, and Freund's complete adjuvant, which consists of mycobacterial products contained within a water-in-paraffin mineral oil emulsion. The third group are gel-based adjuvants and include the only adjuvant to be fully licensed for human use in America, potassium aluminium sulphate or alum. Particulate adjuvants include lipid based liposomes and immunostimulatory complexes (ISCOMS), microparticles and virosomes. Synthetic adjuvants include copolymers, MPL analogues and polyphophazene. In addition, cytokines have been used to experimentally enhance immune responses to vaccines and skew the subsequent immune response to the desired phenotype. Examples include IFNγ, IL-2, IL-12 and type I IFNs. Some of these adjuvants will now be considered in more detail.

## 17.8 Alum

Most adjuvants in human use induce a dominant Th2-type immune response. This was thought to be preferential as Th2 cytokines are potent inducers of antibody production. Alum exerts its adjuvanticity in one of two ways; the first is by providing a depot effect for the co-administered antigen so that the immune cells receive a prolonged antigenic stimulation, thereby enhancing the magnitude and longevity of the response; the second effect is through immunomodulation. The immune system is modulated by alum so that a Th2 response is induced. Alum activates the complement cascade so that the complement products iC3b and C5a bind to their receptors on macrophages and cause the inhibition of IL-12 (Figure 17.6). The inhibition of IL-12 enables the development and proliferation of Th2 cells, which produce Th2 cytokines such as IL-4, IL-5 and IL-13. Alum also induces the upregulation of MHC class II molecules on APCs therefore favouring the expansion of CD4 T helper cells. Due to the Th2 cytokine environment and enhancement of T helper cells, antibody-producing B cells predominantly produce high levels of IgG1. Alum is very effective at enhancing antibody responses to bacteria and is suited to vaccines against extracellular pathogens. However, it is less well suited to vaccines aimed at protecting against intracellular pathogens, particularly intracellular virus infections.

## 17.9 Freund's complete adjuvant

Freund's complete adjuvant (CFA) consists of cell wall components from killed *Mycobacterium tuberculosis* within an oil-in-water mineral oil formulation. The mineral oil enables co-delivered antigens to be presented to the immune system over a prolonged period of time and in a more concentrated form at the vaccine delivery site. The mycobacterial-derived products contain PAMPs that stimulate the innate immune system and increase the magnitude of the adaptive immune response. CFA increases the rate of phagocytosis, enhances antigen uptake by DCs and macrophages and results in increased cytokine secretion, especially TNF production. This causes an inflammatory infiltrate at the site of injection,

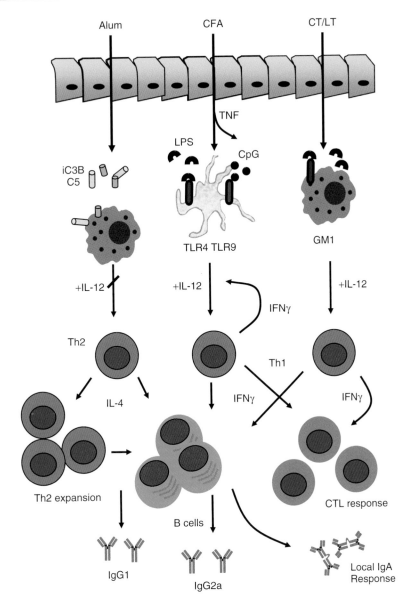

**Figure 17.6** The mechanism of immune cell activation by adjuvant. Alum induces the production of complement proteins, which inhibit the production of IL-12. This allows for the induction of Th2-mediated immune responses. Complete Freund's adjuvant (CFA) contains mycobacterial components that stimulate TLRs on APCs, which induce the production of IL-12. CD4+ T Th1 cells produce IFN-γ resulting in the induction of CD8+ CTL and IgG2a B cell responses. Labile toxin (LT) and cholera toxin (CT) are thought to activate APCs through ganglioside receptor (GM1), which activates Th1 cells and induces CTL responses but at the same time mucosal administered LT/CT enhances the production of local IgA.

which is often very painful. Binding of PAMPs to their corresponding TLRs on APCs induces the production of IL-12 by DCs and macrophages. DNA containing CpG motifs in the mycobacterial component signal through TLR9, while cell wall components such as lipoarabinomannans bind to TLR2 activating the MyD88 signalling pathway. Heat shock proteins (HSPs) released by damaged cells at the site of injection are recognized by TLR4 and signal to APCs in a similar manner to LPS. These danger signals trigger the production of IL-12 by

DCs and macrophages, which enhances the development of Th1 T cell clones and cell-mediated immunity. CD4+ Th1 cells produce high amounts of IFN-γ and this in turn increases IL-12 production by DCs and stimulates the development of CD8+ CTLs. The Th1 cytokine environment also induces antibody producing B cells to secrete predominantly IgG2a (Figure 17.6).

Incomplete Freund's adjuvant (IFA) is also frequently used to enhance immune responses. It differs from CFA in that it contains no mycobacterial components, only paraffin and surfactant. However, IFA is still immunogenic as it stimulates innate immunity and provides a depot for the co-administered antigen. Although very powerful adjuvants, CFA and IFA are unsafe for human use due to the inflammation following injection and the potent polyclonal T cell responses they induce, which may also heighten the risk of developing autoimmune disease.

## 17.10 Mucosal adjuvants and vaccine delivery

Most pathogens infect humans across mucosal surfaces. Therefore, stimulation of a local immune response at the site of infection would be highly desirable and be the overriding goal of a mucosal vaccine. For example, the generation of a mucosal IgA response is considered to be the milestone of a successful mucosal vaccine. However, immunization at mucosal surfaces has been problematic due to the difficulties in stimulating mucosal immune responses. There is therefore a need to develop effective vaccine adjuvants and vaccine delivery systems that transport antigen into MALT and provide protective immunity. One of the problems faced by oral or nasal delivery of vaccine is the natural mechanism of immune tolerance to ingested or inhaled antigens. An effective mucosal vaccine therefore has to overcome oral or nasal tolerance in order to stimulate a protective immune reaction to the delivered antigen.

## 17.11 Prospects in adjuvant design

In Europe, an oil-in-water adjuvant called MF59 has recently been given a licence for human use in combination with influenza vaccine. Critically this adjuvant enhances the immune response to influenza in elderly individuals, who are most susceptible to disease and whose immune system has been compromised with age. It increases anti-influenza antibody titres in a predominantly Th2 manner, although the precise mechanism of adjuvanticity remains poorly understood. It is thought that MF59 enhances the recruitment of macrophages and other APCs into the site of injection and induces the upregulation of chemokines, thereby augmenting antigen processing and presentation.

CT and LT from *E. coli* have been extensively studied due to their potential use as adjuvants. Moreover, they are highly immunogenic at mucosal sites, a characteristic that is becoming an important consideration for future vaccine design. However, both CT and LT are far too toxic for use in humans – therefore efforts have been made to produce mutants that are non-toxic but retain immunogenicity. For example, the LT mutants LTK63 and LTR72 enhance mucosal immune responses by stimulating Th1 and Th2 responses respectively (Figure 17.6). Likewise, the non-toxic CT B subunit, or enzymatially inactive forms of CT, have been demonstrated to be particularly immunogenic when delivery to mucosal tissues. The mechanism of LT and CT adjuvanticity is multifaceted. They increase the immunogenicity of most antigens that are co-delivered with them by various means. They increase the permeability of epithelial layers, which increases the availability of co-delivered antigens to APCs located in sub-epithelial compartments and augments the recruitment of leukocytes into the area of administration. LT and CT increase the antigen processing and presentation abilities of DCs, for example by upregulating MHC class II, CD80/CD86 and CCR7 expression. In addition, LT and CT influence cytokine production, including increasing IL-1 expression by DCs, which act as a pro-inflammatory mediator. Lastly, CT and LT are potent stimulator of antibody class-switching to IgA and therefore enhance IgA secretion across mucosal tissues (Figure 17.6).

By selecting an appropriate adjuvant (in combination with vaccine), the immune response can be tailored to a particular pathogen in order to obtain the most effective protection. For example a Th1-inducing adjuvant would be more appropriate than a Th2-inducing adjuvant at inducing CD8+ CTL responses against intracellular viruses. Conversely a Th2-inducing adjuvant would be more appropriate at inducing CD4+ T cell help in order to promote antibody responses against extracellular bacteria. The route of administration is also an important consideration for the delivery of an efficacious vaccine. Different routes of administration tend to promote subtly different immune responses. For example, oral and intranasal administration tends to promote IgA responses, while parenteral administration enhances IgG responses.

**Table 17.3** Protective immune responses depend on appropriate Th1/Th2 T cell activation. The phenotype of the immune response to a particular pathogen is critical in determining whether an individual is protected or not. The table illustrates the appropriate protective immune responses against selected pathogens.

| Th1 | Th2 |
|---|---|
| Influenza virus | Measles virus |
| RSV | *Borrelia burgdorferi* |
| *Mycobacterium tuberculosis* | *Trichuris muris* |
| *Chlamydia pneumoniae* | *Schistosoma mansoni* |
| *Bordetella pertussis* | *Nippostrogylus brasiliensis* |
| *Leishmania major* | *Brugia malayi* |

## 17.12 Th1/Th2 polarization and vaccine development

In order to elicit protective immunity a successful vaccine must induce an appropriate immune response (Table 17.3). Even the antibody isotype is critical as some are useful for neutralizing extracellular pathogens, while others are efficient at eliminating intracellular viruses or bacteria. As mentioned in Chapter 3, T helper cell responses can be divided into either Th1 or Th2 depending on the cytokines that are produced. The success of a vaccine critically depends on inducing the correct subset, exemplified in the murine model of *Leishmania*. BALB/c mice develop an inappropriate Th2 response to *Leishmania* and are susceptible to disease, while C57BL/6 mice develop an appropriate Th1 response and are therefore resistant to infection. Differentiation of T cells into the wrong phenotype can lead to pathogen persistence, disease and even death.

There are many parameters that influence the differentiation of T cells into either Th1 or Th2. The first critical parameter is whether the vaccine is alive or dead. Live vaccines replicate in their normal environmental niche and therefore induce appropriate immunity. Inactivated vaccines or particulate antigens do not replicate and are therefore poor inducers of CD8+ T cell responses. The dose of a vaccine or its half-life can also influence Th1/Th2 development, as can the amount of TCR engagement and co-stimulation. The immediate cytokine environment is also critical. Vaccines delivered into an environment rich in IL-12 and IFN-γ will direct T cells into a Th1 phenotype, while IL-4 and IL-5 will induce a Th2 phenotype. The outcome of an adaptive immune response will therefore depend on the initial pathways triggered following vaccination.

The CD4+ T cell phenotype will also determine the subsequent antibody isotype. Naïve and primary effector B cells express IgM, which has high avidity but low affinity. Primary effector B cells undergo isotype switching to IgG. In mucosal sites further isotype switching occurs resulting in the production of IgA or IgE. Many vaccines promote good antibody production but it is important to induce the appropriate isotype. The production of IgA is often an important criterion for vaccination against mucosal infections, while some vaccines aim to promote complement fixing antibody isotypes. Some of the main vaccine formulations and the type of immune responses they induce will now be examined.

## 17.13 Live-attenuated vaccines

The basis of any live-attenuated vaccine is to induce protective immunity without causing disease. The most widespread method of achieving attenuation in a live vaccine is to passage the microorganism (bacteria or virus) many times in tissue culture (Figure 17.7). This passage selects for mutants that are more suited to growth in tissue culture than growth in human cells and results in an attenuated organism with very low pathogenicity. A further method of attenuation is by temperature sensitive adaptation, usually by cold adaptation. For example, virus is grown in tissue culture at a low temperature so that it becomes adapted for growth at that temperature but ill suited to growth at body temperature. Live-attenuated vaccine strains still retain a modest ability to invade, replicate and grow in human cells, a feature that allows for the induction of a strong, lasting immune response but without the undesired effect of causing pathology and disease. Probably the most famous and successful live-attanutaed virus vaccine is the vaccinia virus vaccine that completely eradicated global smallpox (Box 17.1: Smallpox). The major disadvantage of live-attenuated vaccines is the potential of reversion to a pathogenic state. For example, the Sabin polio vaccine has a 1 in 24 million chance of reverting to a pathogenic state and in many countries the inactivated Salk polio vaccine is preferred. Live-attenuated vaccines do have the advantage of activating the immune response in a natural way. Both the MHC class I and MHC class II antigen presentation pathways are engaged, through transient intracellular replication and uptake of exogenous antigen respectively.

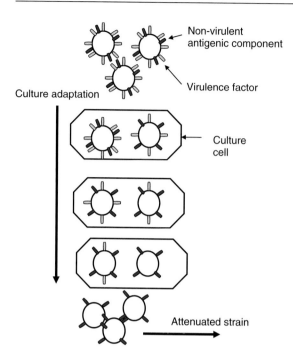

**Figure 17.7** Tissue culture attenuation of live virus vaccines. As an example, wild type viruses are highly pathogenic and contain certain molecules on their surface that confer them with their virulence. Attenuation by repeated passage through tissue culture mutates the molecules of the virus so that it is no longer virulent to its host. However, successful live-attenuated vaccines retain the antigenic components necessary for the induction of strong immune responses.

---

## Box 17.1. Vaccination Against Smallpox.

Edward Jenner, more than 200 years ago, pioneered the field of rational vaccination following his realization that farm workers infected with the mildly pathogenic cowpox virus were protected from the dreaded smallpox. Jenner embarked on a cognisant immunization protocol and empirically demonstrated that protective immunity to smallpox (*Variola major*) could be transferred by the related cowpox virus (*Variola vaccinae*). This method of smallpox vaccination continued until the mid-nineteenth century when cowpox was replaced by vaccinia virus, a related orthopoxvirus of unknown origin. The procedure for vaccination is termed variolation (now known as scarification) and involves repeated puncture of the skin with infectious virus.

Vaccinia virus was used during the WHO sponsored smallpox eradication program and remains the virus vaccine of choice today. Essentially it is an attenuated virus vaccine produced by scarifying bovine calves on the flank and collecting infectious liquid from productive pustules.

---

Vaccination introduces infectious virus into the skin where it invades keratinocytes and spreads from cell to cell causing necrosis and oedema at the site of administration. An inflammatory infiltrate migrates to the vaccination site by day 7 post vaccination, characterised by an influx of macrophages, neutrophils and lymphocytes. Pro-inflammatory cytokines, such as IFN-γ, IL-12 and TNF, are released activating NK cells and inducing the differentiation of CD4+ Th1 cells and CD8+ CTLs. Cell-mediated immunity is vital for the control of vaccinia infection, while B cell-mediated antibody responses are less important. In order to sustain an effective memory T cell population and hence for continued protection against smallpox, revaccination every 5–10 years is ideal, although protective efficacy may last 30 years.

---

Recognition of native PAMPs by the innate immune system tend to drive the adaptive immune system toward an IL-12 and IFN-γ dominated, Th1-type response, with the generation of long-lasting cell mediated and humoral immunity.

One of the most important live-attenuated vaccines is the bacille Calmette-Guerin (BCG) vaccine used to protect against *Mycobacterium tuberculosis*, the causative agent of TB. It is estimated that approximately 2 billion individuals are infected with TB at any one time worldwide, with the highest rates of infection in developing nations. BCG vaccination has been used extensively since the 1940s, most effectively against childhood TB infections. BCG is a tissue culture adapted strain of *Mycobacterium bovis*, which elicits a protective immune response to the closely related species *Mycobacterium tuberculosis*. Intradermal injection is the preferred method of delivery, resulting in a local inflammatory response that forms a red papule and the expansion of lymphocytes in the regional lymph nodes. Although the precise immune mechanisms that result in protective immunity following BCG vaccination are poorly understood, both activated DCs and TB reactive CD8+ CTLs are required for protection. A Th1-type immune response, characterized by IFN-γ production, is necessary for the control of TB infection. Despite the efficacy of the BCG vaccine against childhood TB, it is far less effective against adult pulmonary TB or against new, emerging strains of TB. This is a particular problem in developing countries with high TB burdens; therefore there is a critical need to either enhance the immunogenicity of current BCG vaccination regimes or develop novel vaccines. Vaccination against TB is a clear example where a Th2-dominated immune response would not be successful. IFN-γ from CD8+ T cells and CD4 Th1 cells is vital for the control and containment of this intracellular infection.

An example of an attenuated live viral vaccine is the triple combination measles, mumps and rubella (MMR) vaccine. Measles virus alone is estimated to infect nearly 40 million people, causing one million deaths worldwide each year, with children under the age of 5 being the most susceptible. The mumps virus was the leading cause of viral meningitis in the UK before the MMR vaccine was introduced in 1988. The rubella virus is largely a threat to the unborn child, causing developmental abnormalities such as congenital heart disease, defective growth and deafness as a result of the pregnant mother being infected. The introduction of the MMR vaccine has significantly reduced the incidence of all three virus infections. As with many live virus vaccinations, there is always a risk of developing side-effects, such as fever, rash or convulsions associated with the measles component of the vaccine. However, the risks of such side-effects are low and far less serious than the disease caused by infection from wild-type virus (Figure 17.8).

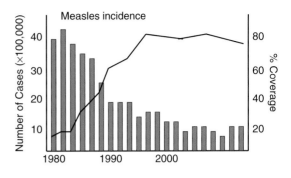

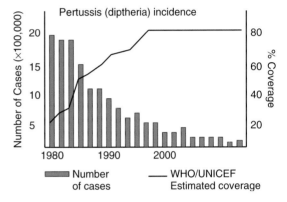

Figure 17.8 Incidence of measles and diptheria following vaccination. The incidence of measles and diptheria cases dramatically declines when vaccines against each of these infections were introduced. Although some side effects occur infrequently in the vaccinated population, the prevention of disease associated with infection far outweighs any risks. Graphs reproduced courtesy of the World Health Organization (http://www.who.int).

## Box 17.2. The Bioterrorist Threat.

A modern development in social awareness to the dangers of terrorism is no more apparent than in the interest generated by the threat posed by biological agents. From a list of possible pathogens, including *Bacillus anthracis* (anthrax), *Yersinia pestis* (plague), *Mycobacterium leprae* (leprosy) and Ebola virus, the most evocative and possibly most likely biological threat comes from the smallpox virus. Although eradicated worldwide in 1977, two internationally accepted stocks of smallpox virus remain; one at the Centres for Disease Control and Prevention (CDC) in the United States and one at the State Centre of Virology and Biotechnology in the former USSR. However, concerns exist that certain clandestine individuals have access to smallpox for use as a bioweapon.

As an example, the United States discontinued national smallpox vaccination in 1982. Therefore the majority of the population have never been immunized or immunity to smallpox has waned sufficiently in those vaccinated that they are no longer protected. Smallpox is an ideal bioweapon due to its virulence, pathogenicity and ease of spread. It has been estimated that only 50–100 infected individuals could lead to a national epidemic. Only one smallpox vaccine is currently licensed in the United States. It was manufactured in the 1970s by the calf scarification method using the New York City Board of Health (NYCBOH) vaccinia strain and comprises just 15 million undiluted doses in total (held at the CDC). Efforts are now being made to manufacture a tissue culture vaccine using the NYCBOH strain as seed virus and are expected to be licensed in the near future.

## 17.14 Inactivated vaccines

Inactivated vaccines are based on whole pathogens that are either UV, heat or chemically treated so that they are no longer able to replicate in the host. Certain methods of inactivation, for example heat inactivation, can permanently denature the proteins that are required by the immune system to generate an appropriate immune response. Therefore chemical inactivation, usually with formaldehyde, is the preferred method. Inactivated vaccines are not able to invade or replicate inside host cells and are therefore very poor at inducing cell-mediated immune responses. However, they do induce strong antibody responses to the antigenic components of the pathogen. Unfortunately, most inactivated vaccines need to be administered several times (repeated boosters) and require the presence of an effective adjuvant. Another disadvantage of inactivated vaccines is that chemical treatment sometimes fails to kill the pathogen completely, allowing development of serious disease.

The current influenza virus vaccine, administered to high-risk populations in order to prevent epidemics of the flu, is based on inactivated virus. The influenza vaccine is trivalent and consists of three different virus strains, which are chemically inactivated. The influenza A virus strain(s) (highly pathogenic) likely to circulate in the following season are predicted and incorporated into the vaccine together with an influenza B virus strain (mildly pathogenic). Influenza viruses are identified by the type of haemagglutinin and neuraminidase (H and N) proteins they possess. For example, the current vaccine contains H1N1 and H3N2 virus haplotypes plus a strain of influenza B virus (such as the Beijing or Hong Kong strain). However, the inactivated influenza vaccine is not always protective, due to changes in the influenza virus strain as a result of antigenic drift (single mutations in the H and N genes) or antigenic shift (re-assortments of whole gene segments, e.g H3N2 to H5N2). The vaccine is then rendered non-protective as the immune system has never encountered the new virus strain and so the threat of an influenza epidemic or even a pandemic remains. Recent concerns have arisen due to geographically isolated human outbreaks of H5N1 virus, a haplotype thought only to be a problem in birds. Efforts are now being made to produce a H5N1 virus vaccine, although this virus is toxic to chicken embryos, the preferred medium for vaccine virus growth.

An example of an inactivated bacterial vaccine is the formaldehyde inactivated whole-cell pertussis vaccine (WCPV) administered in childhood for protection against whooping cough. This was first administered in 1914 in the USA and consists of inactivated *Bordetella pertussis* absorbed onto aluminium salt as an adjuvant. WCPV is usually given in combination with diptheria toxoid and tetanus toxoid (DTwP). A toxoid is the chemically inactivated form of the toxin that causes disease. Although the DTwP vaccine was responsible for a massive decline in the incidence of whooping cough, diptheria and tetanus, some individuals had adverse reactions to the vaccine. The preferred trivalent diptheria, tetanus, pertussis (DTP) vaccine contains an acellular pertussis vaccine (DTaP) containing one or more inactivated components *B. pertussis*, including the toxin and filamentous haemagglutinin antigen. However, the DTaP vaccine is less efficacious than the DTwP vaccine and requires repeated immunizations.

## 17.15 Polysaccharide vaccines

Polysaccharide vaccines are also known as conjugate vaccines and consist of bacterial polysaccharides conjugated to proteins. They have been very successful in preventing diseases caused by certain bacteria that possess a polysaccharide capsule, such as *Neisseria meningitis* (meningitis), *Haemophilus influenzae* type b (meningitis) and *Streptococcus pneumoniae* (pneumonia). The capsule surrounding these bacteria protect them from the hosts immune system but antibodies generated against the polysaccharide component of the capsule are protective and hence have formed the basis for a new generation of vaccines.

Polysaccharide antigens are not very immunogenic by themselves and induce only weak T cell-independent (TI) antibody responses (stimulate B cells directly with no T cell help). By chemically conjugating the polysaccharide to a protein antigen Th cells are activated and can provide help to B cells effectively transforming the polysaccharide from a TI antigen to a T-dependent (TD) antigen. This enhances the B cell antibody response, induces class-switching and affinity maturation and elicits an effective memory response. These properties make conjugate polysaccharide vaccines ideal for the immunization of infants. The Hib vaccine consists of polysaccharide from *H. influenzae* type b conjugated to a protein carrier, most effectively either diptheria toxoid or tetanus toxoid. Both require several rounds of immunizations but produce good immune memory and strong antibody responses in infancy and result in high levels of protection against Hib-related disease. A major advantage is that they are extremely safe for human use, comprising only basic molecules.

## 17.16 Peptide vaccines

A novel method of vaccination utilizes specific peptides containing the immunodominant T cell epitopes from pathogen-derived antigens. This represents a way in which an immune response can be rationally directed to molecularly defined components of a specified pathogen, without the interference from poorly immunogenic pathogen components. The selection of antigenic peptides is therefore an important consideration when designing peptide-based vaccines. They have the advantage of being safe for human use, easily manufactured and stable and they can be administered via non-invasive routes such as on bare skin or intranasally. These peptides are endocytosed by APCs and are presented directly to T cells in conjunction with MHC class I or class II molecules. They predominantly induce CD4+ and CD8+ T cell responses and are therefore ideal vaccine candidates against intracellular pathogens. However, they induce weak antibody responses and are poorly immunogenic, thus requiring co-administered adjuvants. Current peptide-vaccine design has focused on developing vaccines against diseases where a protective vaccine is not available, such as malaria, leishmaniasis and hepatitis.

One other disadvantage is that each person binds peptides via slightly different anchor residues depending on their particular MHC molecules (MHC haplotype). A peptide sequence that binds the MHC molecules of one person may not bind those of another person. It is possible to predict epitopes that will bind to the more common MHC gene products, which can then be incorporated into a generic peptide vaccine.

## 17.17 DNA vaccination

The recent discovery that plasmid DNA encoding an antigen is able to induce an immune response has opened an exciting and fast-growing area of vaccine research. DNA vaccination is also known as plasmid, nucleic acid, polynucleotide or genetic vaccination and refers to the use of bacterial plasmid DNA engineered to express a gene encoding an antigen. Advances in molecular biology techniques have enabled the construction of plasmid DNA that optimally expresses the gene of interest in cells (Figure 17.9). DNA vaccines have a number of advantages over existing live attenuated or killed vaccines; they are stable over a broad temperature range (therefore they do not require refrigeration); they are able to induce both CTL and antibody responses; they are inexpensive to manufacture; there is no danger of reversion to a virulent state; and there is prolonged expression of antigen that is beneficial for the development of a long-lasting memory response.

The most favourable routes of administration are injection of plasmid DNA into muscle or by propelling DNA attached to gold microparticles into the skin by a method known as gene gun immunization. DNA vaccines elicit an immune response following the direct transfection of tissue cells such as myocytes in muscle or keratinocytes in the skin (Figure 17.10). Following uptake of plasmid DNA by these cells, the encoded protein is expressed inside the cell and presented on MHC class I molecules.

However, presentation of antigen by these somatic cells alone may not be enough to induce a cell-mediated immune response. Direct transfection of professional APCs (such as DCs) by plasmid DNA may result in direct presentation of peptide on MHC class I molecules to CD8+ T cells. In addition, indirect priming of DCs may take place, whereby secreted protein is endocytosed by APCs. This allows peptide to enter the exogenous processing pathway, be presented on MHC class II molecules and hence activate CD4+ T helper cells. In addition, cross-presentation of exogenous antigen on MHC class I molecules contributes to the induction of cell-mediated responses. The ability of DNA vaccines to induce both cell mediated and be able to express protein in its natural form is a great advantage for future vaccine design.

Immune responses generated following DNA vaccination are usually Th1 dominated and the inclusion of immunostimulatory motifs enhances their immunogenicity. The immunostimulatory properties of plasmid DNA are due primarily to the presence of CpG oligodeoxynucleotide (ODN) motifs. Bacterial CpG motifs are short sequences of DNA that contain properties that are distinct from eukaryotic DNA. They are rich in unmethylated cytosine-guanine resides flanked by certain nucleotide motifs, with an optimal configuration requiring the presence of two $5'$ purines and two $3'$ pyrimidines. Eukaryotic DNA does not contain unmethylated CpG motifs (most CpG islands in eukaryotic DNA are methylated) and hence the immune system has evolved a mechanism by which it can recognize bacterial DNA. Bacterial CpG motifs are recognized by TLR9 expressed on APCs and certain lymphocytes. Recognition of CpG motifs in bacterial DNA provides plasmid DNA with an adjuvant effect. TLR9 signalling results in the induction of the MyD88 pathway and the production of IL-12 by APCs. This results in the generation of Th1 cells secreting IFN-$\gamma$, TNF and IL-2. However, delivery to mucosal surfaces has been shown to induce a Th2-mediated IgA response, which is beneficial for the neutralization of pathogens, while at the same time augmenting Th1-meiated systemic responses. These properties may make CpG-containing adjuvants and DNA vaccines ideal candidates for mucosal delivery.

## 17.18 Immuno-stimulatory complexes (ISCOMs)

Immuno-stimulatory complexes (ISCOMs) are particulate antigens (about 40 nm in diameter) that have a

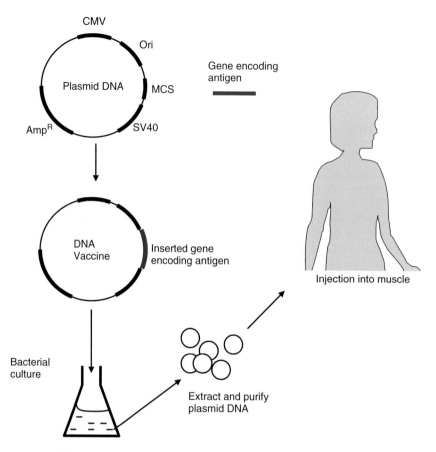

**Figure 17.9** Construction and immunization of a DNA vaccine. Using standard molecular biology techniques a particular gene that encodes an antigen can be introduced into a circular DNA plasmid. These bacterial-derived plasmids are then replicated within bacterial cells (usually laboratory strains of *E. coli*) under the control of the origin of replication (*Ori*) and harvested to give a usable quantity of DNA. The plasmid expresses ampicillin (amp$^R$) so that only those bacteria containing the plasmid are able to grow on ampicillin-selective media. The plasmid DNA, encoding the antigen, is then immunised into muscle for example and the gene expressed in the tissue of the host under the control of the cytomegalovirus promoter (CMV). An immune response is then generated against the expressed antigen.

comparative size and dimension to virus-like particles (VLPs). ISCOMs are therefore recognized by the immune system as if they were a virus, resulting in a potent CTL response characterized by IFN-γ and IL-12 production (similar to an anti-viral response). Most ISCOMs are composed of Quillaja saponin subunits together with cholesterol and phospholipids that when combined form an icosahedral cage-like structure (Figure 17.11). Specific antigens are then introduced by utilizing hydrophobic interactions between the ISCOM and the protein antigen of interest.

Part of the ISCOMs adjuvanticity derives from the Quillaja saponin content. Quillaja saponin is a plant glycoside derived from *Quillaja saponaria* and

its immunostimulatory properties have been known since the early 1900s. Following immunization ISCOMs enhance MHC class II expression on APCs and augment IL-12 production. This promotes T cells to secrete the Th1-type cytokines IFN-γ, TNF and IL-2 and elicits a strong CTL response, which is further enhanced by the delivery of antigen into the cytosol for presentation on MHC class I molecules. In addition, ISCOMs are very proficient at increasing antibody responses, mainly due to the efficient targeting of antigen to APCs. In general systemic IgG2a and mucosal IgA is upregulated in response to many antigens. Although immune responses to ISCOMs are dominated by IL-12 and Th1-type responses, they also enhance Th2-type cytokine

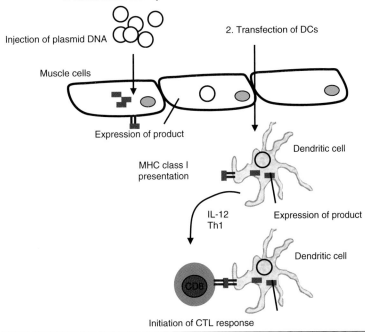

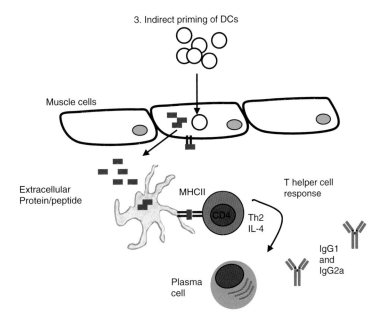

**Figure 17.10** Mechanisms of DNA vaccination. DNA vaccination involves direct transfection of muscle cells (or keratinocytes in the skin). The antigen encoded by the plasmid is expressed in the muscle cell and peptides presented in conjunction with MHC class I. The second mechanism involves direct transfection of APCs by the plasmid DNA. The antigen is then expressed within the APC and peptide loaded onto MHC class I (cross-presentation on MHC class II molecules may also occur). The third mechanism is indirect priming of APCs.

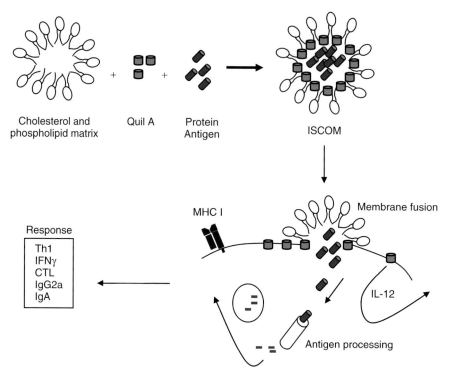

**Figure 17.11** Mechanism of immune activation following ISCOM immunisation. ISCOMS are composed of a cholesterol and phospholipid matrix together with Quil A and the desired protein antigen. ISCOMS are efficient at fusing with the cell membrane, thus releasing protein antigen into the cytoplasm. Antigen enters the endogenous processing pathway and peptide is presented on the cell surface in conjunction with MHC class I. This, combined with the induction of IL-12 expression drives a Th1 dominated cell-mediated immune response and the production of Th1-type antibodies.

production under certain circumstances invoking a more balanced immune response. Most importantly ISCOMs generate protective immunity to a number of different acute and chronic viral, bacterial and parasitic infections. Although not licensed for human use, several promising ISCOMs containing antigens from influenza virus, hepatitis B virus and HIV are under development. In addition, ISCOMS are immunogenic when delivered by the oral route. They are though to enhance the uptake of antigen across epithelial layers, where resident DCs are located. Activated DCs then migrate to mesenteric lymph nodes where they present antigen to T cells and B cells and induce an immune response, rather than inducing oral tolerance.

## 17.19 Dendritic cell vaccines

DCs are responsible for initiating the immune response to microorganisms by presenting captured antigens, and through the recognition of pathogen-associated signals direct the type and magnitude of the adaptive immune response. DCs therefore have the potential to form the basis of a new generation of vaccines. By loading DCs with antigen *ex vivo* and injecting them back into the host, they can act both as a natural adjuvant and as a vaccine carrier. Exposure to microbial products during the antigen loading process educates DCs to become either a Th1 or a Th2 inducer depending on the nature of the stimulus. This maturation is necessary for DCs to efficiently act as an adjuvant and drive an effective adaptive immune response. For example, DCs loaded with *Leishmania major* lysate effectively induce a protective immune response to the pathogen. This protective immunity is Th1-driven, requiring the production of IL-12 and IFN-γ. The induction of an appropriate adaptive response can mean the difference between effective vaccination and vaccine failure. For instance DCs loaded with the yeast form of *Candida albicans* produce IL-12 and induce a Th1 response that is protective, but DCs loaded with the hyphal form produce IL-4 that is not protective.

The *ex vivo* loading of DCs with inactivated pathogens or pathogen products has several advantages. DCs can be directly activated by the pathogen itself, thereby inducing an appropriate immune response; a single DC is able to present all the pathogen's antigenic epitopes so as to activate the entire T cell repertoire; both cell-mediated and humoral responses are induced and peptide is loaded onto both MHC class I and class II molecules. However, whole pathogen lysates often contain toxins that are dangerous for human use and therefore methods of loading only specific antigens are favoured. Another approach is to direct antigen specifically to DCs *in vivo* by various cell-targeting methods. In this way, the antigen is complexed to a sensor molecule that specifically binds a molecule on the surface of DCs. For example MHC class II, CD40 and the endocytic receptor DEC-205 (CD205) have all been utilized for the targeted delivery of antigen to DCs.

## 17.20 Mucosal administration of vaccines

Most vaccines in current use are administered intramuscularly (into the muscle) or subcutaneously (under the skin). However, the majority of pathogens enter, infect and cause disease at mucosal sites such as the lung, intestine or urogenital tract. Therefore, an effective mucosal vaccine delivered to the site of infection would be a considerable advantage. The only effective mucosally-delivered vaccine for humans is the oral polio vaccine (OPV), which is a live-attenuated Sabin strain of polio virus. The effectiveness of this mucosal vaccine (together with the injected Salk strain) has been well demonstrated following the near-global eradication of polio. The OPV provides long lasting immunity, even by the standards of conventionally injected vaccines. As well as establishing long-term local immunity, mucosal vaccines are easily administered, especially orally (mouth) administered vaccines such as the OPV. Orally administered vaccines also reduce the risk of needle contamination associated with contemporary parenteral administration (administered into part of the body not involving the alimentary canal, usually intramuscular or intravenous). The induction of an immune response at the mucosa has been shown to result in the production of secreted IgA, which provides a high level of protection against pathogen adhesion, invasion and colonization. In addition, the induction of mucosal immune responses results

in the systemic synthesis of IgG, thereby providing further protection against infection and inhibiting the spread of the pathogen. It is also hoped that a vaccine delivered to a mucosal surface will generate a cell-mediated immune response, as well as an antibody response. In particular, CD8+ CTLs would provide effective immune defence against intracellular pathogens such as viruses.

There is much interest in developing a mucosal vaccine against *Vibrio cholerae*, the bacteria that cause the debilitating and life-threatening disease cholera, which infects the gastrointestinal tract of as many as 7 million people each year. Although several parenterally administered vaccines have been developed, they only provide short-term protection against cholera and do not prevent the transmission of the bacteria from person to person. Subcutaneous or dermal administration of cholera vaccines also causes local immune reactions and adverse events in many individuals. There is therefore a particular requirement for developing an orally administered cholera vaccine. Two oral cholera vaccines have been developed and are available for human use. The first vaccine is based on the non-toxic B subunit of cholera toxin (CtxB), which is administered orally in 2 or 3 separate doses. The second is a live attenuated *V. cholerae* vaccine that is devoid of an active cholera toxin A subunit, which is given as a single oral dose. However, neither vaccine provides protection in all vaccinated individuals and both are short lived, having a markedly reduced efficacy after only two years. There is still a need for a more effective mucosal cholera vaccine to be developed.

Seasonal influenza virus infection is a serious global health issue. In the Unites States alone, more than 200,000 hospitalizations are reported each year. Considering the number of people who are likely to be at risk of infection each year, there is a need for an easily delivered and inexpensive seasonal influenza virus vaccine. A cold adapted, live-attenuated influenza virus vaccine has been developed that can be administered intranasally in two separate doses. It has been reported that delivery into the nasal passages is more efficacious than parenteral delivery of conventional influenza vaccines. The advantage of nasal delivery is that the vaccine induces a protective immune response at the same mucosal site as the virus infects.

However, the efficacy of orally administered vaccines have generally been very low due to the acid and enzymatic environments of the stomach and gut, which tend to degrade vaccine formulations before uptake; and the mucous lining of the gut, which also interferes with antigen uptake. Furthermore, the immune environment within mucosal tissues tends to be immunoregulatory, rather than inflammatory, with a propensity towards tolerance. As a result of these restrictions most mucosal vaccines are based on replicating microorganisms, while killed or subunit vaccines require adjuvants to enhance the mucosal immune response. Current focus is therefore on developing a vaccine delivery system that is suitable and efficacious for use across mucosal surfaces. The most effective mucosal adjuvants tested are the bacterial toxins CT and LT. These toxins have powerful adjuvant effects but are far too toxin for human use. Therefore, several mutants of CT and LT have been engineered to be non-toxic but to retain their immunogenicity.

## 17.21 Nasally administered vaccine against genital infections

As discussed in chapter 9, the mucosa of the urogenital tract is inherently immunoregulatory, which has made immunization at this site problematic. However, recent evidence suggests that the nasal route of vaccination may provide immune protection within the genital tract. This has important implications in the generation of vaccines against sexually transmitted pathogens that infect the urogenital tract, such as HIV and *Chlamydia trachomatis*. Some studies suggest that nasally delivered vaccines induce strong antibody responses in both the respiratory tract and the urogenital tract, although high titres of antibodies were detected in the vagina but not in the cervix. The induction of an immune response in one mucosal tissue, together with a protective immune response at a distal mucosal tissue, relies on the concept of the common mucosal immune system (CMIS). Lymphocytes within the respiratory tract have also been shown to share chemokine receptors and cell adhesion molecules that allow them to migrate to both the respiratory and the urogenital tract following immunization. Some studies suggest that nasally delivered vaccines induce strong antibody responses in both the respiratory tract and the urogenital tract, although high titres of antibodies are often only detected in the vagina and not the cervix. The induction of an IgA response in the genital tract is also possible following nasal immunization, which is beneficial for the neutralization of pathogens and the prevention of invasion across the epithelium. This maybe particularly relevant for a HIV vaccine, as IgA is considered to provide far greater protection than IgG.

Another possibility for the induction of protective immunity in the genital tract is rectal administration. This route of vaccine delivery has been shown to induce IgA

and IgG responses in the genital tract, as well as the colon and large intestines. In addition, sublingual administration (directly under the tongue) is able to induce antibody responses in several mucosal tissues including the genital tract, gastrointestinal tract and respiratory tract. This route of delivery may also be safer than the intranasal route, with fewer adverse reactions. The efficacy of sublingual administration is thought to be related to the higher number of CCR7+ DCs located within the surrounding tissues. This chemokine receptor enables DCs to migrate to the cervical lymph nodes where they can present antigen to T and B cells. As well as establishing mucosal IgA responses, sublingual vaccines my also stimulate CTL responses, an advantage for vaccines directed against intracellular pathogens such as HIV or influenza virus.

## 17.22 New strategies for vaccine development

Vaccination has dramatically reduced the burden of infectious disease since its inception more than 100 years ago but the methodology used for vaccine development has changed little, even though it remains ineffectual against many diseases. There is no effective vaccine against pathogens such as *Plasmodium falciparum* (malaria), HIV (AIDS) and *Leshamania major* (leshmaniasis), therefore new approaches to rational vaccine design are needed (Box 17.2). The advent of modern genome-based technologies and bioinformatics has revolutionized vaccine design and allowed for the analysis and identification of a large array of potentially protective antigens.

The recent sequencing of whole pathogen genomes has allowed the identification of gene products that may provide suitable vaccine candidates. This has been facilitated by powerful computer analysis for the prediction of such antigens. Modern molecular biology techniques then permit the isolation and screening of individual proteins in immunological and protection based assays. Candidate genes can be manipulated further and inserted into vector vehicles such as attenuated viral or bacterial vectors, or DNA plasmids in order to express the gene of interest *in vivo*. Such vectors include recombinant vaccinia virus, poxviruses, BCG and *Salmonella* species, all adapted so as to provide optimal expression of inserted genes. Such vector vaccines have been incorporated into prime-boost vaccination regimes. Immunization with a recombinant vector is followed by a booster immunization with a

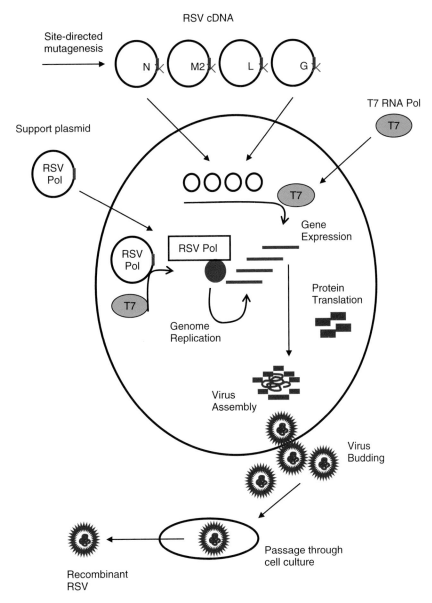

**Figure 17.12** Reverse genetics and the development of novel vaccines. Reverse genetics involves the generation of whole virus from separate genetic components. Individual genes of the virus (RSV has 10 genes in total) inserted into a DNA expression vector (only four genes are shown for simplicity: N (nucleoprotein); M2 (second matrix protein); L (polymerase); and G (surface glycoprotein)). These plasmids are then transfected into a suitable host cell. In order for plasmid gene expression T7 RNA polymerase is also inserted into an expression vector. Plasmid gene expression generates many copies of mRNA, which are translated into viral proteins. In order for whole virus particles to be generated a support plasmid that contains the RSV polymerase is inserted into the cell. The RSV polymerase then replicates the viral RNA. The recombinant RSV (rRSV) is then passaged in cell culture and attenuated for use as a vaccine.

heterologous antigen in order to enhance cell-mediated immunity, antibody production and memory.

Paralleling genome analysis is proteomics, whereby the function and antigenicity of previously undefined proteins can be identified. In a similar way bioinformatics has enabled the prediction of T cell epitopes that form the basis of peptide vaccine design. Another exciting development is the reverse genetics approach (Figure 17.12), whereby specific genetic alterations are inserted into known genes of negative-strand RNA viruses. The cDNA from these modified genes are co-transfected with the remaining genes of the virus and the new virus is then isolated for phenotypic analysis. The prevention of infectious diseases has therefore entered a new and promising era.

## 17.23 Summary

**1.** Vaccination, or immunization, induces a protective immune response to a microorganism.

**2.** Vaccination induces the expansion of antigen specific T cells and B cells that establish a pool of memory cells.

**3.** Vaccines often require an adjuvant, which is co-administered, in order to induce an immune response to the desired antigens.

**4.** The route of immunization is often important in order to induce the most appropriate immune response to a pathogen.

**5.** Although most pathogens infect mucosal tissues, very few vaccines are delivered to mucosal tissues.

**6.** New developments in vaccine and adjuvant design are establishing new possibilities for successful mucosal vaccines.

# Glossary of Terms

**accessory cell** Any non-lymphocytic cell required for the initiation of an immune response. Often used to describe an antigen-presenting cell (APC).

**accessory molecules** Molecules that participate in the activation and effector functions of lymphocytes, other than antigen receptors and major histocompatibility complex (MHC) molecules. Examples include CD4 and CD8 co-receptors on T lymphocytes.

**activation-induced cell death** The process by which a cell undergoes apoptosis, or programmed cell death, following a stimulatory signal.

**acute-phase proteins** A group of plasma proteins that participate in the early phases of innate host defense against infection, known as the acute phase response. Acute phase proteins are predominantly produced in the liver in response to IL-6, interferons and TNF. Acute phase proteins include C-reactive protein and complement factors.

**adaptive immune response** The response of antigen-specific T lymphocytes and B lymphocytes to antigen. Activation involves the presentation and specific recognition of non-self antigen. Includes the development of immunologic memory so that a stronger response is mounted against subsequent antigen encounter (immunity). Also known as acquired immunity.

**adenoids** Organized secondary lymphoid structures that contribute to the Waldeyer's ring, located in the upper airways and at the opening to the oro-pharynx.

**adenosine deaminase (ADA)** An enzyme that catalyzes the deamination of adenosine to produce inosine. A mutation or deficiency in the ADA gene results in a form of severe combined immunodeficiency (SCID).

**adhesion molecules** Molecules that are responsible for the binding of adjacent cells or to the extracellular matirx. They are important in cell migration during an immune response. Examples include integrins, selectins, and members of the immunoglobulin gene superfamily.

**adjuvant** A substance that enhances the immune response to an antigen when administered together. Many vaccines rely on adjuvants to elicit effective immune responses.

**affinity** A measure of the dissociation constant between a protein and its ligand. For example, a measure of the strength of binding of an antibody to its antigen through a single bond.

**affinity maturation** The process of Darwinian selection by which B cell clones that express high affinity antibody to a specific antigen are preferentially selected. Those B cells that express antibody with the highest affinity to antigen are selected to undergo proliferation and expansion, while those with low affinity antibody are deleted.

**agglutination** The aggregation of cells or particulate antigen in the presence of antibodies. Antigen is usually present on the surface of a cell, for example red blood cells or bacteria. Agglutination of particulate antigen can lead to immune complex formation.

**allele** A variant of a single genetic locus within the genome, usually describing two or more forms of the genetic sequence of a gene. Alleles can differ within a population and are considered polymorphic. As the human genome contains two sets of chromosomes (diploid genome), one allele of a gene is normally inherited from the mother and one from the father.

**allelic exclusion** The selective expression of one allele over the other. This occurs in lymphocytes whereby only one allelic form of the antigen-specific receptor (BCR or TCR) is expressed.

**allergen** A type antigen responsible for the sensitization and initiation of allergic immune responses.

**allergic asthma** An immune disease of the airways associated with smooth muscle contraction and bronchoconstriction that is mediated by contact with an allergen.

*Immunology: Mucosal and Body Surface Defences*, First Edition. Andrew E. Williams.
© 2012 John Wiley & Sons, Ltd. Published 2012 by John Wiley & Sons, Ltd.

**allergic rhinitis** An allergic reaction to an allergen in the upper airways that causes runny nose, sneezing and tears. Otherwise known as hay fever.

**allergy** An immune reaction to an allergen, which is associated with inflammation and disease.

**allogeneic** Genetic differences between members of the same species.

**allotypes** The endogenous antigen repertoire that may vary between members of the same species. Transplant recipients are given tissues with the same or similar allotype as the donor in order to avoid rejection.

**alternate complement pathway** The pathway of complement activation that relies on the intrinsic activation of C3, rather than C1q complexes.

**alveolar macrophage** A macrophage population found in the alveoli of the lung that may be involved in maintaining lung homeostasis.

**anaphylatoxin** Proteins, such as the complement fragments C3a, C4a and C5a, that are involved in the initiation of inflammatory reactions, often associated with degranulation during allergic responses.

**anaphylaxis** A life-threatening immediate hypersensitivity reaction to an allergen involving IgE-mediated degranulation of mast cells.

**anergy** The inability of T cells or B cells to mount an effective immune response to a particular antigen, usually as a result of immunological tolerance.

**antibody** A protein produced by B cells that is able to bind to an antigen. Also known as an immunoglobulin.

**antibody-dependent cell-mediated cytotoxicity (ADCC)** A cell-mediated immune mechanism by which a target cell is lysed following the binding of antibody to its cell surface. NK cells, eosinophils and neutrophils participate in ADCC through the recognition of antibody via Fc receptors.

**antigen** Any substance that can be bound by an antibody. The term antigen is also used in a more general sense and can mean any substance that a B cell or T cell recognises through their antigen receptors.

**antigen presentation** The display of processed antigenic peptides by MHC molecules on the surface of cells, which are subsequently recognized by T cell receptors expressed on T cells.

**antigen processing** The degradation of proteins into small peptides so that they can be presented in conjunction with MHC molecules. Antigen processing is necessary for peptide recognition by T cells.

**antigen receptor** Any receptor that is able to recognize an antigen, although the term specifically applies to B cell receptors and T cell receptors expressed by B and T cells respectively.

**antigen-binding site** The specific part of an antibody or T cell receptor that forms a molecular interaction with an antigen.

**antigenic determinant** The particular molecular structure that interacts with an antibody. Antigens tend to have several antigenic determinants, otherwise known as epitopes.

**antigen-presenting cell (APC)** Specialized leukocytes that are particularly adept at initating T cell responses following the processing and presentation of antigen. Dendritic cells and macrophages are examples of APCs.

**antiserum** The protein-rich fraction of clotted blood (serum) that contains antibodies against a particular antigen. Antisera can be used to transfer passive immunity to pathogens or neutralize toxins.

**apoptosis** The highly orchestrated process of programmed cell death, which cells undergo in response to an external stimulus.

**appendix** A gut-associated lymphoid tissue located at the junction between the ileum and the colon.

**atopy** The predisposition of an individual with abnormally high levels of IgE, which makes them more susceptible to allergic reactions.

**autoimmunity** An immune response directed to ones own molecules (endogenous antigens) that often results in tissue damage

**autologous** The description of a tissue or substance that is derived from the same individual.

**autophagy** The sequestration and degradation of protein from the cytosol by lysosomes.

**avidity** The combined synergy of all molecular interactions (different to affinity). Usually used to describe the interaction between an antibody and an antigen that has multiple bonds.

**B lymphocyte (B cell)** A differentiated lymphocyte that is characterized by the synthesis of antibody. B cells terminally differentiate into antibody secreting plasma cells.

**bacteria** A diverse domain of prokaryotic microorganisms. Bacteria comprise the vast majority of the commensal microflora. In addition, pathogenic bacteria cause various diseases in humans.

**basophils** A type of granulocyte that stains with basic dyes. They have a similar function to mast cells.

**BCG (bacille Calmette-Guerin)** A vaccine that has long been used against human tuberculosis, caused by *Mycobacterium tuberculosis*. The vaccine comprises the related *Mycobacterium bovis* strain.

**bone marrow** A primary lymphoid tissue located in the cavities of the long bones and pelvis, where haematopoiesis and early B cell differentiation take place.

**bronchus associated lymphoid tissue (BALT)** Relating to the mucosal-associated lymphoid tissue located in the bronchioles of the respiratory tract.

**bursa of Fabricius** A primary lymphoid organ found in birds, located in the cloacal sac. It is the primary site of B cell development in birds. The bursa gives its name to the B cell.

**C3** A complement component important for the activation of the classical pathway. C3 fragments also act as anaphylatoxins, opsonins and chemoattractants.

**C3 convertase** Another name for the complement proteins C4b2a or C3bBb, which split C3 into C3b and C3a in the classical and alternative pathway respectively.

**C5** A complement component that links complement pathway activation to components that form the membrane attack complex.

**C5 convertase** A complement component that splits C5 into C5a and C5b in both the classical and the alternative pathway.

**caspases** A family of cysteine proteases that play important roles in apoptosis, by mediating the cleavage of proteins.

**cell-mediated cytotoxicity (CMC)** The mechanism by which leukocytes kill target cells.

**cell-mediated immunity (CMI)** An arm of the immune system that involves the activation and effector function of leukocytes such as macrophages, NK cells and T cells. Cell-mediated immunity can be separated from humoral immunity.

**chemokines** A family of low molecular weight cytokines that are involved in the migration and activation of leukocytes. They play a central role in chemotaxis and homing of cells to tissues.

**chemotaxis** Migration of cells up a concentration gradient of a chemoattractant, often involving chemokines.

**clonal deletion** The deletion of B cells or T cells with a particular specificity to an antigen. T cell clonal deletion occurs during thymic selection and B cell clonal deletion occurs during affinity maturation.

**clonal selection theory** The selective proliferation of T cells or B cell within a larger population as a result of specific antigenic stimulation. Those cells that possess superior antigen recognition will proliferate over those cells with a weaker antigen recognition and eventually dominate the lymphocyte repertoire.

**cluster of differentiation (CD)** The nomenclature given to host antigens, which are used to phenotype cells based on expression.

**colony-stimulating factors (CSF)** Proteins that are similar in function to cytokines that play a role in cell differentiation, growth and activation.

**combinatorial joining** The genetic mechanism responsible for the ligation of two gene segments during B cell receptor (antibody) and T cell receptor recombination.

**common-mucosal immune system** This term refers to an integrated system of separate mucosal tissues, whereby cells induced in one mucosal tissue can migrate to others. However, in reality there are distinct compartmentalizations between the different mucosal tissues that require discrete homing receptors.

**complement pathway (classical)** A series of protein interactions that ultimately result in the formation of a membrane attack complex and the release of pro-inflammatory mediators. Initiated following binding of the C1 complex to antibody-antigen complexes on cell surfaces (classical pathway).

**complement receptors (CR)** Proteins that are expressed on the surface of cells that recognise complement proteins and induce cell activation and effector functions.

**complementarity-determining regions (CDRs)** The molecular regions of antibodies and T cell receptors that make contact with antigenic determinants (epitopes) on the antigen. These regions possess the most receptor diversity and are known as variable domains.

**conjunctiva associated lymphoid tissue (CALT)** The organized mucosal lymphoid tissue located around the eye.

**constant region (C region)** The invariant domain of an antibody or T cell receptor. The constant region of antibodies largely determines their immune effector function, while the constant region of T cell receptors are involved in intracellular signalling.

**contact dermatitis** A type IV delayed-type hypersensitivity reaction in response to an allergen that has made contact with the skin.

**costimulatory molecules** Receptors or ligands that are expressed by interacting cells, which are necessary for full cell activation. For example, CD28 is an important costimulatory molecule expressed by T cells and is required for full activation following T cell receptor/MHC molecule ligation.

**C-reactive protein (CRP)** An acute phase protein produced by the liver following an inflammatory episode. It can act as an opsonin that enhances bacterial phagocytosis.

**cross-reactivity** The ability of an antibody to react to two or more separate antigens.

**CTLA-4** Cytotoxic T lymphocyte antigen 4, which is expressed by T cells, is a high-affinity receptor for CD80/CD86 molecules expressed by APCs. It functions as an inhibitory receptor during T cell responses.

**cytokine** Soluble proteins that are secreted by various cells, which have a number of effects on other cells.

**cytokine receptors** Proteins expressed on the surface of cells that recognize cytokines and activate intracellular signalling events that cause alterations in the function of the target cell.

**cytotoxic T lymphcytes (CTLs)** A sub-population of T cells that have potent cell lytic effects on virally infected and abnormal cells. Most CTLs are CD8+.

**damage associated molecular pattern (DAMP)** A molecule or metabolite that is released by host cells in response to stress, damage or death. DAMPs are recognized by pattern recognition receptors (PRRs) and induce cell activation and cytokine release.

**decay-accelerating factor (DAF)** A glycoprotine involved in regulating the complement cascade by initiating the dissociation of C3 convertase (C4b2a) into C4b and C2a.

**degranulation** A process whereby intracellular granule fuse with the plasma membrane and release their contents into the extracellular environment; often associated with granulocyte activation.

**delayed-type hypersensitivity (DTH)** A cell-mediated immune response to an antigen that occurs days after initial contact, and one which is associated with Th1 activation.

**dendritic cells** A heterogenous population of haematopoietic cells characterized by numerous protruding dendrites found throughout all peripheral tissues and secondary lymphoid organs. They can function as professional APCs and initiate T cell and B cell responses

**desensitization** A reduction in the intensity of an allergic response to an allergen over the course of time.

**diapedesis** The migration of leukocytes from the blood, across endothelial cell walls and into tissues.

**double-negative thymocytes** Immature T cells found within the thymus during T cell development that lack expression of CD4 and CD8 molecules.

**double-positive thymocytes** Immature T cells found within the thymus during T cell development that express both CD4 and CD8 molecules.

**effector cells** Lymphocytes that have been activated by antigen and have differentiated from naïve or memory cells. They perform specific biological functions in peripheral tissues so as to remove pathogens or antigens. Activated CTLs or antibody secreting plasma cells are examples of effector cells.

**endocytosis** A mechanism by which molecules are taken into a cell from the extracellular environment.

**endotoxins** Endotoxins are derived from bacteria and are capable of stimulating the immune system. The archetypal endotoxin is lipopolysaccharide (LPS) derived from gram negative bacteria.

**eosinophils** A type of granulocyte that contributes to allergic responses through the recognition of IgE. Eosinophils are also thought to be important in controlling parasitic infections.

**epithelial cell** A cell type that forms the cellular barrier at mucosal surfaces.

**epitope** The part of an antigen that forms a molecular interaction with an antibody, also known as an antigenic determinant. Antigens often have many individual epitopes.

**exocytosis** The mechanism by which intracellular molecules are released into the extracellular environment.

**exon** The region of a gene that is transcribed into mRNA, which forms the template for protein translation.

**F(ab)' 2** The resultant fragment of an immunoglobulin following cleavage by the enzyme pepsin, which possesses two antigen binding sites.

**Fab** The resultant fragment of an immunoglobulin following cleavage by the enzyme papain, which possesses one antigen binding site.

**Fas** A cell surface protein, which is a member of the TNF receptor super-family, that induces the cell to undergo apoptosis following binding to Fas ligand.

**Fas ligand (FasL)** A cell surface protein, which is a member of the TNF receptor super-family, that induces cells expressing it receptor (Fas) to undergo apoptosis.

**Fc** The resultant fragment of an immunoglobulin following cleavage by the enzyme papain, which corresponds to the invariant constant region.

**Fc receptor (FcR)** A cell surface protein that binds to antibody molecules. Cross-linking of FcRs by numerous antibodies results in cell activation.

**follicle associated epithelium (FAE)** The section of epithelium that contains high numbers of M cells and overlies sub-epithelial domes and lymphoid follicles.

**follicular dendritic cells** A type of non-haematopoietic dendritic cell located in the germinal centres of lymphoid follicles. They present fixed antigen to B cells during B cell differentiation and affinity maturation.

**Freund's complete adjuvant** An adjuvant used in experimental research comprising of an oil containing killed mycobacteria and an emulsifier. The co-administration of an antigen enhances the immune response to that antigen. Freund's complete adjuvant (FCA) is such a potent immunogen that it is not safe for human use.

**genetic immunization** Otherwise known as DNA immunization or vaccination. A gene that encodes the antigen of interest is injected into a recipient. Host cells transcribe the gene and manufacture the protein, against which an immune response in generated.

**genotype** The particular genetic makeup of an individual.

**germ line** Those genes contained within germ line cells as opposed to somatic cells. In immunology often referring to unrearranged B cell and T cell receptors.

**germinal centers** Central structures contained with lymphoid follicles where B cells undergo rounds of proliferation and selection during an antibody response. They are the sites of dendritic cell, T cell and B cell interaction.

**goblet cell** A specialized epithelial cell located in the small airways, which secretes mucus from granules stored in its cytoplasm.

**granulocyte** A type of leukocyte characterized by intracellular granules. Neutrophils, eosinophils, mast cells and basophils are all granulocytes.

**granulocyte-macrophage colony-stimulating factor (GM-CSF)** A type of cytokine known as a growth factor that is involved in leukocyte differentiation and activation.

**granuloma** A structure that forms in tissue as a result of a delayed-type hypersensitivity reaction to an antigen or pathogen. It is characterized by a central sphere of monosytes and macrophages, surrounded by T cells and fibrous sheath.

**gut associated lymphoid tissue (GALT)** GALT comprises organized and diffuse lymphoid aggregates along the entire length of the gastrointestinal tract, including the oesophagus, stomach, small intestine, large intestine, colon and appendix. Peyer's patches are part of GALT, which form in the sub-epithelia of the intestine.

**haplotype** A set of alleles on a chromosome that are inherited together from a single parent. Haplotype is often used to describe the inheritance of MHC alleles, one from each parent.

**hapten** A small molecule that elicits an antibody response only when bound to a larger carrier protein.

**heavy chain (H chain)** The larger of the two chains that form immunoglobulins, comprising of variable heavy ($V_H$) and a constant heavy ($C_H$) domain.

**helper T cells** A sub-population of T cells that express the CD4 co-receptor. Their main function is to elicit the induction of antibody and cell-mediated immune responses through co-stimulatory receptor ligation and cytokine production.

**haematopoiesis** The development and differentiation of the cells of the blood, including red blood cells, platelets and leukocytes. Much of haematopoiesis takes place in the bone marrow.

**hematopoietic stem cell** A multi-potent stem cell in the bone marrow that is capable of differentiating into any number of different blood cells.

**hemolytic disease of the newborn** A disease of a newborn child that occurs when maternal antibodies cross the placenta and bind to antigens of the surface of fetal red blood cells. The disease is characterized by anaemia and jaundice.

**heterodimer** A protein complex that is formed of two different protein molecules. Examples include T cell receptors and MHC complexes.

**heterozygous** Individuals that possess two different alleles of a gene.

**high endothelial venules (HEVs)** Specialized structures of the vasculature that are located predominantly in lymphoid tissues, and which provide a favourable environment for leukocytes to migrate out of the blood and across endothelium.

**hinge region** Part of antibody molecules that provide a degree of flexibility in the variable domains.

**histamine** A molecule stored in mast cell, basophil and eosinophil granules. It is released upon degranulation following IgE cross-linking and causes vasodilation, smooth muscle contraction and inflammation and is associated with the symptoms of allergy.

**histocompatibility** The immunological compatibility of organs or cells derived from separate individuals, based on the expression of the same cell surface antigens. Histocompatibility often specifically refers to MHC antigens.

**homodimer** A protein complex composed of two identical proteins.

**human immunodeficiency virus (HIV)** The retrovirus that is the causative agent of acquired immune deficiency syndrome (AIDS).

**human leukocyte antigen (HLA)** The term used to describe the human MHC complex, which is encoded by a large genetic locus that contains many genes related to the immune system, including MHC class I and MHC class II genes.

**humoral immunity** The arm of the immune system that involves soluble proteins, including complement factors, acute phase proteins and especially antibody.

Humoral immunity can be differentiated from cell-mediated immunity.

**hyperimmune** The elevated state of immunity to an antigen, characterized by high titres of antibody.

**hypersensitivity** A heightened reactivity to an antigen or irritant that is associated with a deleterious outcome. Hypersensitivity reactions can be antibody-mediated or cell-mediated.

**hypervariable regions** The part of an antibody or T cell receptor that makes molecular contact with an antigen and one that contains the most sequence variation between molecules.

**immediate-type hypersensitivity** Otherwise known as type I hypersensitivity that involves an immediate reaction to an antigen. Immediate-type hypersensitivity is due to previous sensitization and reexposure to antigen and is mediated by antibody.

**immune complex** A complex of molecules formed by the cross-linking of antigens and antibodies. Immune complexes enhance phagocytosis of antigen but deposition in tissues can cause immune damage.

**immune modulators** Molecules that can alter the magnitude of an immune response, usually referring to a molecule that downregulates an immune response.

**immunity** The generation of a protective immune response to a pathogen, which is long-lasting and is associated with the maintenance of a memory lymphocyte population.

**immunodeficiency** A defect in normal immune function that compromises host defence.

**immunogen** A molecule that is capable of inducing an immune response in the absence of an antigen. Adjuvants used to promote immune responses to vaccines are examples of immunogens.

**immunoglobulin (Ig)** A protein complex composed of two light chains and two heavy chains, which is responsible for binding to antigens. The term immunoglobulin is synonymous with antibody and also describes the central protein of the B cell receptor.

**immunoreceptor tyrosine-based activation motif (ITAM)** An amino acid domain contained within the cytoplasmic tail of many cell surface receptors, which recruits adaptor proteins and initiates intracellular cell signalling events that lead to cell activation.

**immunoreceptor tyrosine-based inhibitory motif (ITIM)** An amino acid domain contained within the cytoplasmic tail of several cell surface receptors, which results in the inhibition of intracellular signalling events.

**inflammasome** A highly regulated protein complex within the cytosol that recruits caspase-1 in order to cleave the pro-inflammatory cytokines IL-1β and IL-18.

**inflammation** A protective response to infection or tissue injury that involves a cascade of events involving the vasculature, epithelia, leukocytes and plasma proteins. Acute inflammation describes the immediate response to tissue injury, while chronic inflammation involves progressive tissue remodelling.

**innate immunity** The early, non-specific response to infection induced by the recognition of danger signals and involving macrophages, granulocytes, NK cells and factors of humoral immunity. No immune memory is induced.

**integrins** A family of cell-surface adhesion molecules important for the migration and trafficking of leukocytes.

**intercellular adhesion molecules (ICAMs) 1, 2, and 3** Adhesion molecules expressed on the surface of several cell types, typically endothelial cells, that bind to integrins.

**interdigitating dendritic cells (IDC)** A population of dendritic cell found in the thymus, which present endogenous antigens during the negative selection of developing thymocytes.

**interferons (IFN)** A family of cytokines that possess potent antiviral properties and are involved in modifying immune responses.

**interleukins (IL)** A family of cytokines that have important functions in cell signalling and orchestrating immune responses.

**intra-epithelial compartment** Describes the location of cells within the epithelial lining of mucosal tissues. For example, T cells located within this compartment are known as intra-epithelial lymphocytes (IELs).

**intron** The sections of DNA within genes that are not transcribed into mRNA and therefore do not code for protein.

**isotype switching** The genetic recombination event that results in antibody isotype (class) switching during B cell differentiation. The VDJ unit is recombined with a different constant domain, for example IgM is replaced with IgG.

**isotypes** Otherwise known as antibody classes, each of which has a different function within the immune system Five basic isotypes exist, IgM, IgD, IgG, IgA and IgE, each of which binds to different Fc receptors.

**J chain (joining chain)** A small polypeptide the links the constant heavy chains of IgM and IgA enabling the formation of polymeric immunoglobulins.

**killer-cell immunoglobulin-like receptor (KIR)** A set of receptors expressed on the surface of NK cells that function by surveying host cells for signs of abnormality or infection. KIRs recognize MHC molecules, which are downregulated by viruses or tumour cells.

**killer T cells** A sub-population of T cells, synonymous with cytotoxic T lymphocytes (CTL), that express the CD8 co-receptor and possess potent cytolytic capabilities.

**lamina propria** The subepithelial compartment of the gastrointestinal tract where immune effector mechanisms take place and where Peyer's patches develolp.

**Langerhan's cells** A population of dendritic cells characterized by the expression of langerin. They were first discovered in the epidermis of the skin where they survey the tissue environment for signs of infection or stress.

**leukocytes** A term used to describe white blood cells that originated in the bone marrow and have undergo haematopoiesis. Leukocytes encompass all the bone-marrow derived immune cells and therefore have a central role in all immune responses.

**ligand** A molecule that binds to a receptor.

**light chain (L chain)** The smaller part of an immunoglobulin molecule that only has a variable domain and no heavy domain. Two light chains, either κL or λL, combine with two heavy chains to form a complete immunoglobulin molecule.

**lipopolysaccharide (LPS)** An endotoxin derived from gram-negative bacteria that has immunostimulatory properties. It is recognized by TLR4.

**lymph** A circulatory fluid contained with the lymphatic system, which transports molecules and cells contained within tissue fluid.

**lymph nodes** Organized secondary lymphoid nodules associated with the lymphatic system. They are important in the generation of adaptive immune responses to antigens derived from regional tissues. They act as inductive sites for T cell and B cell activation.

**lymphatic system** A circulatory system driven by the movement of the body that transports molecules and cells in lymph. The lymphatic system is important for the transport of immune cells and antigens from tissues and into draining lymph nodes.

**lymphocytes** A type of white blood cell that originated in the bone-marrow. Lymphocytes are small leukocytes that include NK cells, T cells and B cells. They have many important functions in the immune system.

**lymphoid follicle** An organized aggregation of lymphoid cells found in secondary lymphoid tissues such as lymph nodes, Peyer's patches and tonsils. Lymphoid follicles are where dendritic cells, T cells and B cells interact during the development of an immune response. Mature lymphoid follicles contain germinal centres, in which B cells proliferate and differentiate into antibody-producing cells.

**macrophages** A type of bone marrow-derived leukocyte, thought to be differentiated from a monocyte and found in many tissues and organs. They are highly phagocytic cells involved in the killing and clearance of pathogens and in the maintenance of tissue homeostasis.

**major histocompatibility complex (MHC)** A large genetic locus containing many genes associated with the immune system, including MHC class I and MHC class II molecules. Polymorphisms in the MHC are responsible for allogenicity and a lack of histocompatibility during transplant rejection.

**mast cell** A type of bone marrow-derived granulocyte located within connective tissue and peripheral tissues. Cross-linking of antibody causes degranulation and release of histamine, which are associated with allergic reactions.

**membrane attack complex** A protein complex formed from complement factors (C7-C9), which form a pore on the plasma membrane of target cells, resulting in cell lysis.

**memory** In immunology, refers to a population of long-lived T cells and B cells which respond rapidly to a secondary challenge with the same antigen. Memory T cells and B cells form the basis of adaptive immune responses and provide immunity to pathogens.

**MHC class I molecule** A heterodimer formed of an α-chain and β$_2$-microglobulin, which presents antigenic peptides to CD8+ cytotoxic T cells.

**MHC class II molecules** A heterodimer formed of an α-chain and a homologous β-chain, which presents peptide antigens to CD4+ T helper cells.

**MHC class III molecules** Proteins encoded by the major histocompatibility complex (MHC), which play no role in peptide presentation, but instead are secreted proteins involved in immune responses such as complement factors and TNF.

**MHC restriction** The mechanism by which T cells can only respond to antigen when presented in the context of an MHC class I or MHC class II complex.

**microfold cell (M cell)** A specialized epithelial cell located throughout the epithelium of the gastrointestinal and respiratory tracts, which is responsible for the transport of antigen from the lumen into the sub-epithelial compartment.

**minor histocompatibility antigens** Molecules that are encoded by genes outside the MHC locus but can still participate in histocompatibility reactions and transplant rejection.

**mitogen** A molecule that is able to stimulate many different lymphocyte clones independently of a specific antigen.

**molecular mimicry** The serendipitous similarity of molecular structures between host molecules and pathogen molecules, which may be associated with autoimmune responses.

**monoclonal** A population of cells that have been derived from a single cell and therefore have the same antigen specificity. This can apply to both T cell and B cell clones. Antibodies can also be monoclonal if they are derived from the same B cell clone or share epitope specificity.

**monocytes** A type of leukocyte found in the blood, which can differentiate into macrophages or myeloid dendritic cells in tissue.

**mucosal associated lymphoid tissue (MALT)** A tissue that is able to secrete mucus across an epithelial membrane and participate in immune defence. Examples include the gastrointestinal, respiratory and urogenital tracts.

**mucous** The adjective of mucus. Relating to the secretion of mucus, such as a mucous membrane or mucous lining.

**mucus** A sticky substance formed predominantly of mucin glycoproteins, which is secreted into the lumen of mucosal tissues, forming a protective barrier.

**nasal associated lymphoid tissue (NALT)** In humans, this includes the organized lymphoid structures associated with the nose and the pharyngeal ring (Waldeyer's ring) and is therefore often referred to as the nasopharynx-associated lymphoid tissue. In rodents, this describes the lymphoid tissue located in the nasal cavity.

**natural killer (NK) cells** A large cytotoxic lymphocyte that contributes to innate immunity by secreting IFN-γ and executing ADCC. They possess potent antiviral and anti-tumour capabilities.

**negative selection** The apoptotic process that occurs in the thymus that deletes self-reactive T cells. B cells in the bone marrow that sythesize self-reactive antibodies are also deleted.

**neutralization** The mechanism by which antibodies block the effects of an antigen or toxin, or inhibiting the invasion and dissemination of pathogens.

**opsonization** The binding of an antibody or complement factor to the surface of a target cell in order to tag it for recognition by an effector cell. Opsonization enhances phagocytosis and ADCC by macrophages and NK cells, respectively. The antibody or complement factors are referred to as opsonins.

**pathogen associated molecular pattern (PAMP)** A molecular motif derived from a microorganism that is recognized by pattern recognition receptors (PRRs) expressed on host cells. Recognition of PAMPs usually induces an immune response following cell activation and cytokine release.

**paneth cell** A specialized epithelial cell located in the crypts of villi of the gastrointestinal tract, which is responsible for the secretion of mucus into the gut lumen.

**paracortex** The area surrounding a lymphoid follicle in a secondary lymphoid structure, which is dominated by T cells.

**passive immunization** The transfer of protective antibodies from one person to another, which does not involve the induction of an immune response.

**pathogen** An infectious microorganism that causes disease in its host, such as a pathogenic virus, bacterium, fungus or parasite.

**pattern recognition receptor (PRR)** A receptor expressed on a variety of host cells that recognizes pathogen associated molecular patterns (PAMPs) derived from microorganisms. Members of the Toll-like receptor (TLR) family are examples of PRRs.

**perforin** A molecule secreted by NK cells and cytotoxic T cells that forms a pore in the plasma membrane of target cells and causes cell lysis.

**peripheral lymphoid organs** Secondary lymphoid tissues such as the spleen, lymph node and organized lymphoid structures of MALT.

**peripheral tolerance** The induction of anergy or unresponsiveness to an antigen in T cells and B cells that is independent of the thymus and central tolerance.

**Peyer's patches** Organized secondary lymphoid structures along the length of the intestine that can act as inductive sites for local T cell and B cell responses. Peyer's patches typically contain several lymphoid follicles.

**phagocytosis** The endocytosis, or engulfment, of large particles by phagocytic cells such as macrophages and neutrophils.

**phenotype** The observable manifestation of the genotype, combining aspects of physiology, biochemistry and behaviour.

**pinocytosis** The engulfment of small particles or liquids by the formation of cellular vesicles.

**plasma** The liquid fraction of the circulating blood, which contains numerous proteins and other metabolites.

**plasma cell** A type of mature lymphoycte. A terminally differentiated B cell that sythesizes large quantities of antibody.

**platelets** Bone-marrow derived cells, otherwise known as thrombocytes, which are important in the blood clotting process.

**poly-Ig receptor** A protein expressed on the basolateral surface of epithelial cells responsible for the binding and transport of IgA across the epithelium and into the lumen. Referred to as secretory component when physically attached to an antibody molecule or when free in the lumen.

**polymorphism** Differences in the genetic sequence of a gene, known as an allele, between different people. Polymorphisms within a population can influence an individual's immune response to a pathogen.

**polymorphonuclear leukocytes (PMN)** Circulating leukocytes that have distinctive multi-lobed nuclei: includes neutrophils, basophils and eosinophils.

**positive selection** The process by which developing B and T cells receive signals in the primary lymphoid organ in which they are developing to continue their differentiation; in the absence of these signals the cells die.

**primary follicle** An organized region of a secondary lymphoid organ that contains immature B cells. A primary follicle develops into a secondary follicle upon antigen stimulation.

**primary lymphoid organs** These are the sites of early lymphocyte development, otherwise known as central lymphoid organs. In humans, the bone marrow is responsible for haematopoiesis and B cell development, while the thymus is responsible for T cell development.

**primary response** The adaptive immune response initiated following the first exposure to an antigen. Often requires a lag phase in order for antigen processing, presentation and maturation of naive lymphocyte and is characterized by the production of relatively high levels of IgM.

**priming** The exposure to antigen and subsequent activation of naïve lymphoctes.

**proteosome** A protein complex located in the cytoplasm that degrades proteins into small peptides. An important piece of cell machinery required for the processing of peptide for MHC class I presentation.

**pyrogen** A molecule that is capable of inducing fever (high temperature), such as TNF.

**RAG-1 and RAG-2 (recombination activating genes)** Genes that encode enzymes responsible for orchestrating the recombination events during B cell receptor and T cell receptor synthesis.

**receptor** A protein usually expressed on the cell surface that binds to a ligand and induces an intracellular signalling cascade that modifies the activity of a cell.

**receptor editing** The process by which immunoglobulins undergo secondary genetic modifications in order to alter the specificity of their antigen-binding domain.

**repertoire** The total population of B cells or T cells that are capable of responding to antigen.

**reticuloendothelial system (RES)** The population of cells involved in phagocytosis, including monocytes and macrophages, that accumulate in tissues and secondary lymphoid organs.

**rheumatoid arthritis** An immune-mediated autoimmune disease that primarily affects the joints.

**secondary lymphoid organs** Organized lymphoid structures, other than the bone-marrow or thymus, that can act as inductive sites for the generation of antigen-specific immune responses. These include lymph nodes and lymphoid aggregates associated with mucosal tissues.

**secretory component** The cleaved component of the poly-Ig receptor.

**selectins** A family of proteins expressed on the cell surface that contribute to cell adhesion and migration.

**sensitization** The first encounter with an antigen, usually an allergen, that subsequently leads to an antibody response following secondary encounter.

**sepsis** A severe form of systemic inflammation induced by bacterial infection of the bloodstream.

**serum** The protein-rich fraction of clotted blood, which contains high titres of antibodies.

**severe combined immune deficiency (SCID)** A general term for a genetic disorder of the immune system, which normally results in a defective T and B cell system.

**signal transducers and activators of transcription (STATs)** Intracellular signalling molecules involved in cytokine signalling, some of which can act as transcription factors.

**signal transduction** The process of delivering a signal received from a receptor to an intracellular signalling cascade that ultimately results in transcription.

**skin associated lymphoid tissue (SALT)** The extensive leukocyte network associated with the epidermis and dermis of the skin. Although not a mucosal tissue, SALT provides widespread immune defence across the body surface.

**somatic gene conversion** A process whereby genetic elements are swapped between different immunoglobulin genes in order to provide variability in antigen specificity.

**somatic hypermutation** Genetic modification of the variable gene segments of antibody genes, which provides variability to the synthesized antibody and increases the repertoire of B cell clones.

**spleen** A large secondary lymphoid organ situated in the upper abdomen, which filters antigens from the blood, acts as a storage site for monocytes and eliminates obsolete erythrocytes from the circulation.

**superantigen** A molecule that is able to activate T cell receptors bearing certain β-chains independently of antigen specificity, and is therefore able to act as a mitogen.

**suppression** A mechanism by which immune cells are modulated through the release of active mediators that downregulate the activity of immune cells.

**switch region** The genetic region within immunoglobulin genes where recombination takes place between the VDJ unit and the constant region, facilitating isotype-switching.

**syngeneic** Genetically identical, for example, like syngeneic twins or inbred mouse strains.

**systemic lupus erythematosus (SLE)** An autoimmune disease that can infect many organs of the body, mediated by the deposition of autoantibodies directed against components of the nucleus, including DNA and histone proteins.

**T cells** A type of lymphocyte that undergoes development in the thymus. Several subpopulations of T cell exist including T helper and cytotoxic T cells, which form an important component of the adaptive immune system

**Th1** A subset of CD4+ T cells that induce cell-mediated immunity through the secretion of the cytokines IFN-γ and TNF.

**Th2** A subset of CD4+ T cells that induce antibody-mediated immunity through the secretion of the cytokine IL-4, IL-5 and IL-13.

**TAP-1 and TAP-2** Proteins that are associated with the transport of peptides into the endoplasmic reticulum for loading onto MHC class I molecules.

**T-cell receptor (TCR)** A heterdimeric protein expressed on the surface of all T cells composed of an α-chain and a homologous β-chain, which recognizes pepide antigens presented by MHC molecules. Two types of heterodimers exist, αβTCR and γδTCR, giving rise to separate populations of T cell.

**T-dependent antigen** An antigen that requires the help of Th cells in order to induce an antbody response against it.

**terminal deoxynucleotidyl transferase (TdT)** An enzyme that introduces nucleotides at the junctions of V, D, and J gene segments of immunoglobulin genes.

**thymocytes** The name given to immature T cells within the thymus.

**thymus** A primary lymphoid organ responsible for the differentiation of T cells. The organ where negative and positive T cell selection takes place.

**T-independent antigen** An antigen that does not require the help of Th cells in order to induce an antibody response.

**titer** A measure of the amount of antibody in body fluids.

**tolerance** The mechanism by which B cells and T cells are made unresponsive to an antigen. Central tolerance is vital in order to prevent T cells and B cells reacting with self-antigens. Peripheral tolerance is necessary in order to prevent T cells and B cells reacting to harmless substances such as food antigens or allergens. Tolerance is an important mechanism for the maintenance of immune homeostasis.

**transplantation** The transfer and engraftment of an organ or cells from a donor individual to a recipient individual.

**tumor necrosis factor (TNF)** A cytokine that possesses potent anti-tumour effects and can augment inflammatory processes.

**urogenital tract associated lymphoid tissue.** The organized and diffuse lymphoid tissue associated with male and female genital, urinary and reproductive tracts. This includes the vagina, cervix, uterus and ovaries in females. An important mucosal tissue for the prevention of sexually transmitted diseases and for the induction of tolerance to a foetus.

**V(D)J recombination** The genetic mechanism that joins V, D and J segments of BCR and TCR genes in order to produce variability in the antigen-biding domain.

**vaccination** The process of generating protective immunity to a pathogen by the injection of a product containing antigen from that pathogen, otherwise known as immunization. Originally described as providing protection against smallpox by vaccinating with cowpox (vaccinia).

**variable (V) regions** The regions of antibody molecules and TCRs that bind to specific antigens. The variability on antigen binding is the result of VDJ recombination.

**villus-crypt unit** A structure formed from enterocytes in the gastrointestinal tract that increases the surface area for the absorption of nutrients.

**virion** A complete virus particle.

**virus** A microscopic organism that requires a host cell in order to replicate. Many viruses are pathogenic and can cause disease in humans.

**Waldeyer's ring** The organized lymphoid structures located at the opening of the pharynx and larynx, including the palatine tonsils and the adenoids.

# Index

acquired immunodeficiency syndrome *see* HIV/AIDS
active immunization, 344
acute lung injury (ALI), 156, 157, 168
acute respiratory distress syndrome (ARDS), 157, 175
adaptive immune system, 5–6, 41–67
  activation and co-stimulation, 60–3, 65–6
  antibody structure, 42–5
  BCR diversity, 46
  constant region and antibody isotypes, 45–6
  dendritic cells, 38–40
  Fc receptors, 42–5, 51–4
  fungal infections, 268–70, 272
  genetic recombination of BCR genes, 46–51, 54, 57
  immunoglobulin class switching, 50–1, 173, 190
  innate immunity, 38–40
  junctional diversity, 48–9
  MHC processing pathways, 59–60
  parasites, 287–8, 289–90, 293–4, 295–6
  somatic hypermutation and affinity maturation, 49–50
  T cells and B cells, 41–2, 55–8, 60–7
  urogenital tract, 186–7, 190
  vaccines, 347
adenoids, 161–3
adjuvants, 347–50
affinity maturation, 49–50
airway surface liquid (ASL), 160–1
airways *see* respiratory tract
allergic reactions, 105–6, 131–2, 194–5, 302–4, 307–13
alloimmunization, 189
alternative complement pathway, 29
alum, 348
anaphylactic shock, 303
aneuploidy, 324
angiogenesis, 326–7
anterior chamber associated immune deviation (ACAID), 193
antibodies
  adaptive immune system, 42–6
  bacterial infections, 249–50
  constant region and isotypes, 45–6

cross-linking, 52, 53–4
domains and fragments, 43
effector functions, 45
eye and conjuctiva, 191
fungal infections, 273–5
gastrointestinal tract, 149–51
immune disorders, 304–6, 308, 312, 319
mucosal immunology, 113–15, 124–6
parasites, 290, 293, 294
respiratory tract, 161–4, 173–4
skin, 204
structure/Fc receptor structure, 42–5, 51–3
urogenital tract, 179–80, 181–2, 189–91
vaccines, 343–4, 347
viral infections, 232–3
antibody-dependent cellular cytotoxicity (ADCC), 13, 274
  adaptive immune system, 45–6, 54
  bacterial infections, 248–9
  immune disorders, 304
  innate immune system, 33–4
  mucosal tumour immunology, 336
  parasites, 293–4, 296
anti-fungal molecules, 265–6
antigen presenting cells (APC), 12–13, 15, 17
  adaptive immune system, 57–8
  eye and conjuctiva, 193
  fungal infections, 270–2
  mucosal immunology, 123
  mucosal tumour immunology, 328–30, 339–40
  parasites, 289–90, 294, 299–300
  respiratory tract, 163–4, 172
  skin, 198–9, 210
  urogenital tract, 182–3
  vaccines, 344–7, 350, 357–8
antigens
  adaptive immune system, 41–6
  bacterial infections, 251–3
  evasion by shift and drift, 235–6
  gastrointestinal tract, 140–6
  immune disorders, 302
  mucosal immunology, 111–12, 114–15, 121–3
  mucosal tumour immunology, 328–33
  respiratory tract, 161–3
  skin, 203–5

transepithelial transport, 121–3
vaccines, 342
viruses, 232–3
anti-inflammatory drugs, 109
antimicrobial peptides, 160–1, 181, 200–1, 242–3
arthroconidia, 261
*Ascaris* spp., 285–7
asthma, 108–9, 131–2, 160, 175, 303, 307–10
autocrine function, 68–9
autoimmune diseases, 189, 213–14, 313–20
autophagy, 251–3
autoreactive T cells, 211–13

B cell receptors (BCR), 6, 8, 15, 41–2, 46–9
B cells, 3, 4–9, 15–19
  adaptive immune system, 41–2, 63
  chemokines, 101–2
  eye and conjuctiva, 192–3
  gastrointestinal tract, 136, 140–1, 143–4, 149–51
  immune disorders, 308, 313–15, 318
  mucosal immunology, 114–17, 127–8
  mucosal tumour immunology, 328
  respiratory tract, 161–4
  skin, 210, 213
  urogenital tract, 177–81, 188–9
  vaccines, 344–5
bacille Calmette-Guérin (BCG) vaccine, 256, 352–3
bacteria, 238–57
  antibodies, 249–50
  antimicrobial peptides, 242–3
  autophagy and intracellular bacteria, 251–3
  barrier functions, 241–2
  classification, 238–40
  complement pathway, 244–5
  cross-presentation of antigens, 252–3
  dendritic cells, 250–1
  DTH response and granuloma in TB, 253–4
  evasion strategies, 239, 247–8, 251
  eye and conjuctiva, 194
  gastrointestinal tract, 137–8, 154
  immune diseases/disorders, 241, 314
  mucosal immunology, 126, 241–2, 250
  neutrophils, 245–7

NK cells and ADCC, 248–9
phagocytosis, 244–5, 247–8, 251
recognition by TLRs, 243–4
respiratory tract, 161, 175–6
skin, 215–16
structure, 240–1
T cells, 254–6
urogenital tract, 187–9
vaccines, 256, 343, 354
basophils, 3, 9–11
bioinformatics, 360–2
bioterrorism, 353
bone marrow, 2–3, 11, 101–2
breastfeeding, 111, 204
bronchus associated lymphoid tissue (BALT),
    156–7, 165–6, 172, 174
bystander effects, 231–2

C-type lectins, 299–300
cadherins, 117–18
calcium fluxes, 61–2
candidal vulvovaginitis (CVV), 264
capsids, 217–18
carcinoembryonic antigen (CEA), 331–2
cathelicidins, 160–1, 201–2
CD3 complex, 55, 61–2
CD4+/CD8+ see T cells
cell adhesion molecules (CAM), 71, 76, 98,
    102
    bacterial infections, 247
    gastrointestinal tract, 152–3
    mucosal immunology, 115–16
    urogenital tract, 187–8
cestodes, 284
chemokines, 93–110
    anti-inflammatory drugs, 109
    chemotaxis, 97, 99
    expression and function, 97
    extravasation of leukocytes, 97–9
    families, 69–70
    G-protein coupled receptors, 94–7,
        99–100, 104–6
    gastrointestinal tract, 143, 152–3
    homeostasis, 104
    ligands, 95–6
    lymphocyte migration, 101–2
    lymphoid structure formation, 102–4
    mucosal immunology, 115, 117, 121, 126–7
    pathogenesis of human diseases, 108–9
    receptor signalling cascade, 99–100
    redundancy, 106–8
    respiratory tract, 163–4, 169, 172–3
    skin, 199–200, 207–9
    structure and nomenclature, 93–4
    T cells and B cells, 101–2, 104–6
    tissue specific homing, 100–1
    urogenital tract, 179
    vaccines, 345–7
    viral mimicry, 107
chemotaxis, 97, 99

chitin, 258–9
chronic obstructive pulmonary disease
    (COPD), 156, 157, 160, 168, 175
Clara cells, 159, 167
class switching, 50–1, 173, 190
classical complement pathway, 28–9
clonal selection theory, 49–50
coeliac disease, 214, 317
coiling phagocytosis, 269
collectins, 160–1, 265–6
colon, 134
colony stimulating factors (CSF), 77–8, 92
colorectal cancer, 324–5, 331–2
commensals, 111–12, 135–6, 137–8, 146, 250,
    314
common γ-chain family, 73–4
common mucosal immune system, 115–16
complement pathway, 27–30, 244–5, 266–8,
    293
complementarity determining regions (CDR),
    50
complete Freund's adjuvant (CFA), 348–50
conidia, 261
conjunctiva associated lymphoid tissue
    (CALT), 191–5
constant region, 45–6
co-stimulation, 60–3
Crohn's disease, 316–17
cross-linking, 52, 53–4
cross-presentation of antigens, 203–5, 228,
    252–3
cryopatches, 141–2
cutaneous lymphocyte antigen (CLA), 100–1,
    117, 206–7
cysteine residues, 94–7
cytokines, 4, 9–11, 68–92
    fungal infections, 272–5, 277
    gastrointestinal tract, 138, 141, 143, 151–2
    IFN-γ signalling pathway, 84–5
    IKK complex, 87–8
    IL-33 and ST2 signal regulation, 91
    immunopathology and regulatory
        cytokines, 79
    JAK/STAT signalling pathway, 80–4
    mucosal immunology, 124, 128–31
    mucosal tumour immunology, 339
    NF-κB activation, 87–90
    plasticity in type I signalling, 83
    receptor signalling, 79–86
    respiratory tract, 167–8, 170–2
    skin, 199–200, 211
    soluble decoy receptors, 90–1
    structure of cytokine families, 69–71
    suppressor of cytokine signalling, 83–4
    T helper cells, 74–5, 78–9
    TGF-β/SMAD signalling pathway, 85–6
    therapeutic potential, 91–2
    type I and type II receptors, 79–81, 83
    type III receptors, 86–7
    type IV receptors, 88–90

see also individual cytokine types
cytopathic viruses, 233–5
cytotoxic T lymphocytes see T cells

damage-associated molecular patterns
    (DAMP), 25–7, 36
    gastrointestinal tract, 137–8, 146
    mucosal tumour immunology, 337
    respiratory tract, 157, 168
    skin, 206
    vaccines, 346
    viral infections, 235
Darwinian selection, 338–9
decoy receptors, 90–1
defensins, 138–9, 160–1, 181, 200–1, 242–3,
    265–6
degranulation of mast cells, 305–6, 308, 310
delayed-type hypersensitivity (DTH), 253–4,
    272–3, 292, 304–7
dendritic cells (DC), 3, 12, 18–19, 36–40
    bacterial infections, 250–1
    fungal infections, 268–70
    gastrointestinal tract, 136, 140, 143–4, 149
    immune disorders, 310
    mucosal immunology, 122–4
    mucosal tumour immunology, 333–5,
        337–8
    parasites, 294–5
    respiratory tract, 161–4, 168–9
    skin, 196, 200, 203–5
    urogenital tract, 182–3
    vaccines, 344–7, 358–9
    viral infections, 227–8
dermatitis, 105–6, 109, 214, 311
dermis, 196–7, 204–7
diabetes mellitus, 318–20
DiGeorge syndrome, 320–1
disulphide bonds, 44, 93
DNA vaccination, 355
Duffy antigen receptor for chemokines
    (DARC), 97
duodenum, 133–4, 136

E-selectin ligand-1 (ESL-1), 98
embrogenesis, 133–5
endocrine function, 68–9
endocytosis, 123, 125
enterocytes, 135
enveloped viruses, 218
eosinophils, 3, 9, 11, 108, 294–5, 307
epidermis, 196–8
epithelium
    gastrointestinal tract, 134–6, 138
    immune disorders, 308–10
    M cells, 121–3
    mucosal immunology, 114, 117–23
    mucosal tumour immunology, 324–6
    respiratory tract, 158–61, 165–8
    structural types, 119–20
    urogenital tract, 178–81

epithelium *(continued)*
  viral infections, 221–2
Epstein–Barr virus (EBV), 220–1, 230, 234
eye associated lymphoid tissue (EALT), 191–5

Fas/FasL expression, 340–1
Fc receptors, 9–11, 13, 18, 42–5, 51–4
foetal immune system, 185–6, 189–90
follicle-associated epithelium (FAE), 114, 135, 140, 165
follicular dendritic cells (FDC), 18–19, 149, 163–4
food allergies, 311–12
fungi, 258–77
  adaptive immune system, 268–70, 272
  anti-fungal molecules, 265–6
  antibodies, 273–5
  barrier functions, 263–5
  complement pathway, 266–8
  cytokines, 272–5, 277
  dendritic cells, 268–70
  dimorphism, 261–2
  DTH response and granuloma formation, 272–3
  evasion strategies, 276–7
  immune diseases, 259, 262–3
  immuno-modulatory products, 276
  innate immune system, 263, 268–70
  instructive signalling, 270
  macrophages, 270–2
  morphology, 258–60
  moulds, 260–1
  mucosal immunology, 263–5
  natural killer cells, 271–2
  neutrophils, 271
  PAMPs and TLRs, 266–70
  phagocytosis, 268–70, 276–7
  skin, 215–16
  vaccines and immunotherapies, 274–6
  yeasts, 260–2, 269
γδ T cells, 16, 146–7, 206–7

G-protein coupled receptors (GPCR), 94–7, 99–100, 104–6, 247
Gastrointestinal associated lymphoid tissue (GALT), 136, 142, 146–7, 152–3
gastrointestinal tract (GIT), 133–55
  anatomy and physiology, 133–4
  bacterial infections, 250
  barrier functions, 136–8, 139
  chemokines and lymphocyte migration to GALT, 152–3
  cryopatches and inducer cells, 141–2
  cytokines in the gut, 151–2
  defensins and trefoil factors, 138–9
  dendritic cells and lumen contents, 143–4
  development, 133–5
  epithelium, 134–6, 138

immune diseases/disorders, 137–8, 153–4, 303
  intestinal production of IgA, 150–1
  intra-epithelial lymphocytes, 140–1, 143–6, 148, 152–3
  lymphoid follicles and germinal centre formation, 140–2
  M cells in intestinal lumen, 143
  mucosal immunology, 135–6
  parasites, 294–6
  Peyer's patches, 133, 136, 139–42
  sub-mucosal B cells and mucosal IgA, 149–50
  T cells, 146–9
  viral infections, 220
genetic recombination of BCR genes, 46–51, 54, 57
genital infections, 360
genome analysis, 360–2
germinal centre formation, 140–2
*Giardia* spp., 291–2
goblet cells, 159–60, 166, 265
granulocytes, 3, 9
granuloma formation, 253–4, 272–3, 294–5
growth factors, 14, 92
  bacterial infections, 255–6
  eye and conjuctiva, 193
  family of cytokines, 69–70, 77–8
  gastrointestinal tract, 147, 150–1
  immune disorders, 309, 313
  mucosal immunology, 124, 128–30
  mucosal tumour immunology, 323, 326, 337, 339
  parasites, 299–300
  respiratory tract, 157, 170–1
  TGF-β/SMAD signalling pathway, 85–6
gut microflora, 111–12, 135–6, 250
gut-associated lymphoid tissue (GALT), 114, 122

H1N1/H3N2 viruses, 235–7, 354
haematopoiesis, 7–10
hair follicles, 202, 207
helminths, 282–7, 292–6, 298–300
herpes simplex virus (HSV), 231
high endothelial venules (HEV), 136, 140, 159, 163–5, 173, 192
HIV/AIDS
  chemokines, 105, 107, 109
  fungal infections, 263, 264–5
  immune cell infection, 221, 224–5
  skin, 216
  urogenital tract, 177, 179–80
  vaccines, 237, 360
homeostasis, 104
human leukocyte antigens (HLA), 58, 184–5, 224–5, 231, 336, 340
human papillomavirus (HPV), 215–16, 332–3
hygiene hypothesis, 296, 313, 314
hyphae, 260–1

idiopathic thrombocytopenia purpura (ITP), 189
IFN-γ signalling pathway, 84–5
IKK complex, 87–8
IL-1R-associated kinase-1 (IRAK1), 88–9
IL-33/ST2 signal regulation, 91
ileum, 133–4, 136
immune complex disease, 304
immune disorders, 302–21
  acute phase responses, 304–5
  allergic asthma, 303, 307–10
  allergic reactions, 302–4, 307–13
  allergic rhinitis, 310–11
  autoimmune diseases, 313–20
  coeliac disease, 317
  disregulation of tolerance, 313–16
  early phase responses, 304–5, 308
  epithelial cells, 308–10
  food allergies, 311–12
  hygiene hypothesis, 313, 314
  immunodeficiencies, 320–1
  inflammatory bowel disease, 316–17
  late phase responses, 304–7, 310
  mast cell degranulation, 305–6, 308, 310
  sensitization, 304–5
  skin allergy and atopic dermatitis, 303, 311
  systemic lupus erythematosus, 317–18
  T cell subsets in allergies, 312–13
immune privilege, 192–3, 202
immune surveillance, 328–9, 333
immunodeficiencies, 320–1
immunogenicity, 329–30
immunoglobulin *see* antibodies
immunological memory, 7–9
immuno-modulatory products, 276
immunoreceptor tyrosine-based activation motif (ITAM), 52, 305–6
immunoreceptor tyrosine-based inhibition motif (ITIM), 52, 54
immuno-stimulatory complexes (ISCOM), 355–8
inactivated vaccines, 353–4
inducer cells, 141–2
inducible nitric oxide synthase (iNOS), 32–3
inflammasome activation, 226–7
inflammatory bowel disease (IBD), 154, 316–17
influenza viruses, 236
innate immune system, 5–6, 20–40
  activation, 36–8
  adaptive immunity, 38–40
  antibody-dependent cellular cytotoxicity, 33–4
  complement cascade, 27–30
  dendritic cells, 36–40
  Fc receptors, 32
  fungal infections, 263, 268–70
  mechanisms, 21
  missing self hypothesis, 35–6
  natural killer cells, 33–6

neutrophils, 32–3
opsonization, 27–8, 30–1
PAMPs and DAMPs, 24–7
parasites, 287, 288, 292–3, 294–5
pattern recognition, 22–3
peptidoglycan and Nod-like receptors, 24–6
phagocytosis, 31–2
receptors and cells, 20–2
respiratory burst, 32–3
respiratory tract, 168
signalling in response to LPS, 23–4
toll-like receptors, 22–4
instructive signalling, 270
integrins, 117, 121, 124, 173
interferons (IFN), 11
adaptive immune system, 63–5
bacterial infections, 248–9
family of cytokines, 69–70, 75
fungal infections, 272–5
gastrointestinal tract, 146–8
IFN-γ signalling pathway, 84–5
immune disorders, 311, 318
innate immune system, 37, 39
mucosal immunology, 128–31
mucosal tumour immunology, 333
parasites, 289–90, 294
respiratory tract, 157, 167–8, 170–2
skin, 204, 206, 208–10, 212–13
therapeutic potential, 91–2
urogenital tract, 186, 189
vaccines, 356–7
viral infections, 222–3, 225–7
interleukins (IL), 10–12, 14, 37, 39
adaptive immune system, 63–5
common γ-chain family, 73–4
eye and conjunctiva, 193–5
families and superfamilies, 69–75
fungal infections, 266, 272–5
gastrointestinal tract, 138, 141, 143, 147–8, 150–2
IL-33 and ST2 signal regulation, 91
immune disorders, 304, 307–14
JAK/STAT signalling pathway, 81–4
mucosal immunology, 124, 128–31
mucosal tumour immunology, 333, 339
NF-κB activation, 88–90
parasites, 287–90, 292–6, 299
respiratory tract, 157, 167–8, 170–2
skin, 198–200, 210–13
therapeutic potential, 92
urogenital tract, 186–7, 189–90
vaccines, 349–50, 356–9
viral infections, 226–7
intestinal lumen, 143
intra-epithelial lymphocytes (IEL), 16
gastrointestinal tract, 140–1, 143–6, 148, 152–3
immune disorders, 317
respiratory tract, 165–6, 171

skin, 206
intracellular bacteria, 251–3

JAK/STAT signalling pathway, 80–4, 99–100
jejunum, 133–4, 136
joining chain (J-chain), 124–6
junctional diversity, 48–9

Kaposi's sarcoma, 230–1
keratinocytes, 196–8, 199–202
killer cell immunoglobulin-like receptors (KIR), 13, 35, 223–4

lactoferrin, 161
lamina propria DCs (LPDC), 143
Langerhan's cells (LC), 182–3, 196–8, 200, 202–3
lectins, 29, 266–8, 293, 299–300
*Leishmania* spp., 281, 290–1, 298
leukocytes, 3, 97–9, 115–16, 159
lipopolysaccharide (LPS), 23–4, 120–1, 240, 244, 247
live-attenuated vaccines, 351–3
lock and key hypothesis, 44
lumen, 143
lung diseases, 156–7, 326
lymph nodes, 16–19
gastrointestinal tract, 153
mucosal immunology, 115, 117
mucosal tumour immunology, 335
respiratory tract, 163–5, 169
skin, 205
urogenital tract, 183
lymphocytes, 3, 7–9, 101–2, 117, 140–6, 148, 152–3
lymphoid follicles, 140–2, 163, 192
lymphoid structures, 102–4, 162, 165–6
lymphoid tissue inducer (LTi) cells, 140–1
lysozyme, 161

M cells, 121–3, 135, 140, 143, 161–3
macrophages, 3, 9, 11–12
fungal infections, 270–2
gastrointestinal tract, 136, 140
mucosal immunology, 123
mucosal tumour immunology, 336–7
parasites, 294–5, 300
respiratory tract, 169–71
vaccines, 344–5
viral infections, 225–6
major histocompatibility complex (MHC), 12–15
adaptive immune system, 41–2, 55, 57–62, 65
bacterial infections, 251–3
eye and conjunctiva, 193
gastrointestinal tract, 144–7
innate immune system, 35–6
mucosal immunology, 128

mucosal tumour immunology, 328–30, 335–6, 339–40
parasites, 288–91
skin, 202–3, 210
vaccines, 346–8, 351, 355–9
viral infections, 223, 228, 230–1
mammalian target of rapamycin (mTOR), 83
mammary glands, 204
mannose binding lectin (MBL), 266–8, 293
mannose receptor (MR), 243–4, 267–8
MAPK signalling pathway, 83, 85, 89–90, 100, 306
mast cells, 3, 9–11, 205–6, 305–6, 308, 310
measles, mumps and rubella (MMR) vaccine, 237, 353
melanocytes, 197–8
melanoma differentiation antigens, 332
membrane attack complex (MAC), 27–8, 266–7
memory T cells, 7–9, 66–7, 153, 172, 229
menstrual cycle, 181–2
metastasis, 327
microvilli, 133–4
missing self hypothesis, 35–6, 335
monocytes, 3, 9, 11–12
moulds, 260–1
mucins, 160
mucocillary elevator/clearance, 159–60
mucosal adjuvants, 350
mucosal immunology, 111–32
antigen sampling at mucosal surfaces, 121–2
bacterial infections, 241–2, 250
common mucosal immune system, 115–16
cytokines, 124, 128–30
dendritic cells, 122–4
epithelium, 114, 117–23
eye and conjunctiva, 191–2
fungi, 263–5
gastrointestinal tract, 135–6
immune diseases, 130–2
J-chain regulation, 124–6
M cells, 121–3
organized lymphoid tissue, 127–9
pathogens, commensals and non-pathogens, 111–12, 124–7, 130–1
respiratory tract, 156
secretory component expression, 126
secretory dimeric antibody at mucosal surfaces, 124–6
sub-mucosa, 125, 126–7
T cells and B cells, 114–18, 127–32
toll-like receptors and NOD proteins, 120–1, 124
transepithelial transport of antigens, 121–3
urogenital tract, 177–8, 182–3
vaccines, 359–60
viral infections, 217, 220

mucosal tumour immunology, 322–41
  aneuploidy and colorectal cancer, 324–5
  angiogenesis, 326–7
  augmentation by immune cells, 337
  cancer types, 323
  carcinoembryonic antigen, 331–2
  chemokines, 108–9
  cytokines, 339
  Darwinian selection, 338–9
  dendritic cells, 333–5, 337–8
  disregulated MHC expression and antigen
      presentation, 339–40
  effector molecules, 333
  evasion strategies, 337–8
  Fas/FasL expression, 340–1
  immune surveillance, 328–9, 333
  immunogenicity of tumour cells, 329–30
  lymph nodes, 335
  macrophages and neutrophils, 336–7
  melanoma differentiation antigens, 332
  metastasis, 327
  NK cell recognition and missing self, 335
  NKG2D receptor on NK cells, 335–6
  p53 mutations, 323, 324, 326, 330–1
  phagocytosis, 336–7
  proto-oncogene activation, 323–5
  Ras proteins and proliferation, 324
  recognition of transformed cells, 330–1
  transformation into cancer cells, 322–4
  tumour associated antigens, 328, 330–1,
      335
  tumour cell escape, 338–41
  tumour reactive T cells, 335
  tumourigenesis, 324–6
  viral tumour associated antigens, 332–3
mucosal-associated lymphoid tissue (MALT),
    2, 9, 11, 15, 136, 151, 165, 177–8
multiple sclerosis (MS), 108–9
myeloid cells, 3, 7–9, 12

nasal associated lymphoid tissue (NALT),
    113, 156–7, 161–2, 165, 172
natural killer (NK) cells, 3, 9, 12–13, 16, 33–6
  bacterial infections, 248–9
  fungal infections, 271–2
  mucosal tumour immunology, 331, 335–6,
      340–1
  parasites, 288–90
  respiratory tract, 171
  skin, 205–6
  urogenital tract, 183–6, 189
  viral infections, 222–5
natural killer T (NKT) cells, 37–8, 147–8, 256,
    271–2
nematodes, 284, 285–7, 294–6
neonatal immune system, 189–90
neoplastic transformation, 324
neutralizing antibodies, 233, 274
neutrophils, 3, 9, 11, 32–3
  bacterial infections, 245–7

fungal infections, 271
immune disorders, 307
mucosal immunology, 126–7
mucosal tumour immunology, 336–7
respiratory tract, 161, 170
urogenital tract, 190
nitric oxide (NO), 32–3
NOD-like receptors (NLR), 24–6, 226
NOD proteins, 120–1, 137
non-cytopathic viruses, 233–5
nuclear factor (NF-κB), 87–90

opportunistic fungi, 262–3, 264
opsonization, 27–8, 30–1, 244–5
oral polio vaccine (OPV), 359
oropharyngeal candidosis (OPD), 264

P-selectin glycoprotein ligand-1 (PSGL-1),
    97–8
p53, 323, 324, 326, 330–1
pancreatic β-cells, 319–20
Paneth cells, 138–9
paracortex, 18–19
paracrine function, 68–9
parasites, 278–301
  adaptive immune system, 287–8, 289–90,
      293–4, 295–6
  antibodies, 293
  classification, 279
  evasion strategies, 296–300
  gastrointestinal tract, 154
  helminths, 282–7, 292–6, 298–300
  immune diseases/disorders, 284, 297, 314
  innate immune system, 287, 288, 292–3,
      294–5
  life cycles, 280–2, 284–5, 286–7, 298
  protozoa, 278–82, 287–92, 298–9
  rare infections, 297
  structure, 278–80, 283–4, 285–6
  Th1 versus Th2 *Leishmania* immunity,
      290–2
  variable surface glycoprotein, 287, 297–8
passive immunity, 181
passive immunization, 344
pathogen-associated molecular patterns
    (PAMP), 4, 12, 24–6
  bacterial infections, 243–4
  fungal infections, 263, 266–70
  gastrointestinal tract, 137, 143
  immune disorders, 309
  mucosal immunology, 120, 124
  mucosal tumour immunology, 337
  parasites, 294–5
  respiratory tract, 168–9
  skin, 209–10
  vaccines, 346, 349
  viral infections, 226–7, 235
pattern recognition receptors (PRR), 4, 12,
    22–3
  bacterial infections, 243–4

fungal infections, 266
gastrointestinal tract, 138, 143
immune disorders, 309
mucosal immunology, 120
parasites, 295
respiratory tract, 167
pelvic inflammatory disease (PID), 187–9
peptide vaccines, 354–5
peptidoglycan, 24–5, 240, 243
Peyer's patches (PP), 133, 136, 139–42, 161
phagocytosis, 31–2, 123, 170–1
  bacterial infections, 244–5, 247–8, 251
  fungal infections, 268–70, 276–7
  mucosal tumour immunology, 336–7
  parasites, 298
  viral infections, 225–6
PI3K/AKT/mTOR signalling pathway, 83, 99
plasma cells, 3, 8
*Plasmodium* spp., 279–80, 281–2, 288–90,
    298–9
plasticity, 83
polysaccharide vaccines, 354
pre-eclampsia, 184–5
pregnancy, 183–6, 189–90
protease-activated receptors (PAR), 247
protease inhibitors, 160–1
protein kinases, 323
proteomics, 362
proto-oncogene activation, 323–5
protozoa, 278–82, 287–92, 298–9
pseudohyphae, 261
psoriasis, 105–6, 109, 211–13

Ras proteins, 324
reactive oxygen species (ROS), 32–3, 232
redundancy, chemokines, 106–8
regulatory T cells (Treg), 14, 105–6, 210–11,
    255–6, 273, 313, 318–20
reproductive system *see* urogenital tract
respiratory burst, 32–3
respiratory syncytial virus (RSV), 175–6
respiratory tract, 156–76
  alveolar macrophages, 169–71
  anatomy and physiology, 158–9, 161–3,
      165–6
  bacterial infections, 245
  BALT structure, 165–6
  barrier functions, 159–60
  cells of the lower airways, 166–7
  defensins and antimicrobial peptides,
      160–1
  dendritic cells, 161–4, 168–9
  development, 156–7
  effector site T cells, 171–2
  epithelium, 158–61, 165–8
  immune diseases/disorders, 156, 157, 168,
      174–6, 302–3
  immune modulation by airway epithelial
      cells, 167–8
  innate immune system, 168

lymph nodes and immune generation, 163–5
migration of T cells into lung tissue, 172–3
mucins and mucocillary clearance, 160
mucocillary elevator, 159–60
mucosal immunology, 156
NALT structure, 165
natural killer cells, 171
pathogens, 174–6
production of IgA, 173–4
surfactant proteins, 167
tonsils and adenoids, 161–3
viral infections, 220, 232
reverse genetics, 361–2
rheumatoid arthritis (RA), 108–9, 318–19

saliva, 111
*Schistosoma* spp., 283–5, 292–4, 299
semi-allogeneic foetus, 183–4, 186
sensitization, 304–5
severe acute respiratory syndrome (SARS), 129, 131, 176
severe combined immunodeficiency (SCID), 320
sexually transmitted diseases, 177, 187–9, 360
signal transduction, 80–3
skin, 196–216
    anatomy and physiology, 196–7
    autoimmune diseases/disorders, 213–14, 303, 311
    barrier functions, 196–8
    cellular immune system, 198–9
    chemokines and migration, 207–8
    cutaneous lymphocyte antigen, 206–7
    cytokines, 211
    DCs and cross-presentation of antigen, 203–5
    hair follicles and immune privilege, 202
    immune response initiation, 208–11
    intra-epithelial lymphocytes, 206
    keratinocytes, 196–8, 199–202
    LCs as immune sentinels, 202–3
    lymphocytes in the dermis, 206
    mast cells and NK cells, 205–6
    psoriasis, inflammation and autoreactive T cells, 211–13
    systemic diseases, 214–15
    viral infections, 220
skin associated lymphoid tissue (SALT), 198–9
SMAD signalling pathway, 85–6
small intestine, 133
smallpox vaccines, 352
soluble cytokine receptors, 90–1
somatic hypermutation, 49–50
spleen, 18, 19
spores, 261
stem cell factor (SCF), 9
sub-mucosa, 125, 126–7, 149–50
superoxide dismutase (SOD), 32–3

suppressor of cytokine signalling (SOCS), 83–4
surfactant proteins, 167, 266
systemic diseases, 214–15
systemic lupus erythematosus (SLE), 214–15, 317–18

T cell receptors (TCR), 6, 8, 12, 15–16, 41–2, 54–7, 145–8, 172
T cells, 3, 6–9, 12–19
    adaptive immune system, 41–2, 55–8, 60–7
    bacterial infections, 250–6
    chemokines, 101–2, 104–6
    cytokines, 74–5, 76, 78–9
    eye and conjuctiva, 192–5
    fungal infections, 264–5, 271–6
    gastrointestinal tract, 136, 137–8, 140–1, 142, 143–9, 153–4
    immune disorders, 304, 307–20
    innate immune system, 37–8, 39
    mucosal immunology, 114–18, 127–32
    mucosal tumour immunology, 328–35, 340
    parasites, 288–95, 299–300
    respiratory tract, 161–5, 171–3
    skin, 199–200, 206–13, 216
    urogenital tract, 177–81, 185–91
    vaccines, 345–7, 350–1, 355–9
    viral infections, 221, 224, 227–31, 233–5
Tamm-horsfall protein (THP), 190–1
TGF-β/SMAD signalling pathway, 85–6
thymus, 2–3, 55–7, 64, 101–2
tissue specific homing, 100–1
TNF receptor-associated factors (TRAF), 86–9
TNFR1-associated death domain protein (TRADD), 86–8
toll-like receptors (TLR), 22–4, 79
    bacterial infections, 243–4, 250
    fungal infections, 266–70
    gastrointestinal tract, 137–8
    mucosal immunology, 120–1, 124
    respiratory tract, 167–8
    urogenital tract, 178–81
    viral infections, 226
tonsils, 161–3
transepithelial transport of antigens, 121–3
transporter associated with antigen processing (TAP), 60
trefoil factors, 138–9
trematodes, 283–5, 292–4
*Trypanosoma* spp., 279–80, 281, 287–8
tuberculosis, 253–4
tumour associated antigens (TAA), 328, 330–1, 335
tumour cell escape, 338–41
tumour necrosis factor (TNF), 11–12
    immune disorders, 307–8, 318
    innate immune system, 37
    ligand superfamily, 69–70, 75–7
    mucosal tumour immunology, 331, 333

parasites, 287–8, 290, 294
respiratory tract, 157, 168, 170–1
skin, 210–13
urogenital tract, 186
viral infections, 225–6
tumour rejection, 108–9
tumourigenesis, 324–6

ulcerative colitis, 316
urogenital tract, 177–91
    adaptive immune system, 186–7, 190
    alloimmunization and autoimmune diseases, 189
    anatomy and physiology, 177–8
    antibodies, 179–80, 181–2, 189–91
    antigen presenting cells, 182–3
    barrier functions, 178–81
    epithelium, 178–81
    immune diseases, 179–80, 184–5, 187–9
    immunity in the urinary tract, 190–1
    maintenance of foetal tolerance, 185–6
    mucosal immunology, 177–8, 182–3
    NK cells and pregnancy, 183–6, 189
    passive immunity, 181
    pregnancy, 183–6
    viral infections, 220

vaccines, 342–62
    active immunization, 344
    adaptive immune system, 347
    adjuvants, 347–50
    bacteria, 256, 343, 354
    delivery systems, 350, 359–60
    dendritic cell vaccines, 358–9
    development strategies, 360–2
    DNA vaccination, 355
    fungi, 274–6
    immuno-stimulatory complexes, 355–8
    inactivated, 353–4
    live-attenuated, 351–3
    passive immunization, 344
    peptide vaccines, 354–5
    polysaccharide vaccines, 354
    principles, 342–3
    processing for immune recognition, 344–7
    Th1/Th2 polarization, 351
    viruses, 235–7, 342, 354
variable surface glycoprotein (VSG), 287, 297–8
variable surface proteins (VSP), 292
varicella zosta virus (VZV), 215
VDJ recombination, 46–50, 54, 57
villi, 133–4
villus-crypt unit, 135, 139
viruses, 217–37
    antibodies, 232–3
    bystander effects, 231–2
    classification, 217, 218
    cytopathic and non-cytopathic, 233–5
    dendritic cells, 227–8

viruses (*continued*)
  epithelial cell infection, 221–2
  evasion by antigenic shift and drift,
    235–6
  evasion of CTL-mediated immunity,
    229–31
  gastrointestinal tract, 154
  immune cell infection, 220–1
  immune disorders, 314
  infection types and forms, 219–20
  inflammasome activation, 226–7

interferons, 222–3, 225–7
macrophages, 225–6
mimicry of chemokines, 107
mucosal immunology, 130, 217, 220
mucosal tumour immunology, 330,
  332–3
natural killer cells, 222–5
replication within host cells, 218–19
respiratory tract, 175–6
skin, 205–6, 215–16
structure, 217–18

T cell responses, 227–31
TLRs and NLRs, 226
vaccines, 235–7, 342, 354

Waldeyer's ring, 156, 161–3
Wiskott-Aldrich syndrome, 320

yeasts, 260–2, 269

zipper-like phagocytosis, 269
zymosan, 266